History of Life

THIRD EDITION

Richard Cowen

University of California, Davis

© 2000 by Richard Cowen

Blackwell Science, Inc.
Editorial Offices:
 Commerce Place, 350 Main Street, Malden,
  Massachusetts 02148, USA
 Osney Mead, Oxford OX2 0EL, England
 25 John Street, London WC1N 2BL, England
 23 Ainslie Place, Edinburgh EH3 6AJ, Scotland
 54 University Street, Carlton, Victoria 3053, Australia

Other Editorial Offices:
 Blackwell Wissenschafts-Verlag GmbH,
  Kurfürstendamm 57, 10707 Berlin, Germany
 Blackwell Science KK, MG Kodenmacho Building,
  7-10 Kodenmacho Nihombashi, Chuo-ku,
  Tokyo 104, Japan

Distributors:
USA
 Blackwell Science, Inc.,
  Commerce Place, 350 Main Street, Malden,
  Massachusetts 02148
  (Telephone orders: 800-215-1000 or 781-388-
  8250; fax orders: 781-388-8270)

Canada
 Login Brothers Book Company
  324 Saulteaux Crescent,
  Winnipeg, Manitoba, R3J 3T2
  (Telephone orders: 204-224-4068)

Australia
 Blackwell Science Pty, Ltd.
  54 University Street, Carlton, Victoria 3053
  (Telephone orders: 03-9347-0300; fax orders:
  03-9349-3016)

Outside North America and Australia
 Blackwell Science, Ltd.
  c/o Marston Book Services, Ltd., P.O. Box 269,
  Abingdon, Oxon OX14 4YN, England
  (Telephone orders: 44-01235-465500;
  fax orders: 44-01235-465555)

Acquisitions: Nancy Anastasi Duffy
Production and Manufacturing: Lisa Flanagan
Cover design by Leslie Haimes
Printed and bound by Braun-Brumfield, Inc.

Cover illustration: Tree reflected in Lochan Na
H-Achlaise, Rannoch Moor, Scotland.
From Corel Premium Photos. Used under license.

Printed in the United States of America
99 00 01 02 5 4 3 2 1

The Blackwell Science logo is a trade mark of
Blackwell Science Ltd., registered at the United
Kingdom Trade Marks Registry

Library of Congress Cataloging-in-Publication Data

Cowen, Richard, 1940-
  History of life/Richard Cowen.—3rd ed.
    p. cm.
  Includes bibliographical references and index.
  ISBN 0-632-04444-6
  1. Paleontology. I. Title.
QE711.2.C68   2000
560—dc21   99-16542
                                   CIP

To Jo, Claire, and Alexandra

Contents

Preface

For Everyone

For 32 years I have taught a course called *History of Life* at the University of California, Davis. This book, now entering its third edition, was written for that course, but it is meant not just for students, but for everyone interested in the history of life on our planet. Paleontology, the study of ancient life, requires some knowledge of biology, ecology, chemistry, physics, and mathematics, but not very much, so the average person can have access to it without deep scientific training. My aim is ambitious: I try to take you to the edges of our knowledge in paleontology, showing you how life has evolved on Earth, and how we have constructed the history of that evolution from the record of rocks and fossils.

However, there is a snag. Human history is never simple, even when we try to describe events that happened last year. It's even worse when we ask *why* events happened. It's not likely that any account of the history of life is going to be simple either. The living world today contains all kinds of creatures that do unexpected things. There are frogs that fly and birds that can't. There are mammals that lay eggs, reptiles that have live birth, and amphibians that suckle their young. There are fishes that breathe air and mammals that never touch the land. We have to expect that there were complex and unusual ways of life in the past, and that evolution took some unexpected turns at times. The challenge of teaching paleontology, and the challenge of writing a book like this, is to present a complex story in a way that is simple enough to grasp, yet true enough to real events that it paints a reasonable picture of what happened and why. I believe it can be done, and done so that you can learn enough to appreciate what's going on in current research projects.

Some of my colleagues believe that we can never know enough about the ways that creatures evolved or the ways they lived to present a coherent picture. Some biologists state flatly that we can never identify the ancestors of anything. Others say that attempts to reconstruct how extinct creatures lived are at best "story-telling" rather than science.

I think that's rubbish. I think paleontologists can identify how evolution happened and how the creatures of the past lived. We can't *prove* it, any more than we can *prove* what motivated George Washington. But we can state clearly what we know and don't know, we can suggest why certain events happened, and we can describe the evidence we used and the thoughts behind our suggestions. Then people can accept the ideas or not, as they wish. In short, I believe that the fossil record is good enough to allow us to know a lot more than the pessimists think. I'd go further: I think we are limited more by a lack of good ideas than by the facts of the fossil record. I have not been shy about offering *explanations* of events as well as descriptions of them. Mostly they are other people's explanations, but now and again I've suggested some of my own. You can accept these or not, as you wish. The question you face is that of a jury member: is this idea sound "beyond all reasonable doubt?" If you don't accept an explanation for an event, you can leave it as an abominable mystery, with no explanation at all, or you can suggest a better explanation yourself.

There is one caution, however. No one is allowed to dream up any old explanation for past events. A scientific suggestion (a hypothesis) has to fit with the evidence, with the laws of physics, chemistry, and engineering, and with the principles of biology and ecology that have been pieced together over the past 200 years of scientific investigation.

There's yet another wrinkle. A jury decides on a case, once and for all, with the evidence available. But in a science, the jury is always out, and new evidence comes in all the time. You may have to change a verdict—without regret, because you made the best (wrong) decision you could based on the old evidence. Some of the ideas in the earlier editions are wrong, and you won't find them here; you'll find better ones. Sometimes the new answer is more complex, sometimes it is simpler. Always, however, the new idea fits the evidence better. That's simply the way science works: not on belief, not on emotional clinging to a favorite idea (even if it is your own), but on evidence. I never expect to be able to write *the final solution* to the major questions about the history of life, but I do expect to be able to provide better answers this year than I could last year. If my lectures are the same this year as they were last year, then something is wrong with our science, or something is wrong with me. Paleontology is exciting because it is advancing so quickly.

Since paleontology is so fast-moving as a science, this book has changed too. I have radically re-written the sections on the origin of life, on the great Cambrian radiation of metazoan animals, on dinosaurs, on birds, and on human evolution. I have recognized that it is impossible to survey paleoclimatology in one chapter, and I have added a brief survey of the functional anatomy of invertebrates as an appendix. Every other section has been fine-tuned to reflect new research.

The book will be out of date before it is published, at least in a few places. So I have begun a continuing project to keep it abreast of current research, using the Internet. There will be a Web page for this book, and it will serve several functions.

First, it will link you to sites that contain illustrations of fossils and of fossil reconstructions, as color images, sometimes animated, that simply cannot be included in the book.

Second, it will allow me to write short updates that will present new research results directly linked to the pages in this book. And I will be able to present lists of references to those new results, to augment the references already presented at the end of each Chapter.

Third, it will link you to news reports, press releases, and some journal articles that describe the new research directly. Some science magazines are very generous about providing free Web access to selected articles they print: *Discover*, *Scientific American*, and *Science News* are good examples as I write. Many scientific journals are suspicious of the Web, often requiring a subscription by the institution before pages can be accessed on the Web. I have made it a personal policy not to use links to these journals, because not everyone will be able to access them.

Fourth, there will be links to the Web pages of individual research scientists who have made the effort to explain and illustrate for Internet users what they are doing, and how, and why. (It takes time and trouble to set up a good Web page, so this is a real service to colleagues and the public in general.)

As I write, the full development of the Internet in publishing and educating and informing is in the early stages of a revolution. It is going to be an exciting few years ahead, and I hope that the fundamental founding principle of free interaction on a global scale will not be compromised by governmental and commercial interests.

The Web site for this book can be reached at

www-geology.ucdavis.edu/~GEL3/   (note the dash - !)

To My Teaching Colleagues

The course for which this book was written serves three audiences at the same time: it is an introduction to paleontology; it is a "general education" course to introduce nonspecialists to science and scientific thought; and it can serve as an introduction to the history of life to biologists who know a lot about the present and little about the past. Therefore, the style and language of this book are aimed at accessibility. I do not use scientific jargon unless it is useful. I have tried to show how we reason out our conclusions—how we choose between bad ideas and good ones. I have not diluted the English language down to pidgin to make my points. In short, I have aimed this book at the intelligent nonspecialist.

I have not covered the fossil record evenly. I've tried to write compact essays on what I think are the most important events and processes that have molded the history of life. They illustrate the most important ways we go about reconstructing the paleobiological past. I've used case studies from vertebrates more than from other groups simply because those are the animals with which we are most familiar. Most fossils are marine invertebrates, and most paleontologists, including myself, are invertebrate specialists. I have tried to write briefly about invertebrates at an introductory level. They are not the easiest vehicles to use at this level, and that thought has controlled my choice of subject matter.

I'm pleased with the text of this book: I believe it communicates a lot about our science in the space available. But it's impossible to communicate paleontology well without a much greater visual component than can be included in a relatively inexpensive book format. So this book is under-illustrated. I use a lot of slides in my classes in an attempt to bring fossil and living organisms into the classroom, and to give life to the words and names.

The references are a careful mixture of important books, primary literature, news reports, and review articles that bring the latest work into this edition as it went to press. I have deliberately skewed the lists to include items likely to be found in small college and city libraries.

If this book contained nothing controversial, it would be very dull and far from representing the state of paleontology as it stands today. I have tried to present arguments for and against particular ideas in case studies that are presented in some detail, such as the K–T boundary controversy. Often, however, space or conviction has led me to present only one side of an argument. Please share your dissatisfaction and/or more complete knowledge with your students, and tell them why my treatment is one-sided or just plain wrong. That way everyone wins by exposure to the give and take of scientific argument as it ought to be practiced between colleagues.

To Students

Several thousand people like you have voted with their comments, questions, body language, and formal written evaluations on the content of my course. Those people have had more influence on the style and content of this book than anyone else. So you and your peers at the University of California, Davis, can take whatever credit is due for the style in which the material is presented.

After all the thanks, however, I do have another point to make. You don't have to take any of the interpretations in this book at face value. Facts are facts, but ideas are only suggestions. If you can come up with a better idea than one of those I've included here, then work on it, starting with the literature references. It would make a great term paper, and (more important) you might be right. The 1960s slogan "Question Authority!" is still valid. Your suggestion wouldn't be the first time that a student found a new and better idea for interpreting the fossil record.

Why do I, and why should you, bother with the past? Henry Ford said, "History is more or less bunk." If we don't understand the past, how can we deal intelligently with the present? We and our environment are reaching such a state of crisis that we need all the help we can get. Nature has run a series of experiments over the last 3.5 billion years on this planet, changing climate and geography, and introducing new kinds of organisms. If we can read the results of those experiments from the fossil record, we can perhaps define the limits to which we can stretch our present biosphere before a biological disaster happens.

The real pay-off from paleontology for me is the fun involved in reconstructing extinct organisms and ancient communities, but if one needs a concrete reason for looking at the fossil record, the future of the human race is surely important enough for anyone.

This Book

I begin this book with the formation of Earth and the great unsolved problem of the origin of life. Then I describe an early Earth populated entirely by bacteria, so strange in its chemistry and ecology that it might well be another planet. Eventually, living things so alter their world that we begin to recognize environments and organisms that seem much more familiar. I describe the evolution of animals and begin to worry about their physical and ecological environment. By now, we are dealing with a world whose geography we can begin to reconstruct, which leads to chapters on plate tectonics and the climates of the past, and how they might have affected living things.

The vast record of invertebrates allows us to measure the diversity of life through time, which shows that there have been times of high diversity, and times of dramatic extinctions. I deal with extinction, mainly to look at the crises or "mass extinctions" that have occurred sporadically through time. Then I turn largely to the history of vertebrates, following some of the great anatomical, physiological, and ecological innovations by which fishes gradually evolved into the major classes of tetrapods on the land, including ourselves.

Some of my colleagues are dubious about evolutionary "progress," but I regard the evidence for it as overwhelming. I have not tried to write a simple historical catalog of fossils. Instead, I have tried to set interesting episodes in Earth history into a global picture. For example, the tragedy that overtook Mesozoic communities 65 million years ago has to be seen in terms of their success until that time, and the radiation of the mammals can only be appreciated against a background of changing planetary geography and climate. Finally, the rise to ecological dominance of humans has its counterpart in the massive changes in land faunas that accompanied it, all set in the context of the ice ages.

I owe a great deal to the people at Blackwell Science who have encouraged and helped me over the years. I would like to thank Simon Rallison and Jane Humphreys for past help, and Nancy Hill-Whilton, Nancy Duffy and Lisa Flanagan for this edition. My friends Norm Gilinsky and Ken McKinney provided very valuable early advice. Doug Eernisse at California State University, Fullerton, constructed the first serious Web pages linked to the second edition of this book, and I used his work as I began to construct the present version at Davis.

The book was produced on my Macintosh™ computer system, because it is so much fun to use and because the contents can be current three months before the book appears on the shelf. This production style may result in a less professional and more personal book, but it cuts down the final cost and makes future revision simple. I composed the text in WriteNow™ and laid out the pages in ReadySetGo!™.

None of these people or programs should be blamed for deficiencies: please complain directly to me at **rcowen@ucdavis.edu**.

Finally, I thank my wife Jo and my daughters Claire and Alexandra for tolerating my neglect of them while all this was in process. If you buy this book, you may well contribute toward their further deprivation by encouraging another new edition.

Davis, California   June 1999

COVER PHOTOGRAPH: Trees appear to be deeply rooted in the human psyche: one need only think of the trees in the Garden of Eden, the great world tree Yggdrasil of the Norse legends. Biologists use the same images: they use the metaphors of trees, roots, stems, and branches as they try to describe evolution. Branches may fail (they are "pruned" by extinction). This tree is obviously old, and has had many failed branches during its lifetime, just like our metaphorical tree of life. To complete the imagery, think of the paleontologist, who does not see the tree itself, but its rather blurred reflection in the fossil record. The real tree in the photograph stands on Rannoch Moor, in the Highlands of Scotland.

The Origin Of Life

Life on Earth is a fact, although we don't know where and how it began. There is no evidence of life, let alone intelligence or civilization, anywhere in the universe except on our planet. The most reasonable hypothesis to explain these two statements is that life evolved here on Earth.

There are complex organic molecules in interstellar space, on interplanetary dust, in comets, and in the meteorites that hit the Earth from time to time. It makes good chemical sense that such compounds form naturally in interplanetary or interstellar space, because gas clouds, dust particles, and meteorite surfaces are bathed in cosmic and stellar radiation. But life as we know it consists of cells, composed mostly of liquid water that is vital to life. It is impossible to imagine the formation of any kind of water-laden cell in outer space; that could only have happened on a planet that had oceans and therefore an atmosphere.

Planets may have organic compounds delivered to them from space, by way of comets or meteorites, but it is unlikely that this process in itself leads to the evolution of life. Organic molecules must have been delivered to Mercury, Mars, Venus, and the Moon as well as to Earth, only to be destroyed by inhospitable conditions on those lifeless planets.

Experiments show that it is fairly simple to form large quantities of organic compounds in planetary atmospheres and on planetary surfaces, given the right conditions. Space-borne molecules thus may add to the

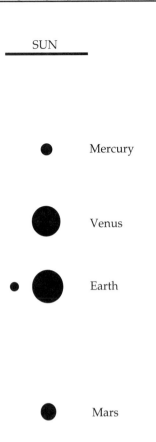

Figure 1.1 The Sun and the terrestrial or inner planets. Their relative sizes and their relative distances are correct, but their sizes are exaggerated by about five billion times compared with their distances.

supply on a planetary surface, but they would never be the only source of organic molecules that led to the origin of life.

In testing the idea that life evolved here on Earth (from nonliving chemicals), we deal entirely with principles of nature that we can study. Geologists and astronomers can use evidence gathered from the Earth, Moon, and other planets to reconstruct conditions in the early solar system. Chemists can determine how complex organic molecules could have formed in those environments. Geologists can try to determine when life became established on Earth, and biologists can design experiments to test whether these facts fit with the idea of evolution of life from nonliving chemicals.

The Formation of the Solar System

The Universe is about 12 or 13 billion years old, according to current estimates. The Sun, planets, asteroids, and comets that make up our solar system are much younger, relative newcomers at an age of 4.55 billion years (that is, 4550 million years), give or take a mere 50 million years!

As astronomers reconstruct it, a cloud of interstellar dust and gas floated in our inconspicuous part of the Milky Way galaxy for several billion years. Then, a nearby supernova explosion blasted new material and a lot of energy into the cloud; as a result, or by coincidence, the cloud began to collapse on itself. Most of the material condensed in the center of the cloud to form a new star, our Sun, but about 1% of the cloud remained in orbit around the new star as dust and gas.

Dust particles collide softly and tend to stick together by electrostatic and gravitational attraction in a process called **accretion**. "Dust bunnies" form under the bed in the same way. Around a new star, the dust bunnies can build up and compact into substantial solid masses a kilometer or so in diameter. Computer models show that in only a few million years, several thousand bodies the size of large asteroids will coalesce into larger units that we now see as **planets**. The new planets continue to be bombarded by asteroid-sized objects for perhaps several hundred million years in an **era of huge impacts**. Sometimes planets may have been shattered in huge impacts, or had fragments splintered off them into space. For example, a body larger than Mars may have hit the Earth just after it formed, knocking its axis into the present 23° tilt that gives us our seasons, and blasting debris into Earth orbit that quickly accreted to form the Moon. Around this time, our Solar System took on its present form, with three or four major terrestrial planets in stable orbits (Figure 1.1), giant gas planets orbiting outside them, and meteorites, asteroids, and comets still orbiting in space as celestial debris.

Earth is one of four **terrestrial** (rocky) planets in the inner part of our solar system. Venus and Earth are about the same size, and Mars and Mercury are significantly smaller. They all formed in the same way, and most likely, they were all bombarded heavily in the era of huge impacts. All the rocky planets would have been very hot for a long time, with many active volcanoes. They melted deeply enough to form planetary cores made of iron and to give off gases to form atmospheres. But there their similarity ended, and each inner planet had its own later history.

Life on Planetary Surfaces

Once a planet survives the era of huge impacts and cooled, its surface conditions are largely controlled by its distance from the Sun and by any volcanic gases that erupt into its atmosphere. The geology of a planet therefore greatly affects the chances that life might evolve on it.

Liquid water is vital for life as we know it, so surface temperature is perhaps the single most important feature of a young planet. Surface temperature is primarily determined by distance from the Sun: too close, and liquid water evaporates to water vapor; too far, and water freezes to ice. But that's not all, otherwise the Moon would have oceans like Earth's. Gas molecules tend to escape into space from the weak gravitational field of a small planet. The smaller the planet, the faster gases are lost and the heavier are the molecules that escape. Gases are also lost from the atmosphere as they react chemically with surface rocks. They can be released again only by eruptions that melt those rocks. A small planet cools quickly, and its volcanic activity stops as its interior freezes. After that, no further volcanic gases will erupt to return or add gases to the atmosphere. Therefore, a small planet quickly evolves to have a very thin atmosphere or no atmosphere at all.

Volcanic gases include large amounts of water vapor and CO_2 (Figure 1.2). Both gases trap solar radiation in the atmosphere (the **greenhouse effect**), and keep the planetary surface warmer than one might expect simply from its distance from the Sun. For example, Earth would have been frozen for most of its history without CO_2 and water vapor in its atmosphere. With these principles in mind, let's look at the prospects for life on the planets of our solar system.

Both **Mercury** and the **Moon** had active volcanic eruptions early in their history, but they are small. They cooled quickly and are now solid throughout. Their atmospheric gases either escaped quickly to space from their weak gravitational fields or were blown off by major impacts. Today Mercury and the Moon are airless and lifeless.

Venus is larger than the Moon or Mercury, almost the same size as Earth. Volcanic rocks cover most of the planetary surface. Like Earth, Venus has had a long and active geological history, with a continuing supply of volcanic gases for its atmosphere, and it has a strong gravitational field that can hold most gases. But Venus is closer to the Sun than Earth is, and the larger amount of solar radiation hitting the planet was trapped so effectively by water vapor and CO_2 that water molecules may never have condensed out as liquid water. Instead, they remained as vapor in its atmosphere until they were dissociated, broken up into hydrogen (H_2), which was lost to space, and oxygen (O_2), which was taken up through oxidization of the hot surface rocks of the planet.

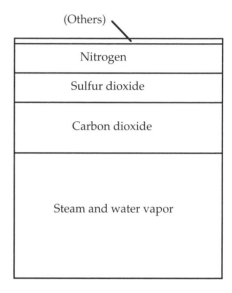

Figure 1.2 A planetary interior may be hot enough to melt rock. If molten rocks reach the planetary surface in volcanic eruptions, the gases they give off may help to form an atmosphere. The mixture of gases shown in this diagram was measured at Kilauea Volcano, in Hawaii; other volcanoes erupt different mixtures, but the basic ingredients are the same. Sulfur dioxide is caught up in raindrops, leaving water vapor, carbon dioxide, and nitrogen as the main constituents of the atmosphere.

Today the dense, massive atmosphere of Venus consists largely of CO_2. Atmospheric pressure at the surface is 100 times that of Earth's. Sulfur gases react in the atmosphere to make tiny droplets of sulfuric acid (H_2SO_4), forming the famous clouds that hide the planetary surface. Water vapor has vanished completely: it has either dissociated or has been bound into molecules of sulfuric acid or carbonic acid. Although the sulfuric acid clouds reflect 80% of solar radiation, CO_2 traps the rest, so the surface temperature is about 450°C (850°F). We can be sure that there is no life on the grim surface of Venus.

Mars is much more interesting than Venus from a biological point of view. It is smaller than Earth, and farther from the Sun. But it is large enough to retain a thin atmosphere, mainly composed of CO_2. Mars today is cold, dry and windswept, with dust storms that can cover half the planet.

No organic material can survive now on the surface of Mars. There is no liquid water, and the soil is highly oxidizing. But while Mars was still young, and was actively erupting volcanic gases from a hot interior, the planet may have had a thicker atmosphere with substantial amounts of water vapor. The crust may still contain ice in cracks and crevices that could be set free as water if large impacts heated the surface rocks deeply enough to melt it, or if climatic changes were to melt it briefly.

Mars had surface water in the distant past. Canyons, channels, and plains look as if they were shaped by huge water floods, and other features look like ancient sandbars, islands, and shorelines. Ancient craters on Mars, especially in the lowland plains, have been eroded by gullies, and sheets of sediment lap around and inside them, sometimes reducing them to ghostly rims sticking out of the flat surface. The flood waters may have drained and dried very quickly, but there may have been temporary oceans. Some estimates suggest that there was water in shallow oceans for as long as 500 m.y. during the early history of Mars.

Mars was too small to sustain geological activity for long. As the little planet cooled, its volcanic activity stopped, and its atmosphere was blasted off by impacts, or lost by slow leakage to space and by chemical reactions with the rocks and soil. The surface is now a dry, frozen waste, and even floods generated by meteorite impact cannot last long enough to sustain life. But life may once have existed briefly on Mars.

In 1996 a team of researchers reported they had found organic compounds and tiny fossil bacteria in ALH84001, a meteorite picked up on the Antarctic ice cap. This meteorite most likely originated on Mars, and was splashed into space by an asteroid impact, to arrive on Earth after spending thousands of years in space. The researchers suggested that the organic compounds and the bacteria were Martian. By 1998 the evidence was looking very weak. The organic molecules are real, but most of them are contamination after the rock reached Earth. The "fossil bacteria" are not real. They are about 1/1000 the size of normal bacteria, physically too small to contain enough genes and chemicals to operate a cell. It remains true that the only planet known to have life is ours, the Earth, and that there is no good evidence for past or present life on Mars.

The **asteroid belt** lies outside the orbit of Mars. Some asteroids have had a complex geological history and may once have been part of larger bodies. There is no question of life in the asteroid belt now. No planet or moon outside the orbit of Mars could trap enough solar radiation to form liquid water on its surface to provide the basis for life. Complex hydrocarbon compounds can accumulate and survive on the surfaces of meteorites, or in the atmospheres of the outer planets or some of their satellites, but those bodies are frigid and lifeless.

So we return to **Earth** as the only known site of life. Earth probably had a molten surface originally, because the catastrophic impact that formed the Moon probably melted Earth completely. After this dramatic event, Earth settled down with a core, mantle, and at least a small area of crust. Impacts and volcanic eruptions continued to release gases to form a thick atmosphere that consisted mainly of CO_2, with small amounts of nitrogen, water vapor, and sulfur gases (Figure 1.2). Any hydrogen quickly escaped to space.

Carbon dioxide in the atmosphere trapped solar radiation, so Earth's surface was warm. But Earth was cool enough to form a crust, and water vapor condensed to form oceans. Oceans in turn helped to dissolve CO_2 from the atmosphere and deposit it into carbonate rocks on the seafloor. This process absorbed so much CO_2 that Earth did not develop runaway greenhouse heating as Venus had (Figure 1.3), and very little of Earth's water vapor was broken down in the atmosphere, to be lost as hydrogen leaked to space. Instead, large, shallow oceans covered most of Earth, with a few crater rims and volcanoes sticking out as islands.

Huge impacts on the early Earth would have wiped out any life or proto-life on the planet. The life forms that were our ancestors could not have evolved until after the last sterilizing impact. Smaller late impacts may have encouraged the evolution of life, as comets and meteorites fell on to Earth from space. All comets and a few meteorites carry a significant proportion of organic molecules, and comets may have been a major source of the organic molecules that made possible the evolution of life on Earth. But processes here on Earth also fostered the formation of important organic chemicals.

Lightning and intense ultraviolet (UV) radiation from the young Sun acted on the atmosphere to form small traces of very many gases. Most of the gases dissolved easily in water, and as they formed they rained out into the early oceans, making them rich in carbon. The gases included ammonia (NH_3), methane (CH_4), carbon monoxide (CO), and ethane. As much as three million tons per year of formaldehyde (CH_2O) could have formed. Nitrates accumulated in seawater as photochemical smog, and the nitric acid produced in lightning strikes rained out. But perhaps the most important chemical of all was **cyanide** (HCN). Cyanide forms easily in the upper atmosphere, from solar radiation and from meteorite impact, and then dissolves in raindrops. Today it is broken down almost at once by oxygen, but early in Earth's history it built up at low concentrations in the ocean. Cyanide seems to be a basic building block for more complex organic molecules. Life probably evolved in chemical conditions that would have killed us instantly.

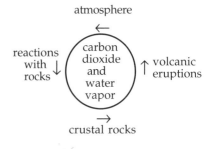

atmosphere

reactions with rocks

carbon dioxide and water vapor

volcanic eruptions

crustal rocks

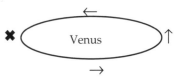

Venus

gases remained in the atmosphere

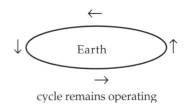

Earth

cycle remains operating

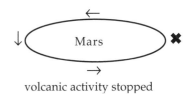

Mars

volcanic activity stopped

Figure 1.3 Atmospheric evolution took completely different courses on Venus, Earth, and Mars. On a model planet (top), water and carbon dioxide cycle around the rocks, the atmosphere, and the ocean. This still happens on Earth. Mars cooled so quickly that volcanic activity soon ceased, stopping the cycle with the gases frozen in the crust. Venus heated so much that carbon dioxide remained permanently in the atmosphere, stopping the cycle with a hot, dense atmosphere.

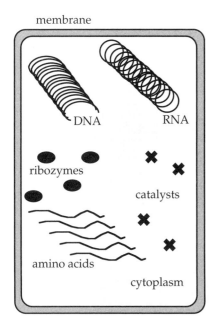

membrane

DNA   RNA

ribozymes

catalysts

amino acids

cytoplasm

Figure 1.4 An attempt at reconstructing the "universal minimal cell," based mainly on suggestions by Lynn Margulis.

By 3.8 b.y. ago, rocks were forming on Earth at surface pressures and temperatures we think of as "normal". Earth had oceans by then, and a fairly stable climate. Therefore the origin of life probably occurred in physical conditions very much like those of today.

Reconstructing the Origin of Life

The simplest cell alive today is very complex. After all, it has evolved through many billions of generations. We must strip away these complexities as we try to imagine what the first living cell might have looked like and how it worked.

What is the simplest living cell one can imagine, based on our knowledge of cell biology? A "universal minimal cell" (Figure 1.4) is surrounded by a **cell membrane**, an envelope that helps to isolate chemical reactions inside the cell from dilution or contamination from outside, but also allows some molecules to pass in and out of the cell as necessary. Most of the cell interior is water, but organic molecules are dissolved in that water to form a rich viscous fluid, the **cytoplasm**. Some of the molecules in the cytoplasm provide fuel for chemical reactions: adenosine triphosphate (**ATP**) is the most important of those. The cell also contains the nucleic acid deoxyribonucleic acid (**DNA**), in the form of a long double chain made up of **bases**, a **sugar**, and **phosphoric acid**. Ribonucleic acid (**RNA**) is formed in a specific pattern laid out by either of the DNA chains, and, in turn, RNA controls the assembly of **proteins**, which consist of long chains of **amino acid** molecules linked end to end in a specific sequence. Some proteins act as organic **catalysts**, which help reactions inside the cell, while RNA sequences called **ribozymes** also act as catalysts for making more RNA. Other proteins are structural units inside the cell and in the membrane.

Such a minimal cell could not have appeared from nowhere. Any theory that advocates the evolution of life, as opposed to its creation by a divine being, must include a time during which **chemical evolution** produced the basic building blocks of life on a lifeless planet, brought them together in the right place at the right time, and molded and perfected the chemical reactions to the point where we could say that a living thing had emerged. It's a tough proposition, because even a minimal cell already contains organic catalysts to help along its internal reactions, without which it would die. Information from geology and organic chemistry can help to clarify which reactions were and were not possible on the early Earth.

The vital part played by phosphorus in cell chemistry is not a cosmic accident. The chemistry of phosphoric acid allows it to hold molecules together along the chains of DNA and RNA and make them very stable. Phosphoric acid is the only commonly available acid that has these properties and could form the basis for evolving a stable biochemistry. If life has evolved anywhere else, it must necessarily be based on water, carbon, and phosphorus. That gives us a starting point for designing experiments on the origin of life from non-living chemicals.

BOX 1.1 NECESSARY CONDITIONS FOR THE EVOLUTION OF LIFE FROM NON-LIVING CHEMICALS

Energy is needed to form complex organic molecules. Some laboratories have used electrical discharges to simulate lightning on the early Earth; others use high-energy particles from a cyclotron in place of radioactivity from rocks and cosmic rays, heat for volcanic activity, shock waves or laser beams for meteorite impacts, or lamps for solar UV radiation (Figure 1.5). All these energy sources were present on the early Earth.

Protection. Continued energy input (especially heat) will destroy any complex organic molecules that form in reactions, so after they form they must quickly be protected from strong radiation. Laboratory experiments are often designed to allow organic molecules to drop into cold water away from the energy source. On the early Earth, molecules may have been protected under shallow water, in sheltered tide pools or rock crevices, under rocks, ice, or particles of sediment.

Concentration. All chemical reactions run better at high concentration, but almost all reactions leading toward life give low yields in the laboratory. Life is water-based, yet too much water dilutes chemicals so that they react slowly. Some process must have concentrated chemicals on the early Earth. Evaporation is one, and there are others.

Catalysis. Catalytic converters in the exhaust systems of cars contain platinum as a catalyst that encourages the breakdown of pollutants. An organic substance that works as a catalyst is called an **enzyme**. All the reactions inside our cells and our bodies are aided by enzymes, which are necessary even in the simplest possible living cell. Suitable catalysts may have encouraged difficult reactions on the early Earth, even at low energy levels and low concentrations. Later, the last stages leading toward life may have been aided by catalysts trapped on or inside membranes.

Amino Acids

Everyone agrees that certain conditions are necessary if life is to evolve from nonlife (Box 1.1). The first experiment to include some of them was done in 1953 by Stanley Miller, who was then a graduate student at the University of Chicago. He passed energy (electric sparks) through a mixture of hydrogen, ammonia, and methane in an attempt to simulate likely conditions on the early Earth (Figure 1.6). Any products fell out of the energy into a protected flask. Among these products, which included cyanide and formaldehyde, were four different amino acids. This result was surprising at the time because amino acids are not simple compounds. Although Miller's experiment used a rather unlikely mixture of starting gases, it encouraged many other experiments on the origin of life, and it showed that chemical experiments related to this problem were not only respectable science but were very exciting and rewarding.

Most of the amino acids found in living cells today could have formed naturally on the early Earth, from a wide range of ingredients over a wide range of conditions. They form readily from mixtures that include the gases of Earth's early atmosphere. One experiment yielded seven amino acids from dilute cyanide mixtures. Moreover, the same amino acids that form most easily in laboratory experiments are the most common in living things today. The only major condition is that amino acids do not form if oxygen is present.

Amino acids form easily on the surfaces of **clay particles**. Clays can attract some organic molecules and thus can act as natural concentrators. Shallow water clays on the early Earth could also have protected organic molecules from breakdown by UV radiation.

Organic molecules occur in meteorites. Again, the most common ones are also the most abundant in laboratory experiments. Nucleic acid bases are also found in meteorites, so all these compounds would have

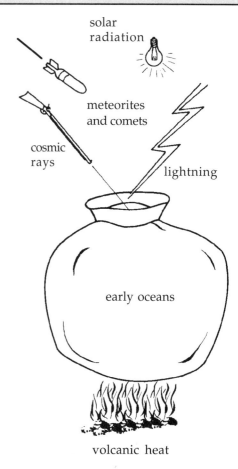

Figure 1.5 Some of the energy sources that would have been available to power chemical reactions on the early Earth. (After Cowen, *History of Life.* © 1976 McGraw-Hill Book Company.)

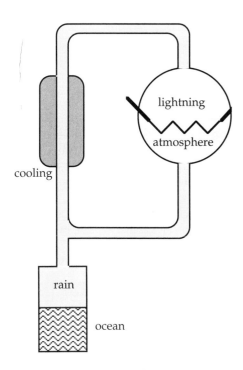

Figure 1.6 Stanley Miller's experiment was designed to simulate conditions on the early Earth. An atmosphere of methane, ammonia, and hydrogen was subjected to lightning discharges, and the reaction products cooled, condensed, and rained out to collect in the ocean. The reaction products included amino acids.

been supplied in quantity to the early Earth. We do not know how much organic matter was supplied to the oceans by natural synthesis on Earth and how much by meteorite infall. That ratio is not as important as the fact that the right materials were present on the early Earth to encourage further reactions.

Larger Organic Molecules

Linking sequences of amino acid molecules into chains to form protein-like molecules involves the loss of water, so scientists have tried evaporation experiments to simulate the process under early Earth conditions. High temperatures help evaporation, but they also pose a problem, because organic molecules tend to break down if they are heated: the longer the molecule, the more vulnerable it is to damage. Here again, natural minerals provide an attractive alternative. Amino acids in water can be adsorbed onto the surfaces of clay or phosphate minerals and linked there into proteinlike substances.

Nucleic acids (RNA and DNA) have structures made up of nucleic-acid bases, sugars, and phosphates. Nucleic-acid bases can be made from cyanide in experiments that simulate lightning strikes on the early Earth. The base **adenine** forms easily from cyanide, but the formation of other bases from cyanide mixtures was a pleasant surprise. Sugars have been formed in laboratory conditions simulating the flow of water from hot springs over beds of clay. Naturally occurring phosphate-bearing acids and minerals are associated with volcanic activity. Thus all the ingredients for nucleic acids were present on the early Earth, and the universal cell fuel ATP could also have formed easily on the early Earth.

Linking sugars, phosphates, and nucleic-acid bases to form fragments of nucleic acid called **nucleotides** is also a dehydration process, and the phosphates themselves can act as catalysts here. As for amino acid chains, long nucleotides form much more easily on phosphate or clay surfaces than they do in suspension in water.

The linear structure of clay minerals encourages nucleotides to arrange themselves naturally in lines, making them likely to link into chains to form short nucleic acid molecules (Figure 1.7). The interesting point about this reaction is that nucleotide molecules are large enough to force clay layers apart, providing new active surfaces for further reactions, which would then open up new active surfaces, and so on. Clays also absorb polyphosphates such as ATP, which might have provided energy for some of the organic reactions taking place on clay surfaces.

Figure 1.7 Clay minerals have long, straight cleavage planes. Linear organic molecules may line up along the cleavages, encouraging reactions that form long-chain organic molecules such as amino acids and nucleic acids.

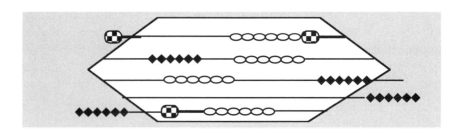

Toward the First Living Cell

The basic organic molecules that make up cell membranes and cell contents may have been present in reasonable amounts in the oceans of the early Earth. We must still explain how they evolved into a cell that could reproduce itself. First, organic chemicals have to be concentrated. Four concentration mechanisms could have occurred naturally: evaporation; freezing; scums, droplets, or bubbles; and mineral grains.

Evaporation is usually associated with warmth, which also speeds up chemical reactions. Chemicals can be concentrated in desert lakes (the Dead Sea is a good example) and in tide pools along shorelines. More rarely, lavas running from active volcanoes into the sea can cause rapid heating and evaporation along the shore.

Concentration by freezing. "Ice beer" is made by freezing ordinary beer for a short time. Some water freezes out, leaving behind a beer with a higher alcohol content. Although chemical reactions are normally slow at freezing temperatures, they would be encouraged by high concentrations, and any complex molecules formed would not be likely to break down easily.

Scums are layers of chemicals that form on water surfaces because they repel water and are light. They are therefore natural concentrates. If they are tossed around in turbulent water, scums may form **droplets** that can then act as tiny, concentrated chemical reaction chambers. Complex molecules could be formed inside such droplets.

Bubbles also form in water as waves break. Most of them rise back to the surface and burst, or gradually disappear as their contained gases dissolve. But water is never pure, and organic substances may line the surface of the bubbles. A rich organic soup may form a foam that persists for a long time, giving chemical reactions more time to work.

Mineral grains may sometimes have a small surface electric charge that acts to repel or attract molecules toward them. Clay minerals are abundant in nature, have long linear crystal structure, and are very good at attracting and adsorbing organic substances: cat litter is clay and works on this principle.

On the early Earth, clay particles could have attracted a wide variety of organic molecules from the surrounding water, concentrating them many thousands of times and lining them up along the natural crystal structure of the clay particles (Figure 1.7). Local surface reactions would then be strongly favored, and the reaction products would tend to stay with the clays and be protected by them. Even inside a living cell, reactions do not take place loose in the cytoplasm. Proteins are made on the surfaces of structures called **ribosomes**, for example.

Putting this all together, we might conclude that the best places of all for concentrating important chemicals might be in sheltered, scummy lagoons around a stormy volcanic island in high latitudes, where clays were forming in a freezing environment!

The Naked Gene

In living cells today, each protein is coded on long sequences of nucleic acid. The long sequences of DNA that specify these protein structures are themselves difficult to replicate, and replication requires many proteins to act as enzymes to catalyze the reactions. Protein synthesis and DNA replication are interwoven in cells today; neither is independent of the other, even though they use very different chemical pathways and probably evolved independently. How could these two processes have begun independently, then evolved to depend on each other?

Some RNA sequences called **ribozymes** can act as enzymes and make more RNA even when no proteins are present. Other RNA sequences act as catalysts that speed up the assembly of proteins. Perhaps then the first living things were pieces of RNA, or **naked genes**, which happened to have the right structure to act as enzymes that helped RNA to replicate itself. In this theory, RNA ribozymes on the early Earth were able to replicate themselves, slowly and inaccurately, and therefore could have been considered alive. This scenario that begins with naked genes is currently the best hypothesis for the origin of living cells.

On the early Earth, early ribozymes that also coded for powerful protein enzymes would replicate faster than other ribozymes. (In experiments, RNA replication is about a thousand times faster if protein enzymes are present.) Increasingly successful early ribozymes would very quickly have outcompeted all other ribozymes to become the ancestors of life on Earth. DNA evolved later, in a major mutation that was superior because the double strand of DNA is more easily repaired after damage than the single strand of RNA.

It's unlikely that the first cells evolved in open water. Most biochemical reactions have to be protected from free exchange with the external environment, because reactions are diluted by water or poisoned by toxins. The first cells probably evolved inside a membrane, on a clay surface, under a membrane on a clay surface, or inside a coating of polysaccharide "slime." Living cells today have an membrane with special proteins embedded in it to control intake and output of substances.

If life evolved by way of naked genes, then it did not do so in hot springs. At such high temperatures, nucleic acid bases are unstable unless they are protected by a cell membrane, and so are RNA and DNA. Naked genes could not have existed (for long) in hot springs.

Many organic membranes are made of sheets of molecules called **lipids**. A lipid molecule has one end that attracts water and one end that repels water (Figure 1.8a). Lipids line up naturally with heads and tails always facing in opposite directions; a sheet of lipid molecules therefore repels water (Figure 1.8b). If the sheet of lipids happens to fold around to meet itself, it forms a waterproof membrane around whatever contents it has trapped. Such packets, called **liposomes**, form spontaneously in lipid mixtures. They are simple spheres with an outer membrane of lipids (Figure 1.8c). Whipping up an egg in the kitchen produces liposomes as the yolk is frothed around. David Deamer discovered that liposomes can

(a)

(b)

(c)

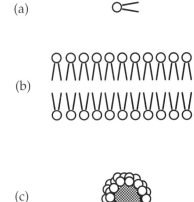

Figure 1.8 (a) The polar structure of a lipid molecule repels water at one end (o) and attracts it at the other. This allows lipids to form either (b) water-repellent sheets and scums or (c) water-repellent spherical containers (liposomes).

form from a mixture of molecules that would have been present on the early Earth—fatty acids, glycerol, and phosphates. He then found fatty acid molecules in meteorites, and he made globules from them by drying them out and then rewetting them. If DNA is added to the original solution, it can be trapped inside the liposomes as they form.

Liposomes could have formed in great numbers in the early ocean as wave turbulence thrashed around lipid layers on the water surface, or as lipid scum washed up on shore. Some globules may have contained important chemicals: amino acids, primitive forms of nucleic acid, and so on (Figure 1.8c). This mechanism allows for the formation of a set of liposomes with greatly variable contents: the "best" would have operated chemical reactions much more efficiently than the "worst" ones.

If simple proteins and DNA are added to mixtures from which liposomes are forming in the laboratory, the amount of DNA trapped inside some of them is multiplied 100 times over Deamer's original results. This suggests that there is a powerful affinity between nucleic acid and some simple proteins, so that the formation of liposomes that had cell-like contents may not have been as difficult as one might think; perhaps one might call such liposomes **protocells**.

Catalysts are needed in cells today to speed up the formation of long chains of molecules, nucleic acids, and protein structures. Clays are natural catalysts. Perhaps the first association between RNA and proteins occurred on the ocean floor on the catalytic surfaces of clay minerals (Figure 1.7). Later, the first successful protocells with lipid membranes evolved to become globules in the water (Figure 1.8c) rather than organic coatings on the surfaces of clay minerals.

Many things are unknown about reactions on the early Earth, but we do know that they took place in natural environments and not in test tubes. Chemical reactions on clay surfaces are often faster, easier, and more efficient than experiments in sterile glass test tubes. Clay minerals are so finely layered that a 1-cm cube has a reactive surface area that is not 6 square centimeters, as one would expect, but 2800 square **meters**. Kitty litter works so well in absorbing smells from urine and feces because it is made from finely divided clay particles.

Clays are not the only natural catalysts. **Zeolites** are very reactive natural minerals that form as basaltic lavas break down. They often occur with clays and share many of their properties. (Oil companies patent the zeolites they use as catalysts in refining petroleum.) Perhaps protein synthesis and RNA replication evolved first on the surfaces of clays, zeolites, or other minerals, and reactions inside membranes followed later.

However it formed, any protocell could have grown by absorbing simple organic substances from seawater; perhaps some divided as they grew. Some protocells might have failed to grow because they had an incompatible set of proteins or nucleic acids, while others grew rapidly and divided frequently. As proteins and/or nucleic acid mixtures changed during internal reactions or during protocell splitting, a few "mutants" could have acquired an increased ability to absorb substances and to grow faster. Simple enzymes possibly aided such reactions in protocells, because many enzymes are proteins.

Successful protocells, which grew fast and split frequently, would have produced many "daughter" protocells that were like their parent in chemistry, and also would have been able to grow and split faster than others. Protocells, then, could have behaved much like living cells, except that their internal reactions, and in particular their splitting, were not totally controlled by DNA. It would have been very difficult to draw the line between living and nonliving protocells, because the change from one state to the other would have been so gradual. A protocell could have been called living as soon as protein formation, nucleic acid formation, and cell division became integrated and reliable.

Where Did Life Evolve?

At least six different major habitats have been seriously suggested as the environment in which life began (Box 1.2). However, some are less likely than others. **Soil surfaces** would not attract large quantities of organic material that would be available in ponds or in the ocean. **Interstellar space** and the **atmosphere** are too dry.

The "warm pond" idea of Charles Darwin, and modern experiments such as those of Deamer, encourage the view that **tropical lagoons** could strongly favor the origin of life. Warm temperatures promote chemical reactions, and an early tropical island would most likely have been volcanic and therefore liable to have intermittent strong heating and a ready supply of fresh minerals and reactive clays.

In the laboratory, cyanide and formaldehyde reactions occur readily in half-frozen mixtures. Perhaps **cold volcanic islands** were natural environments for cyanide reactions on the early Earth. Volcanic eruptions often generate lightning storms, so eruptions, lightning, fresh clays, and near-freezing temperatures (ice, snow, hailstones) could all have been present on the shore of a cold volcanic island. Furthermore, RNA bases are increasingly unstable as temperatures rise above 0°C: normal tropical waters (~25°C) may be about as warm as it could be for the origin of life.

Most theories of the origin of life suggest surface or shoreline habitats in lakes, lagoons, or oceans. Solar radiation or atmospheric phenomena are likely energy sources. But deep in the oceans lie the **midocean ridges**, long underwater volcanic rifts where the seafloor is tearing apart and forming new oceanic crust. Enormous quantities of heat are released in the process, much of it through hot water vents on the floors of the rifts, and myriads of bacteria flourish in the hot water. Perhaps life began nowhere near the ocean surface, but deep below it, at volcanic vents.

Laboratory experiments have implied that amino acids and other important molecules can form in such conditions, even linking into short protein-like molecules, and currently the deep-sea hypothesis is enjoying favor. However, it remains true that nucleic acid bases are unstable at high temperatures, so the idea seems unlikely.

The deep-sea hypothesis has led to speculation that life might have evolved deep under the surface layers of other planets or satellites. (For

---

BOX 1.2 POSSIBLE HABITATS FOR THE ORIGIN OF LIFE

- Soils

- The upper atmosphere

- Space

- Tropical lagoons

- Glacial islands

- The deep sea

example, Jupiter's moon Europa probably has liquid water under its icy crust.) The hypothesis helps to generate money for NASA's planetary probes. But the internal heat energy of such planets and moons is very much less than Earth's, and water-borne organic reactions would be much less likely to work deep under the icy crust of Europa than the surface processes on Earth we have already discussed.

New laboratory experiments are producing organic chemicals in conditions that simulate ices forming on dust grains in the freezing near-vacuum of space; in other words, on comets. Even so, the ices have to be thawed out to a *water*-based chemistry to react further. Processes in space may generate chemicals and deliver them to planets; but if a planet is hospitable, the right conditions for generating organic molecules and life already existed *on* that planet.

Living Things

Living things grow by using the information coded into the nucleic acids they inherited from their parent or parents, so that a successful design for an organism stands a reasonable chance of being passed on. The nucleic acids carry a code for the right combination and sequence of proteins to be formed, so that the growing organism is provided with a sound structure and biochemical system. But the growing organism lives in a world that fluctuates. Conditions may or may not favor its growth and success. A growing organism may not survive to reproduce, or it may have fewer successful offspring than a competing cell.

If the nucleic acid is damaged, it represents a **mutation**. The mutation may cause a change in the protein coded in the nucleic acid sequence. The "wrong" protein may have a major effect on the organism if it is a very important one, or the effect of the change may be so minor as to be insignificant. Most often, but not always, individuals with major changes are unsuited to the environment, and they fail to reproduce themselves and their changed DNA. Small-scale mutations often allow their bearers to survive and reproduce, so variation in DNA and in the form and function of organisms can increase with time in a population. Mutations lead to the appearance of descendants that differ from their ancestors and are therefore likely to be better or worse suited to their environment. Such changes are likely to affect reproductive success.

Living things grow and reproduce. Sometimes an individual may simply make **clones** (identical copies) of itself, in **asexual reproduction**. Sometimes two individuals cooperate to produce a new individual with a mixture of nucleic acids from each parent, in **sexual reproduction**. While the new individual that grows as a result of sexual reproduction is genetically similar to both its parents, it is not identical to either. Sexual reproduction, and the **recombination** of genes that occurs during it, does not introduce any new genetic material into a population, but it does produce new, unique combinations of genes that are coded to produce a new, unique individual.

Therefore, mutation and sexual reproduction, in different ways, result in organisms that are subtly different from one another, and thus are likely to have different success in reproduction. The process by which a particular organism leaves more or fewer successful offspring than another is called **natural selection**. Many chance factors affect the survival of particular individuals in a population. But, in the long run, individuals that are well suited to a prevailing environment survive long enough to leave offspring, while individuals that are not so well suited die before they can reproduce, or die while they are still capable of further reproduction, thus leaving fewer offspring than others. Offspring of comparatively successful individuals will tend to inherit the characters that made that parent successful, so that they in turn are likely to become reproductively successful if environmental conditions remain the same. **Differential reproductive success**, which results directly from natural selection, is the link that connects organisms and their environment.

By now, natural selection has been operating on living things for billions of generations, acting at each generation to fine-tune the relationship between organisms and their environment. As a result, even the simplest living organism is too complex to create in the laboratory. I do not think that is likely to change in the near future.

Energy and Life

Living things use energy; the biological term for that energy use is **metabolism**. Much of biology consists of studying the ways in which living things acquire the energy they need to grow and reproduce.

The first protocells had energy available to them in the form of ATP, amino acids, and other organic compounds that they could absorb from water. The compounds had been accumulating in the ocean over a long period, so for some unknown time protocells would have had fuel available as they gradually perfected their simple biochemical reactions. But as protocells became more numerous and more effective in attracting and using organic molecules, there must have come a time when demand exceeded supply. As simple organic molecules became scarcer and scarcer in the oceans, living cells encountered the world's first energy crisis. Paradoxically, this would have happened first in environments where protocells were most successful and abundant. There were two very different reactions to a shortage of "food," and the results can still be seen among living organisms nearly 4 billion years later.

Organotrophy and Fermentation

Organisms that obtain their metabolic energy by breaking down organic molecules they have absorbed from the environment are called **organotrophs** or heterotrophs. We do not know what the earliest cells were, and we may never know. We do know that the earliest cells evolved in an environment that contained large quantities of naturally formed organic molecules, a "food" supply for those that were able to break down the molecule to release energy.

One of the reactions evolved by early organotrophic cells was **fermentation**, in which cells break down sugars such as glucose. Glucose is often called the universal cellular fuel for living organisms, and it was probably the most abundant sugar available on the early Earth. Humans use fermenting microorganisms to produce beer, cheese, vinegar, wine, and yogurt, and fermentation reactions break down much of our sewage.

Fermentation has by-products which can be broken down still further to yield energy to organisms. Microorganisms called **methanogens** can use by-products of fermentation to gain energy and produce methane gas as their by-product. Methanogens are as different from true bacteria as bacteria are from us, and they have been recognized as part of a special group of microorganisms, the **Archaea**.

Autotrophy: Lithotrophy and Photosynthesis

Organotrophic organisms rely on a supply of organic molecules, and if the first living organisms gained their energy from the organic soup that had built up in Earth's oceans, they would have had an easy time at first. However, organotrophy would quickly have depleted the organic soup (automatically choking off the possibility of any further type of living organism evolving from those organic molecules), and the world's first energy crisis began. A new opportunity now arose: any organism that could generate its own energy, rather than deriving energy from organic molecules, would be strongly favored by natural selection—it would have been able to grow and divide faster than its competitors.

Organisms did evolve ways to generate their own energy (**autotrophy**), either by extracting chemical energy from inorganic molecules (**lithotrophy**), or by trapping heat or light energy (**photosynthesis**).

Lithotrophy can occur when a microorganism rips oxygen molecules off one inorganic compound and transfers it to another, making an energy profit in the process. That energy is then used to build organic food molecules. For example, some methanogens gain energy from lithotrophy by breaking up carbon dioxide and transferring the oxygen to hydrogen, forming water and methane as by-products:

$$4H_2 \ + \ CO_2 \quad \Rightarrow \quad CH_4 \ + \ 2H_2O$$

If this ability evolved very early, it may have been the first time (but not the last) that living things modified Earth's chemistry and climate. By taking the "greenhouse gas" carbon dioxide out of the atmosphere, methanogens might have made the early Earth cooler than it otherwise would have been, perhaps making it more hospitable to living cells.

Photosynthesis is simple in concept, but biochemically complex. Some molecules can capture light energy and store it by raising the energy of electrons within the molecule. **Porphyrins** are the most important of these in living things. They are complex, ring-shaped molecules that perform many functions in cells because they combine easily with metal atoms that fit in the center of the ring. Porphyrins such as **hemoglobin** and **cytochrome** contain iron atoms, **chlorophyll** contains magnesium, **vitamin B12** contains cobalt, and nickel-bearing porphyrins are found in all methanogens.

Porphyrins can be formed from simpler substances that would have been present in Earth's early oceans. A protocell could have incorporated porphyrins inside it, and internal reactions could have been powered by light energy trapped by the porphyrins. Chlorophyll (of various kinds) has become the most widely used light-trapping porphyrin, and organisms use that light energy, carried by electrons, to build up food molecules inside their cells. Chlorophylls reflect green light but absorb red and blue light to power reactions that use CO_2 and H_2 to build up sugars (CH_2O) that are stored to be fermented later.

Photosynthesis produced the first major changes on the Earth that were generated by living things. Immediately, the energy trapped by chlorophyll was used to build more biomass (biological substance), and energy flow through biological systems on Earth increased significantly.

Any early cell that happened to contain chlorophyll could have evolved photosynthesis. For individual bacteria, the advantages of photosynthesis (or lithotrophy) became important as soon as simple organic molecules began to run low, and fermenters began to run short of food. Autotrophic cells now had an energy store, a buffer against times of low food supply, that could be used as needed. In particular, it's easy to see how such cells could come to depend almost entirely on photosynthesis for energy. In many habitats, sunlight is a richer and more reliable energy resource than organic matter that must be sought and captured.

The earliest photosynthetic cells probably used hydrogen from H_2, H_2S, or lactic acid. Later, some bacteria began to break up the strong hydrogen bonds of the water molecule. The step might first have been an act of desperation, but any bacteria that successfully broke down H_2O rather than H_2S,

$$H_2O \;+\; CO_2 \;+\; light \;\Rightarrow\; (CH_2O) \;+\; 2\,O$$

immediately multiplied their energy supply sixfold. There was a penalty, however. The waste product of H_2S photosynthesis is sulfur (S), which is easily disposed of. The waste product of H_2S photosynthesis is monatomic oxygen (O), which is a deadly poison to a cell because it can break down vital organic molecules by oxidizing them. Even for humans, it is dangerous to breathe pure oxygen or ozone-polluted air for long periods.

Cells needed a natural antidote to this oxygen poison before they could operate the new photosynthesis consistently. We and most other organisms still use such antidotes (**superoxide dismutases** or SODs) in our own cells. They are fairly simple compounds that could have been available to an early bacterium with a fortunate mutation among its internal enzymes. SODs absorb monatomic oxygen as soon as it forms and deliver it in strong chemical bonds to be expelled from the cell as a waste product: O_2, ordinary or diatomic oxygen, which is much less active.

Their bacterial plight was pathetic

It's hard to be unsympathetic

Volcanic heat diminished,

Organic soup finished,

The solution was photosynthetic.

Cyanobacteria were the cells that made the first breakthrough to water/oxygen photosynthesis. From then on, we can imagine early communities of bacteria made up of autotrophs and heterotrophs, evolving improved ways of gathering or making food molecules. The total amount of amino acids, sugars, and other nutrients on Earth must have increased greatly, even if they were now tied up in biomass. Autotrophs that died and rotted released nutrients that became available to heterotrophs, and for the first time considerable amounts of energy were being transferred from organism to organism, in Earth's first true **ecosystem**.

Photosynthesizers need **nutrients** such as phosphorus and nitrogen to build up their cells, as well as light and CO_2. In most habitats, the nutrient supply varies with the seasons, as winds and currents change during the year. Light, too, varies with the seasons, especially in high latitudes. Since light is required for photosynthesis, great seasonal fluctuations in the primary productivity of the natural world began with photosynthesis. Seasonal cycles still dominate our modern world, among wild creatures and in agriculture and fisheries.

We can now envisage a world with a considerable biological energy budget and large populations of bacteria. So we can expect an increased chance that a geologist might find a very early bacterium preserved as a fossil in the rock record. In Chapter 2 we shall look at geology, rocks, and fossils, instead of relying on reasonable but speculative arguments about Earth's early history and life.

Review Questions

These review questions are meant only for those of you among my readers who are students eagerly looking forward to tests and examinations! They are meant to give students an opportunity to check whether they have absorbed the most important material in the Chapter. These are examination questions I have used myself. They are not meant to be a guide to **your** exam questions: that is between you and your own instructor.

Sometimes I ask a "thought question" that poses a problem that neither you nor I have a "correct" answer for. In my own exams I give credit for a student's ability to use the facts and ideas he or she has learned in order to come up with a sensible suggestion. At the end of Chapter 24 there are sample thought questions that require you to use material from more than one Chapter at once!

So here, for Chapter 1, are some review questions.

What makes us think that life evolved on Earth?

How did the Earth get its atmosphere?

Name three gases that were abundant in the early atmosphere.

Name one energy source that might have formed amino acids on the early Earth.

Name an energy source that can be used to form amino acids in the laboratory.

List three environments on Earth that have been suggested as likely sites for the origin of life.

What conditions are necessary for the formation of the building blocks of life on the early Earth?

Proteins are formed by linking amino acids together. How does that linkage occur, and how might a scientist try to reproduce it in the laboratory?

How might proteins have formed from amino acids on the early Earth?

The deadly poison cyanide was formed in the early atmosphere. Why is cyanide important for the evolution of life on Earth?

What is the difference between autotrophy and photosynthesis?

Why did Earth's early organisms evolve photosynthesis?

Further Reading

I have space here to explain how I chose the items to list under Further Reading, here and after all the other chapters. Many readers will not have the privilege of access to major research university libraries, but depend on small college libraries or city and county libraries. Under "Easy Access Reading," I have listed widely sold paperback books and articles in journals such as *Nature, Science, Discover, Scientific American, National Geographic Magazine*, and *American Scientist*, perhaps the six most widely distributed journals that deal with all aspects of science.

Under "More Technical Reading", I have listed books and articles in specialized journals. I have also tried to select well-written books of general interest, and I have tried, given the choice, to list articles in widely distributed journals rather than those that are difficult to find. Sometimes I have found scientific gems in obscure places, and I would rather list them than leave them out.

Important earlier work may not be listed directly but is summarized in more recent articles. Always, however, you should be able to work quickly backward to older literature from references in recent articles.

For up-to-date references, updates to the book, and for further reading, extra stories, links with other sciences, and some personal stories about science and scientists, see the World Wide Web page linked with this book:

http://www-geology.ucdavis.edu/~GEL3/

Easy Access Reading

Stars and Planets

Carr, M. H., et al. 1998. Evidence for a subsurface ocean on Europa. *Nature* 391: 363–365.

Flamsteed, S. 1997. Impossible planets. *Discover* 18 (9): 78–83. Extrasolar planets.

Gibbs, W. W. 1998. Endangered: other explanations now appear more likely than Martian bacteria. *Scientific American* 278 (4): 19–20.

Gibson, E. K., et al. 1997. The case for relic life on Mars. *Scientific American* 277 (6): 58–65. But see Gibbs, 1998 and Kerr, 1998.

Golombek, M. P. 1999. A message from warmer times. *Science* 283: 1470–1471. Warmer, wetter ancient Mars.

Ida, S., et al. 1997. Lunar accretion from an impact-generated disk. *Nature* 389: 353–357.

Jakosky, B. M. 1999. Mars: water, climate, and life. *Science* 283: 648–649. Warmer, wetter ancient Mars.

Kerr, R. A. 1998. Requiem for life on Mars? Support for microbes fades. *Science* 282: 1398.

Malin, M. C. 1999. Visions of Mars. *Sky & Telescope* 97 (4): 42–49.

Treiman, A. 1999. Microbes in a Martian meteorite? *Sky & Telescope* 37 (4): 52–58.

Williams, D. M., et al. 1997. Habitable moons around extra-solar giant planets. *Nature* 385: 234–235. Planets are hopeless, but their moons might not be.

Wood, J. A. 1999. Forging the planets: the origin of our Solar System. *Sky & Telescope* 97 (1): 36–48.

Origin of Life

Amend, J. P., and E. L. Shock. 1998. Energetics of amino acid synthesis in hydrothermal systems. *Science* 281: 1659–1662. Amino acids could form readily at ocean vents.

Bernstein, M. P., et al. 1999. Ultraviolet irradiation of polycyclic aromatic hydrocarbons in ices: production of alcohols, quinones, and ethers. *Science* 283: 1135–1138, and comment, pp. 1123–1124. Organic chemistry in outer space.

Benner, S. A. 1993. Catalysis: design vs. selection. *Science* 261: 1402–1403. Predicts that a self-replicating RNA molecule will be made within 10 years.

Cech, T. R. 1986. RNA as an enzyme. *Scientific American* 255 (5): 64–75.

Chyba, C. F., and C. Sagan. 1992. Endogenous production, exogenous delivery and impact-shock synthesis of organic molecules: an inventory for the origins of life. *Nature* 355: 125–132. Review article with large reference list.

Clark, S. 1999. Polarized starlight and the handedness of life. *American Scientist* 87: 336–343.

Cone, J. S. 1991. *Fire Under the Sea*. New York: Morrow. Popular account of the discovery of deep-sea vents.

de Duve, C. 1995. *Vital Dust: Life as a Cosmic Imperative*. New York: BasicBooks.

DeLong, E. 1998. Archaeal means and extremes. *Science* 280: 542–543. Brief summary of current research on the ecology and evolution of Archaea.

Ferris, J. P., et al. 1996. Synthesis of long prebiotic oligomers on mineral surfaces. *Nature* 381: 59–61, and comment, pp. 20–21.

Gaidos, E. J., et al. 1999. Life in ice-covered oceans. *Science* 284: 1631–1633. Life under ice on Europa is unlikely.

Holland, H. D. 1997. Evidence for life on Earth more than 3850 million years ago. *Science* 275: 38–39.

Joyce, G. F. 1989. RNA evolution and the origins of life. *Nature* 338: 217–224.

Joyce, G. F. 1992. Directed molecular evolution. *Scientific American* 267(6): 90–97. Laboratory experiments designed to "evolve" complex organic molecules.

Lee, D. H., et al. 1996. A self-replicating peptide. *Nature* 382: 525–528, and comment, pp. 496–497.

McKay, C. P., and Borucki, W. J. 1997. Organic synthesis in experimental impact shocks. *Science* 276, 390–392.

McPherson, A., and P. Schlichta. 1988. Heterogeneous and epitaxial nucleation of protein crystals on mineral surfaces. *Science* 239: 385–387. Proteins crystallize very well on many different natural mineral grains.

Miller, S. L. 1953. A production of amino acids under possible primitive earth conditions. *Science* 117: 528–529. The paper that started it all.

Miller, S. L., and J. L. Bada. 1988. Submarine hot springs and the origin of life. *Nature* 334: 609–611, and comments, p. 564 and v. 336, p. 117.

Monastersky, R. 1998. The rise of life on Earth. *National Geographic* 193 (3): 54–81.

Nisbet, E. G., et al. 1995. Origins of photosynthesis. *Nature* 373: 479–480.

Orgel, L. E. 1994. The origin of life on the Earth. *Scientific American* 271 (4): 77–83.

Pennisi, E. 1999. Is it time to uproot the tree of life? *Science* 284: 1305–1307.

Schimmel, P., and R. Alexander. 1998. All you need is RNA. *Science* 281: 658–659.

Service, R. F. 1998. A biomolecule building block from vents. *Science* 281: 1936–1937. Ammonia.

Simpson, S. 1999. Life's first scalding steps. *Science News* 155: 24–26. Evidence favoring a deep-sea origin.

Szalai, V., and G. W. Brudvig. 1998. How plants produce dioxygen. *American Scientist* 86: 542–551. Photosynthesis.

Zhang, B., and T. R. Cech. 1997. Peptide bond formation by in vitro selected ribozymes. *Nature* 390: 96–100. Ribozymes that can join amino acids into peptides.

Zimmer, C. 1995. First cell. *Discover* 16 (11): 71–78. David Deamer's work on liposomes and the origin of cells.

More Technical Reading

Forterre, P. 1996. A hot topic: the origin of hyperthermophiles. *Cell* 85: 789–792.

Horowitz, N. H. 1986. *To Utopia and Back: The Search for Life in the Solar System.* New York: W. H. Freeman.

Jay, D., and W. Gilbert. 1987. Basic protein enhances the incorporation of DNA into lipid vesicles: model for the formation of primordial cells. *Proceedings of the National Academy of Sciences* 84: 1978–1980.

Jeffares, D. C., et al. 1998. Relics from the RNA world. *Journal of Molecular Evolution* 46: 18–36. Companion paper to Poole et al. (see below).

Kral, T. A., et al. 1998. Hydrogen consumption by methanogens on the early Earth. *Origins of Life* 28: 311–319. Methanogens cooled the early Earth by using up carbon dioxide.

Lazcano, A., and S. L. Miller. 1996. The origin and early evolution of life: pre-biotic chemistry, the pre-RNA world, and time. *Cell* 85: 793–798.

Levy, M., and S. L. Miller. 1998. The stability of the RNA bases: implications for the origin of life. *Proceedings of the National Academy of Sciences* 95: 7933–7938. RNA bases do not survive well much above 0° C.

Levy, M., and S. L. Miller. 1999. The prebiotic synthesis of modified purines and their potential role in the RNA world. *Journal of Molecular Evolution* 48: 631–637.

Miller, S. L., and A. Lazcano. 1995. The origin of life: did it occur at high temperatures? *Journal of Molecular Evolution* 41: 689–692. Miller and Lazcano argue that it did not..

Orgel, L. E. 1998. The origin of life—how long did it take? *Origins of Life* 28: 91–96. We have no idea. [A crack at Miller, who thinks it was very fast.]

Poole, A. M., et al. 1998. The path from the RNA world. *Journal of Molecular Evolution* 46: 1–17. For biochemists only; for summary, see Pennisi 1998.

Schwartz, A. W., and A. Henderson-Sellers. 1983. Glaciers, volcanic islands and the origin of life. *Precambrian Research* 22: 167–174.

Shapiro, R. 1986. *Origins: A Skeptic's Guide to the Creation of Life on Earth.* New York: Summit. Published in paperback, 1988.

Sleep, N. H., and K. Zahnle. 1998. Refugia from asteroid impacts on early Mars and the early Earth. *Journal of Geophysical Research* 103: 28529–28544. How did they get this stuff published??

Woese, C. 1998. The universal ancestor. *Proceedings of the National Academy of Sciences* 95: 6854–6859. The universal ancestor may be undiscoverable.

Zuckerman, B. and M. H. Hart (eds.) 1995. *Extraterrestrials, Where Are They?* New York: Cambridge University Press.

CHAPTER TWO

Earth's Earliest Life

When we move from the astronomical observatory and the laboratory to the Earth itself to search for evidence about early life, we look for fossils. A **fossil** is the trace of an organism preserved in rock. Most fossils are made of the hard parts of organisms—shells, bones, or wood—that are often more or less unchanged after death. Minerals may crystallize out of groundwater to fill up large or small cracks, crevices, and cavities in the original substance, so fossils may be denser and harder than they were in life. Sometimes the original shell or bone may be replaced by another mineral, making the fossil easier to recognize or easier to extract from the rock (Figure 2.1).

Obviously, the hard parts of an organism are more likely to be preserved than more fragile parts. But occasionally soft parts may leave an impression on soft sediment before they rot. Even more rarely, a complete organism may be encased in soft sediment that later hardens into a rock. Bees, ants, flies, and frogs have been preserved as fossils in fossilized tree resin, **amber** (Figure 2.2), and individual cells have been preserved in **chert**, a rock formed from silica that impregnated the cells and retained their shapes in three dimensions. A fossil may not be part of an organism at all, but a **trace fossil** (a trail, a burrow, a footprint, or a fecal mass) (Figure 2.3).

All kinds of agents may destroy or damage organisms beyond recognition before they can become fossils or while they are fossils. After death the soft parts of organisms may rot or be eaten. Any hard parts may be dissolved by water, or broken or crushed and scattered by scavengers

Figure 2.1 A brachiopod whose original calcite shell was replaced by silica. This made it fairly easy to dissolve the shell out of limestone for study.

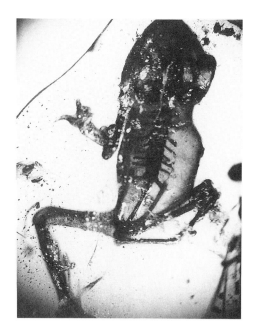

Figure 2.2 A frog preserved in amber. This is a particularly interesting specimen because it is almost undamaged externally, though it has broken internal bones. It was probably seized by a predator (a hawk or an owl?) and carried to a tree where it was covered in resin and preserved in this spectacular fashion. (© George Poinar, University of California, Berkeley.)

or by storms, floods, wind, and frost. Remains must be buried to become part of a rock, but a fossil may be cracked or crushed as it is buried. After burial, groundwater seeping through the sediment may dissolve bones and shells. Earth movements may smear or crush the fossils beyond recognition or may heat them too much. Even if a fossil survives and is eventually exposed at the Earth's surface, it is very unlikely to be found and collected before it is destroyed by weathering and erosion.

Even when they are studied carefully, fossils are a very biased sample of ancient life. Fossils are much more likely to be preserved on the seafloor than on land. Even on land, animals and plants living or dying by a river or lake are more likely to be preserved than those in mountains or deserts. Different parts of a single skeleton have different chances of being preserved. Animal teeth, for example, are much more common in the fossil record than are tail bones and toe bones. Teeth are usually the only part of sharks to be fossilized. Large fossils are usually tougher than small ones and are more easily seen in the rock. Spectacular fossils are much more likely to be collected than apparently ordinary ones. Even if a fossil is collected by a professional paleontologist and sent to an expert for examination, it may never be studied. All the major museums in the world have crates of fossils lying unopened in the basement or the attic.

When we look at museum display cases, it seems that we have a good idea of the history of life. But most of the creatures that were living at any time are not in a museum. They were microscopic or soft-bodied, or both, or they were rare or fragile and were not preserved, or they have not been discovered. Maybe we have found as many as 2% of all the species of animals with hard parts that have lived on Earth, but many people think this estimate is too optimistic. We do have enough evidence to begin to put together a story. But that story is always changing as we discover new fossils and look more closely at the fossils we have found already.

Figure 2.3 A trace fossil—a dung ball 22 cm (nearly 10 inches) across, from Pleistocene deposits in Bechan Cave, Utah. This is a trace fossil because it was not part of the original organism but is evidence that the organism once existed. This dung ball is attributed to a mammoth. (Courtesy of Emily M. and Jim I. Mead, Bilby Research Center and Quaternary Studies Program, Northern Arizona University.)

How Paleobiology Works

Paleobiology deals with the interpretation of fossils as organisms, and it is a part of the larger science of paleontology. Paleobiology works in much the same way as archeology or history: it is the study of past events. Most of us don't study fossils for their intrinsic interest, although some of us do. Their greater value lies in what they tell us about ourselves and our background. We care about our future, which is a continuation of our past. One good reason for trying to reconstruct ancient biology is to manage better the biology of our planet today, so we need to set up some kind of reasonable logic for interpreting life of the past. Some basic problems of paleobiology are much like those of history and archeology: how do we know when we find the right explanation for some past event? How do we know we are not just making up a story?

Anything we suggest about the biology of extinct organisms should make sense in terms of what we know about the biology of living organisms, unless there is very good evidence to the contrary. This rule applies throughout biology, from cell biochemistry to genetics, physiology, ecology, behavior, and evolution.

But suggestions are only suggestions until they are tested against evidence from the fossil itself and from the rock record. Because fossils are found in rocks, we have access to environmental information about the habitat of the extinct organism: for example, the rock might show clear evidence that it was deposited under desert conditions or on a shallow-water reef. Fossils are therefore not isolated objects but parts of a larger puzzle. For example, it is difficult to interpret the biology of the first bird, *Archaeopteryx*, unless we consider environmental evidence from the Solnhofen Limestone in which it is preserved (Chapter 12).

An alert reader should be able to identify four levels of paleobiological interpretation. First, there are **inevitable** conclusions for which there are no possible alternatives. For example, there's no doubt that the extinct ichthyosaurs were swimming marine reptiles. At the next level, there are **likely** interpretations. There may be alternatives, but a large body of evidence supports one leading interpretation. For example, there is good evidence to suggest that ichthyosaurs gave birth to live young rather than laying eggs. Almost all paleobiologists view this interpretation as the best hypothesis available and would be surprised if contrary evidence turned up.

Then there are **speculative** interpretations. They may be right, but there is not much real evidence one way or another. Paleontologists are allowed to accept speculative interpretations as tentative ideas to work with and to test carefully, but they should not be surprised to find them wrong. For example, it seems reasonable to me that ichthyosaurs were warm-blooded, but it's a speculative idea because it's difficult to test. If new evidence showed that the idea was unlikely, I might be personally disappointed but I would not be distressed scientifically.

Finally, there are **reasonable** interpretations. They may be biologically more plausible than others one might suggest, but for one reason or another they are completely untestable and so must be classed as non-scientific. For example, if I asked an artist to draw an ichthyosaur, I'd have to choose some sort of color pattern for the skin. I might suggest bold black-and-white color patterns, like those of living killer whales, but another paleobiologist might opt for more muted tones like those of living dolphins. Both ideas are reasonable, and are surely better than the green-with-pink-spots that one might find in a TV cartoon. But this is speculation, based on no evidence at all.

You will find examples of all four kinds of interpretation in this book. Often it's a matter of opinion in which category to place different suggestions, and this problem is at the root of many controversies in paleobiology. Were dinosaurs warm-blooded? Some paleobiologists accept that as an inevitable conclusion of the evidence, some think it likely, some think it's only speculative, some think it unlikely, and some think it is plain wrong. New evidence almost always helps to solve old questions but also poses new ones. Without bright ideas and constant attempts to test them against evidence, paleobiology would not be as exciting as it is.

The fossil record gradually gets poorer as we go back in time, for two reasons. Biologically, there were fewer types of organisms in the past. Geologically, there are relatively few rocks that have survived from older periods, and even those that have survived have often suffered heating, deformation, and other changes, all of which tend to destroy fossils. Earth's earliest life was certainly microscopic and soft-bodied, a very unpromising combination for fossilization. Only a truly exceptional set of circumstances would preserve such creatures in rocks a few thousand years old, let alone a few billion. But as a result of perseverance and good luck, the earliest fossils now known are more than 3.5 billion years old.

How Do We Know the Age of a Fossil?

Fossils are found in rocks, and usually geologists try to establish the age of the containing rock or of a layer of rock that is not far under or over the fossil (and so might be close to it in age). The age of rocks is measured in two different ways, known as **relative** and **absolute** dating.

Age dating of rocks can only work if one identifies components of the rocks that change with time or are in some way characteristic of the time at which the rocks formed. The same principles are used in dating archeological objects. Coins may bear a date in years (absolute dating), and one can be certain that a piece of jewelry made from a gold coin could not have been made before the date stamped on the coin. The age of waste dumps can be gauged by the type of container thrown into them: bottles with various shapes and tops, steel cans, aluminum cans, and so on. The age of old photographs and movies can often be judged by the dress or hairstyle of the people or the cars or appliances shown in them.

Absolute geological ages can be determined because newly formed mineral crystals sometimes contain unstable, **radioactive**, atoms. Radioactive isotopes break down at a rate that no known physical or chemical agent can alter (Figure 2.4), and as they do so they may change into other elements. For example, potassium-40, ^{40}K, breaks down to form ^{40}Ar, argon-40. By measuring the amount of radioactive decay in a mineral crystal, one can calculate the time since it was newly formed, just as one reads the date from a coin. The principle is simple, though the techniques are often laborious. For example, ^{40}K breaks down to form ^{40}Ar at a rate such that half of it has gone in about 1.3 billion years (Figure 2.4). If we measure the ^{40}Ar in a potassium feldspar crystal today, and find that half the original amount of ^{40}K has gone, then the age of the crystal is 1300 Ma. (By convention, absolute ages in millions of years are given in megayears [Ma], while time periods or intervals are expressed in millions of years [m.y.].)

Several dating methods use this same principle. Perhaps the most well-known is the radiocarbon method, based on the breakdown of ^{14}C, carbon-14. However, the half-life of ^{14}C is only about 5370 years, and only a minute fraction of the original ^{14}C is left after about 40,000 years. Therefore ^{14}C dating is valuable to archeologists and historians, but not to most geologists. There are other problems with the ^{14}C method, too.

Absolute dating is not foolproof. Crystals may have been reheated or even recrystallized, re-setting their radioactive clocks to zero well after the time the rock originally formed. Chemical alteration of the rock may have removed some of the newly produced element, also giving a date younger than the true age. Uncertainties in the precise rate of radioactive decay can give uncertainties of up to several million years.

Worst of all, however, most elements used for radioactive age dating are not used by living organisms to build shells or bones. Usually we cannot date fossils directly. Instead, we have to measure the age of a lava flow or volcanic ash layer as close to the fossil-bearing bed as possible (Figure 2.5), which does contain crystal we can use. This can be a problem. For example, many of the arguments about hominid evolution in East Africa have resulted from problems in assessing age dates from lavas and ashes close to hominid remains. Absolute methods are also expensive and time-consuming.

Paleontologists more often deal with a relative time scale, in which one says "Fossil A is older than Fossil B" (as in Figure 2.5) without specifying the age in absolute years. This is much the same way that archeologists date Egyptian artifacts. We know which Pharoah followed which, even though we do not know the absolute ages for many of the earlier dynasties. So Egyptian history is scaled according to the reigns of particular Pharoahs, rather than recorded in absolute years. It's possible to work this way with fossils, because it is a fact of observation that fossils preserved in the rock record at particular times are almost always different from those preserved at other times. The facts have been well established over the past two centuries by geologists working upward and downward in rock sequences in which sediments have been laid down in successive layers, each layer lying on and therefore being younger than the one underneath. With each study, a small part of geological time has had its sequence of fossils identified, much as archeologists have identified successive events in each Pharoah's reign.

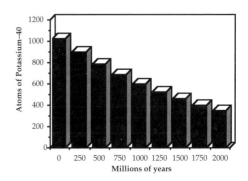

Figure 2.4 The radioactive decay of an isotope proceeds on a logarithmic timetable that is constant under all known conditions. If the decay is recorded in a rock or mineral, we can infer the date when the decay began. Often, but not always, that tells us the age of the rock. This graph shows the atoms of potassium-40 remaining in a crystal on a time scale measured in millions of years, relative to a starting value of 1024 atoms at time zero.

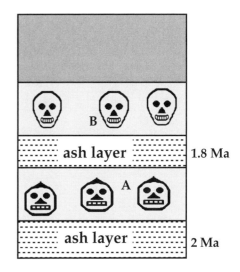

Figure 2.5 The skulls of these two fossil hominids do not contain any radioactive isotopes, but they lie close to two layers of volcanic ash that do. By using relative dating methods, one can say that hominid **A** is older than hominid **B**. Using absolute dating methods, the age of hominid **A** can be fixed closely between 1.8 Ma and 2.0 Ma because there are dated ash layers above and below it. All we know about the age of hominid **B** is that it is younger than 1.8 Ma.

Figure 2.6 The standard relative time scale used by geologists (on the left), and one of several competing absolute time scales (on the right).

Time Divisions		Began (Ma)
CENOZOIC	Quaternary	
	Holocene	0.01
	Pleistocene	1.6
	Tertiary	
	Pliocene	5
	Miocene	23
	Oligocene	35
	Eocene	56
	Paleocene	65
MESOZOIC	Cretaceous	145
	Jurassic	210
	Triassic	250
PALEOZOIC	Permian	290
	Carboniferous	360
	Devonian	410
	Silurian	440
	Ordovician	505
	Cambrian	545
PRECAMBRIAN	Proterozoic	2500
	Archean	3800
	(Age of Earth)	(4550)

With the occasional check from absolute methods in fortunate circumstances, the geological record has been arranged into a standard sequence that allows fast and cheap age determination from the fossils it contains (Figure 2.6). The standard sequence—the **geological time scale**—is divided into a hierarchy of units for easy reference, with the divisions between major units often corresponding to important changes in life on Earth. The names of the eras and periods are often unfamiliar and have bizarre historical roots. For example, the Permian period was so named because Sir Roderick Murchison and the Comte d'Archiac took a stagecoach tour in 1841 to the city of Perm, in Russia, and discovered unfamiliar new rocks there. After a while, however, the names of the eras and periods and their sequence become not a matter for laborious memorization but the key to a vivid set of images of ancient life.

Earth's Oldest Rocks

The first one-third of Earth's history is called the **Archean Era**. The study of Archean rocks requires a special kind of geologist. Most geologists work on the **Principle of Uniformitarianism**, which can be roughly summarized in the saying "The present is the key to the past." A geologist tries to interpret the rock record in terms of well-known processes acting today on the Earth, such as weather and climate, deposition and erosion, tides and currents, volcanic activity and earthquakes, biological and ecological events, and so on.

This approach has been spectacularly successful in interpreting most of the geological record. But the early Earth was not like today's. There was little or no oxygen in the atmosphere. There was much less life in the seas and none on the land. The Earth was young; its interior was hotter, and its internal energy was greater. We can guess that volcanic activity was much greater, but we have no idea whether it was more violent or just more continuous. Reconstructing conditions on the early Earth is difficult and challenging.

Rocks older than 3.5 b.y. are very rare on Earth. There may have been only tiny patches of continental crust on the early Earth; over 90% of the crust would have been oceanic rock that has long been destroyed. The oldest minerals on Earth are zircon crystals over 4 b.y. old, but they have been eroded out of their original rocks and deposited as fragments in younger rocks. The oldest known rocks are in the Isua area of West Greenland, and have been dated by several methods at about 3850 Ma. The rocks include sedimentary formations. Although they have been repeatedly folded, faulted, and reheated, the Isua sediments can still tell us something about conditions on the early Earth when they were formed. They were laid down in shallow, nearshore water, and they include beach-rounded pebbles and weathering products from volcanic lava. Temperatures at the time may have been warm, but they were not extraordinary. The shoreline was probably volcanic and tropical.

The Isua rocks contain quite a lot of carbon in the form of the mineral graphite. When CO_2 is taken up by cells, used in photosynthesis, and incorporated in a cell, the light isotope carbon-12, (C-12, or ^{12}C) is slightly enriched over the heavier isotope ^{13}C in the final organic compounds. The carbon in the Isua rocks is slightly enriched in ^{12}C, enough to convince many geochemists that it once went through photosynthesis inside a cell. Thus there is indirect evidence that there may have been life on Earth at or before 3850 Ma. There is clearer isotopic evidence for biological carbon (life) at 3.7 b.y. The first **fossils** come from later rocks.

Earth's Oldest Cells

Archean rocks are often rich in minerals, and Archean regions have been well explored geologically for economic reasons. An Archean district in northwestern Australia is called North Pole because it is so remote and

Figure 2.7 The oldest trace fossil yet found on Earth, a stromatolite from the Warrawoona rock sequence in Australia. It is associated with wave-affected sediments, and therefore it formed in very shallow water. By comparing this structure with those forming today in Shark Bay (Figure 2.8), it can be interpreted as having been formed by mats of bacteria 3.5 billion years ago. (Courtesy of Stanley Awramik, University of California, Santa Barbara.)

Figure 2.8 Stromatolites are forming today in warm salty water in Shark Bay, Western Australia. The shovel gives the scale. Close study of these modern structures allows us to interpret the stromatolite from Warrawoona (Figure 2.7) as an early trace fossil formed by mats of cyanobacteria. (Courtesy of Paul Hoffman, Geological Survey of Canada.)

inhospitable. It originally attracted geological attention because it is rich in the mineral barite, but it is now famous for more academic reasons.

The local rock sequence is called the Warrawoona Series, and its age is about 3550 Ma. It consists mainly of volcanic lavas erupted in shallow water, or nearby on shore, but there are sedimentary rocks too. The sediments include storm-disturbed mudflakes, wave-washed sands, and minerals that could only have formed by evaporation in very shallow pools. The rocks have not been tilted, folded, or heated very much, and the environment can be reconstructed accurately here, and in rocks of about the same age in southern Africa. The rocks formed along warm, dry, but occasionally stormy tropical volcanic shorelines that physically were just like comparable modern environments. There are two other vital components to the story from Warrawoona: rock structures called stromatolites, and the role of oxygen.

Stromatolites

The Warrawoona rocks contain structures called **stromatolites**, which are low mounds or domes of finely laminated sediment (Figure 2.7). They are usually rich in carbonate, but the carbonate can be replaced by cherty layers. We know what stromatolites are because they can be seen forming today. They are fossilized **microbial mats**, formed mainly by photosynthetic blue-green bacteria (cyanobacteria).

Modern marine stromatolites flourish in Shark Bay, Western Australia, which consists of a set of long shallow inlets, 100 km (60 miles) long but only about 10 m (about 30 feet) deep, along a desert coastline. Summer temperatures are very high, and the salinity of the inlets is 55 to 70 parts per thousand offshore and much higher in places along the shore (normal seawater is 35 parts per thousand). Stromatolites form from the highest tide level down to subtidal levels, but the higher ones, closer to shore, have been better studied (sea snakes, not sharks, are the problem).

Cyanobacteria grow and photosynthesize so luxuriantly that they form dense mats. They thrive in water that is too salty for grazing animals such as snails and sea urchins that would otherwise eat them off. Like most bacteria, they secrete slime, and are also able to move a little in a gliding motion. When the tide comes in at Shark Bay, sediment thrown up in the waves may stick to the slime and cover up some of the bacteria. But they quickly slide and grow through the sediment back into the light, trapping the sediment as they do so. As the cycle repeats itself, sediment is built up in mounds under the growing mats. Eventually the mats grow as high as the highest spring tide, but cannot grow higher without becoming too hot and dry (Figure 2.8).

Some mats just under the water surface grow mainly because carbonate precipitates directly from seawater onto their surfaces, binding the sediment very quickly into a rocklike consistency. Photosynthesis uses up CO_2 and makes it easy for carbonate to crystallize on the mats, building them into wave-resistant structures.

Many different combinations of bacteria can form stromatolites today, especially where salinity is too high for grazing snails. Some cyanobacterial mats are so dense that light may penetrate only 1 mm. The topmost layer of cyanobacteria absorbs about 95% of the blue and green light, but just underneath is a zone where light is dimmer but exposure to UV radiation, heat, and drying is less. Green and purple bacteria grow here and contribute their anaerobic photosynthesis to the growth of the mat. Under the second layer is a zone where light is too low to allow photosynthesis, and here live more bacteria, organotrophs that live off the dying and dead remains of the bacteria above them.

There may be eight different zones in a stromatolite, each with its own microenvironment and each only a few millimeters thick. The zones vary from super-rich in oxygen, with bubbles of pure oxygen trapped in slime, to zones without oxygen or light, saturated with H_2S (hydrogen sulfide). Inside a stromatolite, oxygen levels can vary within fractions of a millimeter and within seconds. Cyanobacteria live in stromatolites immediately next to sulfide-rich, oxygen-poor environments. For example, the cyanobacterium *Oscillatoria* has evolved two photosynthetic systems: one that works in oxygen-rich environments, splitting water molecules and giving off oxygen, and another that works in sulfide-rich environments, splitting H_2S and releasing sulfur. This switch-hitting mechanism allows it to survive episodes of sulfide poisoning. With adaptations like this, it's easy to see how cyanobacteria can live in a stromatolite only fractions of a millimeter from toxic sulfides.

Ancient Stromatolites

Stromatolites are the most conspicuous fossils for three billion years of Earth history, from 3550 Ma to the end of the Proterozoic at about 550 Ma. They are rare in the Archean, probably because Archean geography probably had few clear, shallow-water shelf environments suitable for stromatolite growth. The few Archean land masses were volcanically active, generating high rates of sedimentation that would inhibit mat growth in many shoreline environments.

Stromatolites are easily recognized by their distinctive structure. They are small and rare in the Warrawoona rocks, but much more numerous, complex, and varied forms are found in rocks of the Fig Tree Group in southern Africa, dated at about 3400 Ma (Figure 2.9). Thus, bacterial life was well established by that time, even if stromatolites occurred only in local patches.

Cyanobacteria release a lot of organic material into the water around them. Essentially, their cell walls leak organic slime and other soluble substances into the water, providing food for populations of organotrophic bacteria that live around the mats. Archean stromatolites were not just the remains of simple bacterial mats, but were complex miniature ecosystems teeming with life.

Solar UV radiation was intense in Archean time, with no oxygen (or ozone layer) in the atmosphere. A shield of perhaps 10 m (about 30 feet)

Figure 2.9 Stromatolites from the Fig Tree Group in southern Africa are nearly as old as the Warrawoona stromatolite shown in Figure 2.7. The coin is 2 cm across. (Photograph courtesy of Gary Byerly, Louisiana State University.)

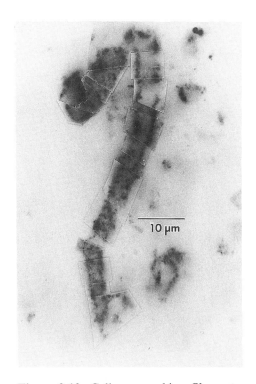

Figure 2.10 Cells arranged in a filament, from the Warrawoona rocks of northwestern Australia, 3.5 b. y. old. (Courtesy of Stanley Awramik, University of California, Santa Barbara, who discovered them.)

Figure 2.11 Stromatolites were more abundant in the Proterozoic than they had been in the Archean. This luxuriant array of stromatolites grew 1.9 b.y. ago on a shoreline where Great Slave Lake now lies on the Canadian Shield. (Courtesy of Paul Hoffman, Geological Survey of Canada.)

of water might have been needed to prevent damage to an early cell by UV radiation. (This estimate may be too high because substances such as ferric iron in the water would have helped to protect cells.) However, the early evolution of the stromatolitic way of life by cyanobacteria may have been a response to radiation. Bacteria were not just existing at Warrawoona. With their layered growth under water, partly protected by slime, sheaths, and sediment trapped on the sticky surface of mats, they were already modifying their microenvironment for survival and success.

Fossil Cells in Chert

Chert is a rock formed of microscopic silica particles. It does not form easily today because all kinds of organisms, including sponges, take silica from seawater to make their skeletons. But silica-using organisms had not evolved in Archean times, and cherts are often abundant in Archean rock sequences. As chert forms, in a gel-like goo on the seafloor, it may surround and impregnate cells with silica, preserving them in exquisite detail. Chert is watertight after it hardens, so percolating water cannot easily dissolve or contaminate the fossils. Some paleontologists expend a lot of effort in looking for, and then looking at, Archean cherts.

Several processes can generate inorganic blobs of chemicals in rocks, and blobs in chert have often been mistaken for fossil cells. Fortunately, the Warrawoona rocks do contain genuine Archean cells (Figure 2.10). By now at least eleven different bacterial cell types have been identified in these cherts and others that are close in age and location. At least some of them were photosynthetic, and they look very similar to cyanobacteria that produce oxygen today. These earliest fossils are not likely to represent the earliest organisms on Earth, however. Cyanobacteria are relatively advanced bacteria, and more primitive organisms are still to be found.

The Late Archean and the Early Proterozoic

The earliest stromatolites are the Warrawoona and Fig Tree examples from about 3500 Ma. By 3100 Ma, there were two distinctly different styles of bacterial mat; by 2800 Ma stromatolites are known from salt-lake environments as well as oceanic shorelines.

There were important geological changes at the end of the Archean, which is defined as 2500 Ma. The Earth had cooled internally to some extent, and the crust became thicker and stronger. The thicker crust affected tectonic patterns—the way the crust moves, buckles, and cracks under stress. Continents became larger and more stable in Early Proterozoic times, with wide shallow continental shelves that favored the growth and preservation of stromatolites. Most Proterozoic carbonate rocks include stromatolites, some of them enormous in extent (Figure 2.11). Proterozoic stromatolites evolved into new and complex shapes as bacterial communities became richer and expanded into more environments.

Finally, stromatolites became less diverse after the end of the Proterozoic as animals appeared in the fossil record. Experiments on living stromatolites suggest that their decline was caused by the rise of grazing metazoans such as snails, limpets, and sea urchins. Today, cyanobacterial mats grow in salty or super-salty shallow lagoons and bays, in desert lakes, and in hot springs. The common feature of such environments is the absence of metazoan grazers.

Banded Iron Formations

Even today, few bacteria are predators. Without the controlling effect of predation, the major checks on Archean populations must have been natural disasters such as storms, heating and freezing, drying, and starvation from lack of nutrients. There must have been very rapid fluctuations in bacterial populations as light levels changed with day and night and summer and winter, altering photosynthetic productivity. Ecologically, the Archean world, populated only by bacteria, was very unstable.

Bacteria can survive rapid environmental fluctuation as long as they are abundant. Although they do not reproduce sexually, evolutionary novelty can appear as a result of mutations. Mutations are rare and usually damaging, but occasionally a beneficial one appears, at a rate proportional to the number of individuals in the population. A favorable mutation can spread very rapidly through a population because bacteria reproduce quickly to form immense numbers of identical clones. In the same way, a population of bacteria can recover from disaster because a few mutant or lucky survivors can quickly repopulate the environment.

Early bacteria lived in a world with little or no free oxygen. So it is a surprise to find massive accumulations of a peculiar rock type that must have been produced by an oxidation reaction of some kind. **Banded iron formations** (BIF) are sedimentary rocks found mainly in sequences older than 1800 Ma. They are alternations of iron ore and chert, sometimes repeated millions of times in microscopic bands. No iron deposits like this are forming now, but we can make intelligent guesses about the conditions in which BIF were laid down.

The chemistry of seawater on an oxygen-poor Earth differed greatly from today's. Today there is practically no dissolved iron in the oceans and little silica because so many organisms extract and use it for their skeletons. But iron dissolves readily in water that has no oxygen, so Archean oceans contained a great deal of dissolved iron as well as silica. Even today, iron is enriched 5000 times above normal levels in oxygen-poor water on the Red Sea floor, where only bacteria can survive.

Silica would have been precipitating much of the time on an Archean seafloor to form thick chert beds in areas that did not receive much silt and sand from the land. But iron can only have precipitated from seawater in such massive amounts by an oxidizing chemical reaction. Therefore, to form BIF, there must have been occasional or regular periods of iron formation against a background of regular chert formation. There is more than one ocean's worth of iron in most BIF formations, so we are not looking at a single event in which the oceans dropped out all their iron. The ocean must have fluctuated between a state in which it held a lot of iron in solution, and a state in which it precipitated iron. Oceanic iron was replenished from erosion down rivers or from deep-sea volcanic vents.

The huge quantities of iron in BIF must have been supplied from a huge reservoir, from the ocean as a whole rather than from seawater in isolated basins. The two main hypotheses about BIF both call for seasonal deposition of iron ore, one by an inorganic process and the other by an organic process. Both may have operated. In the leading inorganic model, UV radiation formed an iron compound in surface waters, which fell out to form a layer of iron ore on the seafloor. This reaction would have been at a peak during the summer, when sunlight was most intense. The inorganic model could explain why there are some BIF in the Isua rocks at 3800 Ma, before there is direct evidence of life or photosynthesis.

However, even this "inorganic" BIF may in fact have been deposited by bacteria. Bacteria can act to form tiny "seed" crystals of iron minerals, which then continue to grow inorganically. It would be extraordinarily difficult to detect that this was in fact begun by bacterial action. Laboratory cultures of bacteria can form iron that looks "inorganic", so it would be a brave person who would say that bacteria were not involved in the formation of BIF.

In the organic model, BIF were formed directly by bacterial photosynthesis. This may have worked in two ways: from cyanobacteria in stromatolites, and from purple, nonsulfur bacteria, much like the bacteria discovered in 1993 by a team of German microbiologists led by F. Widdel. The purple bacteria use photosynthesis to break up iron carbonate dissolved in the water, precipitating oxidized iron as a by-product. Both bacterial reactions would have been at maximum during the summer, when sunlight was most intense. The same seasonal upwelling that brought nutrients to the bacteria also brought iron from deep water, leading to great bursts of iron ore deposition on the seafloor.

Today we probably see in rocks only a fraction of the BIF that once formed on the Archean sea floor, but even the amounts remaining are staggering. BIF make up thousands of meters of rocks in some areas and contain by far the greatest deposits of iron ore on Earth. At least 640 billion tonnes of BIF [this is the metric tonne, 1000 kilograms, that is used internationally; it is very close to an American ton] were laid down between 2500 and 2000 Ma (that's an average of half a million tonnes of iron per year). The Hamersley Iron Province in Australia alone contains 20 billion tonnes of iron ore, with 55% iron content. At times, iron was dropping out in that basin at 30 million tonnes a year. Most of our modern steel industries are based on iron ores laid down by bacteria during that time: Australia and Brazil dominate the world's export trade in iron ore. Cadillacs, Toyotas, and BMWs ultimately owe their existence to Precambrian photosynthetic bacteria.

The Oxygen Revolution

The new research on iron-oxidizing bacteria is so recent that I have to try to reconstruct their Archaean ecology myself. BIF are deposited in bands that can be traced for hundreds of kilometers, so they were laid down uniformly, or at least continuously, over great areas. This suggests to me that the photosynthetic bacteria were floating all across the ocean surface, not simply scattered in a thin ribbon in the shallow water around the edges of the small Archean continents.

Meanwhile, however, stromatolites were flourishing along those same shorelines, containing cyanobacteria using the advanced photosynthesis that split H_2O atoms and produced oxygen.

Oxygen is not released as a pure gas from volcanoes, and only a very small quantity is formed in the upper atmosphere as UV radiation breaks down water molecules. In practical terms, all the oxygen that forms 20% of our modern atmosphere is produced in photosynthesis. Even today, only about 5% of newly formed photosynthetic oxygen is released into the atmosphere and ocean as a free gas; the other 95% reacts with sulfides and iron compounds to form sulfates and iron oxides. The same reactions must have used up almost every oxygen molecule produced on the early Earth, and any other oxygen would have gone to oxidize organic molecules in the water. Since stromatolites did not cover large areas in Archean times, it may have taken literally hundreds of millions of years to satisfy these demands before significant amounts of free oxygen began to accumulate in air and water.

Therefore, the toxic qualities of oxygen would first have affected the cyanobacteria inside which photosynthesis was taking place. As they evolved superoxide dismutase, their antidote to oxygen poisoning (Chapter 1), cyanobacteria could control and then use oxygen in a new process, **respiration** (biological oxidation).

Fermentation leaves byproducts such as lactic acid that still have energy bonded within them. Using oxygen to break the byproducts all the way down to carbon dioxide and water, a cell can release up to eighteen times more energy from a sugar molecule by respiration than it can by simple fermentation.

It looks as if cyanobacteria began to use oxygen in respiration very early. They can photosynthesize in light and respire in the dark. Obviously, such a system must be able to store oxygen in a stable, nontoxic state for hours at a time.

Cyanobacteria, especially those in stromatolites, were the dominant life forms along early ocean shorelines. Their success was largely due to their control over oxygen, which gave them an abundant and reliable energy supply in two different ways: by mastering photosynthesis based on water, and by breaking down food molecules in respiration rather than fermentation.

Stromatolites increase dramatically in the rock record with the beginning of the Proterozoic Era at about 2500 Ma. At the same time, major geological changes resulted in major crustal formation, accompanied by intense volcanic activity that must have provided huge amounts of iron to the ocean water. UV light and/or purple bacteria oxidized iron at the surface of the open ocean, while cyanobacteria gave off waves of oxygen into shallow water that oxidized and precipitated any iron they contacted. Driven by seasonal changes in sunlight, all three reactions worked together to produce BIF, helping to explain why BIF and stromatolites occur together in many rock sequences. BIF production rose to a peak around 2500 Ma, then fell off steadily until 1800 Ma.

The increased oxygen production by stromatolites in shallow water produced the first great masses of BIF, and the gigantic buffering effect of oxidizing the iron allowed time for the biochemistry of oxygen tolerance and utilization to evolve in many bacteria. But eventually the rate of BIF formation slackened and the oceans and atmosphere began to accumulate permanent but very low amounts of free oxygen. BIF are rare after 2000 Ma, as oxygen levels in the oceans reached a permanent level so high that seawater could no longer hold dissolved iron and BIF could no longer form.

Other geological evidence confirms the oxygenation of the oceans around this time. The uranium mineral uraninite cannot exist for long if it is exposed to oxygen, and it is not found in rocks younger than about 2300 Ma. On land, the presence of oxygen in air for the first time would have rusted any iron minerals exposed at the surface by weathering. Rivers would have run red as they flowed across the Earth's surface before vegetation invaded the land. On land and in shallow seas, **red beds**, or sediments bearing original iron oxides, date from about 2300 Ma. An estimate based on carbon isotope work suggests a very rapid rise in global oxygen levels between 2220 and 2060 Ma.

All these changes justify the use of the term **oxygen revolution**: it changed the chemistry of Earth's air, land, and water forever. One of the indirect results was vital for the further evolution of living things. Solar UV radiation acts on any free oxygen high in the atmosphere to produce **ozone**, which is O_3 rather than O_2. Even a very thin layer of ozone can block most UV radiation. Earth's surface has been protected from damaging UV radiation ever since free oxygen entered the atmosphere. After that time, it became possible for organisms to evolve that were longer-lived and more complex than bacteria (Chapter 3).

Stromatolites as Evolutionary Forcing Houses

The success of cyanobacteria didn't just pump waves of oxygen into ocean water and the atmosphere. It also began to affect the anoxic bacteria living within stromatolites. Stromatolites would have become forcing houses of evolution, where massive bacterial populations were stressed by intermittent waves of oxygen that were weak at first but gradually became more intense and longer-lasting. Oxygen tolerance and then oxygen use probably evolved in several lines of bacteria in these microenvironments. Stromatolite microbiology was probably as complex at 2000 Ma as it is today. Certainly the bacterial fossil record became increasingly rich during the oxygen revolution. The Gunflint Chert of the Lake Superior area (about 2000 Ma) contains stromatolites and magnificently preserved bacteria (Figure 2.12).

The Superiority of Slime

Slime is an underappreciated facet of bacterial biology. In nature, all bacteria secrete slime. The bacterial cell membrane may secrete over itself either a rigid outer coating or an extracellular coating of slime, or both.

Slime is very useful to bacteria. It allows limited movement of a cell so that it can adjust a little to the microenvironment, even without obvious locomotory devices. Slime protects against radiation, so cells photosynthesizing in very shallow water are less exposed to harm from UV radiation. Diffusion is slow through slime, so it is easy for a slimy bacterium to survive fluctuations in external chemistry. Slime is chemically active, allowing bacteria to stick tightly to surfaces.

Slime helps to form complex structures. Bacterial cells can divide inside their slime coat, so large colonies can form as large units, well glued together. Slime helps to trap sediment and nutrients. Stable communities of bacteria can resist chemical, physical, or pharmacological disturbance, whether they form slimy photosynthetic mats on the shoreline of salty lagoons or slimy masses of plaque on teeth.

Sometimes bacteria of different kinds are attracted to one another to form a **consortium** inside a common slime coat, where they can act efficiently together. For example, one species will break down organic molecules, releasing hydrogen to methanogenic archaebacteria, which use the hydrogen to produce methane. Consortia of bacteria help to form stromatolites. Bacteria usually grow faster in consortia than they do alone.

Some cyanobacteria today store oxygen in bubbles in their slime coating. The bubbles act as an oxygen store so the bacterium can respire at night; they detoxify any poisonous sulfide that diffuses into the microenvironment; and they may inhibit the growth of nearby anaerobic bacteria. It's easy to imagine early cyanobacteria using their ability to tolerate oxygen, then evolving to use slime, first as a defensive shield and then as an offensive weapon in the competitive environment inside stromatolites.

Slime is thus fundamental to the success of many bacteria. How old is it? Slime was holding together bacterial mats at 3500 Ma at Warrawoona, so it is as old as the earliest known cells. The Warrawoona cells are preserved because they have a rigid outer bacterial cell coat. The rigid coat is a protective layer that is either covered by slime or has slime as an alternative. Perhaps cells used the simpler alternative of slime long before they evolved a rigid outer coat.

Could slime be as old as life itself? Perhaps slime evolved even before the cell membrane. A naked gene coated in polysaccharide slime may have done rather well, nestled inside a cleavage crack in a seafloor clay mineral.

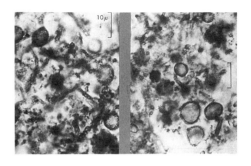

Figure 2.12 The Gunflint Chert of Minnesota contains magnificently preserved bacteria. (Courtesy of J. William Schopf, University of California, Los Angeles.)

Review Questions

When did life evolve? (and what evidence did you use to say so?)

How old are Earth's earliest known fossils? (within 10%)

What kind of organism formed the earliest known fossil?

What are BIF or Banded Iron Formations and how were they formed?

What is a stromatolite (draw or describe). How is it formed?

What is a trace fossil? Give an example.When did the "Oxygen Revolution" take place? Summarize in a few words the evidence used to answer that question.

Why is the word "revolution" used to describe the change in the oxygen content of Earth's oceans and atmosphere?

Further Reading

Easy Access Reading

Barley, M. E., at al. 1997. Emplacement of a large igneous province as a possible cause of banded iron formation 2.45 billion years ago. *Nature* 385: 55–58, and comment, pp. 25–26.

Costerton, J. W., et al. 1999. Bacterial biofilms: a common cause of persistent infections. *Science* 284: 1318–1322. Slime.

Groves, D. I., et al. 1981. An early habitat of life. *Scientific American* 245(4): 64–73. The environment of the Warrawoona rocks.

Mojzsis, S. J., et al. 1996. Evidence for life on Earth before 3,800 million years ago. *Nature* 384: 55–58, and comment, pp. 21–22; also comment in *Science* 275: 38–39.

Nisbet, E. G. 1987. *The Young Earth.* Boston: Unwin Hyman.

Oluwenstein, L. 1995. Death and the microbe. *Discover* 16 (9): 99–103. The biology of bacterial colonies.

Renne, P., et al. 1998. Absolute ages aren't exactly. *Science* 282: 1840–1841.

Schopf, J. W. 1993. Microfossils of the early Archaean Apex Chert: new evidence of the antiquity of life. *Science* 260: 640–646.

Widdel, F., et al. 1993. Ferrous iron oxidation by anoxygenic phototrophic bacteria. *Nature* 362: 834–836, and comment, pp. 790–791.

More Technical Reading

Karhu, J. A., and H. D. Holland. 1996. Carbon isotopes and the rise of atmospheric oxygen. *Geology* 24, 867–870.

Knoll, A. H. 1985. Patterns of evolution in the Archean and Proterozoic Eons. *Paleobiology* 11: 53–64.

Konhauser, K. O. 1998. Diversity of bacterial iron mineralization. *Earth-Science Reviews* 43: 91–121.

Madigan, M. T., et al. 1997. *Biology of Microorganisms*, 8th ed. Englewood Cliffs, N. J.: Prentice Hall.

Schopf, J. W. (ed.). 1983. *Earth's Earliest Biosphere: Its Origin and Evolution.* Princeton: Princeton University Press.

Walter, M. R. (ed.). 1976. *Stromatolites.* Amsterdam: Elsevier.

Sex and Nuclei: Eukaryotes

There are fundamental biological differences between **prokaryotes** (all living Archaea and Bacteria), and **eukaryotes** (all other living organisms) (Figure 3.1, and Box 3.1). Eukaryotes must have evolved from prokaryotes, which were the earliest organisms. Who were those prokaryotic ancestors of eukaryotes, and how did eukaryotes evolve from them? Prokaryotes were and are very successful in an incredible range of habitats, from stinking swamps to the hindgut of termites and from hot springs in the deep sea to the ice desert of Antarctica, and deep in rocks underground. They occur in numbers averaging 500 million per liter in surface ocean waters, 1 billion per liter in fresh water, and about 300 million on the skin of the average human. This diversity makes it difficult to begin to look for potential closest relatives of living eukaryotes!

Symbiosis and Endosymbiosis

Symbiosis is a relationship in which two different organisms live together. Often they act so that they both derive a benefit from the arrangement. Examples range from the symbiosis between humans and cats to bizarre relationships such as that of acacia plants, which house and feed ant colonies that in turn protect the acacia against herbivorous animals and insects. The ultimate state of symbiosis is **endosymbiosis**, in which one organism lives inside its partner. Animals as varied as termites, sea turtles, and cattle are able to live on plant material because they contain somewhere in their digestive system bacteria with the enzymes to break down the cellulose that is unaffected by the host's digestive juices. Many

BOX 3.1. DIFFERENCES BETWEEN PROKARYOTES AND EUKARYOTES

(1) Eukaryotes have their DNA contained within a membrane, and under the microscope this package forms a distinct body, the **nucleus**. Prokaryotes have their DNA loose in the cell cytoplasm.

(2) Prokaryotes have no internal subdivisions of the cell, but almost all eukaryotes have **organelles** as well as a nucleus. Organelles are subunits of the cell that are bounded by membranes and perform some specific cell function. **Plastids**, for example, are organelles that perform photosynthesis inside the cell, generating food molecules and releasing oxygen. **Mitochondria** contain the respiratory enzymes of the cell. Food molecules are first fermented in the cytoplasm, then passed to the mitochondria for respiration. Mitochondria generate ATP as they break food molecules down to water and CO_2, and they pass out energy and waste products to the rest of the cell. They also make steroids, which help to form cell membranes in eukaryotes and give them much more flexibility than prokaryote membranes.

(3) Eukaryotes can perform **sexual reproduction**, in which the DNA of two cells is shuffled and redealt into new combinations.

(4) Prokaryotes do not have flexible cell walls, so cannot expand to engulf other cells. The flexibility of eukaryotic cell membranes allows them to engulf large particles, to form cell vacuoles, and to move freely. Plant cells, armored by cellulose, are the only eukaryotes that have given up a flexible outer cell wall for most of their lives.

(5) Eukaryotes have a well-organized system for duplicating their DNA exactly into two copies during cell division. This process, **mitosis**, is much more complex and precise than the simple splitting found in prokaryotes.

(6) Eukaryotes are almost always much larger than prokaryotes. A eukaryote is typically ten times larger in diameter, but this means that it has about 1000 times the volume of a prokaryote.

(7) Eukaryotes have perhaps a thousand times as much DNA as prokaryotes. They have multiple copies of their DNA, with much repetition of sequences. The DNA content of prokaryotes is small, and they have only one copy of it. There is little room to store the complex "IF... THEN... " commands in the genetic program that turn on one gene as opposed to another. Therefore, genetic regulation is not well developed in prokaryotes, which means that they cannot produce the differentiated cells that we and other eukaryotes can. Multicellular colonies of bacteria are all made up of the same cell type, repeated many times in a clone. Therefore, any species of bacterium is very good at one thing but cannot do others; its range of functions is narrow.

Figure 3.1   The structure of a eukaryote (left) and a prokaryote (right). The eukaryote is much bigger than the prokaryote.

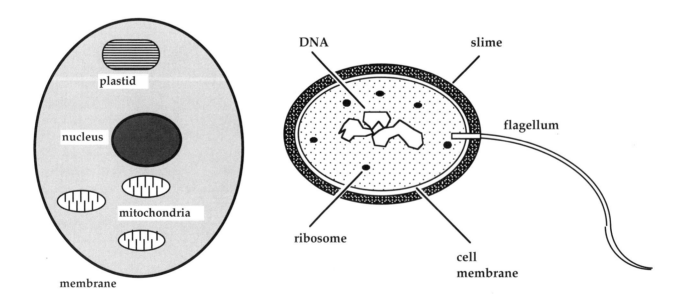

tropical reef organisms have symbiotic partners in the form of photosynthesizing microorganisms. Living inside the tissues of corals or giant clams, these symbiotic partners have a safe place to live. In turn, the host receives a share of their photosynthetic production.

Thanks mainly to Lynn Margulis, it's now clear that endosymbiosis was critical in the evolution of eukaryotes. The first Eucarya were prokaryotic, and are probably as ancient a group of organisms as Bacteria and Archaea. However, at some point, a prokaryotic eukaryan made a dramatic evolutionary breakthrough: it obtained organelles by endosymbiosis with another prokaryote, and in doing so became a eukaryote. Mitochondria and plastids, and perhaps even flagella, were once free-living bacteria. But they became so closely associated with a host prokaryote (the ancestral eukaryan) that eventually they became, for practical purposes, part of it (Figures 3.1, 3.2). At least five major pieces of evidence show that organelles originated by endosymbiosis (see Box 3.2).

Mitochondria and Their Ancestors

How did a proto-mitochondrial bacterium get inside a eukaryan host cell? There are two current scenarios, which do not differ very much. They both assume that the ancestral eukaryan evolved a cell wall flexible enough to engulf other cells.

In one scenario, the eukaryan may have engulfed bacteria to digest them. However, if it engulfed a bacterium that could oxidize its waste products and release some of the calories back to the eukaryan, the eukaryan would benefit by not digesting the bacterium, but keeping it as a living internal guest (or slave, if you like). The bacteria came to reside permanently (and divide) inside the host.

In the other scenario, eukaryan cells and bacteria routinely succeeded better when they lived side by side in very close contact—in external

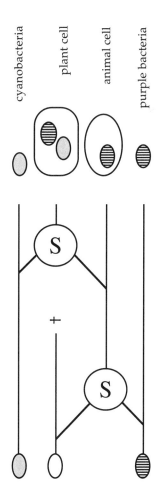

Figure 3.2 Lynn Margulis and others have suggested that plastids and mitochondria were once free-living cells until they became endosymbionts inside other cells. S symbols indicate symbiotic events.

BOX 3.2. EVIDENCE FOR ORGANELLE/EUKARYOTE SYMBIOSIS

1. The DNA in mitochondria and plastids is not like the DNA in the eukaryotic cell nucleus.

2. Mitochondria and plastids are separated from the rest of the eukaryotic cell by membranes; thus they are actually outside the cell. The cell itself makes the membrane, but inside it is a second membrane secreted by the organelle.

3. Plastids, mitochondria, and prokaryotes make proteins by similar biochemical pathways, which differ from those in the cytoplasm of eukaryotes.

4. Mitochondria and plastids are susceptible to streptomycin and tetracycline, and so are prokaryotes; eukaryotic cytoplasm is unaffected by these drugs.

5. Mitochondria and plastids can multiply only by dividing; they cannot be made by the cell cytoplasm. Organelles thus have their own independent reproductive mechanism. A cell that loses its mitochondria or plastids cannot make more.

symbiosis. Each produced substances that the other could use. Eventually, the eukaryan took the bacteria inside its tissues, not as a predator engulfing food, but as a host for internal symbionts.

The host eukaryan and the bacteria eventually came to a stable relationship when they established mutual population control. The host cells grew, flourished, and divided, each daughter cell taking with it a population of bacteria. At some point the symbiont bacteria lost their cell walls and became mitochondria, and we can call the host a true eukaryote.

Early eukaryotes, with their flexible cell membranes, could engulf other cells and gather food for themselves and their internal partners. Eukaryotes came to depend entirely on their mitochondria to provide them with ATP, so the number of mitochondria had to be matched closely to the needs of the host. Therefore, the genes that control mitochondrial reproduction passed to the host nucleus, leaving behind in the mitochondria (as far as we can tell) mainly the genes that control the oxidation they perform for the cell.

Either scenario could have produced **protists**, (single-celled eukaryotes). Protists are amoeba-like **animals**, capable of moving and eating by engulfing other organisms. Food is fermented in the cytoplasm and oxidized in the mitochondria. The same process occurs in our cells today.

> A eukaryote wanting to eat,
>
> Saw bacteria as quite a nice treat.
>
> Increasing closeness
>
> Led to endosymbiosis,
>
> And the modern plant cell was complete
>
> © *Elizabeth Wenk 1994*

Plastids

Plants photosynthesize to accumulate energy. Plant photosynthesis is not performed in the cell cytoplasm but in organelles—in chloroplasts, or **plastids**. Some early protists acquired cyanobacteria as symbiotic partners which evolved into plastids (Figure 3.2), in the same way that even earlier protists had acquired mitochondria. The cyanobacteria benefited more from nutrients in the host's wastes than they would as independent cells. In time, the protist came to rely so much on the photosynthesis of its partners that it gave up hunting and engulfing other cells, gave up locomotion, grew a strong cellulose cell wall for protection, settled in the light, and took on the way of life that we now associate with the word "plant."

This is a scenario for the evolution of the first eukaryotic photosynthesizers (the **algae**) (Figures 3.2 and 3.3). Since that event, the plant-animal dichotomy has been one of the most important in the organic world. We place advanced plants and advanced animals in two different Kingdoms; animals eat plants and one another.

The word **symbiogenesis** is used to describe the appearance of a dramatically new organism by symbiosis rather than mutation. It is a useful term, because we are discovering more and more examples of symbiosis in ecosystems. For example, many trees are successful because they have symbiotic fungi in or around their roots, which help to break down soil debris and make it available to their tree partners: in turn, they take some nutrition from the plant roots. Nevertheless, beware: although symbiogenesis seems dramatic, it is a normal part of evolution. Species must acquire the mutations that give them the capacity to take part in symbioses.

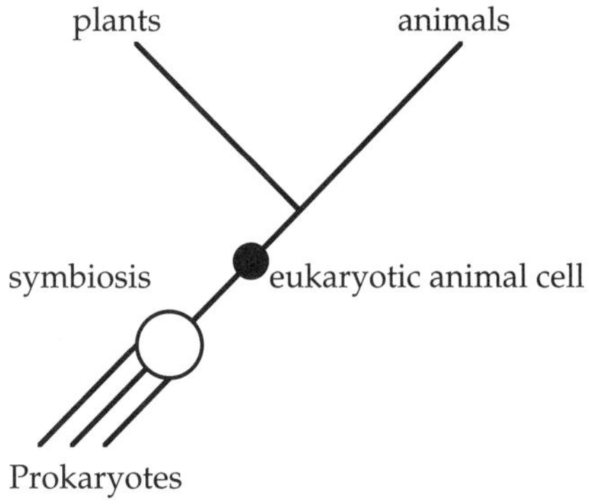

Figure 3.3 This diagram suggests that plants evolved from animal cells, as did Figure 3.2. The ancestors of both groups were single-celled, and the branch between them also occurred among protists.

Each species in the symbiosis continues to evolve under natural selection, and individuals that take part in symbioses do so because they reproduce more effectively than those that do not.

Eukaryotes in the Fossil Record

The endosymbiotic theory is based on biological and molecular evidence. It's not easy to see how it can be tested in the fossil record. It is even difficult to identify the first eukaryotic fossils. Most fossil cells are small spherical objects with no distinguishing features. Although most eukaryotes are larger than prokaryotes, at least one living prokaryote approaches normal eukaryotic size. Experiments in which artificial fossils have been made from rotting prokaryotes have shown that it is almost impossible to distinguish them from eukaryotes after death. After death, the cell contents of prokaryotes can form blobs or dark spots that look like fossilized nuclei or organelles (Figure 3.4). Rotting colonies of cyanobacteria can look like multicellular eukaryotes, and filamentous bacteria can look like fungal hyphae. And finally, early eukaryotes were probably small and thin-walled, and therefore are most unlikely to be preserved as fossils.

We have to take the geological record and interpret it as best we can. First of all, eukaryotes could not have evolved before oxygen became a permanent component of seawater. The symbiotic theory for the origin of eukaryotes requires intimate association between two different prokaryotes with complementary needs. I think that the first eukaryotes evolved in or near stromatolites, in shallow coastal waters. The major problem will be to identify them unambiguously as eukaryotes.

The oldest eukaryote is *Grypania spiralis*, a ribbon-like fossil 2 mm wide and over 10 cm long. It occurs abundantly in rocks around 1400 Ma in China and Montana, and has been described also from banded iron formations in Michigan, dated at 2100 Ma. *Grypania* looks very much like a eukaryotic alga. If so, then eukaryotic algae evolved at or before 2100 Ma, and if Figure 3.3 is correct, simpler eukaryotes had evolved before that, even while major amounts of BIF were still being deposited. Yet living eukaryotes require at least 1% of present oxygen levels. This figure is larger than most geochemists would like for depositing BIF (Chapter 2). No-one would be surprised by *Grypania* at 1400 Ma, but 2100 Ma seems too early. It is not clear how the problem will be resolved.

The Evolution of Sex

Prokaryotes sometimes donate DNA to other cells, but this is not at all similar to sexual reproduction. Essentially, every prokaryote is its own lineage, either dying, budding off, or splitting into daughter cells that are clones of the parent in the process of asexual reproduction.

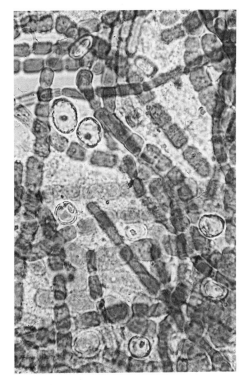

Figure 3.4 A few dead and rotting cells among these living cyanobacteria look as if they have organelles or nuclei inside them. The dead cells have split away from their chains and swollen to a spherical shape, and their rotting contents have formed dark patches and dots. So we cannot identify dots and spots in fossil cells as organelles or nuclei, even if they are. (Courtesy of Andrew Knoll, Harvard University.)

Cell division is simple for prokaryotes. There are no mates to find, organelles to organize. Daughter cells are clones, with the same DNA as the parent cell, so they are already well adapted to the microenvironment. Prokaryotes gamble against a change in the environment: if a change occurs that kills an individual, the change will most likely wipe out all that individual's clones too. Prokaryotes have no way to affect the future of their genes. They can only pass them on unchanged to their offspring.

Many eukaryotes also reproduce by simple fission, cloning identical copies of themselves. *Amoeba* is perhaps the most familiar example, but corals, strawberries, Bermuda grass, and aphids often use this method too. In most living eukaryotes, however, the DNA of two individuals is shuffled and redealt to their offspring in **sexual reproduction**. Offspring are therefore similar but not identical to their parents: in fact, there is an impossibly low chance that any two sexually reproduced individuals are genetically identical, unless they developed from the same egg, as identical twins do.

The offspring of sexual reproduction resemble their parents in all major features but are unique in their combination of minor characters. Sexually reproducing species have built-in genetic variability that is often lacking in clones of bacteria. Individuals vary in the characters of their bodies, which often means that some individuals are slightly better fitted to the environment than others, so stand a better chance of reproducing. The particular sets of DNA in those individuals are thus differentially represented in future populations.

In organisms that reproduce by cloning, a favorable mutation can spread successfully only if it occurs in an individual that outdivides its competitors. The environment selects or rejects the whole DNA package of the individual, which either divides or dies. This is a one-shot chance, and many potentially successful mutations may be lost because they occur in an individual whose other characters are poorly adapted. On the other hand, a favorable mutation may allow one individual such success that it and its clones outcompete all the others, making the population uniform even though it may contain bad genes along with the good one. Uniform populations of yeast may be desirable to a baker or brewer, but in nature a uniform population may easily be wiped out by changes in the environment.

In contrast, a mutation in a sexually reproducing individual is shuffled into a different combination in each of its offspring. For example, a mutant oyster might find her mutation being tested in different combinations in each of her 100,000 eggs. Natural selection could then operate on 100,000 prototypes, not just one. Favorable combinations of genes can be passed on. A sexually reproducing population can evolve rapidly and smoothly in changing environments, and in favorable circumstances evolution can be greatly accelerated by sexual reproduction.

At the same time, sexual reproduction is conservative. Extreme mutations, good or bad, can be diluted out at each generation by recombination with normal genes. The genes may not disappear from the population, but may lurk as recessives, likely to reappear at unpredictable times as recombination shuffles them around.

Eukaryotes are so complex that only approximately similar individuals can shuffle their DNA together with any chance of producing viable offspring. Complex physical, chemical, and behavioral ("instinctive") mechanisms usually ensure that sex is attempted only by individuals that share much the same DNA. Such a set of organisms forms a **species**, which is defined as a set of individuals that are potentially or actually interbreeding. The composite total of genes that are found in a species is called the **gene pool**.

Sexual reproduction has two great flaws. First, a sexual individual passes on only half of its DNA to any one offspring, with the other half coming from the partner. Therefore, to pass on all its genes, a sexual individual has to invest double the effort of an asexual individual. Second, the offspring of sexual parents are not identical. Sets of incompatible genes may be shuffled together into the DNA of an unfortunate individual, which may die early or fail to reproduce. At every generation, then, some reproductive wastage occurs. Yet so many species reproduce sexually that there must be very strong counterbalancing advantages of sex. Although several books have been devoted to this question, there are as yet no convincing answers.

Certainly sexual reproduction has advantages from the point of view of the species or population. It's good for a population to have variability to survive environmental crises; it's good for the future of the species to retain favorable mutations in the gene pool for comprehensive testing. But that's not how evolution really works. Selection operates dominantly, perhaps exclusively, on individuals, and individuals do what's best for themselves and for their genes.

One argument suggests that mutations can be repaired more easily by sexually reproducing organisms because they have a double copy of their DNA. In the same way, computers can be programmed to copy documents more accurately by comparing the new version letter by letter with the original. However, this argument only explains why organisms should be diploid, not why they should reproduce sexually. Diploids could combine the DNA repair mechanisms of their diploid state with asexual reproduction. Almost every animal group has some diploid members that reproduce asexually.

Because the problem of the origin of sex is not solved, let me try a suggestion of my own. The major advantage that I see in sexual reproduction is the ability to manipulate the fate of one's DNA. An individual is born with a certain set of DNA, which cannot be altered. But what is passed on to offspring can be influenced—*by appropriate choice of a mate*. Clearly, one cannot control the details, because recombination is a matter of shuffling and redealing DNA. But while an individual cannot change its own DNA, it can control to some extent the DNA that is to be shuffled with its own. I explain this with a thought experiment.

Orchids, Eukaryotes, and Game Theory: The Origin of Sex

Commercial orchid growers are rewarded for the successful production of striking flowers. They prefer to propagate orchids by cloning the best varieties, so that they can produce large numbers of identical, superior blooms for sale. Yet there is a large but long-term potential reward for successfully breeding new varieties of orchid that in turn can be cloned in high-level production. Therefore, orchid companies devote at least some of their resources to sexual crossing experiments between individual orchids, trying to achieve the right balance between these two alternative styles of producing new orchids. In other words, the best strategy for commercial orchid production includes a mixture of reproductive modes, with most production being asexual cloning and some being sexual.

Now let's think about early eukaryotes that are still reproducing like bacteria, by cloning. What if two early eukaryotes could reproduce sexually? Could sex evolve in the face of its obvious disadvantages?

Suppose a eukaryote (The Clone) has been living successfully in some environment, along with other eukaryote lineages. All of them usually reproduce by cloning, although they can exchange some DNA with other cells, just as bacteria do. The Clone lives alongside its daughter cells, which were all cloned from it and are identical with it—in fact, they all are The Clone. Members of The Clone have no reason to compete with one another, because they share the same DNA. An occasional mutant differs, of course, but most mutants are not successful organisms, and they or their clones die out. Sexual reproduction is not a good idea in these circumstances. Like an established orchid company, The Clone succeeds best by continuing to clone, and so do all other eukaryotes in this environment.

Suppose now that a new lineage of cells appears. Perhaps it is a successful mutant, perhaps it is a strain of cells that represents serious competition to The Clone. What should The Clone do?

The Clone is either superior or inferior to the newcomer and so has perhaps a 50% chance of surviving direct competition with it if both simply continue to clone. But can The Clone increase its chances of survival in the face of new competition? I suggest that The Clone should behave like an orchid company and should devote at least some of its cells to sexual reproduction.

If The Clone devotes one of its identical cells to DNA exchange with the newcomer, the offspring will contain some percentage of The Clone's DNA. If the offspring is inferior and dies out, The Clone and its competitor have each lost an equal, insignificant, investment. If the competitor remains superior, The Clone is doomed anyway, and its investment in sexual reproduction cost it nothing. If The Clone remains superior to its competitors, it continues to clone and dominate the environment. But if the offspring is superior to both, its lineage is the one that will survive. The Clone's genes are at least partly represented in the successful population, which is better than not at all. And if that happens, The Clone would try gene exchange with that successful offspring, to try to increase its genetic representation in the descendant still more. In summary, DNA exchange by a few of its individuals costs The Clone little but could very well mean the difference between genetic survival and genetic extinction. As long as there are many individuals in The Clone, the cost of these attempts is about the same as a typical insurance premium (a small fraction of the potential cost of a disaster).

Each lineage of eukaryotes would have the same best strategy of sex-for-insurance, so occasional DNA exchange would be favored by all lineages that were potential competitors. Each partner in an exchange would try to pass along as much DNA as it could, and that amount would quickly balance out to 50%. Even if a first attempt at DNA exchange did not succeed, the logic behind the attempt would remain the same, and attempts would continue.

This strategy works even if the environment changes. Cells should outcompete other cells if they can; if they can't, they should reproduce sexually with rivals, if possible. This implies that sexual reproduction would have been favored as soon as it was possible, at least as an occasional strategy.

The disadvantages of sexual reproduction do not count in this situation. The reasoning applies to any habitat where two or more potentially interbreeding clones of eukaryotes are competing. Competition between clones is the driving force leading to the evolution of sexual reproduction in this scenario.

This reasoning is not shared by most higher eukaryotes today, where sexual reproduction is normal and groups of clones are rare. In this case, the individual is unique, and no part of it can be spared for a rather speculative insurance premium. Sexual reproduction would not arise in lineages of unique individuals, but it has persisted in evolution for other reasons; that's a different problem.

It is very difficult to pin down the time when sexual reproduction began among eukaryotes. We can only try to identify in the fossil record some creatures whose descendants today reproduce sexually. For certainty, we must look to the appearance of multicellular animals, perhaps 600 Ma. But Andrew Knoll has suggested that a big diversification of protists around 1000 Ma may reflect their evolution of sexual reproduction.

The Classification of Eukaryotes

Eukaryotes occur in the natural world in ecological and evolutionary units called species. As we have seen, species are groups of individuals whose genetic material is drawn from the same gene pool but is almost always incompatible with that of another gene pool. Members of the same species, therefore, can potentially interbreed to produce viable offspring. They tend to share more physical, behavioral, and biochemical features (usually called **characters**) with one another than they do with members of other species. Defining and comparing such characters allows us to distinguish between species of organisms. A species is not an

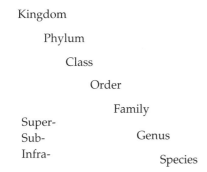

Kingdom

Phylum

Class

Order

Family

Super-
Sub-   Genus
Infra-
                        Species

Figure 3.5 The Linnean hierarchy of taxa, or taxonomic units, as it is used by zoologists. Other categories such as "cohort" can be used, and the qualifying prefixes in the lower left hand corner are used freely. Botanists and microbiologists use rather different categories, but the hierarchical principle remains the same.

arbitrary group of organisms, but a real, or natural, unit.

Biologists use the **Linnean system** of naming species, after the Swedish biologist Carl Linné who invented it in the eighteenth century. A species is given a unique name (a **specific** name) by which we can refer to it unambiguously. Linné gave the specific name *noctua* to the little owl of Europe because it flies at night. Species that share a large number of characters are gathered together into groups called **genera** (the singular is **genus**) and given unique generic names. Linné gave the little owl the generic name *Athene*. She is the Greek goddess of wisdom, and the little owl is the symbol of the city of Athens, stamped on its ancient coins. However, taxonomic names do not have to carry a message, even though a simple and appropriate name is easier to remember. (One must be careful about names: *Puffinus puffinus* is not a puffin, but a shearwater, and *Pinguinus* is not a penguin but the extinct Great Auk!) Thus, Linnean names are only a convenience, but a very valuable one. The bird that the British call the tawny owl, the Germans the wood owl, and the Swedes the cat owl, is *Strix aluco* among international scientists.

Genera may be grouped together into higher categories. For example, many species of owls are grouped together to form the Family Strigidae, or strigids, named after one of its genera, *Strix*. Families may be grouped into superfamilies, and after that into orders, classes, and phyla. Many other subdivisions can be coined for convenience (Figure 3.5).

A division or subdivision that is used to arrange organisms into groups is called a **taxon** (plural, **taxa**). Biologists who try to recognize, describe, name, define, and classify organisms are **taxonomists** or **systematists**, and the practice is called **taxonomy** or **systematics** or **classification**. Although slightly different ranks of categories are used for different kingdoms of organisms, the basic units of classification recognized by all biologists remain the species and genus.

Complex rules have grown up as more and more organisms have been described by taxonomists. Botanists and zoologists have their own rules for describing new species so that names are legally established.

Describing Evolution

Linné did not believe in evolution. But we now recognize that members of a species form a genetically based biological unit that is evolutionarily separate from the rest of the organic world. The recognition and naming of a new species, therefore, is a statement about evolution. It represents the taxonomist's hypothesis that the members of the species share the same gene pool, which is different from the gene pool of any other species because there has been evolutionary divergence over time.

Most taxonomists aim to form genera and higher categories that truly reflect evolution. In evolutionary classification, species are grouped into genera on the hypothesis that genera also share a unique set of characters that are not shared by other genera. In turn, genera are grouped into families, families into orders, orders into classes, and classes into phyla. Decisions about the course of evolution are not always obvious, so taxonomic decisions may have to be revised as new information becomes available. Species are moved around between genera and higher categories as taxonomists refine their classifications to reflect evolutionary history more effectively. The incomplete nature of the fossil record makes classification particularly difficult for paleontologists, and it often leads to uncertainties or arguments about classification.

As organisms evolve through time, their characters change. Characters may change slowly or quickly; they may change gradually or rather suddenly. As changes accumulate, one species may evolve characters that are changed from their original state. The new set of characters may be different enough that a biologist who could examine living specimens at the ends of the series would certainly regard them as separate species. But how can one draw a line between species in time? After all, descendants have always been genetically continuous with their ancestors. There is no discontinuity between ancestor and descendant species like that seen between contemporaneous species in the living world. This is a special question facing paleontologists, and it makes the taxonomy of the fossil record rather difficult, and often viciously argued, as is the case for fossil hominids. The principle is simple, however. At one extreme one could say that all living organisms are the same species, because they all evolved, continuously, from a single ancestor that was the first living cell. But for convenience, and to reflect the reproductive gaps that exist between the species at any given time in Earth's history, a paleontologist must draw lines somewhere between species (and genera, and families, and so on), knowing that the lines are artificial if they pretend to separate ancestors from descendants. Fortunately or unfortunately, the fossil record is spotty enough, and the pace of evolutionary change is rapid enough, that truly intermediate fossils are very rarely found. The fossil record is therefore rather more easily divided into species and higher categories than one might expect.

Cladistics

This book uses a mild form of an approach to evolutionary taxonomy called **cladistics**. The aim of cladistics is to trace the **phylogeny** (the evolutionary pathways) by which species appeared and to use that phylogeny to organize species into **clades**, that is, groups of species that all descended from one ancestral species. To use an analogy, a clade is a branch on a tree that represents all life: if all life on Earth is descended from the first living cell by a series of evolutionary branching events, life as a whole is one clade (the entire tree). Just as trees may branch many times, and branches then branch, and so on, clades of organisms exist in a hierarchy of scales, with the terminal branches representing single species. Every species belongs to a larger clade, which belongs to a larger clade, and so on. Every clade, no matter how large, began with a single branching event that produced the ancestor species of the clade.

Cladists try to identify groups of species that share a set of characters that were evolved as new features in a common ancestor, and then passed

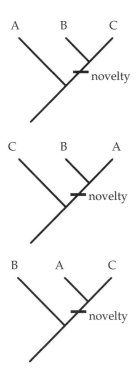

Figure 3.6 Three possible hypotheses (expressed as cladograms) could portray the evolutionary relationship of these three species. Perhaps one might feel unable to work out the relationship. In that case one would portray the three species as separating into three equal branches. Obviously, with more taxa under study, the number of possible hypotheses increases enormously.

on to all descendant species. Such newly evolved characters represent a change from a **primitive** or original state to a novel or **derived** state. For example, all living mammals have fur, but no other living organisms do. Perhaps fur was inherited from a common ancestor of all living mammals that evolved a furry skin as a newly derived character modifying a primitive one (a skin covered by scales). If that is true, then mammals are a clade. Examining the hypothesis, one finds other shared derived characters of living mammals that strengthen the argument: all living mammals are warm-blooded, and suckle their young.

Sometimes problems arise because similar derived characters are found in species outside a clade; those characters have evolved more than once by **parallel evolution**. For example, bats and birds both have wings, and in each group the wing is a derived character that has been modified from some other structure. But bats and birds share very few other derived characters, and the weight of evidence suggests that birds are a clade, bats are probably a clade, but [bats + birds] is not a clade.

Once a clade is established, we then search for characters that are novel in species within the clade. Subclades can be established based on the distribution of such derived characters within the group, until a single best hypothesis emerges about the total evolutionary history of the group of species. The hypothesis can be tested as further characters are examined or existing ones are reassessed, and as new species are discovered and fitted into the evolutionary framework.

As an example, three living species, A, B, and C, could be related along three possible evolutionary pathways (Figure 3.6). Which is correct? Which two of the three species are most closely linked? Two species may look very similar because they share similar characters, but if those are **shared primitive characters** that were also present in a common ancestor, they cannot tell us anything about evolution within the group, because they have not changed within that history. The useful character for solving the problem is the novelty, or **derived character**, which defines the group that has changed the most since the three species all shared the characters of their common ancestor (Figure 3.6).

Cladograms such as Figure 3.6 display the distribution of characters in a visual form, and the cladogram that requires the simplest and fewest evolutionary changes is assumed to represent best the phylogenetic history of the species. A cladogram therefore expresses a hypothesis about the phylogeny of a group. Two of the species are most closely linked, and form sister groups in a clade, while the third species becomes their sister group in a larger clade (Figure 3.6).

Once the preferred cladogram is drawn to portray the best hypothesis, one can make decisions about the best way to classify the species and to describe its evolutionary history. A cladogram in itself does neither of these things.

One could introduce formal names for each clade on a cladogram. It's obvious, however, that this would lead to a great number of names, not all of which might be necessary for everyday discussion around the breakfast table. It conveys more information simply to print a cladogram and to use a minimum of hierarchical names.

A cladogram is always drawn with all the species under study along one edge (Figure 3.7). No species in a cladogram is shown as evolving into another. Some cladists claim that one can never know true ancestor–descendant relationships, and in a strict sense this is correct because we don't have time machines. But sometimes a fossil is known that could well be an ancestor of a later fossil or of a living organism. At present, for example, it seems more reasonable (to me) to suggest that *Archaeopteryx* is the ancestor of owls and penguins than to suggest that these birds are descended from some ancestor that we haven't found yet. Hypotheses like this are expressed on **phylograms** or **phylogenetic trees** that include time information: we are allowed to show a suggested ancestor within the diagram (Figure 3.7). Like cladograms, phylograms are not statements of fact but hypotheses, subject to continuous testing.

Counterintuitive patterns sometimes emerge in cladistics. We are all used to thinking about fishes, amphibians, reptiles, birds, and mammals as classes of vertebrates, in some way equal in rank to one another (Figure 3.8). But this is not a cladistic classification. Tetrapods are actually a clade within fishes, derived from them by acquiring some novel characters, including feet, and amphibians are a clade within tetrapods. Reptiles are in turn a derived subgroup of tetrapods. Mammals and birds are equally clades of derived reptiles.

There's nothing intimidating about this—it simply takes some time to get used to it. The important thing is not to try to force the older taxonomic units into a cladistic framework, but to combine a simple and convenient classification with an explanatory cladogram or phylogram. I have tried to use cladograms and phylograms in this way.

If we classify all living reptiles into one group and draw a cladogram (Figure 3.8), we display the well-known fact that living reptiles and birds are more alike than either is to mammals. The cladogram also carries other information. It shows that warm blood, a derived character that living birds and mammals share, must have evolved independently at least twice, unless living reptiles have lost warm blood.

As we consider smaller subgroups of living and fossil reptiles, we find that the neat picture of reptile classification breaks down, and that we must revise our interpretations. Figure 3.9 shows that "living reptiles" is not a clade. We can define a clade called "reptiles," but we have to include mammals and birds in it. This means that turtles, mammals, birds,

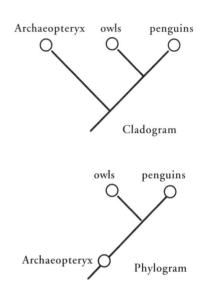

Figure 3.7 A cladogram and a phylogram of three groups of birds. In the cladogram, no group is shown as the ancestor of another, because a cladogram seeks only to show the degree of relationship between groups. Owls and penguins share derived characters that *Archaeopteryx* does not have. A phylogram or phylogenetic tree of the same three groups can be used to show the hypothesis that *Archaeopteryx* was the ancestor of owls and penguins.

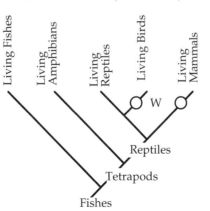

Figure 3.8 A conventional classification and a cladogram of the vertebrates. In the conventional classification, each class has equal rank to the others. In the cladogram, it is clear that all classes are not equivalent in rank: for example, living mammals are the sister group of [living birds + living reptiles]. Note that the novel character W, warm blood, has been independently derived in birds and mammals, according to the hypothesis expressed in this cladogram.

Figure 3.9 A cladogram of some groups of vertebrates that carries more information than the cladogram in Figure 3.8. "Reptiles" is not a clade unless (as shown here) it also includes birds. Living reptiles (*) do not form a clade either. Look also at the evolution of warm blood (W), which occurred some time during the evolution of mammals. Living birds are also warm-blooded. Warm blood may be a primitive character that birds and dinosaurs inherited from archosaurs, or it may be a derived character of birds alone. Drawing a cladogram forces us to look at such questions!

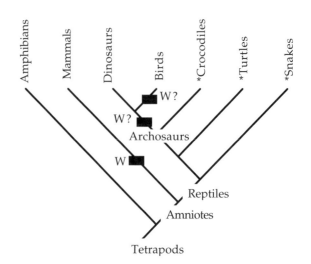

The human species has thrived

In the short time since we arrived.

But the cladist affirms

That we are only worms,

Even though we are somewhat derived.

crocodiles, and lizards are all *reptilian* clades that have diverged from an ancestral, primitive reptile, although some are more derived than others in the sense that they have evolved more novel characters that their common ancestor did not have. In the same way, humans are derived fishes, derived amphibians, and derived reptiles, all at the same time. Once one becomes used to this unusual line of thinking, evolution becomes much more real, and we can see, for example, that humans, tapeworms, and the scums of Shark Bay are all cyanobacteria—though some have accumulated more visible derived characters than others.

Review Questions

How did eukaryotes evolve from prokaryotes? You can express your answer in prose, poetry, or art work.

When did the first living thing evolve that wasn't a bacterium?

Give one possible reason why organisms would evolve sexual reproduction.

Here is a blank cladogram showing the relationship between N groups of animals. Fill in the groups on the cladogram.

Further Reading

Easy Access Reading

DeLong, E. 1998. Archaeal means and extremes. *Science* 280: 542–543. Current research on biology of Archaea.

Fenchel, T., and B. J. Finlay. 1994. The evolution of life without oxygen. *American Scientist* 82: 22–29.

Han, T.-M., and B. Runnegar. 1992. Megascopic eukaryotic algae from the 2.1-billion-year-old Negaunee Iron Formation, Michigan. *Science* 257: 232–235, and comment, v. 259, p. 835. *Grypania*.

Kabnick, K. S., and D. A. Peattie. 1991. *Giardia*: a missing link between prokaryotes and eukaryotes. *American Scientist* 79: 34–43.

Knoll, A. H. 1992. The early evolution of eukaryotes: a geological perspective. *Science* 256: 622–627.

Palmer, J. D. 1993. A genetic rainbow of plastids. *Nature* 364: 762–763. Plastids evolved only once.

Palmer, J. D. 1997. Organelle genomes: going, going, gone! *Science* 275: 790–791. Some organelles lose their genes and others do not.

Riding, R. 1992. The algal breath of life. *Nature* 359: 13–14.

Szalai, V., and G. W. Brudvig. 1998. How plants produce dioxygen. *American Scientist* 86: 542–551. Photosynthesis.

More Technical Reading

Margulis, L. 1981. *Symbiosis in Cell Evolution*. San Francisco: W. H. Freeman.

Sagan, D., and L. Margulis. 1988. *Garden of Microbial Delights*. New York: Harcourt Brace Jovanovich. Charming description of single-celled creatures.

Whitman, W. B., et al. 1998. Prokaryotes: the unseen majority. *Proceedings of the National Academy of Sciences* 95: 6578–6583.

The Evolution of Animals

Eukaryotes come in two grades of organization: single-celled (protists) and multicellular (plants, animals, and fungi). The world today is full of complex multicellular plants and animals: how, why, and when did they evolve from protists?

Proterozoic Protists

A single-celled eukaryote or **protist** can carry chlorophyll (it can be an autotrophic, photosynthetic, "alga"), it can eat other organisms (it can be an organotrophic, "protozoan" "animal"), or it may do both.

In Chapter 3, I suggested that protists probably evolved in or near stromatolites because many kinds of prokaryotes were clustered together there in microhabitats that varied over scales of millimeters. Early protists could have extended into many habitats, however. Beginning about 1850 Ma, we find **acritarchs**, spherical microfossils with thick and complex organic walls. They are probably algae that grew thick organic coats in a resting stage of their life cycle, but spent most of their life floating in the **plankton**, the community of organisms that makes a life in surface waters of oceans and lakes. Many Proterozoic protists seem to have been planktonic, while most prokaryotes still lived in seafloor bacterial mats.

Andrew Knoll suggested that protists became successful planktonic organisms because they could move more efficiently than prokaryotes to find nutrients in the water. Nutrients come from the breakdown of minerals in rock weathering, and so are most abundant near shore. They are also present in seafloor sediment and are recycled upwards toward the surface by upwelling ocean currents. Most protists have flagella that can move them slightly in the water, perhaps into microenvironments richer in nutrients. Early protists would have found rich and untapped, nutrient resources in the surface ocean waters. Photosynthetic protists would have flourished there, and so would carnivorous protists that fed on them.

The fossil record of the Late Proterozoic contains many planktonic protists, but it may not tell us a complete story. We know that a very diverse array of plankton existed by 800 Ma, because they are known as fossils. But many amoebalike protists do not have cell walls made of cellulose and so do not preserve well. It's possible that while the surface layers of Proterozoic oceans had huge numbers of floating plankton, Proterozoic seafloors were crawling with successful populations of protists consuming the rich food supplies available in bacterial mats.

Evolving Metazoans from Protists: Anatomy and Ecology

A flagellate protist is a single cell with a lashing filament, a **flagellum** (plural, flagella), that moves it through the water (Figure 4.1, left). Some species of flagellate protists, the **choanoflagellates**, form clones: they reproduce by budding off new individuals, which then stay together to form a compound animal or **colony** rooted to the seafloor (Figure 4.1, right). The flagella beat to generate a systematic water current around and between the individuals, which filter bacteria from the water flow.

A **sponge** is the simplest multicellular variation on this theme. It contains many similar flagellated cells arranged so that they generate and direct water currents efficiently (Figure 4.1, center; Figure 4.2). Sponges are more advanced than simple colonies of choanoflagellates because

Figure 4.1 Protists and sponges. Left: A choanoflagellate, a protist that uses its flagellum to move water, thus propelling the animal through the water. Food is captured as water is pulled through the collar. Center: A "collar cell" or choanocyte from a sponge. It has almost exactly the same structure and function, except that the cell is anchored in the body of the sponge and does not move through the water. The flagellum beats to produce a water current that is entirely for feeding. Right: A possible evolutionary intermediate, a colony of choanoflagellates. New individuals that bud from a parent cell stay attached to one another and to the seafloor. The colony does not move, but the flagella of all individuals act together to generate a feeding current that draws seawater through the filters of individual cells. Most likely the first metazoan evolved through an organism that looked and functioned like this. (From Barnes et al., *The Invertebrates: A New Synthesis.* © 1993 Blackwell Scientific Publications.)

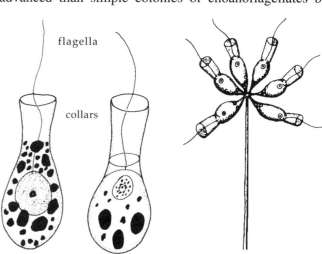

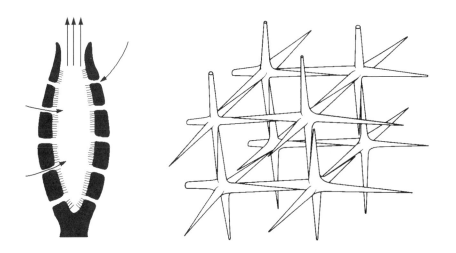

Figure 4.2 Left: The structure of a simple sponge, which is advanced over choanoflagellates in that the filters are more sophisticated and are enclosed inside a body wall. (From Barnes et al., *The Invertebrates: A New Synthesis.* © 1993 Blackwell Scientific Publications.) Right: Many sponges have supporting structures that help them retain their shape, even at large size. For example, some glass sponges have mineralized needles or spicules arranged into a skeletal framework like the steel girders in a skyscraper. (From Boardman et al. (eds.), *Fossil Invertebrates.* © 1987 Blackwell Scientific Publications.)

they also have specialized sets of cells to form a body wall, to digest and distribute the food they collect, and to construct a stiffening skeletal framework of organic or mineral protein that allows them to become large without collapsing into a heap of jelly (Figure 4.2, right). Sponges are thus **metazoans**, not protists. Metazoans are not just multicellular, they have different kinds of cells that perform different functions.

Metazoans are most likely a clade, that is, they all descended from one kind of protist (Figure 4.3). All metazoans originally had one cilium or flagellum per cell, for example. Metazoans also share the same kind of early development. They form into infolded balls of internal cells which are often free to move, and are covered by outer sheets of cells that form an external coating for the animal: a skin, if you like.

The first metazoans were soft-bodied, and we have no fossil record of them. But we can look at the tremendous variety of living animals and at the geologic record to try to reason out what the first metazoans might have looked like and what they might have done.

There are only three basic kinds of metazoans: sponges, cnidarians, and worms (Figure 4.3). Probably the immediate protist ancestor of metazoans was like a colonial choanoflagellate (Figure 4.1). One can imagine scenarios for the divergence of the three great metazoan groups. All of them solved the problem of developing to larger size and complexity, but in different ways. Sponges evolved by extending the choanoflagellate way of life to large size and sophisticated packaging. They continued to pump water (and the oxygen and bacteria they take from it) through their tissues, in internal filtering modules (Figure 4.2).

Cnidarians (or coelenterates), including sea anemones, jellyfish, and corals, are built mostly of sheets of cells, and they exploit the large surface area of the sheets in sophisticated ways to make a living. They consist of a sheet of tissue, with cells on each surface and a thickening layer of jellylike substance in the middle. The sheet is shaped into a baglike form to define an outer and an inner surface (Figure 4.4, top). A cnidarian thus contains a lot of seawater in a largely enclosed cavity lined by the inner surface of the sheet. The neck of the bag forms a mouth, which can be closed by muscles that act like a drawstring. A network of nerve cells

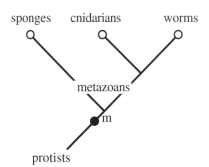

Figure 4.3 The three major groups of metazoans are a clade [Metazoa] that evolved from a protist by acquiring specialized multicellularity (m).

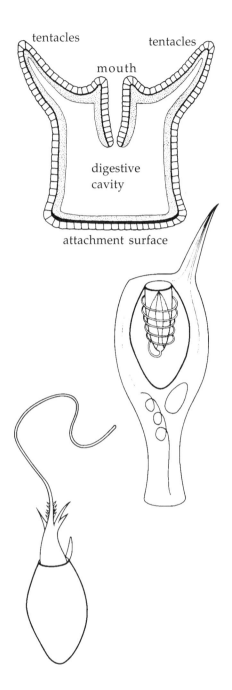

Figure 4.4 Important features of cnidaria. Top: Basic structure of a cnidarian. (From Boardman et al. (eds.), *Fossil Invertebrates.* © 1987 Blackwell Scientific Publications.) Center: Nematocysts are unique features of cnidarians, used in defense and feeding. This cell has not been triggered. Bottom: A discharged nematocyst, showing the range of the weapon. (From Barnes et al., *The Invertebrates: A New Synthesis.* © 1993 Blackwell Scientific Publications.)

runs through the tissue sheet to coordinate the actions of the animal.

In most cnidarians the outer surface of the sheet acts simply as a protective skin. The inner surface is mainly digestive, and it absorbs food molecules from the water in the enclosed cavity. Because cnidarians are built only of thin sheets of tissue, they weigh very little, and can exist on small amounts of food. They can absorb all the oxygen they need from the water that surrounds them, and they absorb all their food molecules too. Digestive cells lining the cavity then leak powerful enzymes into the water inside the animal. The prey is broken down by these enzymes, and the cnidarian absorbs the food molecules through the inner lining.

Cnidarians have **nematocysts** or stinging cells that are set into the outer skin surface (Figure 4.4, center and bottom). The toxins of some cnidarians are powerful enough to kill fish, and people have died after being stung by swarms of jellyfish. Nematocysts are usually concentrated on the surfaces and the ends of **tentacles**, which form a ring around the mouth. They provide an effective defense for the cnidarian, but they are also powerful weapons for catching and killing prey, which the tentacles then push into the mouth for digestion in the cavity.

Hardly any sponges can tackle food particles larger than a bacterium, though there are a few exceptions. Living cnidarians routinely trap, kill, and digest creatures that outweigh them many times. However, there is no guarantee that the first cnidarians had nematocysts: they may have been simple absorbers of dissolved organic nutrients.

The third and most complex metazoan group contains all other metazoans, including vertebrates (Figure 4.3). Here I shall simply call them **worms**. Worms consist basically of a double sheet of tissue that is folded around with the inner surfaces largely joined together to form a three-dimensional animal. In contrast to sponges and cnidarians, worms have evolved complex organ systems made from specialized cells.

All sponges and most cnidarians are attached to the seafloor and depend on trapping food from the water. But many worms, including the most simple group, flatworms, are mobile scavengers and predators. Worms creep along the seafloor on their **ventral** (lower) surface, which may be different from the **dorsal** (upper) surface. They prefer to move in one direction, and a **head** at the (front) end contains major nerve centers associated with checking and testing the environment.

Probably the *mobility* of worms on the seafloor led to the differentiation of the body into **anterior** and **posterior** (head and tail) and into dorsal and ventral surfaces, as the various parts of the animal encountered different stimuli and had to be able to react to them. A well-developed nervous system coordinates muscles so that a worm can react quickly and efficiently to external stimuli.

The same locomotion that gave a worm a front-to-back axis also gave it **bilateral symmetry**. Any other shape would have produced an animal that could not move forward efficiently.

The head usually features the food intake, a mouth through which food is passed into and along a specialized one-way **internal digestive tract** instead of being digested in a simple seawater cavity (Figure 4.5). No sponge cell or cnidarian cell is very far away from a food-absorbing

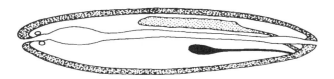

(digestive) cell, so these creatures have no specialised internal transport system. But the digestive system of worms needs an oxygen supply, and the nutrients absorbed there have to be transported to the rest of the body. Worms therefore have a circulation system, and the larger and more three-dimensional they are, the better the circulation system must be.

Flatworms called **acoels** are the most primitive living worms and have a body structure that is simply a double sheet of tissue, with a weakly developed head and flattened dorsal and ventral surfaces. They show what an early worm may have looked like as it evolved from a creature that was more two-dimensional and sheet-like.

Most other worms are more rounded and three-dimensional. They have an internal fluid-filled cavity, the **coelom**. All highly evolved metazoans show modifications of the coelomate wormlike body. In humans, for example, the coelom is the sac containing all the internal organs.

R. B. Clark suggested that the evolution of the coelom in some early flatworm gave it access to new food supplies in organic-rich sediment. Liquid is incompressible, and a flatworm that first evolved any kind of internal fluid pool would have been able to squeeze its internal reservoir by body muscles. Such squeezing would have poked out the body wall at its weakest point, which is usually an end (Figure 4.6). Such a hydraulic extension of the body could have been used as a power drill for burrowing into the sediment. As well as burrowing for food, a wormlike animal with a coelom would have been able to burrow for safety.

This scenario for the evolution of the coelom is very reasonable. Some living flatworms have evolved small hydraulic systems for other purposes, some for hydraulic eversion of the penis in mating (and in food capture), and others for everting an extension of the mouth, a pharynx, for feeding. An early flatworm could have evolved a much larger internal fluid-filled system to become a coelomate.

Figure 4.5 The basic structure of a worm. It is a three-dimensional animal, bilaterally symmetrical, with a head and tail, and a mouth and an anus joined by an internal digestive tract. Other major internal organs are made from specialised cells. It is mobile and has an intricate set of muscles for locomotion. (From Barnes et al., *The Invertebrates: A New Synthesis.* © 1993 Blackwell Scientific Publications.)

The paleontologist's view

Classes worms together with you.

This is based on the claim

That instead of a plane,

A worm's three-dimensional too.

© Elizabeth Wenk 1994

"Complex" is a relative term

That's rarely applied to a worm.

But a worm that's coelomic

Can be rather comic:

It can burrow and wriggle and squirm.

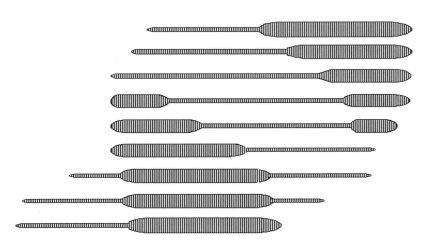

Figure 4.6 How wormlike animals use the coelom as a hydraulic device for burrowing. This worm is burrowing from right to left. It extends the front end by squeezing fluid forward, then forms a bulb at the front to act as an anchor. Then it pulls the rest of the body up to the anchor, and begins the cycle once again. We use much the same logical sequence of actions, but a different mechanism, to winch a stranded truck out of mud. R. B. Clark argued that an advanced worm, partitioned into segments separated by valves, has an even more efficient burrowing action.

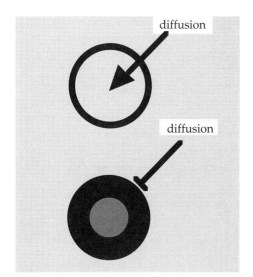

Figure 4.7 Diffusion. Animals with thin tissues (above) can rely on diffusion alone to supply the entire body with oxygen. But diffusion cannot supply oxygen to the interior of animals with thick tissues (below): special respiratory systems are needed so that the inner tissues are not starved of oxygen.

Figure 4.8 Respiration may be a problem as coelomate worms burrow into oxygen-poor sediments. Two solutions: the tube-worm on the left evolved coelom-filled tentacles that project into the water for feeding and respiration. (From Lydekker.) The tube-worm on the right evolved body movements that pump water through the burrow, bringing food and oxygen.

The coelom would have provided another great advantage for worm-like animals. Oxygen must reach all the cells in the body for respiration and metabolism. Single-celled organisms can usually get all the oxygen they need because it simply diffuses through the cell wall into their tiny bodies. Sponges pump water throughout their bodies as they feed, and cnidarians and flatworms are at most two sheets of tissue thick. But larger animals with thicker tissues cannot supply all the oxygen they need by diffusion (Figure 4.7). Oxygen supply to the innermost tissues becomes a genuine problem with any increase in body thickness or complexity. If the animal evolved some exchange system so that its coelomic fluid was oxygenated, the coelom could act as a large store of reserve oxygen. Eventually the animal could evolve pumps and branches and circuits connected with the coelom to form an efficient **circulatory system**.

Most advanced worms have segments: their bodies are divided by septa that separate the coelom into separate chambers connected by valves. Clark suggested that this arrangement is more efficient for the wormlike burrowing action than a simple, single coelomic cavity (Figure 4.6). The segmentation of many animals, including earthworms and caterpillars, is derived from this invention on the Precambrian seafloor.

If Clark is correct, respiration problems were particularly serious for early coelomates because they were burrowing for food in rich organic sediments, which are very low in oxygen. Therefore we might expect a successful worm to evolve some special organs to obtain oxygen from the overlying seawater (at one end?) while the main body of the worm can remain safely below the surface for protection and for gathering food (at the other end?). Many worms and more advanced coelomates that live in shallow burrows have various kinds of tentacles, filaments, and gills that they extend into the water as respiratory organs (Figure 4.8). It is a very short step from here to the point where the coelomate collects food as well as oxygen from the water by **filter feeding** (Figure 4.8, left), as in all bryozoans and brachiopods, in some molluscs, worms, and echinoderms, and in simple chordates.

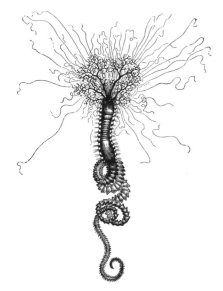

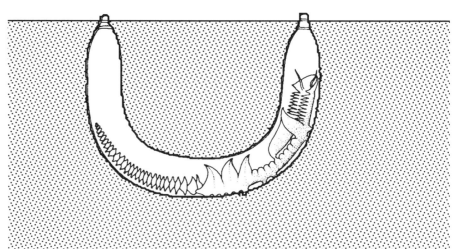

An alternative solution has evolved in many coelomate worms. They burrow so actively that their body movements inside the burrow pump oxygenated water down the burrow over them. In these worms, respiration through the skin surface is sufficient for their oxygen needs as long as they also have an efficient internal circulatory system to distribute the oxygen (as in some worms and in many arthropods such as burrowing shrimp). Some of these animals have also evolved to collect food from the respiratory currents flowing down into their burrows (for example, Figure 4.8, right), but most still are sediment scavengers and predators.

Evolving Metazoans: Regulatory Genes

It is difficult to grow a viable multicellular animal rather than a protist. The DNA has to contain not only the information to build several or many different kinds of cells, but the information to grow them at the right time, to place them accurately within the body, and to drive all the biochemical and biophysical systems that ensure the animal operates as a coherent unit.

The genetic programming that builds an animal works like efficient computer programming. For example, one could instruct a computer to draw a flower, specifying the size, shape, and position of each petal. Given that petals typically have much the same size and shape, however, one could use one shape and size for every petal, and simply tell the computer to move the pen to the right place before drawing the same petal each time (Figure 4.9).

Structural genes build each piece of the animal, and **regulatory genes** make sure the piece is built in the right place at the right time. Thus a set of regulatory genes could be used in combination with a set of "segment" genes to build all the segments along a growing worm. The same sort of regulatory genes could easily be used to build legs on, say, a millipede or a crab, by calling on a "leg" gene the appropriate number of times instead of a "segment" gene. By calling on slight modifications of the "leg" gene as growth developed, regulatory genes could build an animal whose legs were different along its length (as in insects), or build a vertebrate with different bones along the length of a backbone. For example, embryonic snakes have genetically programmed limb buds that show us where the legs were placed on ancestral snakes, but today those buds never develop into legs because the regulatory genes do not send a growth instruction to them.

Developmental geneticists have now identified regulatory genes that control which way up an animal is formed, which is front and back, and how the animal varies along its length or around its edges. The most thrilling discovery is that much the same control box is used throughout the metazoans. Sets of genes that sit close to one another in the nuclear DNA perform much the same job in developing animals, but because they call in a variety of structural genes in a variety of patterns in time and body areas, the results in terms of anatomy are vastly different.

Figure 4.9. One can construct a complex object by judicious placing of very simple units.

Hox genes are such clusters. Sponges have one set of Hox genes (and are simple in structure), whereas mammals have 38 sets in four clusters, and goldfish have 48 in seven clusters. Hox genes control the growth of nerve nets, segments, and limbs throughout metazoans, and their evolution accompanies the divergence in anatomy and physiology and ecology and behavior that gave us all the variety of living animals. Hox genes provide separate, but complementary evidence to accompany the fossil record; however, it is important to remember that we can only study the genes of surviving groups of animals, not those from the 95% of species that have become extinct.

Protists don't need Hox genes, because they don't divide cells in precise patterns to form a multicellular adult. Hox genes evolved in early metazoans, and provided the genetic tool kit to build viable complex animals. Presumably, Hox genes control the lay-out of a sponge that gives efficiency of water currents passing through the body. In the simplest worms, Hox genes lay out the nerve nets that allow the worm to sense the environment all along the body. One can easily imagine that the earliest metazoans, wherever, whenever, and however they evolved, would quickly radiate into a great variety of body shapes and structures, with natural selection acting equally quickly to weed out the shapes that were poor adaptations, and leaving a scrapbook of successful prototypes that proliferated.

Ediacaran (Vendian) Animals

We begin to see a reasonable fossil record of animals from Late Proterozoic rocks, when conditions returned to "normal" around 600 Ma, after at least two great episodes of glaciations. In South Australia the rocks that were laid down at this time have not suffered much damage since. In particular, they bear the traces of soft-bodied animals that are much advanced over any of the protists that dominated the fossil record up to this point (Figure 4.10). This set of animals is called the **Ediacaran fauna**,

Figure 4.10 Ediacaran animals from South Australia. Left: *Glaessneria*, usually interpreted as a fossil sea pen. (Courtesy of Bruce Runnegar, University of California, Los Angeles.) Right: *Mawsonites*, whose similarity in form to modern jellyfish suggests but does not prove that this animal too is a cnidarian. (Courtesy of Mary Wade of the Queensland Museum, Australia.)

named after rocks found in Ediacara Gorge in the Flinders Ranges behind the city of Adelaide, or the **Vendian fauna**, named for fossils of the same type and the same age that have been discovered along the shore of the White Sea in northern Russia.

Thousands of Ediacaran fossils have now been collected worldwide in dozens of different localities. Almost all the fossils occur between 565 and 543 Ma, with the highest abundance and diversity during the last few million years from 550 Ma to 543 Ma. After that the Ediacaran animals seem to have become extinct. Most of them probably left no descendants; others gave rise to some of the Cambrian animals that followed.

There are a few Ediacaran sponges, but most Ediacaran fossils are cnidarians of some sort. Jellyfish and other cnidarians (Figure 4.10, left) floated just like their living relatives. Colonies of sea pens were attached to the seafloor. Sea pens look like plants, but are cnidarians that capture and eat floating animals in the water (Figure 4.10, right). *Dickinsonia* is a very large flattened animal, up to 45 cm long, and there is some debate whether it is a very unusual worm or a very unusual cnidarian. Its "segments" sometimes look offset rather than matched across the central line, which would make it a cnidarian rather than a worm. Certainly that is not true of the specimens shown here (Figure 4.11), but more than one kind of animal may have been called *Dickinsonia*, and squashing during fossilization could affect the symmetry of a soft-bodied animal.

Other Ediacaran fossils are worms that patrolled the seafloor. Some squirmed through the surface sediment; others walked on the tufts of bristles located on their body segments. Shallow fossil burrows in the sediment show that some worms were a centimeter across and were deposit feeders, leaving fecal pellets behind them. Smaller worms left trails on the surface as they wriggled across the sediment. *Spriggina*, an animal that is about 45 cm long, has some sort of head-like structure. It and other Ediacaran animals are probably arthropod ancestors or relatives.

Some Ediacaran fossils are not like any living animals, and research on these is continuing. For example, *Tribrachidium* is not understood at all (Figure 4.12).

Some Ediacaran animals fed on floating plankton, but others were mud eaters. There is no evidence of animals that grazed algae on the seafloor. (They may have been present but were not preserved.) Many Ediacaran animals are small, but some reached astonishing size for such early animals. Since Ediacaran animals were soft-bodied and unprotected, there may have been no large carnivores on the seafloor.

This is the most generally accepted account of the Ediacaran fauna, stressing the similarities that seem to link Ediacaran animals with living animals. This interpretation is attractive because, as we have seen, worms and cnidarians are comparatively simple in body construction and would be expected to have evolved early in metazoan history.

There is a more radical interpretation of the Ediacaran animals. Adolf Seilacher infers from their preservation that Ediacaran animals had much tougher surfaces than any of the soft-bodied animals we are familiar with. He does not see the so-called sea pens as cnidarians, for example, but as extinct animals built rather like air mattresses, lying on the

Figure 4.11 *Dickinsonia*, sometimes interpreted as a large, flat annelid worm, sometimes as a cnidarian. Turn the page to different positions to try to decide which is more probable. (Above: courtesy of the late Martin Glaessner, University of Adelaide; below: courtesy of Bruce Runnegar, University of California, Los Angeles.)

Figure 4.12 The enigmatic Ediacaran fossil *Tribrachidium*. (Courtesy of the late Martin Glaessner, University of Adelaide.)

sand surface. Seilacher and Mark McMenamin suggested that the Ediacaran animals had symbiotic algae in their tissues and gained much of their nutrients from photosynthesis, like most reef corals do today. Seilacher thinks the Ediacaran fossils belonged to a distinct group of animals, and has called them the **Vendozoa**. Others are prepared to accept that these animals may have had symbiotic algae in shallow water, but see no reason to doubt that most of them are cnidarians.

Seilacher believes that the Vendozoa are extinct and left no descendants. In his view, living animals are descended from other ancestors that are barely represented in fossils of this age. Seilacher's idea is losing support as we study new Ediacaran fossils in more detail, and begin to see their relationships to cnidarians, worms, and arthropods more clearly. For example, Ediacaran fossils are probably preserved so well because they had, during life, or after death, or both, a coating of bacteria that allowed the preservation of soft-bodied creatures in a way that was no longer possible once grazing and burrowing animals evolved in the Cambrian. John Gehling has called these bacterial coatings or **biomats** "Ediacaran death masks", and he and others suggest that they take away the basic reason for Seilacher's radical interpretation of Ediacaran fossils. Seilacher continues to maintain his position: time will tell who is right.

The Evolution of Skeletons

One of the most important events in the history of life was the evolution of mineralized hard parts in animals. Various kinds of algal cells and then planktonic protists had evolved tough cell walls and cell coverings by the Late Proterozoic, but they were not mineralized. The Ediacaran animals may have had leathery skins, but no Ediacaran animal had true mineralized hard parts.

Beginning rather suddenly, the fossil record contains skeletons: shells and other pieces of mineral that were formed biochemically by animals. Humans have one kind of skeleton, an internal skeleton or endoskeleton, where the mineralization is internal and the soft tissues lie outside. Most animals have the reverse arrangement, with a mineralized exoskeleton on the outside and soft tissues inside, as in most molluscs and in arthropods (Figure 4.13). The shell or test of an echinoderm is technically internal but usually lies so close to the surface that it is external for all practical purposes. The hard parts laid down by corals are external, but underneath the body, so that the soft parts lie on top of the hard parts and seem comparatively unprotected by them. Sponge skeletons are simply networks of tiny spicules that form a largely internal framework. There is incredible variety in the type, function, arrangement, chemistry, and formation of animal skeletons; biomineralization is a whole science in itself.

With the evolution of hard parts, the fossil record became much richer, because hard parts resist the destructive agents that affect the soft parts of bodies. Almost as soon as geologists realized that fossils marked

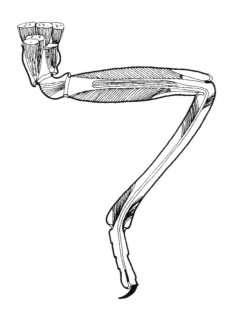

Figure 4.13 Arthropods have exoskeletons secreted from inside by the soft parts. The limbs must be operated from the inside by the muscles and ligaments, unlike the system evolved by vertebrates. (From Barnes et al., *The Invertebrates: A New Synthesis.* © 1993 Blackwell Scientific Publications.)

time periods in earth history, they also recognized that the quality of the fossil record depended on the style, structure, and composition of the hard parts of the organisms that were preserved (Chapter 2). For about a century, in fact, many geologists believed that there was no fossil record before hard parts evolved. The evolution of hard parts defines the beginning of an era in Earth history, the **Paleozoic Era**, and the beginning of its oldest subdivision, the **Cambrian Period**. In contrast, **Precambrian** time (the Archean and Proterozoic Eras) was first seen as a time of no life, and then as a time of soft-bodied, mainly bacterial life. Even today, the base of the Cambrian is defined at a time when major new fossils appear in the record.

Why did hard parts evolve in the first place, and why did they evolve when they did? What difference do hard parts make to the biology of an animal?

Worms are soft-bodied. Sponges are sponges, whether they have tiny mineral spicules forming an internal skeleton, or a soft protein like that in bath sponges. But many metazoan groups have skeletons that are such an integral part of their body plans that they only exist as such when they have hard parts. Although there are molluscs without shells (slugs and squids, for example), it seems impossible to be a clam without a shell. Shells are so important to clams that if a clam evolved to be shell-less, its basic biology would be so changed that we would call it something else. An arthropod without a skeleton is basically a worm (unless it's a caterpillar). Thus any worm that evolved hard parts by definition evolved into some other major group.

Therefore, the evolution of hard parts implies the appearance of new kinds of animals on Earth. When one animal group is radically different from any other and is also considered to be a clade, evolved from some single ancestral species (Chapter 3), it is a **phylum**, defined by its own particular body structure, ecology, and evolutionary history. Mollusca and Arthropoda are familiar phyla that must once have had a common ancestor, but that ancestor wasn't a mollusc or an arthropod. There are arguments about the number of phyla living in the world today, mainly because of the bewildering variety of wormlike creatures, but most people would count about 30 phyla of living animals.

Worms contribute to the fossil record, especially by leaving trace fossils of burrows and trails, and there are various small, puzzling groups of early shelled animals that became extinct without leaving descendants. But for paleontological purposes, only nine or ten phyla are or have ever been important in terms of hard parts (Box 4.1). It is stunning to realize that all but Bryozoa are known from Cambrian rocks, and all but Bryozoa and Chordata are known from *Early* Cambrian rocks. Only two phyla from the list (Cnidaria and Porifera) are found in Ediacaran rocks.

At face value, these facts suggest that a spectacular burst of innovation at the beginning of Cambrian time produced most of the major body plans of animals. Each phylum is radically different from any of the others, so each must have followed a different evolutionary pathway. Even the hard parts they evolved are very different from one another. Sponges evolved an internal skeleton of fine silica needles. Molluscs and most

> **BOX 4.1. THE MAJOR PHYLA OF FOSSIL INVERTEBRATES**
> († **indicates an extinct group**)
>
> - Porifera or sponges
>   (including †Archaeocyatha)
> - Cnidaria
> - Bryozoa
> - Brachiopoda
> - Mollusca
> - Arthropoda
> - Echinodermata
> - Hemichordata
>   (including †graptolites)
> - Chordata (including vertebrates)

brachiopods evolved an external shell made of calcium carbonate, but the two phyla used different minerals, and different crystal structure. Some brachiopods used calcium phosphate for their shells. Arthropods evolved chitin, but different groups of arthropods impregnated the chitin with calcium phosphate or with calcium carbonate. Echinoderms have an internal skeleton just under the skin, made up of small separate plates of calcium carbonate, each one a single calcite crystal. Looking at the variety of minerals involved, it's clear that there wasn't some simple chemical change in the oceans (an increase in phosphate, for example) that triggered the invention of skeletons. Yet the evolutionary event was global, so it was probably triggered by some global biological or ecological factor. The event has come to be called the **Cambrian Explosion**. We need to know why these different creatures evolved different types of skeletons, and why they did it so rapidly. The radiation of the phyla of metazoans was apparently so rapid that we still have no idea of the order of branching. A symposium in 1998 heard several, radically different proposals for metazoan cladograms, and the organizers later reported that ten more years' research would help to begin to solve the question.

However, perhaps the Cambrian Explosion was not so dramatic. The ancestors of Cambrian animals may have lived in Ediacaran times but have not yet been discovered. The earliest metazoans may have been tiny and soft-bodied, for example, shaped more like the larvae of their descendants than the adults. Perhaps they were soft-bodied and wormlike, burrowing between sand grains in the sea floor. If so, metazoans could have had a long history that would not be fossilized. They would have had 150 m.y. or so for evolution within soft parts, after the Hox gene system began to generate diversity. Then suddenly, around 550 Ma, many of them suddenly gained hard parts and large size, and "exploded" into the fossil record. In doing so, they evolved the larger features that allow us to identify them anatomically and ecologically as the metazoan phyla that still survive.

In this scenario, what evolved in the Cambrian explosion was ecological molluscs and ecological arthropods. Their ancestors may have had important differences in the DNA of important genes, but they had all been ecologically similar. This would make the Cambrian explosion an ecological event rather than a phylogenetic one. However, it is also possible that all the divergent evolution that formed the multiple metazoan lineages occurred just before and just into the Cambrian, making the Cambrian explosion genetic as well as ecological, and dramatic indeed. The truth is probably somewhere in between (as it often is).

For some purposes, it doesn't matter whether the Cambrian explosion genuinely represents the evolution of new phyla, or whether it represents the evolution of new characters within pre-existing phyla. What is clear is that the evolution of morphology that reflects new adaptation and novel ecology occurred very rapidly indeed at the beginning of the Cambrian—and that needs explanation. We'll review the evidence, and then look for explanation of the Cambrian Explosion.

The Beginning of the Cambrian: Small Shelly Fossils

In Namibia, where the rock record is excellent, Ediacaran fossils are followed immediately by Cambrian fossils, and there is some preliminary evidence that they overlapped briefly in time. In Siberia, a whole suite of small shelly fossils appears in the rocks as the Cambrian period begins, together with sponges of several different types. Most of the small shells are tiny cones and tubes that we don't understand properly, but at least some of them are complex animals, including the first molluscs (Figure 4.14). Soon, archaeocyathid sponges (Figure 4.15) were forming reef patches. These are dramatic changes that reflect rapidly evolving life.

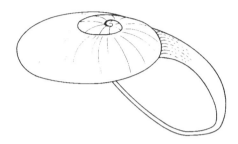

Figure 4.14 A tiny Cambrian animal, *Aldanella*, which may be a mollusc (though some scientists disagree). If it is a mollusc, *Aldanella* is the earliest known snail. The shell is only 1.5 mm across. (From Boardman et al. (eds.), *Fossil Invertebrates*. © 1987 Blackwell Scientific Publications.)

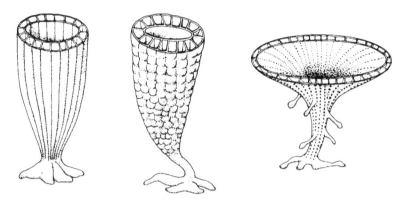

Technically, the base of the Cambrian is marked at a remote spot in a cliff face in Newfoundland, Canada, in rocks that have no fossils except a few worm tracks. As far as most paleontologists are concerned, the excitement begins with the flood of small shelly fossils that marks the base of the Cambrian in Siberia. New research pegs the base of the Cambrian at about 543 Ma, and the world's fauna was completely revolutionized in the 20 m.y. that followed.

The same set of small shelly fossils is now known worldwide, and their discovery shows that there were at least two major steps in the evolution of animal skeletons, because the next stage of the Cambrian sees the appearance of more abundant and more complex creatures.

For 10 m.y. or more, there were no animals larger than a few millimeters long except for the archaeocyathid sponges. Then worldwide, in a few million years after 530 Ma, we see the appearance of a much larger variety of marine life. Dominant among these animals were trilobites, brachiopods, and echinoderms.

Larger Cambrian Animals

Trilobites are arthropods, complex creatures with thick jointed armor covering them from head to tail (Figure 4.16). They had antennae and large eyes, they were mobile on the seafloor using long jointed legs, and they were something like crustaceans and horseshoe crabs in structure. They did not have the complex mouth parts of living crustaceans, so their

Figure 4.15 Archaeocyathids are Cambrian fossils with the same size and shape as sponges, and they may have had much the same biology. The latest interpretation is that they are indeed early sponges. (From Boardman et al. (eds.), *Fossil Invertebrates*. © 1987 Blackwell Scientific Publications.)

Figure 4.16 The Middle Cambrian trilobite *Olenoides*, from the Burgess Shale of British Columbia, Canada. This specimen is preserved with traces of its walking legs still visible under the thick, hard carapace that forms the usual trilobite fossil. (Courtesy of the Palaeontological Association.)

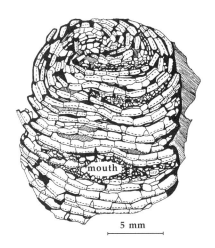

Figure 4.17 Helicoplacoids were the earliest definite echinoderm group. This is a drawing of *Helicoplacus* itself from the Lower Cambrian of California. Helicoplacoids do not have the fivefold symmetry characteristic of living echinoderms—that is a later derived character. Their reconstruction is controversial, and their paleobiology is poorly understood. (Courtesy of Kraig Derstler, University of New Orleans.)

Figure 4.18 Brachiopods. Above: A Cambrian brachiopod. Below: One of the better-known "lamp shell" forms found later in brachiopod history.

diet may have been restricted to sediment or very small or soft prey. They burrowed actively, leaving traces of their activities in the sediment, and they are by far the most numerous fossils in Cambrian rocks. The number of fossils they left behind was increased by the fact that they molted their armor as they grew, like living crustaceans. Thus, a large adult trilobite could have contributed twenty or more suits of armor to the fossil record before its final death. Even allowing for this bias of the fossil record, it is clear that Cambrian seafloors were dominated by trilobites. Other large arthropods are also known from Early Cambrian rocks, although they are much less common.

Echinoderms familiar to us today are sea urchins and starfish, but peculiar plated echinoderms are found in very early Cambrian rocks. The **helicoplacoids** looked rather like twisted deflated footballs covered with small plates (Figure 4.17). They were attached to the seafloor by one pointed end, and three food grooves led to a mouth halfway up on one side. Most echinoderms today are symmetrical, many of them with almost perfect fivefold patterns of arms and plates, but the first known echinoderms were distinctly unlike that, showing that fivefold symmetry is a derived character that evolved later in echinoderm history.

Brachiopods are relatively abundant Cambrian fossils, creatures that had two shells protecting a small body and a large water-filled cavity where food was filtered from seawater pumped in and out of the shell (Figure 4.18). Brachiopods lived on the sediment surface or burrowed just under it.

These animals are large, and they are much more easily assigned to living phyla. For the first time, the seafloor would have looked reasonably familiar to a marine ecologist. Trilobites probably ate mud, and echinoderms and brachiopods gathered food from seawater. Yet some ecological puzzles remain. There are no obvious large predators among these earliest skeletonized Cambrian fossils, no obvious grazers unless trilobites ate algae, and no swimmers, only floating plankton.

The Burgess Fauna

I have discussed the "Cambrian explosion" as if it related entirely to the evolution of skeletons. While this is basically true in terms of fossil abundance, it is clear that there was also dramatic evolution at the same time among animal groups with little or no skeleton. The frequency of worm tracks, trails, and burrows increases at the beginning of the Cambrian, and soft-bodied animals appeared with some amazingly sophisticated body plans.

These soft-bodied animals are found in the Burgess Shale, a Middle Cambrian rock formation high in the Canadian Rockies. Charles Walcott quarried thousands of fossils from the Burgess Shale 90 years ago, and expeditions in the 1960s and 1970s added new specimens. More than 120 species of animals are now known from the Burgess Shale. Similar fossils are now known from Early Cambrian rocks in Greenland, China,

Canada, and Poland, and from Middle Cambrian rocks in Utah. I shall call them all the **Burgess fauna**.

More than half the Burgess animals burrowed in or lived freely on the seafloor, and most of these were deposit feeders. Arthropods such as *Marrella* (Figure 4.19) and worms dominate the Burgess fauna. Only about 30% of the species were fixed to the seafloor or lived stationary lives on it, and these were probably filter-feeders, mainly sponges and worms. Thus, the dominance of most Cambrian fossil collections by bottom-dwelling, deposit-feeding arthropods is not a bias of the preservation of hard parts: it occurs among soft-bodied communities too. Trilobites are fair representatives of Cambrian animals and Cambrian ecology.

The main delights of the Burgess fauna are the unusual animals, which have provided fun and headaches for paleontologists. *Aysheaia* (Figure 4.20, left) looks like a caterpillar, with thick soft legs. It is called a **lobopod** because of the strange shape of its limbs, and may be related to living velvet worms like *Peripatus*, which now live on land on wet rainforest floors. *Aysheaia* has stubby little appendages near its head that may be slime glands for entangling prey (that's how *Peripatus* feeds).

Hallucigenia (Figure 4.21), named for its bizarre appearance, is now recognized as a lobopod with spines, related to *Aysheaia* (Figure 4.20, left) and to another new animal from the Early Cambrian of China.

Some Burgess trilobites have appendages at the top of their limbs that could have been used to shred soft-bodied prey such as worms.

Figure 4.19 *Marrella*, one of the commonest fossils in the Burgess Shale, is a very strange-looking arthropod. Soft parts are preserved: traces of long antennae are visible, legs can be seen under the strangely open carapace, and its guts were squeezed from the body and preserved as a dark stain on the rock surface. (Courtesy of the Palaeontological Association.)

Figure 4.20 Two animals from the Burgess Shale. *Aysheaia* (left) is a lobopod (see text), and *Naraoia* (right) is probably a trilobite that has a head and a tail but no thorax. (Courtesy of the National Museum of Natural History.)

Figure 4.21 The bizarre-looking Burgess Shale animal *Hallucigenia*. It is now recognized as a lobopod with spines. Look at the picture, then at *Aysheaia* (Figure 4.20 left). (Courtesy of the Palaeontological Association.)

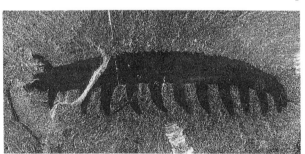

Figure 4.22 A priapulid worm, *Ottoia*, from the Burgess Shale. This predator stabbed animals from underneath with a spiked proboscis that forms part of its mouth (left). This specimen has the gut preserved. (Courtesy of the Palaeontological Association.)

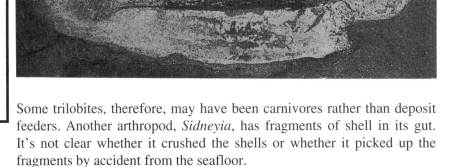

Ottoia just sat there and waited

With spikes that were sharp and serrated

   A worm wriggling by

   Would certainly die

In the U-shaped tube she'd created.

        © *Elizabeth Wenk 1994*

Some trilobites, therefore, may have been carnivores rather than deposit feeders. Another arthropod, *Sidneyia*, has fragments of shell in its gut. It's not clear whether it crushed the shells or whether it picked up the fragments by accident from the seafloor.

There are undoubted predators in the Burgess fauna. Priapulid worms today live in shallow burrows and capture soft-bodied prey by plunging a hooked proboscis into them as they crawl by. There were at least seven species of priapulids in the Burgess Shale, including *Ottoia* (Figure 4.22).

Anomalocarids are the most spectacular Cambrian predators. *Anomalocaris* itself is the largest well-known Cambrian animal, close to half a meter long. (Fragments of larger, unnamed anomalocarids suggest body lengths up to 2 meters!) Anomalocarids had a pair of appendages designed to catch prey that was then crushed in a circular jaw. They are an

Figure 4.23 *Opabinia* had five eyes and one central grasping organ on its front. (Courtesy of the Palaeontological Association.)

extinct group of animals related to arthropods. *Opabinia* (Figure 4.23) is a highly evolved anomalocarid. It is long and slim, with a vertical tail fin, so it probably swam about. It has five eyes and one large grasping claw on the front of its head.

Wiwaxia (Figure 4.24) is a flat creature that crept along the seafloor under a cover of tiny scales that were interspersed with tall strong spines. Halkieriids are best known from the Burgess fauna of Greenland. They are flattened creatures averaging about 5 cm long. They look like flat-tened worms, with perhaps 2000 spines forming a protective coating em-bedded into the dorsal surface. Yet two distinct subcircular mollusc-like shells are embedded in the upper surface close to each end. These crea-tures may be like *Wiwaxia*; if so, they too are worms—with armor.

The Early Cambrian fossils from China (the Chengjiang fauna) were discovered in 1984, and already include at least 70 species. Arthropods dominate here too, then sponges, and other animals as enigmatic as any-thing in the Burgess Shale.

The most significant of the Burgess animals in evolutionary terms are two wormlike creatures, *Pikaia* from the Burgess Shale and *Cathay-myrus* from Chengjiang. They are very much like the amphioxus *Bran-chiostoma*, a creature recognized today as the nearest living primitive rel-ative of the chordates. *Cathaymyrus* and/or *Pikaia* may be the remote ancestor of ourselves and all other vertebrates (Chapter 7).

Altogether, the Burgess faunas give us a good idea of the sorts of ex-citing but extinct soft-bodied creatures that may always have lived along-side the trilobites but were hardly ever preserved.

Figure 4.24 *Wiwaxia* apparently crept along the seafloor (like a mollusc or a worm?) but was covered by an array of overlapping plates and spines. (Courtesy of the Palaeontological Association.)

Solving the Cambrian Explosion

The Cambrian explosion remains a puzzle. The waves of evolutionary novelty that appeared in the seas during the Early Cambrian have few parallels in the history of life. Many groups of fossils appeared quite sud-denly in abundance, thanks to their evolution of skeletons, sometimes at comparatively large body size. This was not some ordinary event in the history of life.

Skeletons

A skeleton may support soft tissue, from the inside or from the outside, and simply allow an animal to grow larger. Therefore, sponges could grow larger and higher after they evolved supporting structures of protein or mineral, and they could reach further into the water to take advantage of currents and to gather food. Large size also protects animals from predators large and small. A large animal is less likely to be totally con-sumed, and in an animal like a sponge that has little organization, dam-age can eventually be repaired if even a part of the animal survives at-tack. As skeletons evolved, even for other reasons, they helped animals to survive because of their defensive value.

Early echinoderms had lightly plated skeletons just under their surfaces (Figure 4.17), and the most reasonable explanation of their first function is support, accompanied or followed by the function of defense.

For other animals, skeletons provided a box that gave organs a controlled environment in which to work. Filters were less exposed to currents, so perhaps they would not clog so easily from silt and mud. A box-like skeleton would also have given an advantage against predation. Molluscs and brachiopods may have evolved skeletons for these reasons.

In yet other animals, hard parts may have performed more specific functions. We have already seen that worms tend to burrow head-first in sediment. But after penetrating the sediment they squirm through it. A worm that evolved a hardened head covering could use a different and perhaps better technique, shoveling sediment aside like a bulldozer. Richard Fortey has suggested that the large headshield of trilobites was evolved and used in this fashion.

But arthropods, and especially trilobites, are strongly armored all over their dorsal surfaces, not just in the head region (Figure 4.16). Most likely their armor served for the attachment of strong muscles. Muscles pull and cannot push. Worms move by using internal hydraulic systems, as we have seen. On the other hand, walking demands that limbs push on the sediment, and that is very unrewarding if the other end of the leg is unbraced. Arthropods evolved a large, strong dorsal skeleton against which their jointed legs were firmly braced (Figure 4.12), allowing them to move much more efficiently than worms do. Later, the strong skeleton would have been effective against predators. It's unlikely that trilobites evolved skeletons for defense. Cambrian trilobites had small tails that did not protect them very well when they rolled up. (Later trilobites had larger tails and could roll up to protect themselves almost completely.)

Skeletons seem to have evolved for many different reasons, in many different chemistries, in many different animals, but why did they evolve in a very short geological time and in two abrupt waves? The only common factor is the dramatic invasion of new ways of life on the Early Cambrian seafloor, into ecologies that were impossible without support or sheltering of internal organs or muscular bracing.

Despite all the discussion of skeletons, the Burgess fauna shows that dramatic evolution took place also in animals that did not have strong skeletons. However, many of these animals had outer coverings that were tough, but lightly mineralized—the Burgess arthropods are particularly good examples.

The common factor along successful groups of Cambrian animals is larger body size. All of this suggests that in some way the world had become ready for large animals, and in turn that tells us that the Cambrian event was driven by worldwide ecological factors—but we do not yet know what they were. They could have been related to a change in food supply in the sea, which in turn depends on upwelling, which in turn depends on climatic and geographic patterns on a global scale. We don't yet know enough about Cambrian geography and climate to say anything sensible about these factors, but it's here that the answer probably lies and where future research should be focused.

Whatever the underlying global cause was, some specific mechanisms have been suggested to explain the Cambrian explosion.

Predation

The predation theory has two aspects. The first is a general ecological argument. The ecologist Robert Paine removed the top predator (a starfish) from rocky shore communities on the Washington coast and found that diversity dropped. In the absence of the starfish, mussels took over all available rocky surfaces and smothered all their competitors. Paine suggested that a major ecological principle was at work: effective predators maintain diversity in a community. If a prey species becomes dominant and numerous, the top predator eats it back, maintaining diversity by keeping space available for other species.

Unfortunately, Paine's original observations may have been made in an unusual community. Along other rocky shores (for example, in South Africa) the top predators are seabirds. They keep the shores so clean of herbivores such as limpets, and so well fertilized with guano, that seaweeds take over and choke out the huge diversity of animals that would otherwise live on the rocks. Thus, the action of predators may lower or raise diversity, depending on the ecological food chain of the particular community. There may be no general principle linking predation to high diversity.

Nevertheless, Steven Stanley used Paine's work to suggest that the evolution of predation triggered the Cambrian radiation. Stanley made an intellectual jump to suggest that predators can *cause* additional diversity in their prey. He argued that if predators first appeared in the Early Cambrian, they may have caused the increase in diversity at that time. Perhaps predators also encouraged the evolution of many different types of skeletonized animals.

Geerat Vermeij provided critical support for Stanley's idea, suggesting how new predators might indeed cause diversification among prey (at any time). In response to new predators, prey creatures might evolve large size, or hard coverings made from any available biochemical substance, or powerful toxins, or changes in life style or behavior, or any combination of these, all in order to become more predator-proof. And as the new predators in turn evolve more sophisticated ways of attacking prey, the responses and counter-responses might well add up to a significant burst of evolutionary change.

Are the characteristics of early Cambrian fossils consistent with a predation theory? Predation as a way of life probably appeared long before the Cambrian. There are predatory protists, and some early tiny metazoans were probably predators too—but all on a tiny scale. The rules of the predator/prey game may have changed radically as large multicellular creatures evolved. Many Early Cambrian fossils have hard parts that look defensive. For example, some sharp little conical shells called sclerites may have been spines that were carried pointing outwards on the dorsal and lateral sides of animals, to fend off predators.

More direct evidence comes from little tubes called *Cloudina* that were the first hard parts ever evolved by any animal. Chinese specimens from late Precambrian rocks occasionally have holes bored into them, presumably by some unknown predator. There are armored and spined Early Cambrian animals, and some Early Cambrian trilobites have healed injuries that may indicate damage by a predator. Defensive structures made of hard parts could therefore have contributed to the increase in the number of fossils in Early Cambrian rocks.

Present evidence suggests that predation played an important part in generating the Cambrian event. It's difficult to be certain, because the only major predators we have discovered are the anomalocarids, and we have no evidence of what they ate—anything they could catch, probably! However, predation does not explain the timing of the Cambrian explosion: why not 100 m.y. earlier, or later? Certainly predation alone cannot account for all the variety of skeletons that we see.

Oxygen Levels

Perhaps the Cambrian explosion is related to global oxygen levels. In one scenario, the evolution of large bodies and skeletons was made possible by high oxygen concentrations. Shells and thick tissues cut off the free diffusion of oxygen into a body, so respiration cannot occur unless there is a high enough oxygen level to push oxygen into the few remaining areas of exposed tissue—gills, for example. This cannot be the whole story, because sponges and cnidarians could have evolved their skeletons (which do not inhibit respiration) in low oxygen conditions.

If the oxygen idea is true, however, it could explain much of the Cambrian explosion. Where did the increased oxygen come from? Oxygen is produced in photosynthesis, but usually the plant tissue that is produced at the same time eventually is eaten and digested (using up oxygen), or it rots (using up oxygen). Only if the organic matter is buried does the oxygen stay free in the sea and the atmosphere. What would increase the amount of carbon buried in the seafloor?

Graham Logan and his colleagues pointed out that the evolution of metazoans ("worms," let us say) also involved the evolution of guts, and therefore of feces, usually in the form of compact fecal pellets. If carbon-rich fecal pellets are buried quickly, organic matter is removed from oxidation effectively, and this raises oxygen levels in the sea and in the atmosphere. In this view, then, the rise of metazoans large enough to produce reasonable quantities of fecal pellets was responsible for the rise in oxygen, which then permitted even more (and larger) metazoans to evolve, and so on, until it became advantageous for those metazoans to evolve skeletons. This is a very attractive idea, especially as "worms" would characteristically produce their fecal pellets in or on the seafloor, where they would be buried easily and quickly.

Even so, it is not clear that oxygen, or predation, or any other single parameter can be identified as the reason for the nature, the scale, and especially the timing of the Cambrian explosion. One could argue (and people have) that the world is full of complex creatures, so complexity must have evolved sometime. Whenever it evolved, it was bound to cause a visible "burst" in the fossil record, but perhaps there was no "trigger" for the Cambrian explosion we see in the record. The first large animals evolved at that time, and it is hardly surprising that they spread rapidly and diversified into many body plans, with different groups evolving hard parts of different chemistry and structure. After the dramatic events early in the Cambrian, the increase in numbers and diversity of fossils later in the period seems anticlimactic. Cambrian fossil collections are not very complex ecologically; they are dominated by trilobites, most of which lived on the seafloor and were deposit feeders. Filtering organisms are very much secondary, and although there are large carnivores, they are represented only by anomalocarids.

The Cambrian explosion is spectacular, but it is not unique; in my view the spectacular diversification of the diapsid reptiles, especially the archosaurs, in the Late Triassic is an analogous case (Chapter 11), as is the diversification of the mammals in the Paleocene (Chapter 18). Such radiations stand out from "normal" evolutionary events just as "mass extinctions" stand out from the rest (Chapter 6). On a real planet inhabited by real organisms, evolutionary rates are likely to vary in time and space, and evolutionary events are likely to vary in magnitude, duration, and frequency. We should not expect that ideal rules we might propose for an ideal planet would be followed by the natural world; instead, we have to find out from that natural world what the rules actually were.

Review Questions

There really are only three kinds of animals — what are they?

What is special about the fossil record during the Cambrian explosion?

What major global geological event coincides with the Cambrian explosion?

Describe in very general terms the ecology of Cambrian animals: how did they live?

What major event is used to define the end of the Precambrian and the beginning of the Paleozoic era?

What biological event marks the base of the Cambrian period?

Why is the base of the Cambrian period such an important marker level for paleontologists?

Diversity changed rather a lot among the animals of the early Paleozoic? How, and why?

What is special about the animals found fossilized in the Burgess Shale? Draw one of them. You don't have to name it, but I have to be able to recognize it. If naming it will help, then do so.

Further Reading

Easy Access References

Bengtson, S., and Y. Zhao. 1992. Predatorial borings in Late Precambrian mineralized exoskeletons. *Science* 257: 367–369. *Cloudina*.

Bengtson, S., and Y. Zhao. 1997. Fossilized metazoan embryos from the earliest Cambrian. *Science* 277: 1645–1648.

Chen, J.-Y., et al. 1994. Evidence for monophyly and arthropod affinity of Cambrian giant predators. *Science* 264: 1304–1308, and comment, pp. 1283–1284.

Davidson, E. H., et al. 1995. Origin of bilaterian body plans: evolution of developmental regulatory mechanisms. *Science* 270: 1319–1325, and comment, pp. 1300–1301.

Erwin, D. H., et al. 1997. The origin of animal body plans. *American Scientist* 85: 126–137.

Gore, R. 1993. The Cambrian period: explosion of life. *National Geographic* 184 (4): 120–136.

Gould, S. J. 1989. *Wonderful Life: The Burgess Shale and the Nature of History*. New York: W. W. Norton. Beautifully written and illustrated book on the Burgess Shale, with Gould's view of the way evolution works.

Grotzinger, J. P., et al. 1995. Biostratigraphic and geochronologic constraints on early animal evolution. *Science* 270: 598–604, and comment, pp. 580–581.

Jensen, S., et al. 1998. Ediacara-type fossils in Cambrian sediments. *Nature* 393: 567–569.

Knoll, A. H., and S. B. Carroll. 1999. Early animal evolution: emerging views from comparative biology and geology. *Science* 284: 2129–2137.

Li, C.-W., et al. 1998. Precambrian sponges with cellular structures. *Science* 279: 879–882.

Logan, G. A., et al. 1995. Terminal Proterozoic reorganization of biogeochemical cycles. *Nature* 376: 53–56, and comment, pp. 16–17.

McMenamin, M. A. S. 1997. *The Garden of Ediacara: Discovering the First Complex Life*. McMenamin loves flashy hypotheses.

Monastersky, R. 1993. Mysteries of the Orient. *Discover* 14 (4): 38–48. The Chengjiang fauna.

Palmer, D. 1996. Ediacarans in deep water. *Nature* 379: 114. Ediacaran animals were tough-skinned.

Ramsköld, L., and X. Hou. 1991. New early Cambrian animal and onychophoran affinities of enigmatic metazoans. *Nature* 351: 225–228, and comment, 184–185. Solution of the *Hallucigenia* mystery.

de Robertis, E. M. 1997. The ancestry of segmentation. *Nature* 387: 25–26.

Shu, D.-G., et al. 1996. A *Pikaia*-like chordate from the Lower Cambrian of China. *Nature* 384: 157–158. *Cathaymyrus*.

Wray, G. A., et al. 1996. Molecular evidence for deep Precambrian divergences among metazoan phyla. *Science* 274: 568–573, and comment, pp. 525–526. See Conway Morris, 1998.

Wright, K. 1997. When life was odd. *Discover* 18 (3): 52–61. Ediacaran fossils.

More Technical Reading

Aguinaldo, A. M. A., and J. A. Lake. 1998. Evolution of the multicellular animals. *American Zoologist* 38: 878–887. As the organizers of the symposium said, ten more years of research might help to solve this question!

Ayala, F. J. et al. 1998. Origin of the metazoan phyla: molecular clocks confirm paleontological estimates. *Proceedings of the National Academy of Sciences* 95: 606–611. See Conway Morris, 1998.

Barnes, R. S. K., et al. 1993. *The Invertebrates: A New Synthesis*. 2d. ed. Oxford: Blackwell Scientific Publications.

Bengtson, S. (ed.) 1994. *Early Life on Earth*. New York: Columbia University Press. Important review papers on Ediacarian times by Fedonkin, Hofmann and Sun (the fossils) and Knoll (evolution and environment).

Brasier, M., et al. 1997. Ediacarian sponge spicule clusters from southwestern Mongolia and the origins of the Cambrian fauna. *Geology* 25, 303–306.

Buss, L. W., and A. Seilacher. 1994. The phylum Vendobionta: a sister group of the Eumetazoa? *Paleobiology* 20: 1–4.

Chen, J-Y., et al. 1991. The Chengjiang fauna: oldest soft-bodied fauna on Earth. *National Geographic Research & Exploration* 7: 8–19.

Clark, R. B. 1964. *Dynamics in Metazoan Evolution*. Oxford: Clarendon Press.

Conway Morris, S. 1997. *The Crucible of Creation: The Burgess Shale and the Rise of Animals*. Oxford: Oxford University Press. New account of Burgess Fauna, critical of Gould's views.

Cooper, A., and R. Fortey. 1998. Evolutionary explosions and the phylogenetic fuse. *Trends in Ecology & Evolution* 13: 151–156. Section on Cambrian.

Gehling, J. G. 1999. Microbial mats in terminal Proterozoic siliciclastics: Ediacaran death masks. *Palaios* 14: 40–57.

Knoll, A. H., et al. 1995. Sizing up the sub-Tommotian unconformity in Siberia. *Geology* 23: 1139–1143.

Morris, P. J. 1993. The developmental role of the extracellular matrix suggests a monophyletic origin of the Kingdom Animalia. *Evolution* 47: 152–165.

Seilacher, A. 1989. Vendozoa: organismic construction in the Proterozoic biosphere. *Lethaia* 22: 229–239.

Seilacher, A. 1999. Biomat-related lifestyles in the Precambrian. *Palaios* 14: 86–93.

Waggoner, B. M. 1996. Phylogenetic hypotheses of the relationships of arthropods to Precambrian and Cambrian problematic fossil taxa. *Systematic Biology* 45: 190–222.

Waggoner, B. M. 1998. Interpreting the earliest metazoan fossils: what can we learn? *American Zoologist* 38: 975–982.

Life in a Changing World

Life has not evolved on its own; it has evolved on a planet that has experienced changing geology, geography, and climate. Life did not evolve in random patterns, either; in order to understand how life has been affected by its global environment over geological time, we have to look at the relationships between the physical and biological world on which it lives.

The Global Diversity Gradient

The diversity of life over the globe can be measured by the number of species that occupy a given habitat or area. Many factors affect diversity, including the available energy supply in the area, the number and variety of the different microenvironments that can be occupied by different species, and the severity of the physical environment. But some factors seem to affect diversity levels more strongly than others.

On the Earth today there is a strong diversity gradient from pole to equator among communities of organisms living in the shallow seas and on land. Many more species live in tropical communities than in high-latitude communities, and the average number of species in any community falls off toward the poles even over a few degrees of latitude. This principle has been documented for many groups of organisms, from birds to bivalves and from foraminiferans to orchids (Figure 5.1).

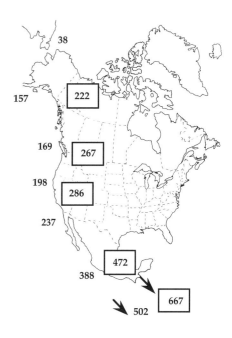

Figure 5.1 The diversity of organisms increases from pole to equator. The map shows the latitudinal gradient in the species diversity of bivalves (in the sea) and of birds (nested in boxes) along the Pacific coast of North America. Arrows point to the maximum diversity further south along the coast, closer to the equator. (Data from Valentine, 1971, and McArthur and Connell, 1966.)

Many explanations have been suggested for the diversity gradient, but generally they relate to the fact that the environment is much more seasonal toward the poles. At the equator, the day length never changes much from a twelve-hour day-night cycle; at each pole there is for practical purposes only one day and one night per year, with three months of full daylight, three months of complete darkness, and two three-month periods of twilight in between. Environmental conditions fluctuate greatly in polar regions, especially the supply of food that comes directly from photosynthetic plant production on land and in the sea. Only relatively few organisms have adapted to live successfully in such an extreme set of conditions. What is the connection between seasonal light intensity and low diversity?

Light and temperature are not in themselves important controls on diversity. Polar waters are actually very stable in temperature, never varying much from freezing all year round. Cold water temperatures do not in themselves foster communities with low diversity, because there is a great diversity of life on the ocean floors at depths of 2–3 km, where it is permanently cold. Nor does lack of light automatically result in low diversity, since the ocean floors are permanently dark.

Perhaps, then, physical extremes of any sort are not important factors. If some species can adapt successfully to difficult environments, why can't many other species do so? Land environments must have seemed incredibly forbidding from the point of view of marine creatures of the Silurian, yet since that time organisms have successfully invaded the land and evolved into myriad forms of life.

To explain low diversity, we need to identify a factor that limits diversity. If a community already has a certain number of species, why can't it include more? The most important of the diversity limiting factors that operate in nature seems to be food supply. If a field will feed only a certain number of animals, the introduction of another individual means that one inhabitant will die of malnutrition, possibly but not inevitably the invading stranger.

The real world is much more complex, but it operates on the same principle. The tropics have a fairly uniform climate, and food supply is stable, available at about the same level all year round. A species can specialize on one or two particular food sources and can rely on them always being available. As each species comes to depend on a narrow range of food sources, it adapts so well to harvesting them that it cannot easily switch to alternatives. Thus, a great variety of specialized species may evolve, competing only marginally with one another, at least for food. In the Serengeti plains of East Africa, for example, several species of African vultures are all scavengers on carcasses. But one species has a head and beak adapted for tearing through the tough hide of a fresh carcass, another is adapted only for eating the soft insides from an opened carcass, another is adept at cleaning bones, and another eats the scraps. Generally, each tropical species lives in a world of stable food resources, and the great diversity of tropical communities is a reliable reflection of this kind of complex ecosystem.

In high latitudes, on the other hand, food supplies may vary greatly from season to season and from year to year. Overall, food supply may be high. Tundra vegetation blooms in spectacular fashion in the spring. There are rich plankton blooms in polar waters during spring and summer, and millions of seabirds and thousands of whales migrate there to share in the abundant food that is produced. Antarctic waters teem with millions of tons of tiny crustaceans (krill) that eat plankton and in turn are fed on by fish, seabirds, penguins, whales, and seals. The Arctic tern migrates almost from pole to pole, timing its stay at each end of the world to coincide with abundant food supply. Yet for organisms that live all year in polar regions, spring abundance contrasts with winter famine, and food variability is the problem, not average food supply (Figure 5.2).

Many polar species switch from one food to another as the seasons progress, taking advantage of whatever opportunities are available. The Arctic gray wolf eats mammals and small birds in summer but attacks weakened caribou in winter. Where food supplies vary, few species can be specialists on only one food source; they must be versatile generalists. Generalists share some food sources, and probably compete more than specialists do. If so, fewer generalists than specialists can coexist on the same food resources. In seasonal or variable environments, where organisms must be generalists, diversity is lower. This factor probably underlies the global diversity gradient from equator to pole.

The argument about food variability is powerful because it can also explain diversity patterns in other ecosystems. For example, much of the deep sea has a small but steady food supply in the form of debris that drops from the surface waters, and flows down from the continental shelves. In response, the fauna of the deep sea floor is very diverse, even though it is made up largely of tiny arthropods digging and scavenging from seafloor mud.

There is yet another factor. When many food sources vary with the seasons, animals tend to concentrate on supplies that are reliable. The most reliable food source in polar regions is seafloor mud, even though it is not particularly nutritious. Clams, worms, and arthropods living on

Figure 5.2 A diagram that summarizes the great seasonal variability of food resources in polar seas. Low light and a cover of sea ice in winter inhibit primary productivity by phytoplankton. When those constraints are removed there is a great surge in food supply in the polar springtime. Similar effects occur on land.

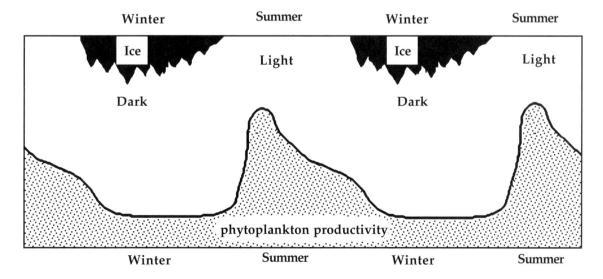

soft seafloors are the most diverse of all groups of polar organisms. Mud grubbing is practiced by many marine animals in high latitudes, and they simply do not perceive any period when food supply fluctuates. In a different strategy, some Antarctic marine animals stop feeding altogether in winter: krill actually shrink!

In slightly lower latitudes, food may be more reliable. Plankton living in the water can be an alternative food source, and filter feeding can become a viable way of life. Here, a new ecological level of marine animals can be added to a community, given reliable food supplies.

Both warm and cold deserts are land environments where food supply is routinely low, but where periods of great abundance occur after a brief rainy season. One would predict that diversity should be comparatively low in such environments, and it is. The most diverse groups of desert organisms are those that are able in one way or another to buffer themselves from variations in their food supply so that they don't perceive periods of famine. In cold deserts, many animals hibernate when food is low. In hot deserts, many animals estivate. Desert plants have an analogous lifestyle, which is to spend long periods as seeds, with plant growth and reproduction fitted within brief damp periods. This buffering strategy is the ecological equivalent of mud grubbing on polar seafloors.

Islands and Continents

The gradient from equator to pole dominates the pattern of diversity on the Earth's surface, but another factor is significant for our understanding of the fossil record. Island groups tend to have more equable climates—referred to as maritime or oceanic climates—compared with nearby continents, no matter what their latitude. Thus the British Isles and Japan have milder climates than Siberia; the West Indies have milder climates than Mexico; and Indonesia has a milder climate than Indochina.

Large continental areas have severe climates for their latitudes. Asia, for example, is so large that extreme heat builds up in its interior in the northern summer, forming an intense low-pressure area. Eventually the low pressure draws in a giant inflow of air from the ocean, the **summer monsoon**, that brings a wet season to areas all along the south and east edges of the continent, from China to Pakistan (Figure 5.3, left). In winter the interior of Asia becomes very cold, a high-pressure system is set up, and an outflow of air, the **winter monsoon**, brings very chilly weather to India, China, and Korea (Figure 5.3, right). Land organisms respond to the great seasonality of the monsoon climate, and organisms in the shallow coastal waters are affected strongly too. As nutrient-poor water is blown in from the surface of the open ocean, food becomes scarce; as water is blown offshore, deeper water upwells and brings nutrients and high food levels (Figure 5.4). As a result, the diversity of marine creatures along the coasts of India is far less than it is in the Philippines and Indonesia, which are far enough away from the Asian mainland that they feel the effects of the monsoons much less strongly.

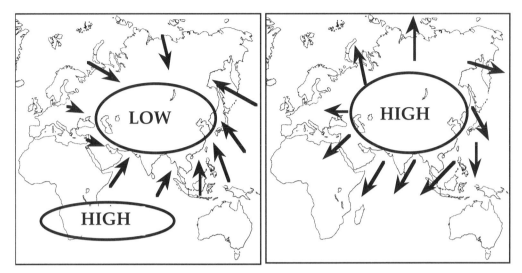

The effects of continental geography as opposed to oceanic geography thus have an important effect on the general pattern of diversity in major regions of the Earth. Their effects, however, are still directly linked to the seasonality of food supply.

Figure 5.3   The monsoons of southeast Asia. Left: In summer, heat builds up over the continent and generates low pressure that draws in moist air from the surrounding oceans. Right: In winter, high pressure over the continent generates cold winds that blow offshore.

Provinces

Organisms often occur in characteristic sets of species called communities, living together in certain types of habitat—rocky shore communities, mudflat communities, and so on. The northwest coast of North America, bathed by a southward flow of cool water, has a characteristic rocky-shore community of plants and animals that occurs from British Columbia to Central California. The coastal communities of the world can be arranged into geographically separate provinces, with each province containing its own set of communities, such as the Oregonian and Californian Provinces of western North America (Figure 5.5).

Provinces are real phenomena, not artifacts of the human tendency to classify things. There are natural ecological breaks on the Earth's surface, usually at places where geographic or climatic gradients are sharp, so that one may pass from one environmental regime to another in a short distance. A classic example is at Point Conception on the California coast. Here, the ocean circulation patterns cause a sharp gradient in water

Figure 5.4   Monsoons and food. Left: Winds blow surface water onshore from the blue water of the open ocean. That water is poor in nutrients. Right: Offshore winds blow water away from the shore. That water is replaced by deep water, which generally contains nutrients and favors growth in the marine food chain.

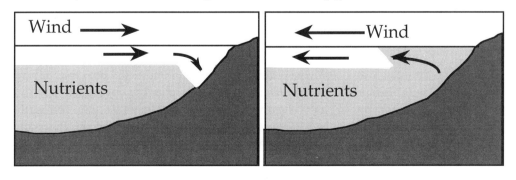

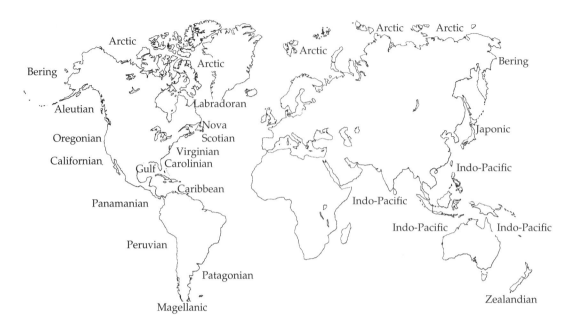

Figure 5.5 The marine biosphere can be divided into biotic provinces. This figure does not show all 31 of them, but I have included all the provinces around the Americas, to stress the differences between east and west coasts and the strong latitudinal gradient that produces many provinces along north-south coastlines. In contrast, the Arctic province is very large, because organisms migrate easily along latitudes around the Arctic Ocean, and the Indo-Pacific province is enormous, because marine organisms migrate easily along east-west coastlines and island chains. The Zealandian province is small because it can occupy only a restricted area of shallow shelf. (Modified from Valentine, 1973.)

temperature (in human terms, Point Conception marks the northern limit of west coast beaches where one can surf without a wet suit). The communities on each side of Point Conception are very different, so a provincial boundary is drawn here, with the Oregonian province grading very sharply into the Californian province (Figure 5.5).

As provinces are identified around the coasts of the world, it seems that the number of species in common between neighboring provinces is usually 20% or less. About 30 provinces have been defined along the world's coasts, mostly on the basis of molluscs, which are obvious, abundant, and easily identified members of coastal communities. Some provinces are very large because they inhabit long coastlines that lie in the same climatic belt (the Indo-Pacific, Antarctic, and Arctic Provinces); some are small, like the Zealandian Province, which includes only the communities around the coasts of New Zealand (Figure 5.5).

Each province contains its own communities and therefore carries unique sets of animals that fill various ecological niches. For example, the intertidal rocky-shore community in New Zealand has its ecological equivalent in British Columbia, even though the families and genera of animals are quite different in the two communities. Multiplying the number of provinces automatically multiplies the diversity of animals around the coasts of the world. The total diversity of the world's shallow marine fauna directly reflects the number of provinces, which in turn reflects climate and geography. Today, for example, there is a steep temperature gradient from equator to poles, so there is a strong climatic zonation of the oceans and continents. Each north-south continental coast has several provinces along it, each different enough in climatic conditions to hold a unique fauna. If Earth's heat were distributed more uniformly, so that the ice caps melted, there would be a lower temperature gradient, fewer climatic zones between equator and pole, and fewer provinces. Perhaps there would be no polar communities as we know them, and the world would have a lower diversity of life. Other things being equal, climatic diversity increases biological diversity.

The more oceanic the general climate of the world is, the greater its diversity. In an oceanic world, each community in a province tends to have more stable food supplies and greater diversity within it. Therefore, the more the continents are fragmented into smaller units, the more oceanic the world's climate becomes, and the more diverse its total biota.

Furthermore, there are severe barriers to free migration by organisms when continents are greatly fragmented and widely separated. Land organisms and shallow marine organisms find it difficult to cross large stretches of ocean. Therefore, unique provinces of land plants and animals and unique provinces of marine organisms evolve along the coastline of each continent.

Climatic barriers limit the distribution of organisms in a north-south direction, so migration is easier in an east-west direction, and the generally expected geographic distributions of organisms occur in belts that run east-west (Figure 5.5). Furthermore, many of the wind and ocean current patterns on Earth have strong east-west components. An Earth with the maximum diversity of life would have a large number of barriers to prevent east-west distribution of organisms, such as north-south mountain chains and deep oceans (Figure 5.6). Such an arrangement would also produce a good deal of climatic variation, setting up geographic situations that would encourage many small provinces, each with its own set of communities. Our present global geography conforms well to that recipe, and the diversity of life on Earth is probably higher now than it has ever been, particularly among shallow-water molluscs.

Once we understand the reasons behind present diversity patterns, as well as the effect of continental movement on geography and climate, we can make some sense of the fossil record. We can suggest how continental movements affect the reliability of food supply, and therefore the diversity of life. When continents are clumped into large masses, they experience great monsoons, so their coastal regions have severe seasons and low biological diversity. When continents are split, the small pieces have oceanic climates, food supply is more even, and communities and provinces are diverse. When there are many continents, especially north-south continents, there are many geographic barriers, and many provinces of animals can evolve, each contributing to a very high total diversity of life on Earth.

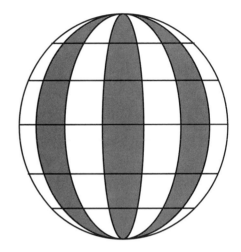

Figure 5.6   The perfect planet Petunia has maximum diversity of life in its oceans. Its many long land masses and its deep north-south oceans form barriers to easy east-west migration by marine animals, and there are many climatic zones and marine provinces along each north-south coast. With no tilt to its axis, it has no seasons.

Diversity Patterns in the Fossil Record

Jack Sepkoski estimated the global diversity of fossils through time, plotting the number of families of marine fossils from Ediacaran (Vendian) to Recent times (Figure 5.7). The data show clear and reasonably simple trends. Few families of animals existed in Vendian times, but the beginning of the Cambrian saw a dramatic increase that followed a steep curve to a Late Cambrian level of at least 150 families. A new, dramatic rise at

Figure 5.7   Jack Sepkoski's 1984 compilation of the diversity of families of marine organisms through time. The lower curve shows organisms with well-preserved hard parts; the upper curve shows all organisms. The similarity between the two curves shows that this bias of the fossil record is not very important, and the basic pattern has not changed with 15 years' worth of new data. (Data from Sepkoski, 1981, 1984.)

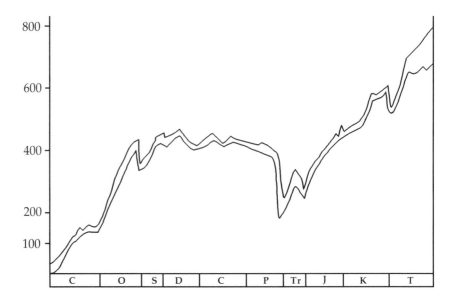

the beginning of the Ordovician raised the total to over 400 families, a number that remained comparatively stable through the Paleozoic. In the Late Permian there was a dramatic drop down to 200 families, but a steady rise that began in the Triassic has continued to the present, with a small and short-lived reversal only at the end of the Cretaceous Period, which marks the end of the Mesozoic Era. This general pattern has been known ever since John Phillips defined the Paleozoic, Mesozoic, and Cenozoic Eras in 1860 (Figure 5.8), but Sepkoski put the pattern in quantitative terms that can be further analyzed.

It's easy to think of possible problems with Sepkoski's approach. For example, only some parts of the world have been thoroughly searched for fossils; some parts of the geological record have been searched more carefully than others; older rocks have been preferentially destroyed or covered over by normal processes such as erosion and deposition. Can we believe Sepkoski's numbers, even if we agree that the general trends he sees are real? His data may in fact be more reliable than one would

Figure 5.8   John Phillips's curve showing diversity through time was published in 1860, before the fossil record was well known, but its general shape remains accepted today (compare Figure 5.7).

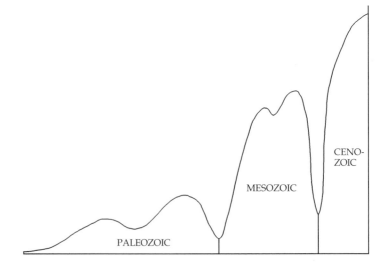

think, especially as a more recent compilation by Michael Benton has shown much the same patterns of diversity change.

Paleontologists have been searching the world for fossils for 200 years. The best-sampled fossil communities are shelly faunas that lived on shallow marine shelves, and our estimate of their diversity through time is likely to be a fair sample of the diversity of all life through time. Larger groups of animals are harder to miss than smaller groups, so we have probably discovered all the phyla of shallow marine animals with hard skeletons. Perhaps we have only found a few percent of the species in the fossil record, but we've probably discovered many of the families. In any case, if the search for fossils has been approximately random (and there's no reason to doubt it) the shallow marine fossil record as we now know it is a fair sample of the fossil record as a whole. And since we now know the biases of the record, we can make intelligent inferences not just about the fossil record but about the living world too.

Global Tectonics and Global Diversity

We have known for over 30 years that the Earth's crust is made up of great rigid plates that move about under the influence of the convection of the Earth's hot interior. As they move, the plates affect one another along their edges, with results that alter the geography of the Earth's surface in major ways. Two plates can separate to split continents apart, to form new oceans, or to enlarge existing oceans by forming new crust in giant rifts in the ocean floor. Two plates can slide past one another, forming long transform faults such as the San Andreas Fault of California. Plates can converge and collide, forming chains of volcanic islands and deep trenches in the ocean, volcanic mountain belts along coasts, or giant belts of folded mountains between continental masses. These movements and their physical consequences are studied in the branch of geology called **plate tectonics**.

James Valentine and Eldridge Moores suggested in 1970 that because plate tectonic movements affected geography, they could in turn affect food supply, climate, and the diversity of life. Over the past 250 m.y., they argued, changes in world geography encouraged the diversification of the world's fauna because continents separated, making the world more oceanic. At the same time, the number of provinces increased, as climatic zonation became stronger and continents separated to create north-south land and sea barriers such as the Americas and the Atlantic Ocean. Does the idea apply to earlier times?

Jack Sepkoski documented three dramatic expansions of diversity: around the beginning of the Cambrian; at the beginning of the Ordovician; and after the Permian extinction (Figure 5.7). If Valentine and Moores are correct, the first two episodes of expansion would fit the idea that the continents were clumped together in the Late Proterozoic, that they began to split by the Early Cambrian, and that they became well separated in the Ordovician. This theory explains the early radiations of

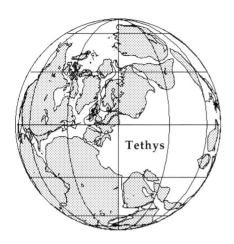

Figure 5.9 The paleogeography of the Permian. Two very large continents, Gondwanaland and Laurasia, combined to form the supercontinent Pangea. This map depicts the Early Permian, about 280 Ma; Pangea is almost complete by this time. A major sea-level drop and the great Permian extinction took place about 30 m.y. later.

animals and their diversity patterns very nicely. Is it true?

It's difficult to make reliable reconstructions of Late Proterozoic and Early Paleozoic geography. The limited evidence suggests, however, that continental movement really was the major control on diversity. Apparently, a Late Proterozoic supercontinent—**Rodinia**—split progressively during the Cambrian and Ordovician, to form a set of small continents that were generally distributed in lower latitudes.

Can we explain decreases in diversity (extinctions), using the same idea? Continental collisions should decrease diversity, just as continental splitting increases it. There were several continental collisions from the Middle Paleozoic through the Permian, and large land masses were formed. The great extinction at the end of the Permian took place shortly after the continents finally merged into a giant supercontinent, **Pangea** (Figure 5.9), composed of a large northern land mass, Laurasia, and a southern land mass, Gondwanaland. Pangea was largely complete in the Early Permian, and the Late Permian extinction coincided with a severe global drop in sea level.

The Permian extinction did not occur gradually over the 150 m.y. of the later Paleozoic, as the continents collided and assembled piece by piece. It's not clear whether the dramatic nature of the Late Permian extinction is a problem for the theory. Most likely, the continental assembly set up the world for extinction, then an "extinction trigger" was pulled. We shall look at the Permian extinction in more detail in Chapter 6.

The rise in diversity that began in the Triassic and continued into the Cenozoic coincides very well with the progressive break-up of Pangea. The break-up was under way by the Jurassic, and reached a climax in the Cretaceous (Figure 5.10). The continental fragments have continued to drift, and today the continents are perhaps as well separated as one could ever expect, even in a random world.

The overall pattern of diversity through time data can receive a first-order explanation from the suggestions of Valentine and Moores. The timing and direction of the changes in diversity correlate well with plate tectonic events. But that cannot be the whole story, for several reasons.

1. Changing faunas through time. If plate tectonics were the only control on diversity, much the same groups of animals

Figure 5.10 The paleogeography of the Cretaceous at about 80 Ma. Pangea is now fragmented, with the Atlantic open from New England southward. The map on the right shows that Africa and India have broken away from Antarctica and are drifting northward.

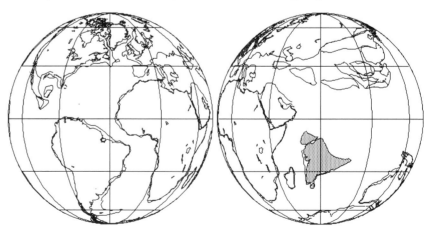

should rise and fall with plate geography. Iinstead, we see dramatic changes in the different animal groups that succeed one another in time.

2. Increase in Global Diversity. An overall increase in global diversity from Ediacaran to Recent times is not predicted on plate tectonic grounds.

3. Mass Extinctions. Major extinctions are much more dramatic than radiations. There are too many sudden "mass extinctions" in the fossil record for a plate tectonic argument to be completely satisfactory. Even if plate tectonic factors set the world up for an extinction, we seem to need some separate theory to explain the extinctions themselves.

Changing Diversity Through Time

Three Great Faunas

Jack Sepkoski sorted his data on marine families through time to see if there were groups that shared similar patterns of diversity. A computer analysis helped him to distinguish three great divisions of marine life through time, which accommodate about 90% of the data (Figure 5.11). He called them the Cambrian Fauna, the Paleozoic Fauna, and the Modern Fauna. The faunas overlap in time, and the names are only for convenience. But they do reflect the fact that different sets of organisms have had very different histories.

The histories of the three marine faunas are shown in Figures 5.12, 5.13, and 5.14. The **Cambrian Fauna** contains the groups of organisms, particularly trilobites, that were largely responsible for the Cambrian in-

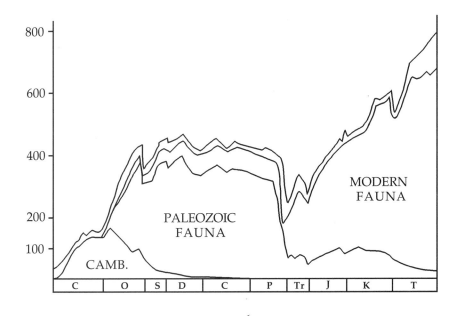

Figure 5.11 The three great faunas defined by Jack Sepkoski in his analysis of the marine fossil record at the family level. They are subsets of the data shown in Figure 5.7. (Data from Sepkoski, 1981, 1984.)

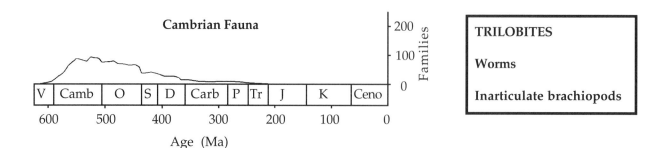

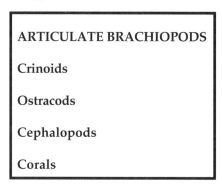

Figure 5.12 The Cambrian Fauna is dominated by trilobites and has generally low diversity (about 100 families). (Data from Sepkoski, 1981, 1984.)

crease in diversity. But after a Late Cambrian diversity peak, the Cambrian Fauna declined in diversity in the Ordovician and afterward (Figure 5.12), even though other marine groups increased dramatically at that time (Figure 5.11). In the same way, the **Paleozoic Fauna** was almost entirely responsible for the great rise in diversity in the Ordovician, and slowly declined afterward (Figure 5.13). The Paleozoic Fauna suffered severely in the Late Permian extinction, and its recovery afterward was insignificant compared with the dramatic diversification of the **Modern Fauna** (Figure 5.14).

The composition of the three faunas is shown in Figures 5.12, 5.13, and 5.14. Their definition is approximate, because Sepkoski's families are grouped at the level of classes or subphyla. There is no zoological affinity between the members of the three faunas, so we might ask whether there are ecological factors at work.

The Cambrian Fauna clearly represents a major addition of animals to the world fauna that existed in the Precambrian. But the Cambrian Fauna is dominated (77%) by trilobites, which were primarily surface-digging deposit feeders (Figure 5.12), and many of the others are also surface diggers, including many of the other arthropods and soft-bodied worms of the Burgess fauna.

Figure 5.13 The Paleozoic Fauna is dominated by suspension feeders and "lie-in-wait" predators and has higher diversity than the Cambrian Fauna (close to 400 families). The Paleozoic Fauna would have been even more sharply defined if Sepkoski had chosen to subdivide crinoids and corals into smaller units, because there are characteristic Paleozoic and post-Paleozoic subgroups of these animals. (Data from Sepkoski, 1981, 1984.)

The Paleozoic Fauna is much more diverse, dominated (80%) by suspension-feeding animals, especially articulate brachiopods, crinoids, and bryozoans, plus corals and cephalopods, which are mainly stationary or slow-moving "lie-in-wait" predators (Figure 5.13). The Paleozoic Fauna did not simply replace the Cambrian Fauna but was richer and added new ecological components to seafloor ecology.

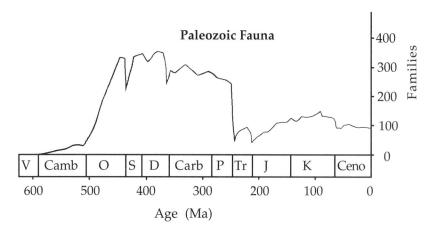

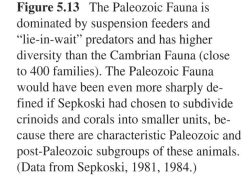

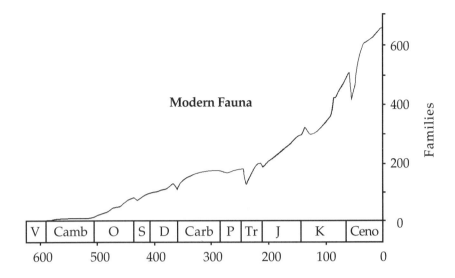

GASTROPODS

BIVALVES

Bony Fishes

Crabs and Lobsters

Echinoids

Figure 5.14 The Modern Fauna is dominated by molluscs and active predators and has even higher diversity than the Paleozoic Fauna (over 600 families). The Modern Fauna was largely unaffected by the Permian extinction. (Data from Sepkoski, 1981, 1984.)

The Modern Fauna is even more diverse. It is rich in molluscs and is dominated (83%) by swimming predators such as fishes and cephalopods, strongly burrowing groups including most bivalves, crustaceans, and echinoids, and versatile opportunists such as crustaceans and gastropods (Figure 5.14). The Modern Fauna again represents not just a replacement for the Paleozoic Fauna but another great addition of ecological novelty to the marine world.

Cambrian communities were dominated by sea-floor deposit feeders and by filter feeders that either fed low in the water or lived in the sediment. The Paleozoic Fauna lived in more tightly defined communities, with a more complex trophic structure. Filter feeders reached higher in the water and fed at different levels, and there was more burrowing in the sediment. The Modern Fauna has many more infaunal, burrowing animals in its communities, and many more predators.

Altogether, marine animals seem to have subdivided their ways of life more finely through time. The overall trend has been to add new ways of life, or **guilds**, to marine faunas through time, a generalization that describes most of the diversity increase, even if it doesn't explain it.

The diversity patterns imply that ecological opportunities in the world's oceans somehow changed through time to favor one particular ecological mixture and then to allow the diversification of others. Obviously there can be many different explanations of the facts, and I have room to discuss only a few suggestions. The diversity patterns have been known in outline for some time, so some of the explanations predate Sepkoski's analysis.

Faunas and Food Supply

Valentine, for example, pointed to the different ways of life that are encouraged under different types of food supply. In the Cambrian, he argued, the continents were not widely separated, food supplies were variable, and the most favored way of life would have been deposit feeding:

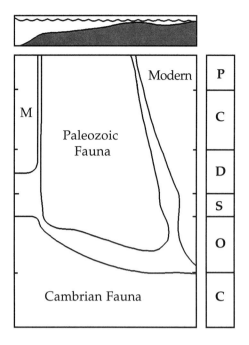

Figure 5.15 An interpretative diagram of the relationship between the three great Faunas and habitat at increasing distance from shore. As Sepkoski sees it, the Paleozoic Fauna displaced the Cambrian Fauna away from shore and then off the shelf, in turn being displaced by the Modern Fauna. Ecospace between the Faunas is occupied by mixed or transitional animals. The story may not be so simple, however—see Figure 5.16. (After Sepkoski, 1991.)

there is always some nutrition in seafloor mud. Thus Cambrian animals are, as Valentine said, "plain, even grubby." The Burgess fossils may not be plain, but many of them were certainly mud grubbers. Even among soft-bodied animals, arthropods dominate Cambrian faunas in numbers and diversity, and most of them were deposit feeders.

In contrast, if the continents were more widely separated in the Ordovician, one might expect much more reliable food supply in the plankton, which would have favored the addition of filter feeders to marine communities. One would also expect that a larger food supply in the form of stationary benthic filter feeders would have allowed slow-moving carnivores to become more diverse. Geerat Vermeij and Philip Signor documented the Ordovician rise of these ecological groups, which went on to dominate later Paleozoic seafloor communities.

If the Permian extinction was induced by continental collisions, one would predict that Paleozoic filter feeders and the predators that depended on them would have suffered a greater crisis than did other groups, because the food supply in the world's oceans would have become much more variable. In general, this prediction is correct: corals, brachiopods, cephalopods, bryozoans, and crinoids felt the Permian extinction most acutely.

But it is more difficult to explain the rise of the Modern Fauna. Other things being equal, one would predict that as continents split again in the Mesozoic, Paleozoic-style predators and filter feeders would again have been favored. They had not become completely extinct, and could surely have been expected to rediversify. In fact, they did, but in a very subdued fashion. Most of the Mesozoic diversification was achieved by other groups that stand out in Sepkoski's analysis as the Modern Fauna. Swimming and burrowing animals were added to marine ecosystems in great numbers.

The Rise of the Modern Fauna

Many people favor explanations that suggest a competitive advantage for the animals of the Modern Fauna, compared with their Paleozoic and Cambrian counterparts, though it is difficult to identify any compelling advantages. Sepkoski and his colleagues suggested some kind of **ecological displacement**. Each great fauna seems to have displaced its predecessor in shallow-water communities, "pushing" it toward the edge of the continental shelf (Figure 5.15). In this model, the Cambrian Fauna and then the Paleozoic Fauna declined as they were pushed offshore. It's not clear that the patterns really imply *competitive* displacement, or, if they do, why shallow-water habitats in particular should have favored more modern kinds of ecological communities. The answers are not likely to be simple, and in fact the patterns are not as clear-cut as Sepkoski argued (Figure 5.16).

Steven Stanley and Geerat Vermeij suggested **predation** as a major factor in the rise of the Modern Fauna, in what Vermeij has called the **Mesozoic Marine Revolution**. The new predators that appeared in the

Middle Cretaceous seem to have been more effective than their predecessors at attacking animals on the seafloor. Modern gastropods evolved, capable of attacking shells with drilling radulae backed with acid secretions and poisons. Advanced shell-crushing crustaceans became abundant, as did bony fishes with effective shell-crushing teeth. Perhaps around this time the filter feeders of the Paleozoic Fauna, which were largely fixed to the open surface of the seafloor, became too vulnerable to predation. They were replaced by animals that can filter food from seawater that they pump down into their burrows. Burrowing bivalves with siphons and burrowing echinoids make up very important components of the Modern Fauna, together with effective, wide-roaming predators such as gastropods and fishes.

Another variant is the **bulldozer hypothesis**, mostly the work of Charles Thayer. Thayer pointed out a major difference between the Modern Fauna and the Paleozoic Fauna: the relative abundance and diversity of strongly burrowing forms (especially worms, echinoids, crustaceans, and bivalves) in modern soft-sediment communities. Their continual churning of sediment means, among other things, that fixed filter feeders find it difficult to attach as larvae, continually run the risk of being overturned as adults, and are at least occasionally subjected to clouds of disturbed sediment that tend to clog their filters. Most filter feeders that are successful on soft sediments today are mobile forms, and fixed filter feeders are confined to the relatively restricted habitat of hard substrates. Most Paleozoic filter feeders were immobile, and they have not been able to compete successfully with the Modern Fauna even though the modern world is relatively oceanic and encourages filter feeders.

Thayer's work has been criticized on the grounds that there were powerful burrowing animals in the Paleozoic, but this criticism is not valid. Many elements of the Modern Fauna were present in the Paleozoic, but they were not dominant, and their bulldozing was limited. We need to explain that fact, of course, but the bulldozer hypothesis is not affected. The predation hypothesis has run into similar difficulties. For example, occasional shell boring by gastropods has now been recorded as far back as the Devonian, but that does not affect the general validity of the hypothesis.

The predation effect and the bulldozer effect could both have operated. Bulldozing and predation are both reasonable explanations of the failure of the Paleozoic Fauna to recover significantly in the Mesozoic, and both could help to explain the diversification of the Modern Fauna in the later Mesozoic.

Energy, or Nutrients?

If one were to seek some kind of overall change in world ecology that favored both bulldozing and predation in the later Mesozoic and not before, one might turn to a suggestion by Richard Bambach. In 1993 he proposed that the additional energy pumped into marine ecosystems by runoff from the land, as it was covered first by advanced gymnosperms

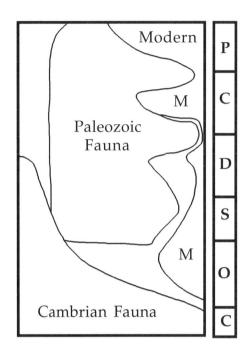

Figure 5.16 Sepkoski's interpretation of Figure 5.15 is based on data that allow this more precise but more complex view of the Paleozoic history of the three great Faunas. The Paleozoic Fauna was not clearly driven offshore until its virtual extinction in the Permian. Indeed, it very nearly drove the Modern Fauna out of the inshore habitat altogether in Late Devonian and Late Carboniferous times. The diversity patterns in Figures 5.12, 5.13, and 5.14 are not closely linked with the habitat patterns, either. It seems once again that the Real Story (when we discover it) is not going to be simple, or even elegant. My diagram is based on data from Sepkoski and Miller, 1986.

and then by angiosperm floras, made it possible to support more complex animals and ecosystems in such diversity.

Geerat Vermeij favors a more general version of this line of reasoning. Most of the world's primary productivity (photosynthesis) occurs in the surface waters of the ocean, and it is likely that variation in ocean productivity through time has been just as significant as the additional run-off from the land. Vermeij argues that variations in the nutrient content of the ocean may have been very important. Nutrients could be high during periods of increased volcanic activity, for example, those that occur as ocean-floor spreading increases as continents separate. In Vermeij's view, this may be the connection between continental splitting and evolutionary radiation, and may be just as important as adding provinces or making environments more equable.

Are There Limits on Diversity? The Logistic Equation

Jack Sepkoski used his data to analyze how fast radiations and extinctions took place. Extinctions were often extremely rapid as well as extremely large. Radiations showed a timetable of their own, in a pattern that has recently been confirmed on an independent data base.

If a few fortunate flour beetles find their way into a very large jar of flour, their food seems unlimited at first, and they grow and breed rapidly. Eventually, however, the population rises to the point where there is some competition for food. The rate of increase slows down, and at some point the population must stop growing when the jar holds as many flour beetles as it can. The population growth in this example follows a mathematically predictable curve called the **logistic equation**.

It turns out that the curve of diversity of life follows a logistic path for most of the time. This may mean there is some limiting factor (presumably ecological), which places some sort of control on the diversity of life at any given time. This is surprising, because one can envisage a control on the *amount* of life more easily than one can explain a control on the *diversity* of life.

What's more, that control seems to change from time to time. Vincent Courtillot and Y. Gaudemer suggested that diversity rose and then levelled out along a logistic path from the beginning of the Cambrian until there was a dramatic extinction at the end of the Permian. After the Triassic period (in which there may also have been logistic growth of diversity followed by a smaller extinction), diversity again followed logistic growth from the end of the Triassic to the present.

If this is real, it implies that there was some sort of global control keeping diversity at a limit of around 460 families for most of the Paleozoic. The rules then changed, to impose a limit of perhaps 360 families during Triassic times, and then changed again to permit a rise to the present diversity of about 1100 families.

Of course, one wonders what that limiting factor might be. If it exists, it is global and ecological. Since we identify families of animals largely by their similar anatomy, we separate out a group of organisms that probably does much the same thing ecologically. There may be only so many ways of making a living in the oceans at any time, which might limit the number of families that can exist—if a new one evolves, chances are that an existing one goes extinct.

Sepkoski, and Courtillot and Gaudemer, agree that extinction events separate periods over which the logistic curves of diversity are different. At various times, Sepkoski has identified such extinction events at the end of the Ordovician, Permian, Triassic, and Cretaceous. Courtillot and Gaudemer think that the Ordovician extinction can be ignored. Either way, these analyses imply that extinction events change the ecological rules of the world ocean. Extinction events are important enough to justify a new Chapter.

Review Questions

Review the logic behind Figure 5.6. Can you improve on that planetary design? if so, explain how your model is better.

THOUGHT QUESTION.—Suppose Earth never had any seasons (it spun straight up on its axis). Think of all the features of life on Earth that seasons affect, and list ways in which the fossil record would be different.

Further Reading

Easy Access Reading

Benton, M. J. 1995. Diversification and extinction in the history of life. *Science* 268: 52–58. Benton uses an independent data base to confirm most of the patterns seen by Sepkoski.

Courtillot, V., and Y. Gaudemer. 1996. Effects of mass extinction on biodiversity. *Science* 381: 146–148.

Erwin, D. H. 1993. *The Great Paleozoic Crisis: Life and Death in the Permian.* New York: Columbia University Press.

Murphy, J. B., and R. D. Nance. 1992. Mountain belts and the supercontinent cycle. *Scientific American* 266 (4): 85–91.

More Technical Reading

Bambach, R. K. 1993. Seafood through time: changes in biomass, energetics, and productivity in the marine ecosystem. *Paleobiology* 19: 372–397.

Smith, A. G., et al. 1980. *Phanerozoic Palaeocontinental World Maps.* Cambridge: Cambridge University Press. A series of maps through time for all continents, now dated.

Sepkoski, J. J. 1981. A factor analytic description of the Phanerozoic marine fossil record. *Paleobiology* 7: 36–53.

Sepkoski, J. J. 1991. A model of onshore-offshore change in faunal diversity. *Paleobiology* 17: 58–77.

Sepkoski, J. J. 1993. Ten years in the library: new data confirm paleontological patterns. *Paleobiology* 19: 43–51.

Tevesz, M. J. S., and P. L. McCall (eds.). 1983. *Biotic Interactions in Recent and Fossil Benthic Communities.* New York: Plenum. Valuable papers by Sepkoski and Sheehan, by Thayer, and by Vermeij.

Valentine, J. W. 1973. *Evolutionary Ecology of the Marine Biosphere.* Englewood Cliffs, N. J.: Prentice-Hall. A pioneering text that laid out and explained the connections between plate tectonics and global paleobiology.

Vermeij, G. J. 1987. *Evolution and Escalation.* Princeton: Princeton University Press. Stresses the importance of predator/prey interactions in evolution. A fine set of essays, with enormous lists of references.

Ziegler, P. A. 1989. *Evolution of Laurussia.* Kluwer: Dordrecht, Holland. Great maps.

CHAPTER SIX

Extinction

Extinction happens all the time. Some species have small populations that depend on a particularly narrow range of food, or habitat, and are vulnerable to even small-scale ecological disturbance. Strictly speaking, if the world has fairly even diversity through time, existing species will become extinct about as often as new species evolve. Extinction is the expected fate of species, rather than a rarity.

However, there are time of extremely rapid extinction, just as there are times of extremely rapid evolutionary radiation. Sepkoski's data of diversity through time show dramatic drops in diversity (Figures 5.7, 5.11). Times of rapid extinction require us to ask whether some special extinction mechanism is at work, exactly the reverse of the questions that are raised by great evolutionary radiations.

We have to proceed in two steps. First, we have to ask whether there really are times of unusually rapid extinction. Then we have to identify the causes for those special events. Were they processes that were just extreme examples of normal (ordinary) Earth processes, or were they catastrophic events that were truly extraordinary?

Mass Extinctions

David Raup and Jack Sepkoski analyzed Sepkoski's data on the fossil record (Chapter 5). They identified extinction events that seemed to be very large, large enough to be called **mass extinctions** (Figure 6.1).

Figure 6.1 Raup and Sepkoski (1986) plotted this graph of extinction rates through time and picked out events in which extinction rates seem to be much higher than background rates: these are then identified as **mass extinctions**. Six of the mass extinctions have been recognized empirically for decades: at the end of the Ordovician; at the end of the Devonian; at the end of the Permian (a double event?); at the end of the Triassic; and at the end of the Cretaceous—the K–T event. The much smaller event late in the Jurassic is not very convincing.

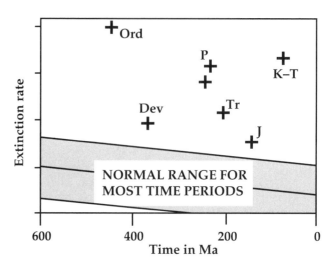

Six mass extinctions have been recognized in a nonquantitative way for decades, and they and others have been studied in some detail:

- At the end of the Ordovician

- At the end of the Frasnian stage of the Late Devonian ("F-F")

- At the end of the Permian period, ? a double event ("P-Tr")

- At the end of the Triassic period

- At the end of the Cretaceous period ("K-T")

These six mass extinctions are alike in some ways. They all represent the extinction of a significant component of global faunas and/or floras, and they are all relatively sudden in geological terms—that is, they occurred over a few million years at most, and in some cases much less than that.

Recovery

Extinctions are followed by recovery, and the patterns of biological and ecological recovery are a new field of study. The larger the extinction, the longer the recovery time, and the more the global ecosystem changes across the event. The rise of the Modern Fauna begins in the Triassic, and the rise of mammals begins in the Paleocene. Groups sometimes show amazing radiations: all living sea-urchins are descended from two Paleozoic genera that survived the P–Tr extinction, for example.

During recoveries we sometimes see **Lazarus taxa**: plants or animals that disappeared at the extinction and are missing from the fossil record for a long time, sometimes millions of years. Corals, for example, are absent from the fossil record of the Lower Triassic, after the P–Tr extinction, and re-appear in the Middle Triassic. The existence of Lazarus taxa implies that somewhere there must have been refuges for scattered survivors; only when they radiate away from the refuges do they appear again in the rocks that we have collected.

Explaining Mass Extinction

Mass extinctions were global phenomena, so they have to be explained by global processes. The first that comes to mind is plate tectonics. Plate tectonic changes could alter global diversity patterns (Chapter 5). However, tectonic changes are relatively slow in geological terms, while mass extinctions stand out because they are relatively sudden. If we are to use plate-tectonic explanations, we have to add a plausible mechanism for a trigger that suddenly fires a deadly "extinction bullet."

If plate-tectonic arguments (with trigger mechanism) do not work, we must look to more extreme suggestions. Some plausible extinction agents that have been examined at various times are:

• A failure of normal ocean circulation affects ocean chemistry enough to cause global changes in climate and atmosphere.

• A rapid change in sea level affects global ecology and climate.

• An enormous volcanic eruption.

• An extra-terrestrial impact by an asteroid.

Only the last is extra-terrestrial in nature. The K-T mass extinction certainly occurs at the same time as a major asteroid impact, so I shall discuss the evidence for impact, and its likely consequences, in Chapter 18. For the purposes of this Chapter, it is important to remember that an asteroid impact large enough to cause a global impact should leave behind physical evidence in the form of a sharply defined layer or **spike** of the element **iridium**, and/or impact-generated glass spherules called **tektites**, and/or quartz crystals with characteristic "shock marks," in the rocks associated with the extinction.

I shall deal with mass extinctions in time order, except for the largest of them all, the Permo-Triassic (P-Tr) extinction, which deserves a more extended discussion of its own.

The Ordovician Mass Extinction

The mass extinction at or near the end of the Ordovician has no iridium spike associated with it, and seems to be closely linked with a major climatic change. A first pulse of extinction happened as a big ice age began, and the second occurred as it ended. Some paleontologists feel that as we collect fossils from more regions of the world, this "mass extinction" may turn out to have been a comparatively minor event.

The Late Devonian (F-F) Mass Extinction

A mass extinction took place, probably in several separate events, at the boundary between the last two stages of the Devonian, the Frasnian and

Famennian (the F–F boundary). There was a major worldwide extinction of coral reefs and their associated faunas, and many other groups of animals and plants were severely affected too. Iridium anomalies, shocked quartz, and glass spherules are reported from China and Western Europe at or near the F–F boundary. However, there are also indications of climatic changes, and major changes in sea-level and ocean chemistry, at the same time. Carbon isotope shifts indicate that global organic productivity was changing rapidly before the boundary.

George McGhee favors an impact scenario. There are three cautions. First, McGhee makes it clear that he always has favored this scenario; second, the geological evidence requires that there were several closely-spaced but medium-sized impacts over perhaps two or three million years, rather than the one tremendous impact that occurred at the K–T boundary; and third, the evidence is incomplete in terms of our understanding of timing, of world geography at the time, and there are difficulties in going from evidence to interpretation. Kun Wang suggests that global ecosystems were already stressed when an impact occurred. There is no "magic marker" of impact phenomena at the extinction event, as there is at the K–T boundary, and that makes the F–F boundary very difficult to work with.

The end-Triassic Mass Extinction

The Triassic–Jurassic boundary marks a turnover of groups on land and in the sea. In 1999 a team of geologists reported that a gigantic eruption took place around this time, in a volcanic episode that marked the first major plate tectonic activity that began to split the Atlantic Ocean (Chapter 5). Critics pointed out that the eruption seemed to have occurred slightly after the boundary, so could not have caused any event at the boundary. To complicate the issue, shocked quartz has been discovered very near the boundary in northern Italy (the large Manicouagan impact crater in Québec, Canada, was formed at some uncertain date around this time). Several craters dated to the late Triassic line up, as if they had been formed by an incoming body that fragmented at the last moment. However, it's not clear whether the craters were formed simultaneously, or if so, whether they were formed at the same time as the extinction. It's not clear that the extinction was large. More research is needed: in fact, it's not clear what the evidence is, let alone the answer!

The Eocene-Oligocene Extinction

An extinction at the Eocene–Oligocene boundary affected marine organisms more than terrestrial animals. It occurred at a time of major climatic change, not obviously linked with any single major impact or any major volcanic event. There were impacts around this time. One struck the east coast of the United States, scattering tektites from a crater that now lies under Chesapeake Bay. Another struck central Siberia, with debris scattered perhaps to Europe. These were clearly separate events. Although it is premature to link them to the extinctions, impacts are not ruled out as contributing to the E-O event.

The Permo-Triassic (P-Tr) Extinction

The extinction at 250 Ma, the end of the Permian, is the largest of all time: the "Mother of Mass Extinctions" according to Douglas Erwin. The extinction was used by John Phillips 150 years ago to define the end of the Paleozoic Era and the beginning of the Mesozoic (Figure 5.8). An estimated 57% of all families and 95% of all species of marine animals became extinct. The Paleozoic Fauna was very hard hit, losing especially suspension feeders and carnivores, and almost all the reef dwellers. The Permo-Triassic (P–Tr) extinction is a major watershed in the history of life on Earth, especially for life in the ocean; the K–T extinction is small in comparison (Figure 5.11).

The P–Tr extinction was rapid, probably taking place in less than a million years, and possibly much faster than that. Although it was much more severe in the ocean, it affected terrestrial ecosystems too. A prolific swamp flora in the Southern Hemisphere had been producing enough organic debris to form coals in Australia, but the coal beds stop abruptly at the P-Tr boundary. No coal was laid down anywhere in the world for at least 6 million years afterward. A large change in carbon isotopes occurred across the P–Tr boundary, which signifies an important and global drop in photosynthesis that lasted a long time.

There is no evidence for an impact at the P–Tr boundary. The continental collisions that formed Pangea in the Permian would account for a major drop in diversity (Chapter 5) but not for a sudden, enormous mass extinction. Perhaps most important of all, the Permian extinction coincides with the largest known volcanic eruption in Earth history.

The Siberian Traps

In addition to plate tectonics (Chapter 5), the Earth also has **plume tectonics**. Occasionally, an event at the boundary between the Earth's core and mantle sets a giant pulse of heat rising toward the surface as a **plume**. As it approaches the surface, the plume melts the crust to develop a flat head of basalt magma that can be 1000 km across and 100 km thick. Penetrating the crust, the plume generates enormous volcanic eruptions that pour hundreds of thousands of cubic kilometers of basalt— **flood basalts**—out over the surface. If a plume erupts through a continent, it blasts material into the atmosphere as well. After the head of the plume has erupted, the much narrower tail will continue to erupt for 100 m.y. or more, but now its effects are more local, affecting only 100 km or so of terrain as it forms a long-lasting **hot spot** of volcanic activity.

Plume events are rare: there have been only eight enormous plume eruptions in the last 250 m.y. The most recent is the Yellowstone plume:

at about 17 Ma it burned through the crust to form enormous lava fields that are now known as the Columbia Plateau basalts of Oregon and Washington, best seen in the Columbia River gorge. North America drifted westward over this "hot spot," which continued to erupt to form the volcanic rocks of the Snake River plain in Idaho (Valley of the Moon and so on), and it now sits under Yellowstone National Park. The hot spot is in a quiet period now, with geyser activity rather than active eruption, but it produced enormous volcanic exposions about 500,000 years ago that blasted ash over most of the mountain states and into Canada.

A massive plume eruption took place exactly at the P–Tr boundary. A new plume burned through the crust in what is now western Siberia to form the "Siberian Traps," gigantic flood basalts that cover 2.5 million sq km in area and are perhaps 3 million cu km in volume. The eruptions coincided exactly with the P–Tr boundary, at 250 Ma, and lasted at full intensity for only about a million years—the largest known, most intense eruptions in the history of the Earth. They lie across the P–Tr boundary and were formed in what was obviously a major event in Earth history.

There is a feeling, particularly among physical scientists, that if we can show that a physical catastrophe occurred at a boundary, we have an automatic explanation for an extinction. But this connection has to be demonstrated, not just assumed.

In 1993 Douglas Erwin felt obliged to suggest the "Murder on the Orient Express" hypothesis for the P–Tr extinction: that is, many factors, all acting together, led to the extinction. This is not a particularly "clean" hypothesis to accept or to test. However, we have six more years' research now, by Erwin and others, and we can do better.

Any explanation of the P–Tr event must account for the severity of the extinctions. The size of the P–Tr extinction is the largest in Earth history—but so is the size of the Siberian Traps eruption. In 1995 Paul Renne and his colleagues suggested a scenario based on the eruption as a primary cause of the extinction.

The plume rises toward the crust and erupts. The tremendous amount of sulfate aerosols would cool the climate enough to form ice-caps, rather quickly, and this in turn would cause a rather rapid drop in sea level along with global cooling, early in the eruptive sequence. In the rock record, we would expect to see changes in carbon and sulfur isotopes, and we do. Furthermore, as the plume erupts, the crust would be raised by the buoyant magma, perhaps enough to form a land footing for the large continental ice sheets that would grow in these high latitudes. Finally, as the eruption dies off, the crust would subside and the aerosols would disperse, making for a rapid end to the volcanically induced glaciation and another rapid change in climate. It is possible, but not calculated yet, that the volcanic gases that had built up during the eruption could have had a greenhouse effect for some time after the eruption ended, taking the earth from a volcanic glaciation to a volcanic hothouse.

Add to this scenario the "usual" effects of a giant eruption, such as acid rain, ozone depletion, a massive dose of carbon dioxide into the atmosphere, or any combination of the above, and the ingredients are in place for a mass extinction.

Though it is easy to imagine that a giant eruption might have caused a catastrophe at the P–Tr boundary, it is not certain that it would. We do not know how much dust, smoke, and aerosols would be produced, even though it is absolutely critical to calculations of their effects that we know those factors rather precisely. We do not know how far volcanic aerosols and stratospheric dust would be carried over the Earth, or in detail what effects they would have. Dust and aerosols in the air can help absorb solar heat rather than reflect it, for example, and some models suggest that parts of the Earth would warm, parts would cool, and parts would stay at about the same temperature.

The most persuasive scenarios of volcanic extinction are quickly summarized. Even a short-lived catastrophe among land plants and surface plankton at sea would drastically affect normal food chains. Large animals would have been vulnerable to food shortage, and their extinction after a catastrophe seems plausible. In the oceans, invertebrates living in shallow water would have suffered greatly from cold or frost, or perhaps from CO_2-induced heating. High-latitude faunas and floras in particular were already adapted to winter darkness, though perhaps not to extreme cold. Thus, tropical reef communities could have been devastated, but high-latitude communities could have survived much better.

These general patterns are observed at the P–Tr boundary, though high-latitude floras were affected worse than one would have predicted. Even so, the patterns do not prove that the eruptions caused the extinction: other factors could have been at work too. Some specific evidence shows that eruptions do not necessarily cause catastrophes. For example, the eruption of Krakatau in 1883 destroyed all life on the island and severely damaged ecosystems for hundreds of miles around. But those ecosystems have completely recovered 100 years later, in a geologically insignificant time. There's no biological trace of the much larger eruption of Toba, 75,000 years ago. No North American extinctions coincided with explosive eruptions from Long Valley caldera, California, from Crater Lake, Oregon, or from Yellowstone, all of which blew ash as far as Canada within the last million years.

Other major plume eruptions are not linked with extinctions: examples include the Jurassic Karroo Basalts of South Africa and the Miocene Columbia Plateau Basalts.

However, one should beware of dismissing catastrophic explanations because small events do not trigger catastrophes. There may be a threshold effect: if the event is not big enough it will do nothing, but if it is big enough it will do everything. Perhaps only two eruptions in the last 500 m.y. were large enough to cause a mass extinction, at the P-Tr boundary and perhaps also the K-T boundary (Chapter 17).

In short, we don't yet know whether an eruption would have catastrophic, severe, or only mild biological and ecological effects, or whether those effects would be local, regional, or global. In any scenario, however, the killing agent is transient: it would have operated for only a short time geologically. Clearly, if such events occur, they are rare. That does not make them impossible, only unlikely. And that means they have to be very persuasive indeed before we accept them!

A gigantic eruption might have caused the P–Tr extinction by inducing climatic changes near and at the boundary. But climatic changes can be caused by more normal agents such as geographic change, and perhaps these could have caused the extinctions. In particular, the Earth's climate is driven dominantly by oceanic factors. An eruption at the P–Tr boundary could have added to a climatic change that was already happening. We should look at some of the data.

In the Late Permian, the continents were clustered together in the supercontinent Pangea (Figure 5.9), which means also that there was a giant ocean, **Panthalassa**, which covered 70% of the Earth's surface. We understand modern ocean circulation rather well, but that does not mean that we can predict how Panthalassa would have worked.

Yukio Isozaki studied sediments from Japan and Canada which were laid down on the deep-sea floor of Panthalassa. Normal oceanic conditions (as we understand them today) deteriorated around 260 Ma, as the deep ocean water became anoxic. Yet the surface waters remained normal: they supported abundant radiolarians, whose bodies were deposited as chert on the sea floor. However, after about 255 Ma the radiolarians become rarer toward the P-Tr boundary, and across the boundary there are none at all, suggesting that Panthalassa became anoxic right to the surface. Long after the extinction, around 245 Ma, the surface waters again became oxygenated enough to support radiolarians, but the deep waters were anoxic until about 240 Ma.

If Panthalassa became anoxic right to the surface, this in itself would cause a catastrophic extinction of marine organisms. The side effects of a marine extinction on atmosphere and climate, not yet spelled out, would in turn have affected land plants and animals.

In Isozaki's model, the extinction is linked with a chemical crisis in the waters of Panthalassa that is symmetrical in time. There is no particular trigger associated with the Siberian Traps eruptions. Instead, the crisis occurs in stages, as he sees it. The first, in which the deep ocean becomes anoxic, can be associated with a major extinction event at the end of the Guadelupian stage of the Late Permian. The second, with full oceanic anoxia, coincides with the P–Tr boundary. The oceanic crisis resolves itself in stages too, with full recovery taking about 10 m.y.: certainly it took this long before reef-dwellers appear again in the Triassic fossil record.

Isozaki admits that the ocean features he infers do not fit our understanding of the way our modern oceans work. But, he argues, he has the data to make his statements, and the fact that we cannot explain how they occurred simply reflects the fact that we do not understand how Panthalassa worked as a superocean. Isozaki also notes that the rapid changes in carbon isotopes and in sea level across the P–Tr boundary itself are not explained by his data: something else is going on at the boundary. (And that something else looks much more sudden.)

In 1996 Henk Visscher and his colleagues reported extreme abundances of fossil fungal cells in land sediments at the P–Tr boundary. There are hints that the fungi-enriched "layer" is the record of a single, world-wide crisis, with the fungi breaking down massive amounts of vegetation that had been catastrophically killed (there were no termites yet). Such a fungal layer is unique in the geological record of the past 500 m.y. The best evidence we have suggests that there were major extinctions among gymnosperms, especially in Europe, and among the coal-generating floras of the Southern Hemisphere. The vegetation of the early Triassic in Europe looks "weedy," that is, invasive of open habitats.

Andrew Knoll and his colleagues have suggested that the extinction was caused by a catastrophic overturn of an ocean supersaturated in carbon dioxide. This would result in tremendous, close to instantaneous, degassing that would roll a cloud of (dense) carbon dioxide over the ocean surface and low-lying coastal areas. An analog might be the recent catastrophic degassing of Lake Nyos, in the Cameroon, where hundreds of people were killed as carbon dioxide degassed from a volcanic lake and cascaded down valleys nearby. The difference is that the proposed P–Tr disaster was global.

In this scenario, the carbon dioxide build-up results from the global geography that included the gigantic ocean Panthalassa. Knoll and colleagues speculated that the abnormal ocean circulation in Panthalassa did not include enough downward transport of oxygenated surface water to keep the deep water oxygenated. With normal respiration and decay of dead organisms, the deep water evolved into an anoxic mass loaded with dissolved carbon dioxide, methane, and hydrogen sulfide. Carbon continued to fall to the sea floor from normal surface productivity, but it was deposited and buried because there was no dissolved oxygen to oxidize it. As carbon dioxide levels fell in the atmosphere, the earth and the surface ocean cooled. Finally, the surface waters became dense enough to sink, triggering a catastrophe as the CO_2-saturated deep waters were brought up to the surface, degassing violently. The event would trigger a greenhouse heating and a major climatic warming.

In 1998 Samuel Bowring and colleagues reported new data from China. They found that the carbon isotope change at the boundary was probably very short-lived: a "spike" only perhaps 165,000 years long. This suggests a major (catastrophic?) addition of non-organic carbon to the ocean, rather than just a failure in the supply of organic carbon. They suggested three possible scenarios. Two of them are variants of the Siberian Traps scenario above, except that in addition the climatic changes could have set off an overturn of Panthalassa and a carbon dioxide crisis. Their third suggestion is an asteroid impact, but there is no evidence for that at all.

Most recently, Greg Retallack and colleagues have found evidence in Australia that suggests a prolonged greenhouse warming set in right at the P-Tr boundary. Several paleoclimatic indicators suggest the same story, which implies that the role of carbon dioxide was the vital link between any environmental disasters and the extinctions. Carbon dioxide in the atmosphere could have been increased by volcanic eruptions, by oceanic turnover, and it would have been accentuated and prolonged if plants were killed off globally. (World floras and oceanic plankton would have to recover before the carbon dioxide could be drawn down out of the atmosphere.) We may be getting close to the answer here!

Repeated, Periodic Mass Extinctions?

Here, I want to dispose of an idea that has garnered a lot of publicity but should be dismissed. In addition to the very largest mass extinctions, Raup and Sepkoski identified smaller ones. They noticed that mass extinctions seemed to have occurred periodically, every 26.2 m.y. since at least the end of the Permian (Figure 6.2). The suggestion was followed immediately by a flurry of claims and counterclaims about cycles in the geological record. Others suggested that the extinction events had a 30 m.y. periodicity; or that the extinctions matched a periodicity of 28.4 m.y. for impact craters on Earth's surface, or that there were periodicities between 30 and 34 m.y. for crater ages, magnetic field reversals, plume eruptions, pulses of mountain-building, and other events.

Then matters got really out of hand. A group from UC Berkeley suggested that the periodic mass extinctions were caused by cometary impacts on Earth (cometary impacts leave no physical evidence!). These periodic cometary impacts were produced by **Nemesis the Death Star**, a small star that is a distant binary companion of the Sun. Every 28 m.y. or so Nemesis approaches close enough to disturb groups of comets that may orbit around the solar system outside the planets. The comets then bombard the inner solar system, including Earth. There is not a shred of evidence for the existence of Nemesis, however. We can't see it, so its supporters say that it must be small and dark. It is, they say, unfortunately (or conveniently) at the other end of its elliptical orbit now, too far away to detect easily.

Even if a Death Star once existed, its orbit would be so distorted by close passages to the Sun that any periodicity would disappear after a few return visits. So even if the extinctions are periodic, Nemesis could not explain them. Altogether, the Nemesis idea, though vivid, was never satisfactory. That didn't prevent it from spawning lots of publicity, several TV programs, two covers of *Time*, and at least five books!

An over-ambitious hypothesis

Called for a Death Star called Nemesis

But one asteroid blast

That occurred in the past

Didn't make for a general synthesis

Figure 6.2 Many of the mass extinctions identified by Raup and Sepkoski seem to occur in cycles of about 26 m.y.

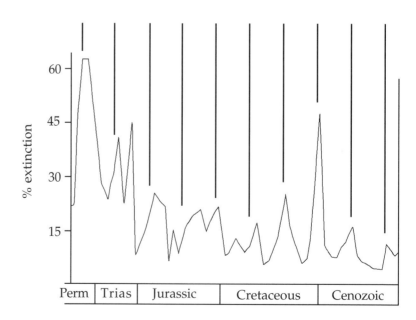

Problems with the Nemesis idea led almost immediately to the Planet X hypothesis. Planet X is a hypothetical planet supposedly orbiting somewhere outside Neptune. If Nemesis the Death Star cannot produce regular comet showers, perhaps X the Death Planet could shower the inner solar system with comets. Planet X must have substantial mass (as much as Earth, or more), yet IRAS, the infrared orbiting telescope, has not been able to find it. In any case, computer models show that Planet X cannot produce periodic showers of comets any more than Nemesis can.

While all the astrophysical arguments were going on, several statisticians argued that the extinction data, and all the other data on physical events in Earth history, contain no cyclic periodicity at all. The so-called cycles are artifacts caused by the wrong methods of statistical analysis, or by undue reliance on absolute dates that have been suggested for geological stages, or both. Most paleontologists are not convinced that the 26 m.y. cycles are real, and a recent analysis found no periodicity at all in a new compendium of data from the fossil record.

The Nemesis idea and its various offshoots are and always were weak scientific hypotheses. They were proposed to explain periodic extinctions that turn out not to have been periodic and therefore did not need special explanation in the first place. Each has relied on the supposed influence of astronomical objects for which there has never been any evidence. We do need to examine major extinction events carefully for evidence of impacts, but we should be very careful about ascribing those impacts to the effects of hypothetical stars, planets, or comets.

So far there is only enough evidence to fix a major impact at one of the mass extinctions in the fossil record, the K–T extinction, and perhaps to suggest another at the F–F boundary. Likewise, there is only enough evidence to connect giant flood basalt eruptions with two mass extinctions, at the P–Tr and the K–T boundary extinctions. It is clear, however, that the largest known impact and the largest known eruption coincide with undoubted mass extinctions. It would be amazing if that was a coincidence. These questions are still open!

Review Question

Construct a set of tables comparing the sizes, effects, and phenomena occurring at each of the "Mass Extinctions." Can you see any pattern? and if you do, write a proposal asking for the research money you would need to test your idea.

Further Reading

Easy Access References

Benton, M. J. 1993. Late Triassic extinctions and the origin of the dinosaurs. *Science* 260: 769–770. Status of the Triassic-Jurassic boundary question.

Benton, M. J. 1995. Diversification and extinction in the history of life. *Science* 268: 52–58. No sign of periodic extinctions in a new data set.

Bowring, S. A., et al. 1998. U/Pb zircon geochronology and tempo of the end-Permian mass extinction. *Science* 280: 1039–1045.

Bottomley, R. et al. 1997. The age of the Popigai impact event and its relation to events at the Eocene/Oligocene boundary. *Nature* 388, 365–368.

Erwin, D. H. 1993. *The Great Paleozoic Crisis: Life and Death in the Permian.* New York: Columbia University Press. Excellent: Erwin clearly labels his thoughts and inferences so that we can participate along with him as co-thinkers. Contains more geology and paleontology than the title suggests.

Farley, K. A., et al. 1998. Geochemical evidence for a comet shower in the Late Eocene. *Science* 280: 1250–1253.

Hut, P., et al. 1987. Comet showers as a cause of mass extinction. *Nature* 329: 118–126. If there's no evidence for asteroid impact, blame comet showers.

Isozaki, Y. 1997. Permo-Triassic superanoxia and stratified superocean: records from lost deep sea. *Science* 276: 235–238.

Quinlan, G. D. 1993. Planet X: a myth exposed. *Nature* 363: 18–19.

Raup, D. M. 1991. *Extinction. Bad Genes or Bad Luck?* New York: W. W. Norton.

Raup, D. M., and J. J. Sepkoski. 1986. Periodic extinction of families and genera. *Science* 231: 833–836.

Renne, P. R., et al. 1995. Synchrony and causal relations between Permian-Triassic boundary crises and Siberian flood volcanism. *Science* 269:1413–1416. Volcanic scenario.

Retallack, G. 1995. Permian-Triassic life crisis on land. *Science* 267: 77–80.

Scotti, J. V., et al. 1991. Near miss of the Earth by a small asteroid. *Nature* 354: 287–289, and comment, pp. 265–267.

Spray, J. G., et al. 1998. Evidence for a late Triassic multiple impact event on Earth. *Nature* 392:171–173. However, there are serious doubts about this one.

Stothers, R. B. 1984. The great Tambora eruption of 1815 and its aftermath. *Science* 234: 1191–1198.

Turco, R. P. et al. 1990. Climate and smoke: an appraisal of nuclear winter. *Science* 247: 166–176. Revised version of their original 1983 suggestion.

More Technical Reading

Clymer, A. K. et al. 1996. Shocked quartz from the late Eocene: Impact evidence from Massignano, Italy. *Geology* 24: 483–486.

Erwin, D. H. 1988. The end and the beginning: recoveries from mass extinctions. *Trends in Ecology & Evolution* 13: 344–349.

Goldsmith, D. 1986. *Nemesis: The Death-Star and Other Theories of Mass Extinction.* New York: Walker.

McGhee, G. R. 1996. *The Late Devonian Mass Extinction: the Frasnian/Famennian Crisis.* New York: Columbia University Press. McGhee covers all serious hypotheses, and favors multiple medium-sized impacts.

McLaren, D. J., and W. D. Goodfellow. 1990. Geological and biological consequences of giant impacts. *Annual Reviews of Earth & Planetary Sciences* 18: 123–171.

Molina, E., et al. 1993. The Eocene-Oligocene planktic foraminiferal transition: extinctions, impacts and hiatuses. *Geological Magazine* 130: 483–499.

Muller, R. A. 1988. *Nemesis: The Death Star.* New York: Weidenfeld and Nicolson. A pop book by a notable member of the Berkeley group.

Rampino, M. R., et al. 1988. Volcanic winters. *Annual Reviews of Earth and Planetary Science* 16: 73–99.

Retallack, G. J., et al. 1998. Search for evidence of impact at the Permian-Triassic boundary in Antarctica and Australia. *Geology* 26: 979–982. They did not find any.

Retallack, G. J., et al. 1999. Postapocalyptic greenhouse paleoclimate revealed by earliest Triassic paleosols in the Sydney Basin, Australia. *Bulletin of the Geological Society of America* 111: 52–70.

Rose, W. I. and C. A. Chesner. 1990. Worldwide dispersal of ash and gases from earth's largest known eruption: Toba, Sumatra, 75 ka. *Global and Planetary Change* 3: 269–275.

Sharpton, V. L., and P. D. Ward (eds.). 1990. *Global Catastrophes in Earth History. Geological Society of America Special Paper* 247. Papers by most of the participants in the controversy, many of them updating previous research.

Stanley, S. M. 1991. Delayed recovery and the spacing of major extinctions. *Paleobiology* 16: 401–414.

Visscher, H., et al. 1996. The terminal Paleozoic fungal event: evidence of terrestrial ecosystem destabilization and collapse. *Proceedings of the National Academy of Sciences* 93: 2155–2158.

Wang, K., et al. 1991. Geochemical evidence for a catastrophic biotic event at the Frasnian/Famennian boundary in south China. *Geology* 19: 776–779.

Wang, K. et al. 1997. Carbon and sulfur isotope anomalies across the Frasnian-Famennian extinction boundary, Alberta, Canada. *Geology* 24, 187–190. An impact occurred when ecosystems were already stressed.

The Early Vertebrates

Vertebrates dominate land, water, and air today in ways of life that combine mobility and large size (more than a few grams). Only arthropods (insects on land and crustaceans in the sea) come close to competing for these ecological niches. As vertebrates ourselves, we have a particular interest in the evolutionary history of our own species and our remote ancestors. It's hardly surprising that vertebrates should receive special treatment in this and almost every other book on the history of life.

It is easier for us to identify with vertebrates than with invertebrates. We can feel how ligaments, muscles, and bones work. We feed by using our jaws and teeth. We have sensory skin and good vision, and we sense vibrations in our ears. We walk, run, and swim. We have special bodily sensations as we thermoregulate, and we understand by experience the bizarre system we have for getting oxygen and circulating it around the body. All vertebrates share some of these systems, and many vertebrates have them all. On the other hand, most invertebrates have quite different body systems that are much more difficult for us to identify with and to understand.

Our familiarity with vertebrate biology helps to make up for the rarity of vertebrate fossils. Vertebrates are rare even today in comparison with arthropods or molluscs, and vertebrate hard parts are held together only by skin, muscles, cartilage, and ligaments that rot easily after death. Even bones crumble and dissolve rather easily once they lose the organic matter that permeates them in life. Land vertebrates in particular live in a habitat that offers little chance of preservation. Bones are scattered and destroyed rather than buried and sheltered by sediment. Only the special interest in vertebrates shown by professional and amateur fossil collectors alike has compensated for the intrinsic poverty of vertebrate preservation. By now we have a very good idea of the major events in vertebrate history.

Vertebrate Origins

Vertebrates must have evolved from invertebrates, which are simpler in structure and have a longer fossil record. Vertebrates, of course, have a backbone, a bony column that contains a nerve canal and a **notochord**. The notochord is a specialized structure that looks like a stiff rod of dense tissue. It is a more fundamental character than the backbone that surrounds it. It is a shared derived character that places vertebrates in the phylum **Chordata**, together with some soft-bodied creatures that have a notochord but do not have a bony skeleton.

By using the stiffness of the notochord, a chordate without a backbone can give its muscles a firm base to pull against, while retaining enough flexibility to allow a push against the water for efficient swimming. The notochord can store elastic energy that is released at the right moment to help swimming. I suspect that the evolution of the notochord, with this mechanism for energy storage and release, is the evolutionary novelty that promoted the success of soft-bodied chordates. It preceded by a long time the evolution of the bony skeleton of a typical vertebrate.

Urochordates and **cephalochordates** are two living groups of soft-bodied chordates that help to show us what a vertebrate ancestor might have looked like. Urochordates include **tunicates** (sea squirts), small boxlike creatures that live as adults in colonies fixed to the seafloor. But their larvae swim actively, using the notochord and muscle fibers in a tail-like structure that is lost when they settle as adults (Figure 7.1).

Cephalochordates (Figure 7.2) are marine creatures that feed by filtering small particles from seawater in a special body chamber that it also uses as a gill chamber for breathing. The notochord runs along the dorsal axis and is surrounded by packs of body muscle arranged in V-shaped chevrons. Alternate contractions of the muscle packs flex the body from side to side in a wave-like pattern that allows it to swim. Sensory organs at the anterior end of the notochord mark the position of a primitive brain. In most of these characters, cephalochordates are much like fishes, even to the pattern of V-shaped muscle fibers that is so obvious when one dissects a fish carefully in a laboratory or a restaurant.

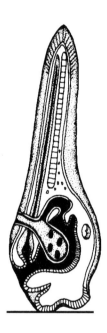

Figure 7.1 The larva of a urochordate has a notochord and swims freely. After a short time it attaches to the seafloor, as shown here, and metamorphoses into an adult that looks more like a sponge than a typical chordate. The adult is a sessile filter feeder. (From Barnes et al., *The Invertebrates: A New Synthesis.* © 1993 Blackwell Scientific Publications.)

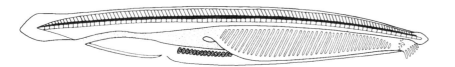

Figure 7.2 Above: *Branchiostoma*, a living cephalochordate, the amphioxus. (From Barnes et al., *The Invertebrates: A New Synthesis*. © 1993 Blackwell Scientific Publications.) Below: *Pikaia*, a fossil cephalochordate from the Middle Cambrian Burgess Shale. (Courtesy of the Palaeontological Association.)

Branchiostoma, the **amphioxus** (Figure 7.2), is a typical cephalochordate, a small animal that lives among sand grains. It can move actively between sand grains and in open water, squirming and swimming in vertebrate fashion with its muscle packs and notochord acting against one another.

These living animals are soft-bodied, so it once seemed very unlikely that we would ever have the fossil evidence to decide which one, if either, was the vertebrate ancestor. But the problem is now solved, because we have found fossil cephalochordates from the Burgess Fauna: *Cathaymyrus* among the Lower Cambrian fossils from Chengjiang, China, and *Pikaia* from the Middle Cambrian Burgess Shale itself. *Pikaia* is extraordinarily similar in size, shape, and anatomy to *Branchiostoma* (Figure 7.2). *Pikaia* is long, with a narrow but deep body; it has a notochord running along its dorsal axis, with V-shaped muscle packs surrounding the notochord in exactly the right arrangement to allow efficient swimming. Furthermore, the Cambrian age of the Burgess cephalochordates is reasonable for a vertebrate ancestor, because the earliest definite vertebrate is Late Cambrian and the earliest definite fish is Ordovician in age. From *Pikaia*, it would have required only some further evolution of the head to produce a primitive fish (Figure 7.3).

> An old flattened worm like Pikaia
> Shouldn't spark intellectual fire
>   But it's related to us!
>   Well now, what a fuss!
> Its relative importance seems higher.

Conodonts

Conodonts are microfossils, tiny tooth-like phosphate pieces that are so abundant and distinctive in marine rocks that they are used as guide fossils for subdividing geological time in Paleozoic rocks. We know now that conodonts were carried by soft-bodied wormlike creatures. After much controversy, the conodont animal was recognized as a chordate. The earliest conodonts are Late Cambrian in age and are therefore the first fossil chordates, but two things are clear. First, we will never know much about conodont biology, however much we know about the phosphate pieces they carry; second, conodonts are not the ancestors of any other vertebrate group.

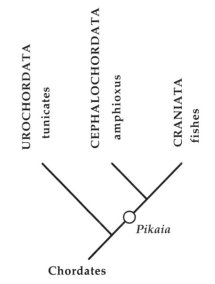

Figure 7.3 A phylogram showing the evolution of fishes from primitive chordates. The Cambrian fossil *Pikaia* has the morphology to be the ancestor of the living amphioxus as well as of all fishes.

Phosphate

The notochord probably evolved as a structure that aided in swimming. The physics of hydrodynamics dictates that swimming efficiency increases with body length. As early chordates explored various ways of life, the more actively swimming species may perhaps have increased in body size. But there must be a body size at which efficient swimming requires more stiffness than a notochord can give. Some kind of cartilaginous or mineralized skeleton then becomes a cheap way of increasing efficiency. (I used the same argument in Chapter 4 in discussing the evolution of hard parts in arthropods.)

In retrospect, one obvious solution would have been the construction of a backbone to stiffen the notochord. But apparently that was not the path that evolution took among the early chordates: Ordovician fishes did not have a bony internal skeleton. Instead, they had mineralized plates that covered some or all of their bodies, adding stiffness and giving rise to the term **ostracoderm** ("plated skin") for these fishes (Figure 7.4). In the same way, sharks today lack an internal bony skeleton but instead have a tough skin with strong fibers that stiffen the body considerably. The plates of ostracoderms would have provided protection too, from possible predators and from abrasion by sand and rock surfaces.

Why did vertebrates use *phosphate* as their hard-part mineral, rather than a more common mineral such as calcite? We, and living fishes, and all other vertebrates can produce powerful bursts of energy, but to do that we break down sugars in our muscle cells faster than we can absorb the oxygen required to do so. In this process (called **anaerobic glycolysis**) we build up an **oxygen debt** as we break down sugar without oxygen. The price we pay is that lactic acid builds up in the blood, making us feel tired, often painfully so. As we pay back the oxygen debt, blood acid levels fluctuate a lot, and calcium may be dissolved temporarily out of the calcium phosphate of our bones. An animal with a calcium *carbonate* skeleton couldn't handle such blood chemistry, because too much of the skeleton would dissolve. That's why living vertebrates have phosphate rather than calcite skeletons. John Ruben suggested that that's also why the first vertebrates had phosphate skeletons. This implies that they needed at least occasional bursts of energy.

Phosphate is a necessary nutrient for living animals, but it is often in short supply in nature. External plates in early fishes may therefore have given them an emergency phosphate supply as well as improving locomotion. Phosphate is dense, however, and a heavy plated coat makes a fish relatively clumsy, with slow acceleration. Thus the earliest fishes must have swum slowly near the bottom, inside heavy phosphate boxes.

All the earliest fishes are found in marine rocks, suggesting that fishes evolved in the sea. Though some isolated plates from Late Cambrian rocks may be the earliest fishes, *Astraspis* from the Ordovician of Colorado is the best-preserved early vertebrate (Figure 7.4), along with *Sacabambaspis* from the Ordovician of Bolivia. A headshield protected the anterior nerve center (which from this point can be called a **brain**), and

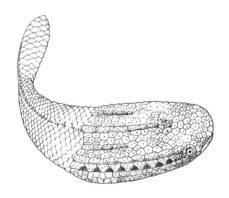

Figure 7.4   An ostracoderm. *Astraspis*, the best-preserved Ordovician fish, from the Harding Sandstone of Colorado. Above: A major piece of the headshield. Below: A reconstruction of the entire fish, which was about 13 cm, or 5 inches, long. (Courtesy of David K. Elliott, Northern Arizona University.)

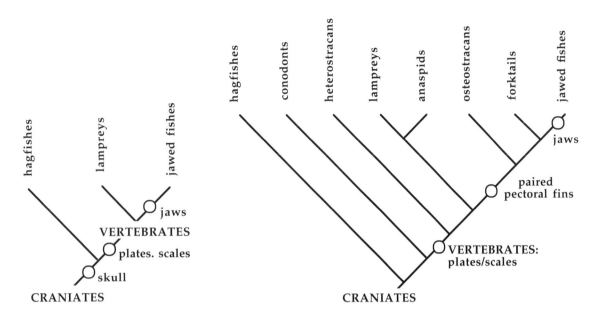

Figure 7.5 Two cladograms showing the relationships between groups of jawless fishes. The cladogram of living groups (left) gives limited information about their history. When extinct groups are included (right), the cladogram shows that they are only a remnant of the former diversity of jawless fishes. (The cladogram on the right is simplified from Janvier, 1996, and Young et al., 1996.)

also provided a stout nose cone for cutting through the water without flexing, and for probing into soft sediment. Behind the eyes were plates with openings to allow water to flow out past the gills. The tail was short, stubby, symmetrical, and small, and these fishes probably swam well but not fast.

None of the early fishes had jaws that opened and closed on a hinge. Instead, they were all jawless or **agnathan**: they had a simple slit or opening somewhere near the front and must have fed on small, easily digested objects.

The few jawless fishes living today (hagfishes and lampreys) have parasitic or otherwise strange ways of life, and they cannot tell us much about the ecology or evolution of their earliest ancestors. Hagfishes have a skull but no bony vertebrae, so are thought to be more primitive than the lampreys and jawed fishes, which are thus the real vertebrates. (Hagfishes therefore count as **craniates** but not vertebrates). Two simplified cladograms show the relationships of the most primitive fishes (Figure 7.5). The living jawless fishes are survivors from a much broader range of vertebrates.

Heterostracans

Heterostracans were the earliest abundant fishes (in the Silurian and Devonian). They are not directly descended from *Astraspis* (Figure 7.4) or *Sacabambaspis*, but probably from some relative of one of them. Heterostracans had flattened headshields with eyes at the side, and they look well adapted for scooping food off the seafloor (Figure 7.6). Some had plates around the mouth that could have been extended out into a shoveling scoop. The rigid head and the stiff, heavy-plated body imply that propulsion came mainly from the tail in a simple swimming style, with none

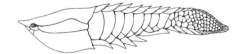

Figure 7.6 Cyathaspids were the earliest abundant heterostracan fishes. They were adapted to scoop food from the seafloor, so the mouth lay on the under side of the head shield. This cyathaspid is *Anglaspis*. (After Kiaer.)

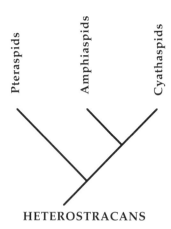

HETEROSTRACANS

Figure 7.7 Cladogram of the major groups of heterostracans. The three major groups evolved largely around different continental coastlines. Simplified from Janvier, 1996.

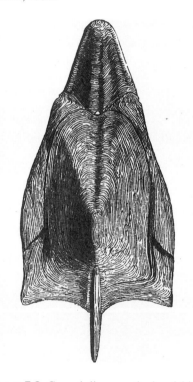

Figure 7.8 Growth lines on the headshield of a pteraspid show how it grew from several independently forming units. (From Lankester.)

Figure 7.9 A reconstruction of *Pteraspis* itself shows the strong plated headshield and the more flexible trunk and tail. (After White.)

of the control surfaces provided by the complex fins of modern fishes. Nevertheless, the heterostracan way of life was successful, and their fossils are found all across the Northern Hemisphere.

Heterostracan fishes diverged quickly in Silurian times into three major groups (Figure 7.7), and they became mechanically more efficient in their movement (swimming, in this case) as they evolved.

The earliest abundant heterostracans, **cyathaspids,** had stout headshields made of a few large plates (Figure 7.6). They were slightly flattened in front view, with a mouth just below the midline and a wide nostril opening above the mouth. Cyathaspids were particularly successful in shallow marine environments in Canada, which was then in a tropical climate. Late in the Silurian, continental pieces collided to form a new, large northern continent, **Euramerica**. As heterostracans colonized its shores and radiated inland into brackish and freshwater environments in Devonian times, they evolved into **pteraspids**. Those heterostracans that reached the island continent of **Angaraland** (now Siberia and Central Asia), radiated into an incredible array of body types as **amphiaspids**.

Pteraspids were most abundant in Devonian times. They had beautifully streamlined armored headshields formed from several plates that each had complex shapes formed by nonuniform growth on their edges (Figure 7.8). The headshield had a sharp nose cone and a smooth curved shape that gave an upward motion to counteract the density of the armor. A spine projected backward over the lightly plated trunk, partly for protection and partly for hydrodynamic stability. Pteraspids had tails with the lower half longer than the upper; other things being equal, this too would have helped the fish to counteract the weight of the headshield (Figure 7.9). The mouth lay under the head, and a ventral plate covered the gills. Water was taken in through the mouth, and the exit passages were neatly tucked toward the back of the headshield, much like the exhausts of a twin-jet fighter aircraft (Figure 7.8, 7.9). In some forms the headshield was very flattened, for gliding through water as a delta-wing aircraft glides through air. Reconstructions of pteraspid and amphiaspid gill systems look like the exhaust systems of 1930s racing cars, sharing their design for efficient passage of fluids (Figure 7.10).

Amphiaspids brought the heterostracan design to new levels in Angaraland. The headshield was almost entirely one unit, which provided more protection but meant that it could not be secreted until the young fish reached final size. On the other hand, the final shape could be more

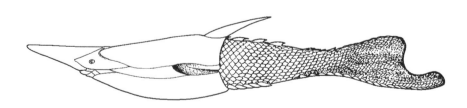

versatile because it didn't depend on modifying a set of plates through several growth stages. The mouth was usually at the front of the headshield, with the nostrils set a little higher and to the side. The eyes were small and far forward, and some amphiaspids were blind (Figure 7.11). Many amphiaspids were well adapted for movement in open water in three dimensions, and they had delta-wing versions too. Other body shapes were apparently adapted for swimming close under the water surface, sucking plankton from it through a mouth extended into a long tube (Figure 7.11). Bottom dwellers also evolved long snouts, which they may have used for probing into the sediment for worms.

In all this successful evolution, heterostracans never evolved fins. Their swimming power came entirely from the trunk and tail, with perhaps a little help from the gill exhaust (Figure 7.10).

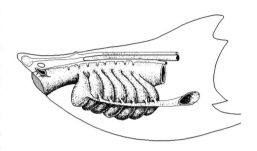

Figure 7.10 A diagram simplified from an internal reconstruction of amphiaspid gills by Larisa Novitskaya, showing their similarity to the exhaust system of an old supercharged multicylinder racing car of the 1930s.

Osteostracans

Other things being equal, any swimming creature would benefit by evolving powerful swimming and better maneuverability. We have seen this already among the heterostracans. A completely new innovation came with the evolution of paired fins, leading to a Late Silurian radiation of new jawless fishes, the **osteostracans**. Osteostracans were like heterostracans in that they had a strongly plated headshield and a comparatively flexible body and tail that provided most of the propulsion.

The most important osteostracans, the **cephalaspids**, lived from Late Silurian to Late Devonian times. They are named for their large solid headshields, which often had a large spine projecting forward and two spines or horns extending backward at each corner. Powerful paired fins were attached at the back corners of the headshield, just inside the protective spines. The body behind the headshield was laterally compressed, as in most living fishes, and small dorsal fins were probably passive aids to stability. The cephalaspid tail was more versatile than the heterostracan tail, and it had some horizontal flaps that added new control surfaces.

Cephalaspids were bottom dwellers. The mouth was on the flat underside of the headshield. The eyes were small and set close together on the top of the headshield. In addition, cephalaspids had large sensory areas on each side of the headshield, covered with very small plates (Figure 7.12). These organs may have served as pressure sensors in murky water, although they may have sensed electrical fields, as living sharks do.

Cephalaspids had so much calcified material in the headshield that it surrounded all the soft tissues; it thus had imprinted on it the tiniest details of the head anatomy, including nerves and blood vessels. The headshield was solidified in place in one piece and was therefore incapable of further growth. This means that cephalaspids did not form a headshield until they reached adult size, at which point they solidified the headshield once and for all. It makes one wonder what a juvenile cephalaspid looked like. There are good reasons for any fish having scales, and juvenile cephalaspids ought to be preserved in some fashion in the fossil record, probably classed under a different group.

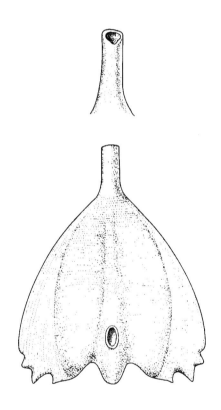

Figure 7.11 Amphiaspids are unusual fishes, but this one, *Eglonaspis*, is spectacular. It was evidently blind, and the headshield is broad, flat, and in one piece. It must have probed downward into soft sediment or upward into surface scum to feed through its tubelike mouth, shown enlarged and turned over in the upper drawing. *Eglonaspis* must have located food by senses other than sight, possibly by detecting electrical signals. (After Obruchev.)

Figure 7.12 The cephalaspid *Hemicyclaspis* has paired fins and a large sensory area on each side of the solid headshield. (After Stensiö and Heintz.)

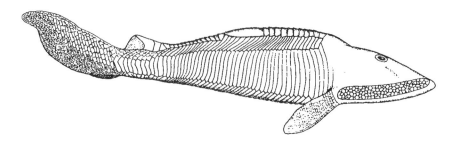

Forktails

In 1998 a new group of jawless fishes was defined on the basis of some small but well-preserved fossils from northern Canada. They are the Furcacaudiformes, but I shall call them **forktails** because that is what the formal name means. Forktails were small and deep-bodied, unusual because mud is preserved inside them, filling a distinct stomach cavity (Figure 7.13). Filter-feeding fishes, which sort their food carefully before swallowing it, do not need large stomachs; mud-eating or carnivorous fishes do. Fork-tails appear to have been mud eaters, but the evolution of a distinct, specialized stomach may have been required before fishes could evolve jaws and eat large prey. In terms of jawless fish evolution, forktails sit nicely as the sister group of jawed fishes (Figure 7.5).

Figure 7.13 A generalized forktail from the Early Devonian of Canada. This diagram combines elements of more than one species of forktails. Note the bulging outline of the big stomach, unexpected in a jawless fish. (After Wilson and Caldwell, 1993.)

The Evolution of Jaws

By the late Silurian, the jawless fishes filled many ecological roles, even without jaws. They were confined to eating small particles, such as plankton from the surface, sediment on the seafloor, or soft, easily swallowed food such as worms or jellyfish. But somewhere among them were fishes in the process of evolving jaws. The evolution of jaws was a major breakthrough for the feeding ecology of fishes; it apparently led to the decline of the surviving agnathans in competition with the newcomers.

Studies of anatomy and embryology suggest, but do not prove, that the bones that form the vertebrate jaw evolved originally from the gill arches of jawless fishes. In living fishes, water is taken in at the mouth and passes backward past the gills, where oxygen and CO_2 are exchanged with the blood system (Figure 7.14).

Figure 7.14 How gills work in most living fishes. Gills are arrays of thin plate-like structures set in rows supported on a strong axis (left). Oxygen-poor blood is pumped along a one-way system through each platelike structure (right). Water is pumped the other way, exchanging gases with the blood by shedding oxygen and taking up carbon dioxide.

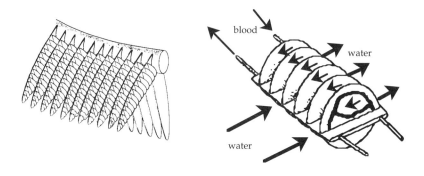

Because gills are soft, they must be supported in the water current by thin strips of bone or cartilage called **gill arches**. The more water passing the gills, the more oxygen can be absorbed and the higher the energy the fish can generate. Living fishes usually have pumps of some kind to increase and regulate the flow of water passing the gills. Most fishes use a pumping action in which they increase and decrease the volume of the mouth cavity by flexing the jaws. In sharks, muscles attached to the gill arches rhythmically flex skin flaps that pump water past the gills. Tuna swim so fast that they create a ramjet action that forces water past the gills, just as the airscoops of some jet fighters funnel air into the turbines.

If jaws evolved from a gill arch, the evolution of the jaw was probably connected originally with respiration rather than feeding. Water flow over the gills of jawless fishes may have been impeded by their small mouths and by a slow flow of water past the gills: therefore, their swimming performance may have been limited by oxygen shortage. Perhaps a joint evolved in the forward gill arch so that it flexed to open the mouth wider, pumping more water backward over the gills (Figure 7.15). In the process, this gill arch was transformed into a true jaw. Opening up the jaws of a medium-sized bony fish such as a trout or salmon shows the similarity between the jaw and the internal gill arches.

Both food intake and oxygen exchange would have been increased by the evolution of the jaw, even if fishes continued to eat microscopic particles. I propose that jaws evolved in actively swimming fishes that were at the edge of their performance envelope (perhaps among forktails, Figure 7.13).

Jawed fishes are **gnathostomes**, as opposed to the agnathans or jawless fishes. There is no general agreement on their classification and early evolution, mainly because they evolved and radiated so quickly in the Late Silurian and Early Devonian that it's difficult to tell which groups are most closely related. A simplified conventional classification and a cladogram (Figure 7.16) will help in discussing the various groups.

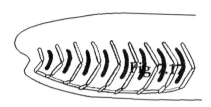

Figure 7.15 How gills probably worked in some early jawless fish. Water was taken in at the mouth, and passed out through several gill slits (shown in black). Between each gill slit was a gill arch (shown in white), a thin strip of bone that supported the gill. Possibly in an adaptation that increased water pumping, the first gill arch became hinged in the middle and evolved eventually into a feeding structure, the jaw.

Figure 7.16 A conventional classification and a cladogram of the jawed fishes or gnathostomes.

GNATHOSTOMES

  †ACANTHODIANS

  CHONDRICHTHYES
  (sharks and rays)

  †PLACODERMS

    †Arthrodires

    †Antiarchs

  OSTEICHTHYES (bony fishes)

    Actinopterygians (rayfins)

    Sarcopterygians (lobefins)

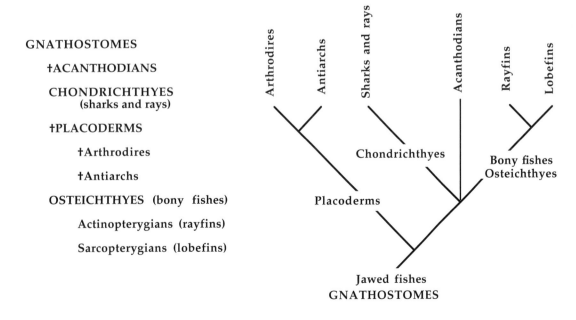

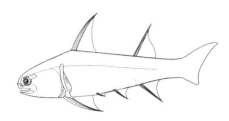

Figure 7.17 *Diplacanthus* is an acanthodian with strong, deeply set spines. (After Watson.)

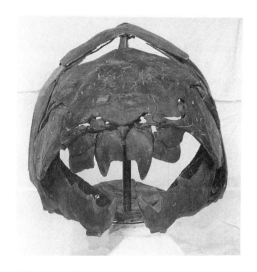

Figure 7.18 The head of *Dunkleosteus*, a gigantic arthrodire from the Late Devonian. Its full body length was about 6 meters (20 feet); the headshield alone was 2 meters (6 feet) long. Tooth plates gave powerful cutting and slicing edges along the jaw line. (Negative 333858. Courtesy of the Library Services Department, American Museum of Natural History.)

Acanthodians

The earliest jawed fishes were small Silurian forms called **acanthodians**. They are lightly built, not well preserved, and not very well known; and they may not be the most primitive jawed fishes (Figure 7.16). I will use them simply as examples of early jawed fishes.

In small acanthodians, the gill arches remaining behind the jaw bore spikes called gill rakers to filter out debris and microscopic food from the respiratory current. Many living fishes, from anchovies to basking sharks, respire and feed by swimming along with their mouths open, filtering plankton on gill rakers.

Acanthodians had paired fins, but they swam only with tail and body. Most of the fins look like passive structures, supported on heavy spines (Figure 7.17). The dorsal spines may have been protective, and the ventral spines could have been used as props on the bottom. Sometimes the spines were firmly fixed, so that the fin acted mainly as a water-deflecting surface for adding maneuverability rather than speed or lift. In other acanthodians, the spines could be folded down, or erected and locked in place. Similar spines in a living triggerfish are defensive. Some acanthodians had deep bodies with many large fins. They probably swam slowly but with high maneuverability, rather like living reef fishes.

The (unknown) ancestor of jawed fishes lived in the Silurian and was most likely a fish that was like an acanthodian in ecology, already filter feeding in open water, not a heavily plated bottom feeder. Fork-tails are plausible candidates (Figure 7.13).

The evolution of jaws and a resulting extension of the potential food supply were the keys to the tremendous ecological expansion and evolutionary success of jawed fishes. But teeth and jaws are only weapons: they must be applied to targets by a delivery system. Paired fins that were active, propulsive units were essential to give the maneuverability and speed needed by active predators. The history of fishes since the Devonian has been largely one of increasing effectiveness in the mounting and hinging of the jaws, in the speed of strike, and in the hydrodynamics of propulsion and maneuverability. The acanthodians, with their non-propulsive spiny fins, were perhaps capable of sophisticated maneuvering in the water, but not great speed or acceleration; this may have been the reason for their eventual extinction.

Placoderms

Placoderms were dominant, worldwide fishes during Devonian times. They are probably the sister group of all other jawed fishes (Figure 7.16). Many different groups of placoderms have been recognized, but only two groups are important, the **arthrodires** (Figure 7.18) and the **antiarchs** (Figure 7.19).

Most placoderms had a well-developed headshield made of several plates, jointed to an armored girdle surrounding the front part of the trunk. The rest of the trunk was lightly scaled, and presumably the trunk

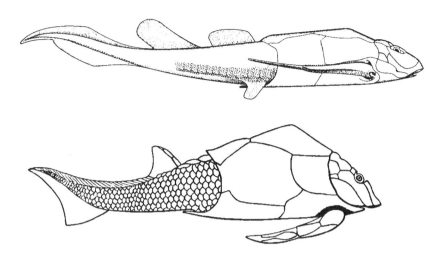

Figure 7.19 Antiarchs. Top: *Bothriolepis*, a Devonian antiarch. (After Stensiö.) Bottom: *Pterichthys* offers a contrast in body shape to *Bothriolepis*. Its headshield is higher, and it looks more like a boxfish. (After Traquair.)

and the long tail were very flexible. There were several pairs of fins, indicating better control over movement than in acanthodians. But the body was usually flattened to some extent, and the eyes were usually small and set on the upper side of the headshield.

Altogether, the body shape and the great forward weight of placoderms suggests that they lived on or near the seafloor, and although they may often have been powerful swimmers, they cannot have been agile or capable of rapid maneuvers. The small eyes imply that they probably used other senses to a large extent, just as living sharks do. The jaws vary quite a lot, but some advanced placoderms had vicious tooth plates set into the jaw; these were large carnivores up to 6 meters (20 feet) long (Figure 7.18). Others had large crushing teeth, perhaps for eating molluscs or arthropods.

Arthrodires were powerful, streamlined fishes, but their great weight of armor and their generally flattened body shapes may have limited their swimming performance. They were the dominant fishes of the Devonian, and included the most powerful predators of the time. Their distribution was worldwide, in both salt and fresh water. Giant arthrodires include *Tityosteus*, the largest known Early Devonian fish, with an length of about 2.5 meters (8 feet). But *Tityosteus* is dwarfed by a Middle Devonian freshwater fish *Heterosteus* and by the Late Devonian *Dunkleosteus*, both of which grew to 6 meters (20 feet) long (Figures 7.18, 7.20).

The major differences among arthrodires are in body shape and tooth type, which probably reflect habitat and diet, and in the proportion of the trunk covered by armor, which reflects swimming capability. Arthrodires with short body armor had a greater length of flexible trunk, and presumably more of the trunk muscles could be recruited for swimming. Large pectoral fins aided stability and provided lift for the heavy armored head. The joint between headshield and trunk armor was well developed. The head could be levered upward on that joint while the lower jaw dropped at the same time, opening a wide gape for effective hunting (Figure 7.20), and possibly aiding water flow over the gills.

Antiarchs are much more specialized and difficult to understand, but these "grotesque little animals" (as one famous paleontologist called

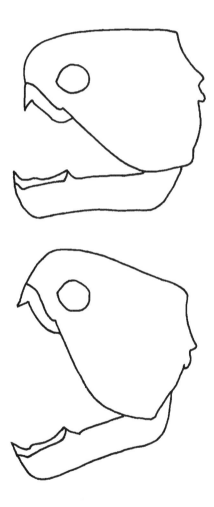

Figure 7.20 The head of *Dunkleosteus*, showing how its jaws worked. The head could hinge upward on a special joint at the back of the skull, while the lower jaw dropped, opening a very wide gape.

them) were successful worldwide, mostly in freshwater environments. They were small, with headshields up to 50 cm (20 inches) long and a maximum known length just over a meter (3 feet). Their headshields were flattened against the bottom, with the eyes set close together high on the headshield. The mouth lay just under the snout. The body armor was long. Instead of pectoral fins, antiarchs had long, jointed appendages that look as if they were used for poling the fish along the bottom rather than swimming (Figure 7.19). Antiarchs had small mouths and probably ate mud, filling an ecological role that had been taken by earlier jawless fishes. It's clear that they were slow, rather clumsy swimmers.

Placoderms evolved at the beginning of the Devonian and were practically extinct by its close, even though they dominated the time between. It seems clear that eventually they were handicapped by their weight of armor, and they never seem to have conquered the problem of agile maneuvering in water. The cartilaginous fishes and the bony fishes, each in their own way, solved the problem and now dominate living fish faunas.

Cartilaginous Fishes (Sharks and Rays)

Sharks and rays, and all their ancestors we have been able to identify, have cartilaginous skeletons rather than bone. This distinction dates back to the Early Devonian, when this group of fishes was just one of the many early successful lines that had recently evolved jaws (Figure 7.16).

The fossil record of sharks and rays is poor, because they do not preserve well as fossils. They have cartilage rather than bone, and a tough skin rather than heavy scales. They do have formidable teeth, which are often well preserved as fossils, but teeth alone give only a vague idea about the entire fish. Occasionally a rare find of a body outline allows us to see that sharks have not changed a great deal in overall body shape during their evolution (Figure 7.21).

Sharks are very successful fishes. They have excellent vision and smell and an electrical sense, all of which combine to equip them well for hunting in all kinds of environments. They all have internal fertilization, and some have live birth. They are certainly not primitive. Sharks are simply a group of fishes that discovered a successful way of life several hundred million years ago.

Figure 7.21   The body outline of *Cladoselache*, a Devonian shark. Even at this early date, the typical shark body had evolved. (After Dean and Zangerl.)

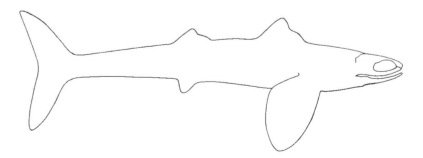

Bony Fishes

Actinopterygians (Rayfin Fishes)

Actinopterygians, or rayfins, have very thin fins that are simply webs of skin supported by numerous thin, radiating bones (called rays). Typically, rayfins are lightly built fishes that swim fast or maneuver very well. They have dominated marine and freshwater environments of the world since the decline of the placoderms at the end of the Devonian. It is tempting to suggest that their evolutionary success largely reflects their mastery of swimming and feeding in open water.

In general, the evolution of the rayfin fishes resulted in a lightening of the bony skeleton and the scaly armor, both of which improved locomotion. Increasing sophistication and variation in the shape and arrangement of the paired fins led to patterns that were optimum for specialized sprinters, cruisers, or artful dodgers. In the most advanced rayfins, swimming has come to depend more and more on the tail fin rather than on body flexing, while the other fins are modified as steering devices and/or defensive spines. Even flying fishes had evolved by Triassic times.

The jaws and skull of rayfins were gradually modified for lightness and efficiency. In particular, intricate systems of levers and pulleys allow advanced fishes to strike at prey more effectively by extending the jaws forward as they close. The same system also allows more efficient ways of browsing, grazing, picking, grinding, and nibbling, all encouraging the evolution of the tremendous variety of living fishes.

Sarcopterygians (Lobefin Fishes)

Sarcopterygians, or lobefins, are distinguished as a separate group of fishes (Figure 7.22) because they evolved several pairs of fins stronger than any found in a rayfin fish. Other differences in scale and skull structure confirm the separate evolution of these groups. In 1999 a fish was described from the late Silurian or Early Devonian of China, which had characters of both, so it probably lies near the evolutionary split between the groups. The specimen, named *Psarolepis*, is not well preserved, and we need more information about it.

The central part or lobe of a lobe fin is sturdy and contains a series of strong bones, while the edges have radiating rays as in ray fins (Figure 7.23). A lobe fin must beat more slowly than a ray fin of the same area because there is more mass to be accelerated and decelerated, but the resultant stroke is more powerful. Furthermore, and fundamental to later vertebrate history, a lobe fin that imparts a powerful stroke to the water has to have some kind of support at its base, just as an oar has to be stabilized in a rowlock. Therefore, lobefin fishes have internal systems of bones and muscles that help to tie one dorsal and two ventral pairs of lobe fins to the rest of the skeleton (Figure 7.23). These ventral linkages evolved to become the pectoral and pelvic girdles of land vertebrates, but

Osteichthyes (bony fishes)

Actinopterygii (rayfins)

Sarcopterygii (lobefins)

Dipnoi (lungfishes)

Actinistia (coelacanths)

Rhipidistia

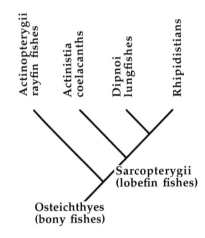

Figure 7.22 A conventional classification and a cladogram of the bony fishes. Rhipidistians are not marked as extinct because they have living descendants (all tetrapods, including humans).

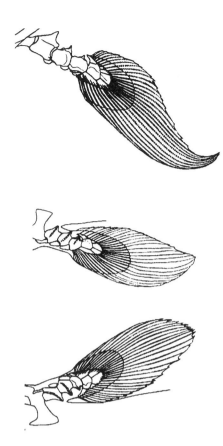

Figure 7.23 From top to bottom: the anterior ventral, the posterior ventral, and the posterior dorsal lobe fins of the coelacanth *Latimeria*. The anterior (or pectoral) fin is significantly larger than the other two, but the posterior dorsal fin is just as large as its ventral counterpart and has just as strong an internal bony skeleton. (After Millot and Anthony.)

Figure 7.24 The coelacanth, *Latimeria chalumnae*. This is the South African species, not the newly discovered Indonesian one. (Drawing by Bob Giuliani. © Dover Publications Inc., New York. Reproduced by permission.)

of course that was not why they evolved: they evolved originally to allow early lobefin fishes to swim more effectively. All Devonian lobefins seem to have been effective swimmers and predators.

Lungfishes and coelacanths both have a long history dating back to the Devonian. but today these two living groups of lobefins are rare and unusual fishes. Two species of coelacanth survive only as scattered small populations off the coast of South Africa, and in Indonesia; and one species of lungfish lives in each of the three southern continents: Australia, South America, and Africa.

All living lobefins have such an unusual biology and ecology that they must be interpreted with caution. They are much evolved from their Devonian ancestors in structure and in habits. so they may not be very good guides to the biology of those ancestors.

Coelacanths

The living coelacanth, *Latimeria* (Figure 7.24), was unexpectedly discovered 50 years ago off the coast of South Africa. Coelacanths had been known as fossils for decades, but it was thought that they had died out after the Cretaceous. In the late 1930s a single specimen was brought in by fishermen and recognized as a lobefin fish by an astute scientist. But it was decades before others were caught much further north around the Comores Islands between Madagascar and the African coast, and 1998 before a second species was reported from Indonesia.

The living coelacanth lives at considerable depth in sea water, is a lazy swimmer, and does not have lungs. It bears live young, as many as 26 at a time, which hatch internally from very yolky eggs.

Lungfishes

Living lungfishes are medium-sized, long-bodied fishes found in seasonal freshwater lakes and rivers in tropical areas. They seem best designed for rather slow swimming. Living lungfishes, or Dipnoi, are so called because they can breathe air, allowing them to survive periods of drought or low oxygen in seasonal lakes and rivers in tropical climates. Lungfishes probably survive today because they can tolerate environments that would kill most other fishes. The African lungfish can even tolerate a dry

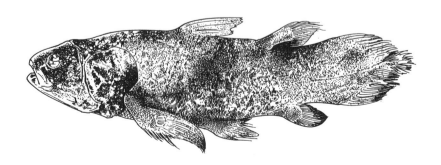

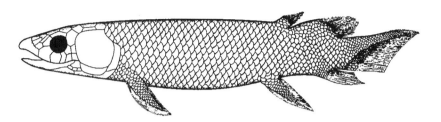

Figure 7.25 Devonian lungfishes such as *Dipterus*, shown here, had no lungs and were more active than their living descendants. (After Traquair.)

season in which its river dries up. It digs a burrow, seals itself inside, and estivates (turns its body metabolism to a very low level) until the rainy season sends water down the river and into the burrow, reviving it.

Lungfishes have evolved considerably to their present anatomy, biology, and ecology. The first lungfishes had no lungs. They were marine, breathed from water through gills, and look as if they were much more active swimmers than their living descendants (Figure 7.25). Late in the Devonian a clade of lungfishes evolved the ability to breathe air in and out of internal sacs that we now recognize as lungs. These lungfishes lived in fresh water, where they evolved changes in teeth and jaws that mark a shift in feeding from other fishes to molluscs and crustaceans. (Living forms have flattened teeth shaped like plates, for crushing their prey.) Early lungfishes also evolved a way of dealing with drought that has not changed much either, it seems. Carboniferous lungfishes have been found fossilized in burrows 300 m.y. old.

Rhipidistians

Rhipidistians (increasingly called Tetrapodomorpha these days) are Devonian lobefins, the sister group of lungfishes (Figure 7.23). All three lobefin groups originated in the Early Devonian, however, and at that time were closely similar. They diverged much more during the Devonian in shape, structure, and ecology. Rhipidistians include the ancestors of amphibians and all other land-going vertebrates, so they are discussed in more detail in Chapter 8. In general, placoderms were the dominant fishes of the Devonian, at least in size. Rhipidistians and the other lobefins were most successful in the shallow waters of coasts and inland waters, but were hardly dominant. After the Devonian the rayfin fishes came to be the most successful group, with their combination of lightness and maneuverability, while the lobefins were gradually confined to unusual habits and habitats. Perhaps in the process of being squeezed, ecologically speaking, the rhipidistian lobefins evolved adaptations that allowed them to expand in an unexpected direction—toward the land.

Review Questions

Describe and draw a very early fish. Label its special features. What did jawless fishes eat, and where did they feed?

In very broad terms, how did jaws evolve? What is a lobe-fin as opposed to a ray-fin? (Draw or describe.)

Further Reading

Easy Access Reading

Erdmann, M. V., et al. 1998. Indonesian 'king of sea' discovered. *Nature* 395: 335. New coelacanth population.

Forey, P., and P. Janvier. 1993. Agnathans and the origin of jawed vertebrates. *Nature* 361: 129–134.

Shu, D. G., et al. 1996. A *Pikaia*-like chordate from the lower Cambrian of China. *Nature* 384: 157–158. *Cathaymyrus*.

Stokes, M. D., and N. D. Holland. 1998. The lancelet. *American Scientist* 86: 552–560.

Thomson, K. S. 1991. *Living Fossil: The Story of the Coelacanth*. New York: W. W. Norton. Not much paleontology, but a lot of science, and a good read.

Thomson, K. S. 1999. The coelacanth: Act Three. *American Scientist* 87: 213–215.

Wilson, M. V. H., and M. W. Caldwell. 1993. New Silurian and Devonian fork-tailed 'thelodonts' are jawless vertebrates with stomachs and deep bodies. *Nature* 361: 442–444.

Young, C. G., et al. 1996. A possible Late Cambrian vertebrate from Australia. *Nature* 383: 810–812, and comment, pp. 757–758.

Zhu, M., et al. 1999. A primitive fossil fish sheds light on the origin of bony fishes. *Nature* 397: 607–610.

More Technical Reading

Aldridge, R. J., et al. 1993. The anatomy of conodonts. *Philosophical Transactions of the Royal Society of London B* 340: 405–421.

Fricke, H., and J. Frahm. 1992. Evidence for lecithotrophic viviparity in the living coelacanth. *Naturwissenschaften* 79: 476–479.

Janvier, P. 1996. *Early Vertebrates*. Oxford: Clarendon Press.

Ruben, J. R. 1989. Activity physiology and evolution of the vertebrate skeleton. *American Zoologist* 29: 195–203.

Schultze, H.-P., and L. Trueb (eds.). 1991. *Origins of the Higher Groups of Tetrapods: Controversy and Concensus*. Ithaca, New York: Cornell University Press. Chs. 1–5.

Thomson, K. S. 1993. The origin of the tetrapods. *American Journal of Science* 293A: 33–62. For discussion of lungfishes and coelacanths.

Wilson, M. V. H., and M. W. Caldwell. 1998. The Furcacaudiformes: a new order of jawless vertebrates with thelodont scales, based on articulated Silurian and Devonian fossils from Northern Canada. *Journal of Vertebrate Paleontology* 18: 10–29. Forktails.

Young, G. C. 1986. The relationships of placoderm fishes. *Zoological Journal of the Linnean Society* 88: 1–57.

Leaving the Water

Plants, invertebrates, and finally vertebrates evolved to live on land in the Middle Paleozoic. There were major problems in doing so (Box 8.1), related not so much to the land surface as to exposure to air. Many marine animals and plants spend their lives crawling on the seafloor, burrowing in it, or attached to it. As a physical substrate, the land surface is not very different. But land organisms are no longer bathed in water. There are predictable consequences for the evolutionary transitions involved, many of them based on the laws of physics and chemistry.

Organisms weigh more in air without the buoyant effect of water, so support is more of a problem. Air may be very humid, but it is never continuously saturated, so organisms living in air must find a way to resist desiccation. Tiny organisms are particularly sensitive to drying out in air, because they have relatively large surface areas but cannot hold large reserves of fluid. Therefore reproductive stages and young stages of plants and animals are very sensitive to drying. Temperature extremes are much greater in air than they are in water, exposing plants and animals to heat and cold. Oxygen and carbon dioxide behave differently as gases than they do when dissolved in water, so respiration and gas exchange systems must change in air. The refractive index of light is lower in air than in water, and sound transmission differs too, so vision and hearing must be modified in land animals.

There are also ecological consequences. Seawater carries dissolved nutrients, but air does not, so some organisms, especially small animals

---

BOX 8.1. PROBLEMS IN ADJUSTING TO LIFE IN AIR

For Plants and Animals

No buoyancy, so support needed
Danger of drying out
Extremes of temperature
Gases behave differently
No nutrients in air

For Animals only

Refraction of light changes
Hearing must be modified

and plants, have a food supply problem in air. It's unlikely that the same food sources would be available to an animal that crossed such an important ecological barrier, so invasion of the land would often be associated with a change in feeding style.

All the major adaptations for life in air had to be evolved first in the water, as adaptations for life in water. Only then would it have been possible for organisms to emerge into air for long periods. We must reconstruct a reasonable sequence of events during the transition, then test our ideas against evidence from fossil and living organisms.

The Origin of Land Plants

We have no idea when plants first colonized land surfaces. Plants must have emerged gradually into air and onto land from water; the first land plants must have been largely aquatic, living in swamps or marshes.

Almost all the major characters of land plants are solutions to the problems associated with life in air. Land plants grow against gravity, so they have evolved structural or hydrostatic pressure supports (hard cuticles or wood) to help them stay upright. They cannot afford evaporation from moist surfaces, so they have evolved some kind of waterproofing. Roots gather water and nutrients from soil and act as props and anchors. Internal transport systems distribute water, nutrients, and the products of photosynthesis around the plant. Even so, all these adaptations for adult plants are useless unless the reproductive cycle is also adapted to air. Cross-fertilization and dispersal require special adaptations in air. All these adaptations must have evolved in a rational and gradual sequence. But because the first stages would have been soft-bodied water plants, the fossil record of the transition is difficult to find.

The following scenario for the evolution of land plants is modified from suggestions by John Raven. Water-dwelling plants, probably green algae, were already multicellular. Green algae grow rapidly in shallow water, bathed in light and nutrients. One might think that cells in a large alga are comparatively independent of one another: in the water, each cell has access to light, water, nutrients, and a sink for waste products. But the fastest growing points of algal fronds need more energy than the photosynthesis of the cells there can supply, so some green algae have evolved a transport system between adjoining cells to move food quickly around the plant. Green algae presumably do this because they can then grow more rapidly.

Raven's scenario begins with green algae living in habitats that were subject to temporary drying. The algae might already have evolved to disperse spores more effectively by releasing them into wind instead of water. Spores, even in algae, are reasonably watertight and could easily have been adapted for release into air from special sporangia (spore containers) growing high enough to extend out of the water on the uppermost tips of otherwise aquatic plants. As plant tissue extended into air, photosynthesis increased because light levels in air are higher than they are in water, especially at each end of the day, and are free from interference by muddy water. Furthermore, CO_2 is more easily extracted from air than it is from water.

As plants grew out into air, some tissues were no longer bathed in the water that had provided nutrients and a sink for waste products. Internal fluid transport systems between cells became specialized and extended. Photosynthesis was concentrated in the upper part of the plant that was exposed to more light. Photosynthesis fixes CO_2, so there had to be continual intake of CO_2 from the air. However, plant cells are saturated with water, but air is usually not, so the same surfaces that take in CO_2 automatically lose water. Sunlight heats the plant, encouraging evaporation. The water loss had to be made up by transporting water up the stem from the roots to the photosynthesizing cells.

Water is transported much more effectively as liquid than as vapor. Early land plants evolved a simple piping system called a **conducting strand** of cells to carry water upward. The conducting strand, found in living mosses, is powerful enough to prevent water loss in small, low plants, if soil water is abundant. But mosses quickly dry up if soil water is in short supply.

Early land plants began to evolve a **cuticle** (a waxy layer) over much of their exposed upper surfaces. The cuticle helps the plant through alternating wet and dry conditions. In wet times it acts as a waterproof coating. It prevents a film of water from standing on the plant that could cut off CO_2 intake. In dry times it seals the plant surface from losing water by evaporation. A cuticle may also have added a little strength to the stem of early plants, and its wax probably helped to protect the plant from UV radiation and from chewing arthropods.

But the cuticle also cut down and then eliminated, from the top of the plant downward, the ability to absorb water-borne nutrients over the general plant surface. Nutrients were taken up more and more in lower parts of the plant, eventually taking place on specialized absorbing surfaces at the base (**roots**) which probably evolved from the runners that these plants often used to reproduce asexually. As roots grew larger and stronger, they helped to anchor and then to support the plant.

The cuticle sealed off CO_2 uptake over the general plant surface. As cuticle evolved, plants also evolved pores called **stomata** where CO_2 uptake could be concentrated. If it is too hot or too dry, stomata can be closed off by **guard cells** to control water loss. As CO_2 uptake was localized, plants evolved an **intercellular gas transport** system that led from the stomata into the spaces between cells, improving CO_2 flow to the photosynthesizing cells. The same system was also used to solve an increasingly important problem. As roots enlarged, more and more plant tissue was growing in dark areas where photosynthesis was impossible; yet those tissues needed food and oxygen (O_2). Soils are low or lacking in O_2, especially when they are waterlogged. The intercellular gas transport system feeds O_2 from the air down through the plant to the roots, sometimes through impressively large hollow spaces.

Later plants evolved the **xylem**, an improved piping system for better upward flow of water from the roots round the plant. The xylem is made

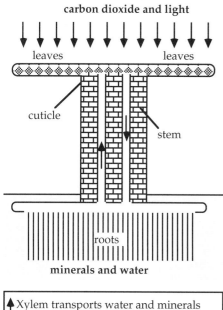

carbon dioxide and light

leaves   leaves

cuticle

stem

roots

minerals and water

Xylem transports water and minerals upward from the roots to the leaves

Phloem transports food around the plant from the leaves where it is made

Figure 8.1 The basic land plant.

of elongated dead cells arranged end to end to form long pipes up and down the stem. Evaporation at the top of the plant essentially sucks water upward through the empty dead cells. Even a narrow xylem can transport water much faster than can normal plant tissue, and, once begun, the evolution of xylem was probably a rapid process that immediately gave plants greater tolerance of dry air. The xylem also carries important dissolved substances as it flows upward, and it is the primary intake and transport system for plant nutrients, particularly for phosphate. Plants that have xylem are called **vascular plants**.

Xylem cells are dead, so xylem transport is passive, driven entirely by the suction—or negative pressure—of evaporation from the upper part of the plant. The forces generated can be very large, so the long narrow walls of xylem cells may tend to collapse inward. Xylem cell walls came to be strengthened by a structural molecule, **lignin**. Once lignin had evolved, it was used later to strengthen the roots and stem as plants grew taller and heavier.

As plants became increasingly polarized, with nutrient and water being taken up at the roots and photosynthesis taking place in the upper parts, the xylem and gas transport systems improved, but neither of them was able to transport liquid downward. This problem was solved by the evolution of another transport system called **phloem** from the cell-to-cell transport system of green algae. Phloem cells transport dissolved substances round the plant in a process that is not yet properly understood. Phloem carries photosynthate from photosynthesizing cells to growing points such as reproductive organs and shoots, and to tissues such as roots that cannot make their own food.

Throughout the process, the advantage that encouraged plants to extend into air in spite of the difficulties involved was the tremendous increase in available light. Marine plants are restricted to the narrow zone along the shore where light has to penetrate sediment-laden, wave-churned water. Growth above water increases light availability. Furthermore, competition for available light tended to encourage even more growth of plant tissues above the water surface, and more effective adaptations to life in air (Figure 8.1). Once plants could grow above the layer of still air near the water surface, spores could be released into breezes. Greater plant height and the evolution of sporangia on the tips of branches were both adaptations for effective dispersal.

The Earliest Land Floras

Pre-Devonian plants all occur fossilized in marine sediments. But plants can be swept downstream in floods and deposited far from their habitats. Mats of floating vegetation can be found today off the mouth of the Amazon, sometimes complete with roots, soil, and sediment. So the fossil record doesn't necessarily imply that pre-Devonian plants lived in the sea. (In fact, Raven suggests that the osmotic systems of land plants imply that they evolved in freshwater marshes.)

The earliest spores that probably belonged to land plants come from Middle Ordovician rocks. They look like the spores of living liverworts. Later spores from Silurian rocks look as if they came from more advanced plants (though we cannot tell which ones). There are no fossils of plant parts other than spores before the Middle Silurian, which means that the spore-producing plants have not been found yet. Ordovician and Early Silurian land plants were probably small and weakly constructed, and difficult to preserve as fossils. Molecular evidence suggests that all land plants today are descended from liverworts, but molecular evidence cannot say anything about extinct groups of plants that might have been the pioneers of plant life in air.

Late Silurian and Early Devonian Plants

A land flora of some sort was certainly well established by Late Silurian times, even though the plants have not been found preserved in their life habitats. Though it is Devonian in age, *Aglaophyton* (Figure 8.2) probably has a grade of structure that evolved in the Late Silurian. It grew to a height of less than 20 cm (about 8 inches). Although it had most of the adaptations needed in land plants—a cuticle, stomata, and intercellular gas spaces—it did not have xylem, only a simple conducting strand.

However, the earliest well-preserved land plants, from Late Silurian rocks of Wales and Australia, almost certainly include vascular plants. *Cooksonia* (Figure 8.3) was only a few centimeters high and had a simple structure of thin, evenly branching stems with sporangia at the tips, and no leaves. But *Cooksonia* also had central structures that were probably xylem rather than simple conducting strands. Later species of *Cooksonia* from the earliest Devonian have definite strands of xylem preserved, and cuticles with stomata, so they probably had intercellular gas spaces and were better adapted for life in air.

Early Devonian land plants were dramatically more diverse. They grew up to a meter high, although they were slender (1 cm diameter). For support they must have grown either in standing water or in dense clusters, aided by the fact that they reproduced largely by budding systems of rhizoids for asexual, clonal reproduction, as strawberries do today (Figure 8.2). This style of reproduction not only gave mutual support to individual stems, but, by "turfing in," a cloned mass of plants could help to eliminate competitive species. Plants like this could have grown and reproduced very quickly, a way of escaping the consequences of relatively poor adaptations for living in air.

Rhynia (Figure 8.4) is like *Cooksonia*: it is a genuine vascular plant. *Rhynia* gave its name to the **rhyniophytes**, an extinct group that includes many early vascular plants. Meanwhile, other major groups of vascular plants also evolved in the Early Devonian. Lycopods are the ancestors of living club-mosses, but they formed trees and forests in the coal swamps of the Carboniferous period. We still cannot identify the rapid evolutionary changes that gave all the other main groups of living vascular plants.

Figure 8.2 *Aglaophyton*, an early land plant from Devonian rocks in Scotland, was nonvascular. It had a simple conducting strand for transporting water up the stem. It budded new plants from a rhizome at the base. (After Kidston and Lang.)

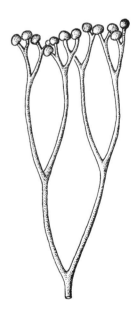

Figure 8.3 *Cooksonia*, an early vascular land plant. (After Edwards.)

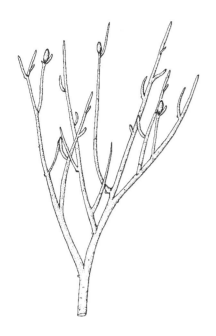

Figure 8.4 *Rhynia*, a Middle Devonian vascular plant from the Rhynie Chert Beds of Scotland with well-developed xylem. *Rhynia* gives its name to a major group of early land plants, the rhyniophytes. (After Edwards.)

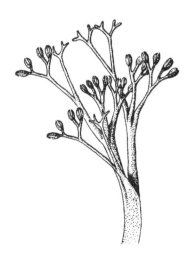

Figure 8.5 The Devonian plant *Psilophyton* represents a new grade of structure in land plants, advanced over rhyniophytes in its strong construction. (After Hopping.)

Later Devonian Plants

New structural advances are seen in later Devonian and Early Carboniferous plants. The successive floras all lived in lowland floodplains, which have a good fossil record. It looks as if we are seeing waves of ecological and evolutionary replacement on all levels, from individual plants to world floras, as structural innovations allowed each plant group to outcompete its predecessor.

For example, *Rhynia* had only 1% of its stem cross-section made of xylem, but *Psilophyton* had 10%, and the whole plant was more strongly constructed (Figure 8.5). So psilophytes were able to grow taller (up to 2 meters high) and photosynthesize more efficiently than rhyniophytes. Other improvements in reproduction and light gathering, through the evolution of leaves rather like those of living ferns and through more complex branching, also aided plant efficiency.

Later plants evolved even more xylem, so the structure became even stronger. Secondary xylem—**wood**—gave great strength and allowed higher growth, producing the first trees (Figure 8.6). Late Devonian trees also evolved true roots, more complex branching patterns, and larger, flatter leaves. Reproductive systems were improved still further: most late Devonian plants had two different kinds of spores, female megaspores and male microspores.

Seed-bearing plants evolved in the Late Devonian. A Late Devonian seedlike structure called *Archaeosperma* looks as if it belongs to a tree very much like *Archaeopteris* (Figure 8.6). This was a great advance: all previous plants had needed a film of water in which sperm could swim to fertilize the ovum, but seed plants can reproduce away from water.

By the end of the Devonian, all the major innovations of land plants except flowers and fruit had evolved. Forests of seed-bearing trees and lycopods had appeared, with understories of ferns and smaller plants.

The dominant process in Devonian plant evolution seems to have been selection based on simple efficiency—in size and stability, photosynthesis, internal transport, and reproductive systems. Plant groups replaced one another as innovations appeared. Perhaps the most interesting part of this story of early plants is the rate at which innovations appeared. There is no obvious reason why the process should not have gone faster, or slower. The innovations we have discussed should have given immediate success whenever they appeared. But it took the length of the Devonian (about 50 m.y.) for rhyniophyte-type plants to evolve to seed-plants. Even "obvious" innovations may take time to evolve and accumulate.

Devonian Plant Ecology

The best-known Devonian plants grew in swampy environments near the equator, where they were very likely to be preserved. The fossil record may not be biased, however; early land plants probably did live in lowlying, damp, tropical regions, where there was little seasonal fluctuation in temperature, light, or humidity.

By Middle Devonian times there were many fernlike plants with well-developed leaves. Fossil tree trunks from the Middle Devonian of New York suggest plants over 10 meters (30 feet) high, with woody tissue covered by bark. Once plants reached these heights, shading of one species by another would have led to fairly complex plant communities.

The evolution of the seed seems to have been the foundation for competitive success in the Early Carboniferous. Seed plants invaded drier habitats, and seed dispersal by wind (rather than water) became important; we have winged seeds from Late Devonian rocks. Seed dispersal allowed some plants to specialize as invaders, avoiding the increasingly dense and competitive habitats in wetlands and along rivers.

The increasing success of land plants, especially their growth to the size of trees, must have produced ever-larger amounts of rotting plant material in swamps, rivers, and lakes, leading to very low O_2 levels in any slow-moving tropical water (O_2 is used up in decay processes). At the same time, the increasing photosynthesis by land plants drew down atmospheric carbon dioxide and increased atmospheric oxygen. All this probably helped to encourage air breathing among contemporary freshwater arthropods and fishes, and it led to better preservation of any fossil material deposited in anoxic swamp water. Coals are known from Devonian rocks, but truly massive coal beds formed for the first time in Earth history in the Carboniferous Period, which was named for them.

Figure 8.6 The first tree *Archaeopteris* shown silhouetted against a Late Devonian sunrise. *Archaeopteris* may very well have been the plant that carried the first seed, *Archaeosperma*. (After Beck.)

Comparing Plant and Animal Evolution

Whether one counts spores or plant macrofossils, there is a striking increase in land plant diversity from Silurian to Middle Devonian time, when a diversity plateau was reached that extended into the Carboniferous. A second increase in Carboniferous land plant diversity was followed by a long period of stability. A third, Late Mesozoic expansion in land plants and animals raised diversity to current levels.

The pattern looks rather like the pattern of Sepkoski's three major faunas in the oceans. But the radiations among land plants and marine animals did not occur at the same times, so they were not directly linked. Extinctions among plants are different from those among animals, which suggests that plants and animals may respond to quite different extinction agents. Specifically, Andrew Knoll suggested three major factors:

- Plants are more vulnerable to extinction by competition;

- Plants are more vulnerable to climatic change;

- Plants are less vulnerable to mass mortality events.

These differences reflect basic plant biology. All plants do much the same thing. They are all at the same primary trophic level, so they cannot partition up niches as easily as animals can. A new arrival in a flora may be competitively much more dangerous than a new arrival in a fauna.

For example, CO_2 uptake must be accompanied by the loss of water vapor, since the plant is open to gas exchange. Many plant adaptations are responses to the problem of water conservation. Because it is so basic a part of their biology, an innovation here could provide a new plant group with an overwhelming advantage. Other plant systems such as light-gathering are equally likely to be improved by innovation.

Plant distributions are sensitive to climate. If climate changes, plants must adapt, migrate, or become extinct. In extreme circumstances, there may be no available refuges. Thus, the tree species of Northwest Europe were trapped early in the ice ages between the advancing Scandinavian glaciers to the north and the Alpine glaciers to the south, and were wiped out. In contrast, similar species in North America were able to move their range south along the Appalachians, then north again as the ice retreated.

On the other hand, plants are well adapted to deal with temporary stress. Plants readily shed unwanted organs such as leaves or even branches in order to survive storms and extreme weather. Many weeds die, to overwinter as seeds or bulbs. The soil is always rich with seeds, so that mass mortality of full-grown plants does not mean the end of the population. Plants are the dominant biomass in communities recovering after volcanic eruptions, tropical storms, or other catastrophes have devastated the local area. In many plant communities, removal of dominant trees by storm, fire, or human agency is followed by the rapid growth of species that are specialized to colonize disturbed areas.

The First Land Animals

As plants extended their habitats into swamps and onto riverbanks and floodplains, they would have provided a food base for animal life evolving from life in water to life in air. The marine animals best preadapted to life on land were arthropods. They already had an almost waterproof cover and were very strong for their size, moving on sturdy walking legs. The incentive to take on exposure to air might have been the availability of organic debris washed ashore on beaches, or perhaps the debris left on land by the first land plants. Plant debris, whether it's on a beach or on a forest floor, tends to be damp; it provides protection from solar radiation and is comparatively nutritious. Thus it is not surprising that the earliest land animals were arthropods that ate organic debris, and other arthropods that ate them. The first abundant fauna of land arthropods is known from Late Silurian trace fossils of arthropod footprints, rather than the animals themselves. Spiders, centipedes, and mites were the earliest land predators, and many small arthropods ate debris, including millipedes and the primitive sister group of insects, the springtails or collembolans.

Different arthropods probably moved into air by different routes. The easiest transition would seem to have been by way of estuaries, deltas, and mudflats, where food is abundant and salinity gradients are gentle.

Very small arthropods have been recovered from several Early Devonian localities. Most of them (mites and springtails) were eating living or dead plant material, and in turn were probably eaten by larger carnivorous arthropods such as primitive spiders. Larger arthropods are usually found in tiny fragments, but it is clear that some were large by any standard. We have pieces of a scorpion that was probably about 9 cm (over 3 inches) long, and a very large millipede-like creature, *Eoarthropleura*, which probably lived in plant litter and ate it. At 15–20 cm (6–8 inches) long, this was the largest terrestrial animal of the Early Devonian.

As plants grew higher, life in air became three-dimensional, and climbing up and down against gravity eventually led to the evolution of flight among insects. This early terrestrial ecosystem did not include any vertebrates as permanent residents, but no doubt the entire food chain, including fishes in the rivers, lakes, and lagoons, benefited from the increased energy flow provided by plants and their photosynthesis.

Rhipidistians

The invasion of the air by plants, and by invertebrates that exploited them for food and shelter, led to a large increase in organic nutrients in and around shorelines. In the Devonian we see the first signs that fishes were beginning to exploit the newly enriched habitats near the shore and near the surface. But we must not imagine that vertebrates adapted quickly to life in air, or that they readily left the water.

As we saw in Chapter 7, lobefin fishes evolved into different ways of life by the end of the Devonian. Lungfishes came to specialize in crushing their prey, crustaceans and small clams. If we can judge by the last surviving species of coelacanth, this group came to hunt by stealth, followed by a quick dash. The rhipidistians seem to have been the Late Devonian lobefins best adapted to hunting fishes in shallow waters along sea coasts and into brackish shoreline lagoons and freshwater lakes and rivers. They look more active than coelacanths and were probably fast-sprinting ambush predators.

Coelacanths use their powerful tails for the final surge after prey, and the paired lobe fins act mainly to adjust attitude and speed as they stalk. The ventral fins beat rather like wings, up and down in the water. The beats are synchronized, with the right pectoral fin beating at the same time as the left pelvic fin, and vice versa. It's not clear why they do this, but it probably has to do with hydrodynamic eddying: it may be difficult to "fly" underwater with two sets of wings beating in pairs. (Dragonfly wings beat in air in the same pattern as coelacanth fins.) The fascinating aspect of the coelacanth pattern of ventral fin beats is that it is exactly the same one amphibians, reptiles, and mammals use to walk on land. So a pattern of fin or limb movement that is used in deep water today by coelacanths for swimming may have been shared by their Devonian relatives the rhipidistians, who also passed it on to the amphibians because it turned out (by chance) to be a useful pattern for squirming and walking across a substrate.

Figure 8.7 The Devonian fishes *Gyroptychius*, above, and *Osteolepis*, below, are typical rhipidistians with powerful low-set lobe fins. (After Jarvik.)

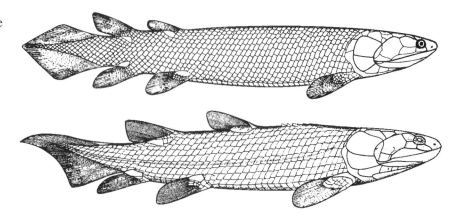

Rhipidistians had long, powerful, streamlined bodies with strong lobe fins and tail (Figure 8.7), adapted for strong swimming. They had long snouts, especially the larger ones. Perhaps as a result, rhipidistians evolved a skull joint that allowed them to raise the upper jaw as well as, or instead of, lowering the lower jaw as they gaped to take prey. This could have had two important effects, both related to life in shallow water. First, the snout movements would have changed mouth volume, perhaps allowing extra water to be pumped over the gills without moving the lower jaw. Second, rhipidistians could have caught prey in shallow water by raising the snout without dropping the lower jaw. Crocodiles do exactly the same thing as they take prey in shallow water. Some rhipidistians may have been able to chase prey right up to or even beyond the water's edge. Their powerful ventral lobe fins, set low on the body, may have allowed them to drive after prey on, over, or through shallow mud banks, thus making rapid trips over surfaces that could be called "land."

The main sprinting propulsion in rhipidistians came from the tail. The lobe fins were set on the dorsal side of the body as well as ventrally (Figure 8.8). In deep water, rhipidistians could attack prey from any angle. But in shallow water the ventral fins took on additional importance. The pectoral ventral fins could be used against the bottom as supports, strengthening the posture of the anterior trunk and acting as props in chasing; the pelvic ventral fins acted to grip and push on the substrate so that maximum effort could be expended against it, adding to the thrust. In some advanced rhipidistians, the lobe fins evolved toward limbs, not as an adaptation for walking but to become a more efficient fish.

Figure 8.8 In the Devonian rhipidistian *Eusthenopteron*, the dorsal lobe fins were attached to the spine just as firmly as the ventral fins were, and presumably played just as important a role in swimming. In those rhipidistians that evolved toward an amphibious life, the ventral fins evolved to become limbs because they happened to be placed where they interacted with the substrate. (After Jarvik.)

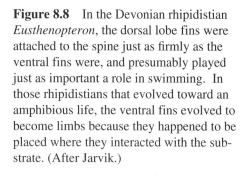

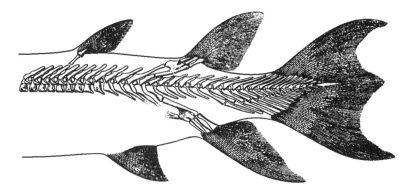

Rhipidistians are usually pictured as living exclusively in freshwater lakes and rivers, but that is not necessarily true. All Devonian lobefin groups were initially marine fishes. Rhipidistians radiated into fresh water mostly in the Middle and Late Devonian, at the same time that land plants were radiating abundantly.

We have a picture, then, of a varied group of rhipidistians, all hunters but some adapted to shallow-water habitats. All of them were fishes; none was adapted to be active out of water for any length of time.

Air Breathing

By Late Devonian time, a group of rhipidistians typified by *Eusthenopteron* (Figure 8.8), had evolved the nostrils that are now used in all tetrapods for air breathing, and they had evolved the internal air passage, the **choana**, that allows air breathing through the nostrils in all later tetrapods. These advanced rhipidistians were therefore breathing air, and they must have evolved lungs at about the same time that lungfishes did. But these two lobefin groups evolved air breathing independently, because the air passages take a different pathway in the two groups. Why did both these lobefin groups evolve the ability to breathe air?

Animals take in oxygen to burn their food in respiration, and produce CO_2 as a waste product. Carbon dioxide is toxic because it dissolves easily in water to form carbonic acid. Animals can tolerate only a small buildup of CO_2 before passing it out of the system. (It is high CO_2 in our lungs that makes us want to breathe out, not shortage of oxygen.)

Gases are exchanged with the environment as body fluids are passed very close to the body surface—for example, blood to the lung surfaces in our own breathing. The process of diffusion lowers any high concentrations of body fluids (say of CO_2) and raises any low ones (say of O_2). The process depends on several factors: the surface area and the thickness of tissue through which the gases must diffuse; the rate at which the external and internal fluids pass across the surface; and the concentrations of gases, both in the internal fluid and in the external medium. In normal fishes, CO_2 and oxygen diffuse in opposite directions across the gill surface.

Oxygen Intake

It may be easier and cheaper to extract oxygen from air rather than water. Water is hundreds of times denser and more viscous than air, and even at best it contains less oxygen. Many gill-breathing animals have to pump external water across their gill surfaces at ten times the rate they pump their internal blood. Gills have to be designed to resist the leakage of dissolved body salts, and the tissues across which oxygen is exchanged cannot be as thin as they can in air, so gas exchange is rarely anywhere near 100% efficient.

Because oxygen diffuses 100,000 times more quickly in air than in water, oxygen-poor air is rare. But oxygen may often be in short supply in water, so that it cannot diffuse inward into the gills. This is particularly likely to happen in tropical regions, wherever warm freshwater or saltwater lakes, ponds, or lagoons are partly or completely isolated, especially in a hot season. Warm, rotting debris can quickly use up oxygen, especially if there is little or no natural water flow. Even if the effect is only seasonal, it is still critical for fishes and other organisms living in the water. The water is stagnant, hot, and full of rotting debris, often teeming with bacteria that may also release toxic substances.

Why would fishes live in oxygen-poor water, where gill breathing is difficult? The food supply may be rich for fishes that can tolerate it, and there are situations in which fishes might benefit from swimming into areas of oxygen-poor water near the surface.

Many carnivorous fishes today are bottom feeders, hunting for small prey that live on or in the surface of the sediment. In warm latitudes, the bottom waters are often much cooler than the surface waters, which are heated by the sun. Digestion can be very slow in cold-blooded animals, especially if they live in cold environments. It may be the critical factor holding back growth and development. In such cases, increasing the digestive rate by swimming into the warm surface water can produce faster growth, earlier maturity, and more successful reproduction.

But what happens to a fish that swims into surface water because it is warm, only to find that it is also oxygen-poor? Even if surface waters are generally low in oxygen, there is always a thin surface layer of water, about a millimeter thick, that gains oxygen from the air by diffusion. Many living fishes in tropical environments come to this surface layer to bathe their gills in the surface oxygen layer. They can breathe, but they have to solve other problems too. If they break the surface, their bodies extend out into the air, losing some buoyancy. Some fishes in this situation bite off bubbles of air and hold them in their mouths for positive buoyancy, to remain at the surface without active swimming. Some living species of gobies breathe this way. Once they have an air bubble, they can extract oxygen from it in the back of their mouths much more efficiently than at the gills. When the oxygen level in the mouth bubble falls, reducing its size and its buoyant effect, the fish must then get rid of the bubble and bite off another. Rhythmic air breathing might evolve as a result of this action, as fishes begin to get rid of CO_2 from their mouth bubbles while they are still losing it at gill surfaces as well.

Oxygen intake in the mouth enriches the blood supply there. If that oxygen-rich blood is then circulated to the gills, oxygen can actually be lost back to the oxygen-poor water. In this situation, some fishes slow down their pumping at the gills. Some air-breathing fishes have reduced gills, but circulate some blood there to get rid of CO_2 because it dissolves so readily in water. This system would apply strong selection to evolve the ability to withstand high CO_2 concentration in the blood supply.

Because air is so rich in oxygen, a fish can store oxygen in an air bubble. An air bubble that takes up only 5% of body volume can increase oxygen storage by 10 times compared with a fish without a bubble.

An oxygen-low situation

Is bad for gill respiration

For some fishes, no trouble

They bite off a bubble

Inspirational improvisation!

Therefore, bubble breathing doesn't mean that a fish is tied to the surface; it can make extended dives to the bottom. This is true today of all air-breathers with low metabolic rates, including crocodiles and turtles.

Once air breathing has evolved, it can dominate respiration, even among fishes that never leave the water. This is true for the African lungfish and for the South American electric eel, which lives in the warm and often stagnant waters of the Amazon.

Air breathing probably evolved in many groups of early fishes that lived in environments where oxygen was seasonally low. In the Late Devonian, the northern continents were clumped together to form Euramerica, a little north of the Devonian equator (Figure 8.9). Rhipidistians and lungfishes lived in shallow lagoons along the shores, or in tropical rivers, deltas, and lakes, where they could have evolved at least the part of respiration that involved oxygen intake.

The evolution of air breathing has usually been linked with seasonal oxygen lows in tropical freshwater bodies, mostly because freshwater air-breathing fishes have been well studied. But one can make a similar argument for shallow marine lagoons. Oxygen is actually up to 10% less available from seawater than from fresh water, so any oxygen shortage would become more acute more quickly in salt water.

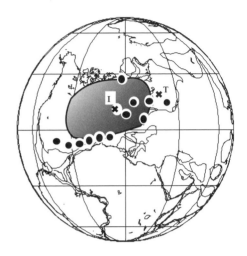

Figure 8.9 The paleogeography of the evolution of air breathing among lungfishes and rhipidistians in Late Devonian times. The black line marks the edge of the continent of Euramerica (North America plus Greenland and much of Europe). Black circles mark sites where lungfishes and rhipidistians have been found. Black crosses mark the sites of *Ichthyostega* in Greenland, and *Tulerpeton* in Russia.

Shedding Carbon Dioxide

It is usually easy to get rid of CO_2 in normal seawater. CO_2 dissolves and diffuses readily in it. But as we have seen, once oxygen is taken up from air rather than water, gills tend to be reduced, and internal CO_2 concentration in the blood tends to be higher. Other ways of getting rid of CO_2 might be favored, even in fishes that remain in the water. But some living fishes can and do leave the water. The walking catfish of Florida is one example (it's actually an invading African fish), and many mudskippers move around on tropical mud flats at low tide. In this situation, gills become inefficient at shedding CO_2, because their filaments stick together in air and collapse. It's difficult to lose CO_2 by diffusion into an air bubble in the mouth, because carbonic acid builds up rapidly as CO_2 and water react. But if gills don't work, another method must be found.

There are two possible alternatives: to lose some CO_2 through the skin surface rather than the gills (living amphibians lose most of their CO_2 this way), or to breathe rapidly and rhythmically, getting rid of the air bubble in the mouth quickly, before it gets too acidic. Many people think that early amphibians used skin surfaces for shedding CO_2, but I doubt it. Carbon dioxide loss at the skin is efficient only in small animals with a large area of skin relative to a small body volume, but the early amphibians were large and scaly. Furthermore, a tough, relatively impermeable, "leathery" skin would have been very helpful in avoiding loss of water and dissolved salts.

Early amphibians probably lost CO_2 by rapid breathing. The evolution of rapid breathing would have been more likely in animals with a large mouth pouch that had evolved some way toward a lung, and a pump to force air in and out, replacing the bite-and-spit method that works for smaller bubbles.

If lungfishes and rhipidistians had spent a long time in seasonally oxygen-poor waters, they might well have evolved a mouth pouch or a lung that allowed them to breathe air, at least for short periods. Rhipidistians had gills as well as nostrils, so they probably did have alternative ways of breathing.

Limbs and Feet: Why Become Amphibian?

Why would rhipidistians, as fast-swimming predators in the water, have evolved lobe fins that increasingly came to look and operate like the tetrapod limb? How would a rhipidistian have benefited from an ability to push on a resistant substrate, rather than using a swimming stroke in water? Why take excursions out into air, rather than simply breathe air at the water surface? In other words, why would a rhipidistian become amphibian? Evolutionarily, it can have resulted only from a chain of events that produced an improved rhipidistian.

The old story about this transition was that an ability to withstand air exposure helped a rhipidistian find another pool of water if the one it inhabited dried up in a drought. This idea is probably wrong. Research in the Florida Everglades has shown that animals around drying waterholes stay with the little supply there is, rather than striking off into parched country in the hope of finding more: it's simply a better bet for survival.

Basking?

The evolution of strong, low-slung lobe fins on rhipidistians probably helped them to hunt small prey in shallow water by poling their bodies through and over mudbanks. The fins became powerful enough to support the weight of the fish, at least briefly, while it gasped and thrashed its way along. The brief exposures to air would not have been long enough to pose much danger of drying out, but they would have pre-adapted rhipidistians for longer periods of exposure.

If some rhipidistians evolved the habit of sunning themselves on mudbanks to warm up their bodies, their digestion would have been faster than in the water; other things being equal, they would have grown faster, matured earlier, and reproduced more successfully than their competitors did. Basking behavior would have been effective even if the fish exposed only its back at first, supported mainly by its own buoyancy. But such effectiveness would have encouraged longer and more complete exposure. Some fishes, and many amphibians and reptiles (including alligators and crocodiles), bask while they digest (Figure 8.10).

As a basking, air-breathing rhipidistian became more exposed, more of its weight would have rested on the ground, threatening to suffocate it by preventing the thorax from moving in respiration. The pectoral fins in particular would have become stronger, to take more and more of the

Figure 8.10 Basking is important for many animals, including crocodiles, which are perhaps the closest living ecological analogues to the early amphibians. (From Lydekker.)

body weight during basking. Part of the shoulder girdle originally evolved to brace the gill region, and part to link with the pectoral fins. So the pectoral fins of rhipidistians still remained strongly linked with the skull and backbone, retaining the neckless appearance of all fishes.

Basking behavior may have made a more competitive fish, but we would still have recognized it as a rhipidistian. What other factors might have encouraged its evolution into a completely new kind of creature in a completely new environment?

Reproduction?

The most vulnerable parts of the life cycle of a fish are its early days as an egg and hatchling. If some rhipidistians could make very short journeys—even a meter or so to begin with—over land, or over very shallow water, they would have been able to find small, warm pools, lagoons, ponds, and sheltered backwaters nearby to spawn in. There would have been fewer predators in these side pools than in open water, and eggs and young would have survived better there. In much the same way, and for the same reasons, salmon struggle to swim far upstream to spawn, and many freshwater fishes swim into seasonally flooded areas to breed.

Isolated warm ponds would also have been ideal breeding grounds for small invertebrates such as crustaceans and insects, which would have formed a rich food supply for the young rhipidistians. Then, reaching a size at which they could handle larger prey and that would give them some protection against being eaten themselves, the young rhipidistians could make their way back to the main stream and take on their adult way of life as predators on fishes. Among young crocodiles, the greatest cause of death (apart from human hunting) is being eaten by an adult crocodile. Crocodiles provide intensive parental care while their young are small. Iguanas tend to separate juvenile and adult habitats.

Figure 8.11 Cladogram of lobefin fishes, simplified from Ahlberg, 1995; Janvier, 1996; and Ahlberg and Johanson, 1998. I have used the simpler name Rhipidistia for the group that modern workers usually call Tetrapodomorpha.

Figure 8.12 Above: Fishes swim by undulating the body significantly, while the head swings less. Below: The same basic body movements are used by a rhipidistian fish swimming or squirming over a mud bank, by an early amphibian crawling, and by a salamander walking on dry land. No sudden or large shifts in locomotory mechanism were required for the transition, even though the fish has fins and the salamander has feet.

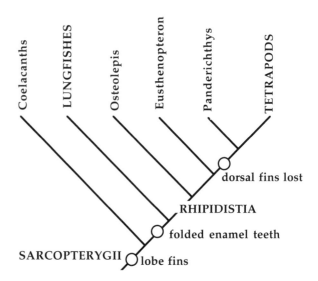

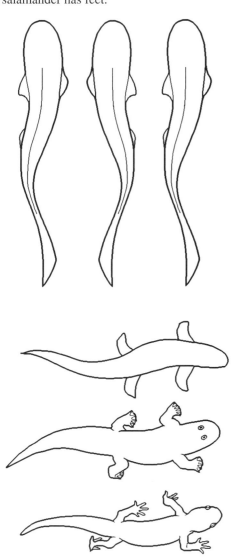

Rhipidistians perhaps solved the same problem in a different way, by arranging for their young to spend their juvenile period away from other adults.

Only one line of rhipidistians took this evolutionary path (Figure 8.11), while the others remained as normal fishes in rivers, lakes, and shallow coastal waters. Rhipidistians such as *Eusthenopteron* have long been recognized as the most likely tetrapod ancestors. The skull bones, the pattern of bones in the lobe fins, and the general size, shape, and geographic distribution of these rhipidistians, called **osteolepiforms**, are close to those of the earliest amphibians. Osteolepiforms have lobe fins with bony elements that correspond to a **1–2–several–many** pattern. Our limbs do the same: in our arms we have humerus; radius + ulna; wrist bones (carpals); hand bones; and in our legs we have femur; tibia + fibula; ankle bones (tarsals); foot bones. We and all other tetrapods share the same pattern, inherited from osteolepiform rhipidistians.

Rhipidistian locomotion in shallow water and on shallow mudbanks would have been improved by stronger fins, especially stronger fin edges. Land locomotion consisted at first of the same undulatory twisting that salamanders still have, with the fins acting simply as passive pivots (Figure 8.12). The fins gradually exerted stronger traction on the substrate, which may have encouraged the multiple rays in the fins to become fewer and stronger until toed feet evolved. In the process, the pectoral fins came to support the thorax, while the pelvic fins came to be better suited to push the body forward. The pelvic fin evolved a hinge joint at a "knee" and a rotational joint at an "ankle," a pattern that persisted into tetrapods. This difference was inherited by all later vertebrates—elbows flex backward, knees flex forward.

As the pectoral and pelvic girdles evolved better linkage with the fins, the fins evolved gradually to become clearly defined limbs. The hindlimbs quickly became as powerful as the forelimbs and evolved strong links with the backbone through the pelvis. Other changes also took place as a line of rhipidistians evolved into amphibians. A leathery skin evolved to resist water loss, fins were strengthened into limbs for lo-comotion and support, and senses improved for an air medium. Ecologi-cally, amphibians and rhipidistians divided up the habitat as they di-verged. Aberrant rhipidistians (evolving amphibians) spent more and more time at and near the water's edge, sunning and basking, while nor-mal rhipidistians remained creatures of open water.

Which creature in this evolutionary progression would we call the first amphibian? I (and many others) would choose the first tetrapod—the first animal that had evolved feet rather than fins. It is becoming clear that toes—digits, to be anatomically precise—are a new structure, *added on* to bones that were present in the fins of earlier fishes. Furthermore, feet, when they evolve, are separated from bones further up the limb by a joint at what we can now call an ankle or wrist.

Even so, if digits are useful to a fish in poling around on a muddy bottom, they may have evolved more than once. For example, the rhipi-distian *Sauripterus* has eleven distinct rays of bone in its lobe fin, though they are not toes, so *Sauripterus* is a fish, not an amphibian.

The First Amphibians

The earliest amphibian (and tetrapod) discovered so far is *Elginerpeton*, from late Devonian rocks of Scotland, perhaps 368 Ma. Its tibia has a well-preserved joint surface at its lower end where an ankle (rather than a fin) would have been (Figure 8.13). Its jaw is also tetrapod-like rather than fish-like. *Elginerpeton* was large. The jaw alone is 40 cm (1 foot or more) long, and the whole animal must have been several feet long. *Elgi-nerpeton* is the sister group rather than the ancestor of later amphibians.

Our knowledge of slightly later Late Devonian amphibians rests on four creatures that are well known enough to discuss: a piece of a shoul-der from Pennsylvania (*Hynerpeton*); two animals known from almost complete skeletons found in Greenland (*Ichthyostega* and *Acanthostega*) and *Tulerpeton* from Russia, which is incomplete. We know that the geo-graphic range, anatomy, habitat, and ecology of early amphibians were varied. *Tulerpeton* and *Hynerpeton* could have walked quite well on land, *Acanthostega* was much more adapted to life in water, and *Ichthyo-stega* was somewhere in between. *Hynerpeton* is around 365 Ma, slightly older than the others, which are very close to the Devonian/Carboniferous boundary (363 Ma).

Hynerpeton died on a muddy tropical delta floodplain, slightly south of the equator (Figure 8.9). It occurs with fishes that include large rhipi-distians, acanthodians, osteolepiforms, and antiarchs. Its shoulder bones suggest that *Hynerpeton* had already lost any internal gills. Also, the bones have places for muscle insertion that suggest a strong muscle sling connected the shoulder girdle to the spine, helping to support the chest and move the shoulder joints effectively in walking.

Acanthostega (Figure 8.14) is the most primitive known tetrapod, with many fish-like characters. For example, it still had functional gills,

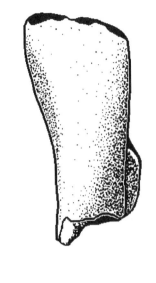
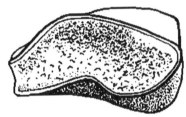

Figure 8.13   *Elginerpeton*, from the Late Devonian of Scotland. The tibia (top, side view) is well enough preserved to show from the ankle joint (below) that this ani-mal could walk. From Ahlberg.

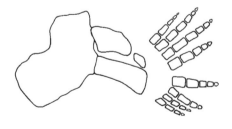

Figure 8.14   The front limb of *Acantho-stega*, a very early amphibian from the Late Devonian of Greenland. Although the limb clearly has toes, it looks like a func-tional flipper rather than a walking foot. There are eight toes. (After Clack and Coates and Thomson.)

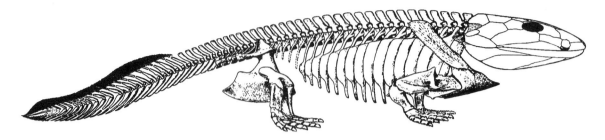

Figure 8.15 *Ichthyostega*, the earliest well-known amphibian, from Late Devonian rocks of Greenland. (After Jarvik.)

which means it was fully capable of breathing underwater as well as in air. Its forelimbs were rather weak, its ribs did not curve round to support its weight well, and its eight-toed lower limbs were still somewhat flipper-like. It may have been best adapted to weed-choked shallows, and it may not have been able to support its weight for long out of water.

Ichthyostega had a massive skeleton but was otherwise very much like Late Devonian rhipidistians in spine, limb, tooth, jaw, palate, and skull structure (Figure 8.15), and probably in diet and locomotion (Figure 8.16). Like rhipidistians, it had a lateral line and a tail fin, but unlike them it had a strong rib cage and powerful limbs and feet rather than lobe fins. These were adaptations for excursions into air. *Ichthyostega* solved the problem of supporting the chest for breathing by having a massive set of ribs attached to the backbone. The body could then be suspended from the frame of the backbone alone, and the skull connection could be less rigid. Amphibians could be described as evolving necks!

Ichthyostega had large conical teeth, suitable for catching large fishes rather than small prey. As an adult, it was probably much like a living

Figure 8.16 Reconstructions of the locomotion of the very early amphibian *Ichthyostega*. Top: Its normal way of life, swimming. Middle: Plowing through shallow water. Bottom: Posing shyly but triumphantly on dry land.

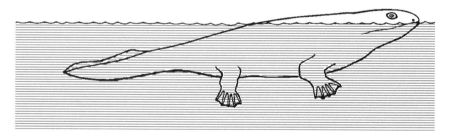

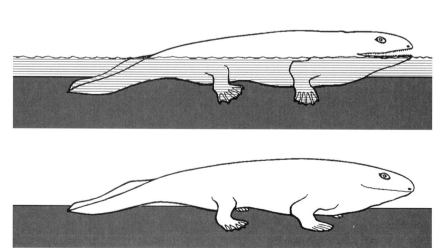

Though Ichthyostega was tired

His imagination was fired

With a gait that was heavy

He'd climbed over the levee

And the tetrapod clade he had sired.

crocodile in appearance (Figures 8.15 and 8.16). Ecologically, the early amphibians resembled reptiles more than frogs and toads. We have not found juveniles yet. Devonian insects and plants were not a suitable food supply for early adult amphibians, which were large (more than a meter long) and had teeth adapted for eating fish. These animals came out into air for reasons other than food, such as the digestive and reproductive advantages I have mentioned. Perhaps only the babies ate crustaceans and insects (and worms and molluscs) in their warm, shallow nursery pools.

All these early tetrapods were polydactylous—they had many digits. Tetrapod toes are not modified rays from the rhipidistian fin, but they are new, "add-on" bones in the lobefin structure. Early tetrapods had rather variable numbers of toes, but not five: *Acanthostega* had eight, *Ichthyostega* had seven (Figure 8.17), and *Tulerpeton* had six.

The loss of toes seems to be linked to the relative adaptation to air exposure: *Acanthostega* was the most aquatic and *Tulerpeton* the most capable of terrestrial travel, with *Ichthyostega* intermediate. This does not mean that *Tulerpeton* was in any way "better" than the others, though it may mean that it basked more and travelled further over mudbanks than the others. In evolutionary terms, *Tulerpeton* is more derived along the clade that leads toward later tetrapods (Figure 8.18).

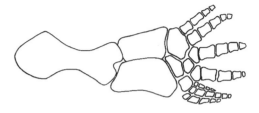

Figure 8.17 The back limb of *Ichthyostega*. This limb is much stronger than that of *Acanthostega*, and looks like an effective walking limb. There are seven toes (well, six and a half). (After Clack and Coates and Thomson.)

There's something I missed, I'm afraid:

The part that the other sex played

SHE contributed most

To the tetrapod host

She LAID the whole tetrapod clade!

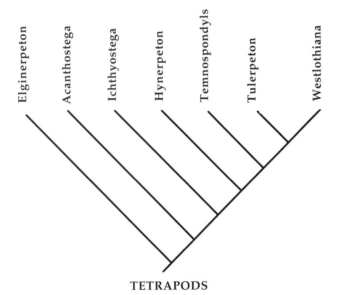

TETRAPODS

Figure 8.18 Cladogram of some early tetrapods, some mentioned in this Chapter and some in Chapter 9. The evolutionary hypothesis in this cladogram suggests that the known early tetrapods show increasing adaptation toward effective movement on land, culminating in three advanced clades: *Tulerpeton*; the temnospondyls, ancestors of living amphibians; and *Westlothiana*, ancestor of all other advanced tetrapods. Simplified from Coates, 1996.

Review Questions

What were the first animals to live on land?

Why would fishes evolve to breathe air?

Why did plants evolve to live in air rather than continuing to live as water plants?

Name a living animal that is probably reasonably close to *Ichthyostega* in ecology, at least as an adult.

Why did amphibians evolve the capacity to leave the water?

THOUGHT QUESTION.—Plants evolved on to land in Silurian times, and there were trees by the Middle Devonian. So land-going animals found that plants were there first. Perhaps it was vital for plants to have been first on to land. 1. Without land plants, would rhipidistians have evolved into amphibians? (Explain) 2. Without land plants, would amphibians have evolved into reptiles? (Explain)

THOUGHT QUESTION.—The first amphibian *Ichthyostega* is BIG. Most evolutionary breakthroughs are made by small creatures. Why was the evolution of amphibians allowed or encouraged by large body size? In other words, why was the first successful amphibian BIG?

Further Reading

Easy Access Reading

Ahlberg, P. E. 1995. *Elginerpeton pancheni* and the earliest tetrapod clade. *Nature* 373: 420–425, and comment, pp. 389–390.

Ahlberg, P. E., et al. 1996. Rapid braincase evolution between *Panderichthys* and the earliest tetrapods. *Nature* 381: 61–64, and comment, pp. 19–20.

Ahlberg, P. E., and Z. Johanson. 1998. Osteolepiformes and the ancestry of tetrapods. *Nature* 395: 792–794, and comment, pp. 748–749.

Clack, J. A. 1994. Earliest known tetrapod braincase and the evolution of the stapes and fenestra ovalia. *Nature* 369: 392–394; see earlier work in 342: 425–427, and comments, v. 344, p. 116 and p. 823.

Coates, M. I., and J. A. Clack. 1990. Polydactyly in the earliest known tetrapod limbs. *Nature* 347: 66–69.

Coates, M. I., and J. A. Clack. 1991. Fish-like gills and breathing in the earliest-known tetrapod. *Nature* 352: 234–235. *Acanthostega*.

Daeschler, E. B., et al. 1994. A Devonian tetrapod from North America. *Science* 265: 639–642. *Hynerpeton*.

Daeschler, E. B., and N. Shubin. 1998. Fish with fingers? *Nature* 391: 133.

Edwards, D., et al. 1992. A vascular conducting strand in the early land plant *Cooksonia*. *Nature* 357: 683–685, and comment, pp. 641–642.

Fricke, H., et al. 1987. Locomotion of the coelacanth *Latimeria chalumnae* in its natural environment. *Nature* 329: 331–333.

Keeley, J. E., et al. 1984. *Stylites*, a vascular land plant without stomata absorbs CO_2 via its roots. *Nature* 310: 694–695, and comment, p. 633.

Shear, W. A. 1991. The early development of terrestrial ecosystems. *Nature* 351: 283–289.

Shear, W. A., et al. 1996. Fossils of large terrestrial arthropods from the Lower Devonian of Canada. *Nature* 384: 555–557.

Shubin, N., et al. 1997. Fossils, genes and the evolution of animal limbs. *Nature* 388: 638–648. Valuable references to previous work by Shubin and others.

Woodward, F. I. 1987. Stomatal numbers are sensitive to increases of CO_2 from pre-industrial levels. *Nature* 327: 617–618, and comment, p. 560.

Wurtsbaugh, W. A., and D. Neverman. 1988. Post-feeding thermotaxis and daily vertical migration in a larval fish. *Nature* 241: 846–848.

Zimmer, C. 1995. Coming onto the land. *Discover* 16 (6): 118–127. Jenny Clack and *Acanthostega*.

More Technical Reading

Ahlberg, P. E., and N. H. Trewin. The postcranial skeleton of the Middle Devonian lungfish *Dipterus valenciennesi*. *Transactions of the Royal Society of Edinburgh, Earth Sciences* 85: 159–175.

Bendix-Almgreen, S., et al. 1990. Upper Devonian tetrapod palaeoecology in the light of new discoveries in East Greenland. *Terra Nova* 2: 131–137. *Acanthostega* carcasses piled up as floating corpses on riverside sandbars.

Coates, M. I. 1996. The Devonian tetrapod *Acanthostega gunnari* Jarvik: postcranial anatomy, basal tetrapod interrelationships and patterns of skeletal evolution. *Transactions of the Royal Society of Edinburgh, Earth Sciences* 87: 363–421.

DiMichele, W. A., and R. W. Hook (rapporteurs) 1992. Paleozoic terrestrial ecosystems. Chapter 5 in A. K. Behrensmeyer et al. *Terrestrial Ecosystems Through Time*. Chicago: University of Chicago Press.

Janvier, P. 1996. *Early Vertebrates*. Oxford: Clarendon Press.

Jarvik, E. 1996. The Devonian tetrapod *Ichthyostega*. *Fossils and Strata* 40: 1–213.

Knoll, A. H. 1984. Patterns of extinction in the fossil record of vascular plants. In M. H. Nitecki (ed.). *Extinctions*, pp. 21–68. Chicago: University of Chicago Press.

Lebedev, O. A., and M. I. Coates. 1995. The post-cranial skeleton of the Devonian tetrapod *Tulerpeton curtum* Lebedev. *Zoological Journal of the Linnean Society* 11: 307–348.

Little, C. 1990. *Terrestrial Invasion: An Ecophysiological Approach to the Origins of Land Animals*. Cambridge, England: Cambridge University Press.

Niklas, K. J., et al. 1985. Patterns in vascular land plant diversification: an analysis at the species level. In J. W. Valentine (ed.). *Phanerozoic Diversity Patterns*, pp. 97–128. Princeton: Princeton University Press.

Randall, D. J., et al. 1981. *The Evolution of Air Breathing in Vertebrates*. Cambridge, England: Cambridge University Press.

Raven, J. A. 1984. Physiological correlates of the morphology of early vascular plants. *Botanical Journal of the Linnean Society* 88: 105–126.

Schultze, H.-P., and L. Trueb (eds.). 1991. *Origins of the Higher Groups of Tetrapods: Controversy and Concensus*. Ithaca, New York: Cornell University Press. The first five chapters deal with amphibians.

Trewin, N. H., and K. J. McNamara. 1995. Arthropods invade the land: trace fossils and palaeoenvironments of the Tumblagooda Sandstone (?late Silurian) of Kalbarri, Western Australia. *Transactions of the Royal Society of Edinburgh, Earth Sciences* 85: 177–210.

Amphibians and Reptiles

Once the first tetrapods evolved from rhipidistian fishes, they radiated quickly into a great variety of sizes, shapes, and ways of life that we now recognize as characteristic of amphibians and reptiles. Although this is the most poorly known part of the terrestrial vertebrate record, it is also one of the most exciting areas in which research is constantly turning up new fossils. In this chapter I will give a progress report on the story as we see it now, and some of the problems that these early land animals faced and solved.

Paleozoic Amphibians

Amphibians can live in air or in water, and most of them spend time in both media. For example, if I have described the ecology of *Ichthyostega* correctly, it spent its adult life in water, with only occasional journeys into air for basking and spawning. The living spadefoot toad spends most of its adult life in air, with only occasional immersion in rain puddles and desert springs for spawning. The biology and ecology of Paleozoic amphibians reflects the fact that they were the first large animals to exploit the environment in and around the water's edge. Their variety reflects different adaptations to different habitats and different ways of life. Some were dominantly terrestrial, some aquatic, and some genuinely amphibian. Naturally, there were variations even within each group.

Living amphibians are all small-bodied and soft-skinned, and they are quite unlike early amphibians. They are the salamanders and newts, the frogs and toads, and the caecilians—burrowing legless amphibians. These living forms are usually classed together as a clade, the "smooth amphibians," **Lissamphibia**. They are very much derived and probably did not evolve until Late Permian or even Triassic times. Though their

Figure 9.1   One possible phylogram for amphibians. There are several competing alternatives, none of them representing a particularly strong hypothesis. The phylogram here is stronger than most, because I have omitted microsaurs, aistopods, and nectrideans, whose position is uncertain.

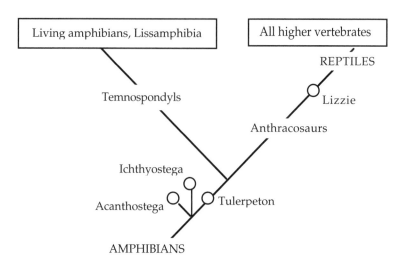

biology is fascinating, they are no guide at all to the origin or the paleobiology of their Paleozoic ancestors.

Early Amphibians

Evolutionary patterns among early amphibians are difficult to work out. Several amphibian groups were evolving in parallel in the Late Devonian, some like *Acanthostega* toward a more aquatic life than *Ichthyostega*, some like *Tulerpeton* toward a more terrestrial life (Chapter 8). One possible evolutionary scheme is shown in Figure 9.1. But it seems that the picture changes each time a new fossil is studied carefully, or an old one is cleaned. Our ideas are likely to change a lot in the next few years.

For example, a new fossil has been named *Eucritta melanolimnetes*, "the creature from the black lagoon," which adds another puzzling animal to the list of early amphibians (and indicates the cinematic taste of its author, Jenny Clack). *Eucritta* has a skull rather like an anthracosaur and a body rather like a temnospondyl. It looks rather like a salamander, but of course it is not one. Clack suggests that *Eucritta* belongs right at the base of amphibian radiation, which means that we are adding complexity to the puzzle with the new fossil, rather than resolving it.

To make matters worse, the first reptiles evolved in the Early Carboniferous, and we have not found their amphibian ancestors. The evolution from fishes to reptiles probably occurred very quickly, by way of a very primitive amphibian. More information on early reptiles will significantly change our view of the ecology of the time, as well as our view of early amphibian evolution.

At present, we have an unsatisfactory view of a diverse set of early amphibians. The best we can do is to distinguish two major lineages. One led (eventually) to living amphibians and one led to living reptiles, birds, and mammals. Both lineages are found among Early Carboniferous fossils, so the split between them was very early.

Early Carboniferous Amphibians

Some early amphibians quickly evolved more terrestrial adaptations than those displayed by *Acanthostega* or *Ichthyostega*. The best-known Early Carboniferous amphibians are from East Kirkton in central Scotland, where there was a complex tropical delta environment at the time (Figure 9.2). The rocks were laid down in shallow pools fed by hot springs. No fishes were found, possibly because the pools were too hot for them. Instead, an early community of terrestrial animals lived in the lowland swamps near the pools, walked by them and sometimes fell into them. These creatures included scorpions and millipedes, the earliest known harvestman, several different amphibian groups including *Eucritta*, and the earliest known reptile.

Ancestors of Living Amphibians: The Temnospondyls

Temnospondyls are the largest and most diverse group of early amphibians: 40 families and 160 genera have been described altogether. There are over 30 skeletons of temnospondyls in the East Kirkton collections. These skeletons are soon after *Ichthyostega* in time, and close to it anatomically and paleogeographically (Figure 9.2). Temnospondyls were clearly descended from an amphibian such as *Ichthyostega*. Many temnospondyls were large, with teeth like those of rhipidistians and *Ichthyostega*. The most common temnospondyl at East Kirkton, *Balanerpeton*, is about 50 cm (20 inches) long. It has heavy bones, and its beautifully preserved feet look strong and well adapted for walking. Like *Ichthyostega*, temnospondyls probably had a biology much like that of crocodiles, and they were fish-eaters as adults (see *Eryops*, Figure 9.3). The jaw was designed to slam shut on prey, and the skull was very strong.

More terrestrially adapted temnospondyls had very massive, strong skeletons capable of supporting them on land, even when they were rather small (Figure 9.4). Some temnospondyls even became more terrestrial during their lifetimes. For example, young *Trematops* had a jaw designed for eating small, soft food items, but adults had a carnivorous jaw and a lightly built skeleton capable of rapid movement, and were probably land-going predators.

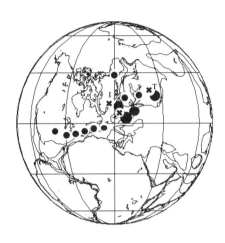

Figure 9.2   The paleogeography of early tetrapods. Black circles mark rhipidistians and Devonian lungfishes, *I* marks the site of *Ichthyostega*, *T* marks the site of *Tulerpeton*, and *R* marks the East Kirkton site in Scotland that yielded many early amphibians and the earliest reptile, Lizzie.

Figure 9.3   Some temnospondyls were capable of land locomotion even though they spent most of their time in water. This heavily built early amphibian, *Eryops*, was much like a crocodile in size and probably in ecology. The limbs were sprawling, the head was massive, and an adult must have been very clumsy on land. (Negative 35632. Photograph by A. E. Anderson. Courtesy of the Department of Library Services, American Museum of Natural History.)

Figure 9.4 The terrestrially adapted Permian temnospondyl *Cacops* was small, with a body length of only 40 cm (16 inches), but massively built. The strength of the front limbs and shoulder girdle is associated with the heavy skull. (After Williston.)

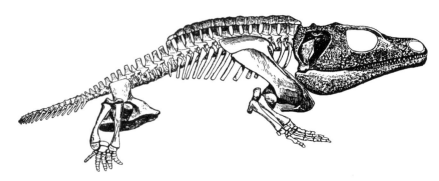

As part of their adaptation to life in air, temnospondyls had an ear structure that suggests they could hear airborne sound. The stapes bone seems to have conducted sound to an amplifying membrane that sat in a special notch in the skull. Other amphibians, including those that are our ancestors, did not evolve such an advanced system.

Temnospondyls were the most common Late Paleozoic amphibians. They survived into the Cretaceous in isolated areas of Gondwanaland. Some Triassic forms were giant marine animals, the largest amphibians ever to evolve, and they ranged as far south as Antarctica (amphibians can live in cold water as long as they spawn in spring or summer). A Late Paleozoic temnospondyl was probably the ancestor of living frogs, toads, and salamanders, though we lack a detailed fossil record of the transition. If the connection is confirmed, temnospondyls are not extinct.

Small but Interesting Groups of Early Amphibians

We do not yet know how to classify some of the small, mainly aquatic early amphibians.

Microsaurs were small, with weakly calcified skeletons. Their remains are usually fragmentary, and they include many juvenile forms. **Aistopods** were small, slim amphibians that had lost almost all trace of their limbs. They probably lived rather like little water snakes (Figure 9.5). They do not preserve well and their fossils are quite rare, so they have not been properly studied.

Nectrideans are better preserved and understood. They had a short body and a long, laterally flattened tail that made up two-thirds of their total length and was probably used for swimming. The vertebrae were linked in a way that permitted extremely flexible bending, and nectrideans probably swam like eels.

Horned nectrideans are fascinating. They had flat, short-snouted skulls with the upper back corners extended backward on each side. In early forms the extensions were quite small, but later they evolved to look like the swept-back wings of a jet fighter (Figure 9.6). The eel-like swimming of nectrideans would have undulated the whole body, including the head, making it difficult to seize prey in a forward strike while swimming. A massive head would have oscillated less than a light, narrow one, perhaps explaining this bizarre skull development. Angela Milner suggested that the backward-pointing skull projections were attached

Figure 9.5 An aistopod, which perhaps lived like a little water snake. A complete specimen is about 75 cm long (30 inches). (After Fritsch.)

by ligaments to the shoulder girdle to cut down head movement in normal swimming. (Note that her suggestion makes sense only if the shoulder girdle oscillated less than the skull does.) Such a connection would have cut down on trunk oscillation too, and some horned nectrideans evolved stiffer, shorter trunks by having fewer vertebrae between shoulder and pelvic girdles. They then began to evolve their hind limbs into flippers, which could now exert some force against a shorter, stiffer trunk. In the later horned nectrideans, the very large delta-winged skulls may have acted as hydrodynamic structures, giving lift and swimming control.

Reptile-like Amphibians: The Anthracosaurs

Anthracosaurs are the other numerous and diverse group of early amphibians. *Tulerpeton* may be the earliest one (Chapter 8), though we need more complete specimens to confirm this. There are certainly two anthracosaurs in the East Kirkton fauna. Most anthracosaurs were adapted for life primarily in water, as long-snouted and long-bodied predators, presumably crocodilelike fish-eaters, with jaws designed for slamming shut on prey. Their limbs were not very sturdy, but they may have been very good at squirming among dense vegetation in and around shallow waters. They must have swum in an eel-like fashion. It's unlikely that they had the speed and power to compete with rhipidistians in open water, even though some were quite large, up to 4 meters (13 feet) long. A few anthracosaurs were smaller, slender animals, adapted to terrestrial life. These amphibians seem to be the closest relatives of the early reptiles, even though their ears remained adapted for low-frequency waterborne sound. *Seymouria* (Figure 9.7) and *Diadectes* are well-known members of this large clade of amphibians.

Figure 9.6 *Diplocaulus magnicornis*: the skull and backbone of a horned nectridean. (Negative 35679. Photograph by Thomson. Courtesy of the Library Services Department, American Museum of Natural History.)

Figure 9.7 *Seymouria* was one of the few anthracosaurs adapted for terrestrial life. It is probably close to all later reptiles. *Seymouria* was about 60 cm long (2 feet). (After White and Romer.)

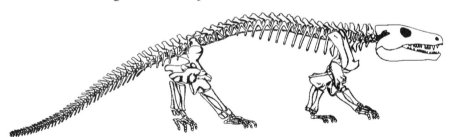

The First Reptiles

The earliest reptile known so far ("Lizzie") was discovered in 1988 at East Kirkton, Scotland. Lizzie, now formally named *Westlothiana*, is a well-preserved skeleton about 20 cm (8 inches) long, with an anatomy that could easily make it the ancestor of all later reptiles (Figure 9.1). In 1999, a new fossil, *Casineria*, was reported from rocks of the same age a few miles away. *Casineria* does not have the head preserved, but its small body is very much like that of later, more advanced reptiles, and it

Figure 9.8 Life reconstruction of *Hylo-nomus*, an early reptile, as an analog of a living lizard. (From Lydekker.)

has five-fingered forelimbs. This underlines the fact that reptiles evolved very early, and at small size.

Why Were the First Reptiles Small?

Rhipidistian fishes and the large early amphibians had a long, heavy skull, with a jaw designed for slamming shut on larger prey, which was then swallowed whole or in large pieces. There was little chewing, and the jaw muscles generated little pressure along the tooth line.

However, some Carboniferous amphibians, and all early reptiles, were *small*. All early reptiles were about the same size as living lizards, and much like them in body proportions, posture, and jaw mechanics (Figure 9.8). They were probably like them in ecology too. They had a notably small skull with a short jaw well suited to hold, chew, and crush small, wriggling prey, and to shift the grip for repeated bites. The small head was set on a neck joint that allowed very swift three-dimensional motion, whereas in many amphibians the long, heavy skull moved mainly up and down and the neck must have seemed very stiff (this adaptation is effective for swimmers). Why this small size?

Animals that spent longer time on land did not do so simply because they were seeking places to breed, but because there were potential food supplies there. Carboniferous forests were rich in worms, insects, and grubs. Young or small animals would have been best suited for foraging after this kind of food. Worms, insects, and grubs are small, though highly nutritious; they are easy to seize, process, and swallow; and they can be found among cracks and crevices in a maze of plant growth in a complex three-dimensional forest setting. Most of these potential prey items are slow-moving, and a successful predator need not have been quick and agile at first. But small and light-bodied animals could have quickly evolved greater agility as their repertoire of prey extended to the expanding number of large insects in Carboniferous forests.

Small body size may also have been favored by a thermoregulatory effect. Animals encounter greater temperature extremes in air than they do in water, and small animals can shelter more easily from chilling or overheating among vegetation, in cracks and crevices, or in hollow tree trunks than can animals the size of *Ichthyostega*. Small bodies are also quicker and easier to heat by basking in the sun. Again, this suggests that terrestrial and/or arboreal excursions would have most benefited juvenile or small animals.

The currently favored scenario of reptile evolution, mostly due to Robert Carroll, is that reptiles evolved on a forest floor covered with rotting material, leaf litter, fallen branches, and tree stumps, ideal places for prey to hide and reptiles to search (Figure 9.9, for example).

This scenario sounds reasonable enough to people used to temperate forests. But tropical forest floors like those of the Amazon today are clean. The shade of the forest canopy is so thick and continuous that no vegetation grows at ground level on the forest floor, except along river banks where water barriers break the continuity of the canopy, or where

storms (or people) have carved an open track of fallen trees. Fallen branches, leaves, bodies, and other pieces of organic debris are broken down and recycled so quickly on the ground by fungi and insects that vertebrates find it hard to make a living there. In contrast, the canopy and the river banks teem with small vertebrates: reptiles, amphibians, mammals, and birds in the canopy, and fishes in the water.

I suggest instead that the first reptiles evolved either on the riverbanks or in the canopy ecosystem of the Early Carboniferous, not on the forest floor. Whichever suggestion one prefers, I would argue that reptiles were living in the canopy forest in the Late Carboniferous. The Late Carboniferous coal forests included giant lycopod trees 30 meters (100 feet) high (Figure 9.9), and dense tree ferns up to 10 meters high formed an understory (Figure 9.10). The lycopod trees were fragile and shallow-rooted, and they may have been hollow in life, as many tropical trees are today. After storms or old age felled them (Figure 9.9), their hollow stumps sometimes remained standing. The fossil forests of Early Pennsylvanian (Late Carboniferous) rocks of Nova Scotia are famous because they include many tree stumps and tree trunks fossilized upright in life position (Figure 9.11). Some early reptiles have been found preserved inside some of the hollow stumps. This may not be a freak of preservation. I think that they *lived* inside the hollow trunks of living trees, as little insectivorous mammals do today in tropical rain forests, but perhaps they sheltered in the hollow stumps or were washed into the stumps in floods.

Figure 9.9 Reconstruction of a Carboniferous swamp. Trees were tall but often had shallow roots and weak structure, and they were frequently felled by storm and flood. A rich fauna of insects, spiders, and other arthropods lived in this ecosystem, and I think it is likely that small early reptiles lived as much in the tree tops as under them. This scene has much growth on and near the forest floor because it is near a natural open space—the water body in the middle distance. (Negative 333983. Courtesy of the Department of Library Services, American Museum of Natural History.)

Figure 9.10 Tree ferns such as *Psaronius* formed a dense understory several meters high in Carboniferous swamp forests, under the 30-meter canopy forest. (After Morgan.)

Figure 9.11 In 1868, the Canadian geologist William J. Dawson described fossils of early reptiles preserved inside hollow tree stumps still standing erect in a fossil forest at Joggins, Nova Scotia. Dawson gave a vivid and sensible word picture of this ancient environment, and this engraving illustrated his account.

Vertical climbing is easy with a small body size, so small Carboniferous animals could have been tree dwellers, as many salamanders are today. Trees offer damp places in which to lay eggs, and rich insect life high in the canopy forest provides abundant food. Even today, salamanders (and spiders) are the top carnivores in parts of the Central and South American canopy forest. The rich fossil record of Late Carboniferous insects, scorpions, spiders, and reptiles may portray the ecosystem of the canopy rather than the forest floor.

Remember that most early amphibians did not evolve toward spending more time on land. If anything, most of them evolved toward spending more time in water. I suspect that this tendency was related to the difficulties they encountered in locomotion on land and in air breathing; I discuss those problems in Chapter 11.

The Amniotic Egg

Living reptiles differ from living amphibians in several characters of the skeleton that can be recognized in fossils, and in other characters that affect the soft parts and cannot be recognized in fossils. The major soft-part character of living reptiles is that they lay amniotic eggs surrounded by a membrane, rather than the little jelly-covered eggs of fishes and amphibians. This fundamental difference in biology needs special attention because it was so important in the evolution of vertebrates into entirely terrestrial habitats. How did the amniotic egg evolve, and who evolved it? (The earliest reptiles probably laid amphibian-style eggs, and the amniotic egg was probably evolved at some later time. Only someone uneasy about the process of evolution would have any problem with this argument.)

Amphibians have successfully solved most of the problems associated with exposure to air. But their reproductive system was and is linked to water, and it remains very fishlike. Almost all amphibians spawn in water and lay a great number of small eggs that hatch quickly into swimming larvae. The eggs do not need any complex protection against drying, because if the environment dries, the larvae are doomed as well as the eggs. Thus, selection has acted to encourage the efficient choice of suitable sites for laying eggs, rather than devices to protect eggs. Both fishes and amphibians may migrate long distances for spawning, and favored sites are often disputed vigorously.

Living reptiles have a different system. Their juveniles hatch into air as competent terrestrial animals, often miniature adults. Yet the stages of embryological development are strikingly similar to those of amphibians. The difference is that reptiles develop for a longer time inside the egg, which in turn means that the egg must be larger and must provide more food and other life-support systems. Reptiles typically lay far fewer eggs than amphibians of comparable body size, so they have evolved more complex adaptations to ensure greater chances of survival for each individual egg.

A reptile egg is enclosed in a tough membrane covered by an outer shell made of leathery or calcareous material. The membrane and shell layers allow gas exchange with the environment (water vapor, CO_2, and oxygen) for the metabolism of the growing embryo, but they also resist water loss. Reptiles lay eggs on land, so eggs are not supported against gravity by water. Instead, the shell gives the egg strength, protects it, holds it in a shape that will allow the embryo room to grow unconstricted, and buffers it against temperature change and desiccation.

Inside the egg, the embryo is nourished by a large yolk, and special internal sacs act as gas-exchange and waste-disposal modules. The most fundamental innovation, however, is the evolution of another internal fluid-filled sac, the **amnion**, in which the embryo floats. Amniotic fluid has roughly the same composition as seawater, so in a sense, the amnion is the continuation of the original fish or amphibian egg together with its microenvironment, just as a space suit contains an astronaut and a fluid that mimics Earth's atmosphere. The rest of the amniotic egg is add-on technology that is also required for life in an alien environment, corresponding to the rest of the spacecraft with its food storage, fuel supply, gas exchangers, and sanitary disposal systems (Figure 9.12). A normal amphibian embryo produces urea as a soluble waste product that is easily flushed from an egg in water; but because water conservation is important in an amniotic egg, a reptile embryo produces uric acid, which is practically insoluble and thus can be stored as crystals inside a special egg compartment without polluting it and without using up precious water that is not easily replaced.

Because the embryo inside an amniotic egg is encased in multiple membranes, and often inside a shell, the female's eggs must be fertilized before they are finally packaged. Internal fertilization must have evolved among early reptiles or their immediate amphibian ancestors.

The evolution of the amniotic egg was the critical event that broke the final reproductive link with water and allowed vertebrates to take up truly terrestrial ways of life. Its evolution demanded changes in behavior patterns and in soft-part anatomy and physiology. The transitional forms either evolved into or were outcompeted by more advanced animals, so they are now extinct and unavailable for direct study.

The amniotic egg was evolved by a creature that we would probably recognize as an early reptile. Its descendants became so abundant, so diverse, and so successful that we now regard them as the first members of a major higher group or clade of vertebrates, the **amniotes**, which include reptiles, birds, and mammals.

I present here a reasonable scenario for the evolution of the amniotic egg among early reptiles. My suggestion could be supported or contradicted by future evidence, or replaced by a simpler or more elegant story.

With the increasing ability of vertebrates of all kinds to make forays onto land, their breeding grounds became much less secure. Sites that had once been safe refuges for young animals gradually became more susceptible to raiders. The same evolutionary pressures that I suggested in Chapter 8 for the origin of amphibians from fishes now drove early reptiles to seek still safer refuges for breeding and the development of

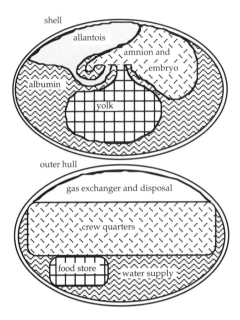

Figure 9.12 The amniotic egg is in many ways analogous to a spacecraft.

their young. In so doing they evolved into the first amniotes.

Forays farther from water became more practicable as the evolution of flourishing plant life provided a myriad of damp hiding places in and around Carboniferous swamps. Small reptiles could have found small, sheltered, hidden places that were damp enough to foster egg development but not obvious enough to attract predators. They may have evolved behavior patterns like those of many living amphibians, particularly tree frogs and tree-dwelling salamanders.

The major problem in laying unshelled eggs away from larger water bodies is drying, at spawning time and during development. A crude sort of egg membrane would have been a partial solution to the developmental problem; further refinement of the system is then fairly easy to imagine. Internal fertilization was probably a preliminary solution to the problem of desiccation while spawning. (Some living amphibians have independently evolved a crude kind of internal fertilization.)

Most living frogs and toads undergo a complex development after hatching. They experience a drastic metamorphosis from tadpole to adult that involves not only a major anatomical reorganization but a major change in life style. The problems associated with this kind of amphibian reproduction can be solved, sometimes in spectacular ways. Tree frog eggs often hatch into tadpoles in places where there is little water. Some frogs carry their tadpoles one by one to little pools in bromeliad plants; some carry them in pouches on their bodies, where the young develop into miniature adults; and in some Australian frogs the females swallow their eggs after they are fertilized and hatch and develop the young internally in the digestive tract (they don't feed while they incubate!).

Early amphibians probably underwent a much more direct development, hatching as miniature adults. A few frogs today lay large eggs, 10 mm across, which hatch miniature adults. These eggs show no sign of evolving toward an amniotic condition, but they show that some living amphibians can lay large eggs that then develop without a complex metamorphosis. Presumably, as the amniotic egg evolved, the reproductive problems faced today by living frogs were avoided by simply allowing the embryo to develop longer and longer inside the egg. Longer development could, of course, have evolved gradually along with increased size and complexity of the egg.

Evolution toward laying eggs in damp air rather than water may have been encouraged by the better oxygen supply available from air. An egg laid in water, especially a small, warm body of water, may be exposed to anoxic conditions, especially in the hot tropical habitats in which early reptiles lived. As long as the egg does not dry out, it may have a better oxygen supply in damp air than in water.

This story provides a unifying theme that links the evolution from fish to reptiles through an amphibian stage. Throughout, evolutionary change is linked with successful reproduction; as a by-product, successful animals are encouraged to enter new habitats. As they do so, they evolve ways of exploiting those habitats, and new ways of life become not only possible but encouraged. Simple themes that explain many facts are always satisfying, and they are often right.

Why did the amniotic egg evolve among small Carboniferous reptiles and not among Carboniferous amphibians? I suspect that here again body size was important. A small female could have hidden her small clutch of eggs more successfully than a large one; she would have invested in only a small mass of eggs; and she probably had a short life span, maybe with only a few reproductive cycles. Egg care would have been particularly important among such creatures, and selection toward an amniotic egg would have been strong.

Carboniferous Land Ecology

Little is known yet about the land ecology of Early Carboniferous times; the East Kirkton fossils are the best-known tetrapod fauna from this time. All the Devonian and Carboniferous tetrapods so far discovered lived in or around the continent of Euramerica, close to the equator (Figure 9.2).

The evidence is much better when we turn to the Late Carboniferous or Pennsylvanian. Late Carboniferous coalfields have been intensely studied for economic reasons, yielding a lot of information that has given a good picture of the flora and global paleoecology of the time. Swamp forests in tropical lowlands were dominated by lycopods, and the vegetation and organic debris that were deposited in oxygen-poor water formed thick accumulations of peat, now compressed and preserved as giant coal beds stretching from the American Midwest to the Black Sea.

Most early vertebrates were carnivorous. The abundant vegetation of the Late Carboniferous was food mainly for insects, but more often raw material for fungal and bacterial decomposition. Fishes, small amphibians, reptiles, and giant dragonflies all ate insects, and in turn provided food for larger carnivores. For vertebrates, then, Carboniferous swamp forests did not provide food directly, but they provided shelter and cover.

We saw in Chapter 8 that there were no herbivores among early land arthropods. The reason may have to do with the evolution of lignin, that universal substance in vascular plants. Lignin is formed through biochemical pathways that include toxic substances which are often stored in cell walls and dead plant tissue. From the Silurian to the Late Carboniferous, lignin and its associated biochemistry probably made vascular plants invulnerable to potential herbivores, especially the very small arthropods that dominated early terrestrial faunas. The terrestrial food chain that began with vascular plants thus led only through detritus.

Eventually, of course, both invertebrates and vertebrates made the breakthrough that allowed direct herbivory. Plant toxins are broken down after death by bacteria and fungi, and it's possible that internal symbiosis with one or both allowed some animals to ingest plant material for internal enzyme-assisted breakdown rather than waiting for natural external breakdown into detritus. Also, the larger sporangia and seeds that were evolved by vascular plants were very nutritious and low in toxins, and therefore more liable to attack by arthropods. Early insects quickly evolved the anatomy to feed on the reproductive tissues of plants. One

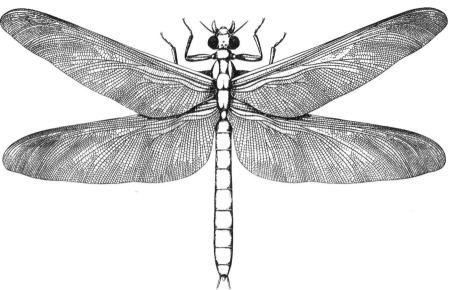

Figure 9.13 One of the giant dragonflies that reached a wingspan of 60 cm (2 feet), from the Late Carboniferous of Western Europe. (Reconstruction after Handlirsch.)

could argue that seeds evolved not only for better waterproofing of the embryo but also to deter insect predation.

Although the first insect is Devonian, the dominant fact of early insect evolution is the explosive radiation of winged insects in the Late Carboniferous, about 325 Ma. Some had mouthparts for tearing open primitive cones, and their guts were sometimes fossilized with masses of spores inside. Others had piercing and sucking mouthparts for obtaining plant juices. Overall, it seems that leaf eating was rare among early insects; instead, they ate plant reproductive parts, sucked plant juices, or ate other insects. Gigantic dragonflies were flying predators on smaller arthropods; Late Paleozoic dragonflies were the largest flying insects ever to evolve, with wingspans up to 60 cm (Figure 9.13).

Explosive evolution had occurred among land-going invertebrates by the Late Carboniferous, much of it linked with the evolution of herbivory among insects: 137 genera of terrestrial arthropods are recorded from the Mazon Creek beds of Illinois, including 99 insects and 21 spiders, with millipedes present also. Most of the living groups of spiders had evolved by the Late Carboniferous, with only the sophisticated orb-web spiders missing. Centipedes were important predators.

Millipedes are important forest recyclers today, feeding on decaying plant material. They include flattened forms that squirm into cracks in dead wood and literally split their way in, reaching new food and making space for shelter and brood chambers at the same time. Carboniferous millipedes reached half a meter in length, and a giant relative, *Arthropleura*, reached 2.3 m (7 feet) long, and 50 cm (18 inches) across. The gut contents of *Arthropleura* suggest that it ate the woody central portion of tree ferns.

Review Questions

What was the novel character that marks the evolution of reptiles from amphibians?

How does a reptile egg differ from an amphibian egg?

What were the first vertebrates to live their whole lives on land?

Describe the feeding ecology of the life in the environment where reptiles first evolved. What was there to eat, and who ate what?

THOUGHT QUESTION—Why were the first reptiles small? Many of their amphibian ancestors were big. What was it about the evolution of reptiles that was allowed, or encouraged, by small body size?

Further Reading

Easy Access Reading

Clack, J. A. 1998. A new Early Carboniferous tetrapod with a mélange of crown-group characters. *Nature* 394: 66–69. *Eucritta.*

Paton, R. L., et al. 1999. An amniote-like skeleton from the Early Carboniferous of Scotland. *Nature* 398: 508–513. *Casineria.*

Monastersky, R. 1999. Out of the swamps: how early vertebrates established a foothold—with all 10 toes—on land. *Science News,* May 22, 1999. *Casineria.*

Shear, W. A. 1991. The early development of terrestrial ecosystems. *Nature* 351: 283–289.

Smithson, T. R. 1989. The earliest known reptile. *Nature* 342: 676–678. Lizzie.

More Technical Reading

Cruickshank, A. R. I., and B. W. Skews. 1980. The functional significance of nectridean tabular horns. *Proceedings of the Royal Society of London B* 209: 513–537.

DiMichele, W. A., and T. L. Phillips. 1996. Clades, ecological amplitudes, and ecomorphs: phylogenetic effects and persistence of primitive plant communities in the Pennsylvanian-age tropical wetlands. *Palaeogeography, Palaeoclimatology, Palaeoecology* 127: 83–105.

Duellman, W. E., and L. Trueb. 1986. *Biology of Amphibians.* New York: McGraw-Hill.

Milner, A. R., et al. 1986. The search for early tetrapods. *Modern Geology* 10: 1–28.

Milner, A. R., and S. E. K. Sequeira. 1994. The temnospondyl amphibians from the Viséan of East Kirkton, West Lothian, Scotland. *Transactions of the Royal Society of Edinburgh, Earth Sciences* 84: 331–361.

Panchen, A. L. (ed.). 1980. *The Terrestrial Environment and the Origin of Land Vertebrates.* London: Academic Press.

Perry, D. 1986. *Life Above the Jungle Floor.* New York: Simon and Schuster. An account of the fauna inside hollow tree trunks reads like a visit to the Pennsylvanian, except for the bats. While the section on the coal forests is good, the sections on dinosaurs are not so sound.

Schultze, H.-P., and L. Trueb (eds.). 1991. *Origins of the Higher Groups of Tetrapods: Controversy and Concensus.* Ithaca, New York: Cornell University Press. Chapters 6–11 deal with amphibians and early reptiles.

Smithson, T. R., et al. 1994. *Westlothiana lizziae* from the Viséan of East Kirkton, West Lothian. *Transactions of the Royal Society of Edinburgh, Earth Sciences* 84: 383–412.

Reptiles and Thermoregulation

The radiation of reptiles was probably encouraged by ecological opportunities away from water bodies. But away from water, microenvironments have lower humidity, more exposure to solar radiation and to colder nights, less vegetation and shelter, and greater temperature fluctuations. Some degree of temperature control or thermoregulation is needed to live in such habitats, and the varied responses of reptiles to environmental and physiological challenges are major themes in their evolutionary history.

Amphibians emerged onto land and the first reptiles evolved in warm, humid, tropical regions along the southern shores of Euramerica. Life away from such swamps and forests demands adaptations for dealing with seasons, where temperature, rainfall, and food supply vary much more and are less predictable than in the tropics. In many ways, such challenges to early land vertebrates were simply extensions of the problems involved in leaving the water. In this chapter we shall follow the early history of reptiles and discuss the adaptations that allowed them their great terrestrial success.

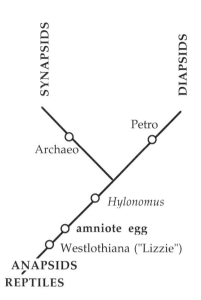

Figure 10.1 Phylogram of early reptiles. Two lineages of reptiles diverged quickly within the amniotic clade. The earliest well-described reptile is *Hylonomus*. Petro is *Petrolacosaurus*, the earliest diapsid; Archaeo is *Archaeothyris*, the earliest synapsid. This phylogram suggests among other things that Lizzie, the earliest reptile, did not lay amniotic eggs (I think Lizzie is too early to have evolved this advanced character), and that synapsids and diapsids represent different evolutionary branches from anapsid ancestors.

Figure 10.2 Three different skull types among reptiles are defined by the number of holes in the skull behind the eye socket. Left: Anapsid, represented by *Captorhinus*. Center: Diapsid, represented by *Petrolacosaurus*. Right: Synapsid, represented by *Dimetrodon*.

The Reptile Radiation

Reptiles came to be dominant large animals in all terrestrial environments in Permian times. We know that the reptile radiation began in Euramerica, because no land vertebrates at all are known from Siberia, from East Asia, or from the whole of Gondwanaland before Middle Permian times. Seaways and mountain ranges may have blocked land migration; alternatively, problems of thermoregulation may have confined land vertebrates to the tropics of Euramerica until the Middle Permian. The reptile invasion of other continents was accompanied by a spectacular evolution of varied reptilian types. Since reptiles rather than amphibians radiated so successfully, perhaps their solution to thermoregulatory problems allowed them to invade regions in higher latitudes.

We know now that the earliest reptile lived in the Early Carboniferous, and that three major groups of reptiles had diverged by the Late Carboniferous and Early Permian (Figure 10.1). The earliest reptiles had **anapsid** skulls (they had no openings behind the eye) (Figure 10.2, left). This character was inherited from fishes and amphibians, so it is primitive. The two major reptile clades have derived or advanced skull types, in which there are one or two large openings behind the eye socket. **Synapsids** (with one skull opening behind the eye socket) (Figure 10.2, right), evolved first, from an anapsid ancestor that we have not yet discovered. Synapsids include the Late Paleozoic pelycosaurs and their descendants, the therapsids and mammals. We know that synapsids evolved first because they never evolved the water-saving capacity to excrete uric acid rather than urea, a character that occurs in all other amniote groups (Figure 10.1). Synapsids dominated Late Paleozoic land faunas.

Some time later the **diapsids** evolved (with two skull openings behind the eye socket) (Figure 10.2, center). Diapsids include the dominant land-going groups of the Mesozoic, including dinosaurs and pterosaurs, and all living amniotes except mammals. Turtles have no skull openings, so are technically anapsid. But their ancestors were most likely diapsid reptiles that **lost** the two skull openings. Turtles evolved in Late Triassic times, from ancestors we have not yet identified.

Late Carboniferous reptiles are best known from sediments laid down in deltas. Some reptiles were already adapting to drier environments, but they are poorly known because their habitats were not as favorable for burial and preservation. Fortunately, a Late Carboniferous

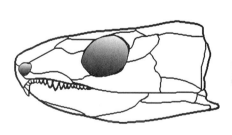
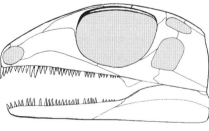
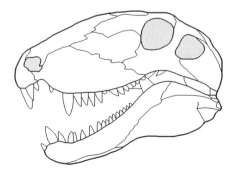

mudflat, on a drier open shore away from the swamp, has been preserved near Garnett, in Kansas. A few lobefin fishes and amphibians are aquatic, but most Garnett fossils are terrestrial, including dozens of specimens of reptiles that were washed down a channel onto the mudflat and preserved there in unusual detail.

The commonest Garnett reptile is *Petrolacosaurus*, known from 50 specimens (Figures 10.2 and 10.3). It was lizardlike in size and general appearance, and it seems to be the best candidate yet for the ancestor of all later diapsids. In turn, *Petrolacosaurus* could easily have evolved from an early anapsid (Figure 10.1). Compared with later diapsids, *Petrolacosaurus* had a heavy earbone, the stapes, that could not conduct airborne sound. As in most early amphibians and reptiles, the massive stapes probably transmitted ground vibrations through the limb bones to the skull.

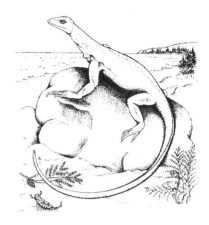

Figure 10.3   A reconstruction of the first known diapsid, *Petrolacosaurus*, from the Late Carboniferous of Garnett, Kansas. (From the cover of *Science*, 3 June 1977. © The American Association for the Advancement of Science, reproduced with permission.)

Pelycosaurs

Although the first diapsid, *Petrolacosaurus*, evolved in the Late Carboniferous, the major radiation of diapsids was a Triassic event. The dominant Late Carboniferous and Permian reptiles were synapsids, including five of the seven reptile species known from Garnett. Early synapsids are classed together as **pelycosaurs**, the most famous of which are the sailbacked Permian forms such as *Dimetrodon* (Figure 10.2, right). They were already the most important group of fully terrestrial tetrapods. Over 50% of Late Carboniferous reptiles were pelycosaurs, and over 70% of Early Permian reptiles. After that they declined, but only because one clade of them had evolved into the dominant therapsids of the Late Permian. Despite their variety, early pelycosaurs are rare except in the fossil forests of Nova Scotia and at Garnett, away from the swamp. Pelycosaurs may have first evolved and lived in habitats that were drier than those favored by amphibians. Later they became much more abundant and widespread.

Archaeothyris, found in the fossil tree trunks of Nova Scotia, was one of the earliest pelycosaurs. It was small, perhaps 50 cm (20 inches) long including the tail, and it was lizardlike in general appearance. But it had the characteristic skull structure of synapsids, with a single opening behind the eye socket instead of the double opening of diapsids. The edges of this opening gave secure attachment for powerful jaw muscles.

Pelycosaur Biology and Ecology

Locomotion

Pelycosaurs are well enough known that we can reconstruct how they walked. The massive front part of the body was supported by a heavy, sprawling forelimb. The lighter hind limb had a greater range of move-

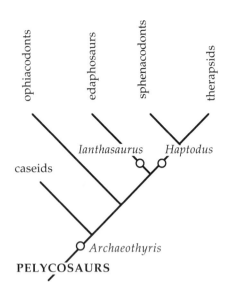

Figure 10.4 A simplified phylogram of pelycosaur evolution that includes some pelycosaurs mentioned in the text.

ment, although it was also a sprawling limb. There was no well-defined ankle joint, and the toes were long and splayed out sideways as the animal walked. Thus, the feet provided no forward thrust but simply supported the limbs on the ground. The forelimbs were entirely passive supports that prevented the animal from falling on its face, while the hind limbs provided all the forward thrust in walking with powerful muscles that rotated the femur in the hip joint. Think of two children playing "wheelbarrow." The propulsion and steering are both from the rear, and the wheelbarrow is stable only as long as the leading child stays stiff. Spinal flexibility is important to many swimming animals, particularly those that actively pursue fishes. But in pelycosaurs, a strong stiff backbone prevented the body from collapsing in the middle under its own weight and allowed thrust from the hind limb to be converted directly into forward motion. Therefore, most pelycosaurs were predominantly terrestrial animals. If they swam, they were slow swimmers that hunted by stealth rather than speed. Only one pelycosaur (*Varanosaurus*) had a really flexible spine, and it may have been almost entirely aquatic.

Carnivorous Pelycosaurs

Early pelycosaurs were all carnivorous. Three major groups are represented at Garnett (Figure 10.4), and they all have the pointed teeth and long jaws of predators. Two groups remained completely predatory. **Ophiacodonts** included the first small pelycosaur, *Archaeothyris,* from Nova Scotia, but later forms became quite large. *Ophiacodon* itself was 3 meters (10 feet) long and probably weighed over 200 kg (450 pounds). Many ophiacodonts have long-snouted jaws with many teeth (Figure 10.5), set in a narrow skull. The hind limbs tended to be longer than the forelimbs. Ophiacodonts may have hunted fishes in streams and lakes of the swamps and deltas of the Late Carboniferous and Permian, although they were perfectly capable of walking on drier, higher ground, and like crocodiles, their prey may well have included terrestrial animals coming down to the water to drink. Their general lack of spinal flexibility (except in *Varanosaurus*) may suggest that they were slow swimmers, possibly eating more amphibians than fishes.

Sphenacodonts were specialized carnivores on land. Many of their skull features betray the presence of very strong jaw muscles, and the teeth were very powerful. They were unlike typical reptile teeth in that they varied in shape and size and included long stabbing teeth that look like the canines of mammals. The sphenacodont body was narrow but

Figure 10.5 The skull of the crocodile-sized pelycosaur *Ophiacodon*, seen from above, is long and pointed, as in many fish-eating animals. (After Romer and Price.)

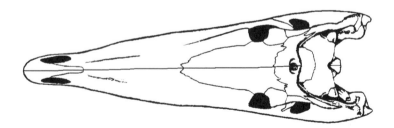

Figure 10.6 *Haptodus*, the earliest known sphenacodont, a little under a meter long (about 3 feet). *Haptodus* probably walked more on its toes than this posture suggests. (From Currie.)

deep, and the legs were comparatively long. Both of these characters suggest that sphenacodonts were reasonably mobile on land.

The earliest sphenacodont was *Haptodus* (Figures 10.4 and 10.6). It was a little less than a meter long and fairly lightly built, but even so it was the largest carnivore in the Garnett community. Similar forms existed throughout the Permian, but later sphenacodonts were much larger. The group is best known from spectacular fossils of *Dimetrodon* (Figure 10.7). *Dimetrodon* had vertebrae extended into spines projecting far above the backbone. (I'll discuss these structures later.)

Evolution within carnivorous pelycosaurs reflected their prey capture. The jaws slammed shut around the hinge, with no sideways or front-to-back motion for chewing. With this structure, a long jaw made it easier to take hold of prey, but the force exerted far from the hinge was not very great. Small prey could perhaps be killed outright by slamming the jaw on them. In ophiacodonts, which may have hunted in water for fish, the difficult part of feeding would have been seizing the prey; their teeth were subequal in length in a long, narrow jaw. Most fish-eaters swallow their prey whole. In sphenacodonts, which were terrestrial carnivores, the head was bigger and stronger. Long, stabbing teeth were set in the front of the jaw (Figure 10.2, right). Struggling prey could be held between the tongue and some strong teeth set into the palate, and could be subdued by powerful crushing bites from the teeth at the back of the jaw. Robert Carroll suggested that the success of pelycosaurs in the Carboniferous and Permian, compared with diapsid reptiles, was due to their massive jaw muscles, which were strong enough to hold the jaws steady against the struggles of large prey. The carnivorous pelycosaurs thus could have been large predators, not simply small insectivores.

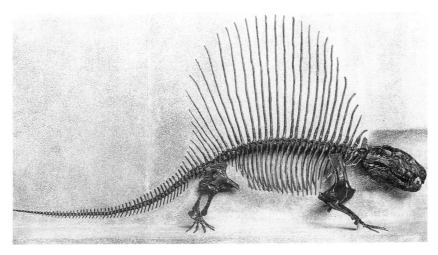

Figure 10.7 The skeleton of the most famous Permian reptile, *Dimetrodon*, a carnivorous pelycosaur with extended vertebrae that formed a "sail" on its back. (Negative 315862. Photo by Julius Kirschner. Courtesy of the Department of Library Services, American Museum of Natural History.)

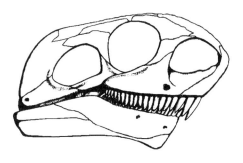

Figure 10.8 A typical caseid, *Cotylorhynchus*, a vegetarian pelycosaur. The teeth are sharper than one might expect, suggesting that food was macerated in the gut or gizzard, not chewed in the mouth. In keeping with this fact, the skull is small for the body size, only about 20 cm (8 inches) long. The nostrils are very large, and one could speculate about the reasons for that. (After Romer.)

Vegetarian Pelycosaurs

Carroll's suggestion cannot be the whole story, because there were also vegetarian pelycosaurs. **Caseids** and **edaphosaurs** were the first abundant large terrestrial animals, and they were among the first terrestrial herbivores. Caseids and edaphosaurs had similar body styles, presumably because they were similar ecologically. They had about the same range of body size as the carnivorous sphenacodonts, but they had smaller, shorter heads that gave more crushing pressure at the teeth. There were no long canines, and the teeth were short, blunt, and heavy. In addition, smoothing of the bones at the jaw joint allowed the lower jaw to move backwards and forwards slightly, grinding the food between upper and lower teeth. Caseids ground their food between tongue and palatal teeth, while edaphosaurs had additional tooth plates in their lower jaw that they used to grind food against palatal teeth. As one would expect, the bodies of all these vegetarians were wide to accommodate a large gut. The limb bones were short but heavy.

Caseids were more numerous than edaphosaurs, and they included *Cotylorhynchus*, which was over 3 meters long (10 feet), and weighed over 300 kg (650 pounds) (Figure 10.8). Caseids had small heads for their size, which perhaps implies that they did not chew very much, and perhaps had powerful digestive enzymes or gut bacteria to help break down plant cellulose.

Edaphosaurs are best known from *Edaphosaurus* itself (Figure 10.9), which had vertebrae extended into spines rather like *Dimetrodon*. The earliest edaphosaur *Ianthasaurus*, however, was a small carnivore found at Garnett. It had a small sail on its back, sharp pointed predatory teeth, and probably ate insects. We find vegetarian pelycosaurs only in the Early Permian, and they must have evolved herbivory as a new ecological way of life. We shall find the intermediate edaphosaurs some day, and perhaps they will show us how herbivory evolved in these early reptiles.

How Did Herbivory Evolve in Tetrapods?

Most plant material is difficult to digest. Vertebrates can break down cellulose only if they chew it well and have some way of enlisting fermenting bacteria to aid digestion, as in the stomachs of modern ruminants.

Figure 10.9 *Edaphosaurus*, a vegetarian pelycosaur that independently evolved a thermoregulatory sail comparable with that of *Dimetrodon*. Total length of body and tail was about 3.5 meters (11 feet). (Skeletal outline after Romer.)

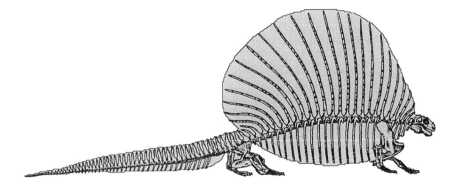

Any vertebrate that begins to eat comparatively low-protein plant material must process large volumes of it, and so must have a rather large food intake at a rather large body size. Some plant material is high in protein or sugar, especially the reproductive parts, but only a small animal can selectively feed on plant parts.

In other words, there are only two possible evolutionary pathways toward herbivory. One of them begins with animals that are small, active, and selective in their food gathering, so high-calorie foods such as plant juices, nectar, pollen, fruits, or seeds can be collected from the plant. Examples are little mammals, hummingbirds, and insects. Later, as gut bacteria are evolved, the diet can contain more and more cellulose and a larger vegetarian can evolve, as in many mammal groups including leaf-eating monkeys and gorillas.

The other pathway begins at rather large body size with rapid and rather indiscriminate feeding, possibly omnivory, so that a large volume of low-calorie food can be processed. Bearlike mammals are examples of a group in which some members have evolved away from a carnivorous way of life toward omnivory and then to a completely vegetarian diet, as in pandas. Later, as more efficient chewing and digestive systems are evolved, large vegetarians can survive at midsize.

Because vegetarianism depends so much on body size, diets must change with growth. Most living reptiles and amphibians change their diet as they grow. Food requirements and opportunities change as they reach greater size and can catch a different set of prey. Among living reptiles, small and young iguanas are carnivorous or omnivorous, while large iguanas are largely vegetarian but take meat occasionally. Living amphibians today are almost all small and carnivorous. I suggest that herbivory evolved among Late Paleozoic tetrapods only after they reached large body size.

The giant Carboniferous coalfields (Chapter 9) contain rock sequences in which many beds consist almost entirely of carbon formed from plant debris such as leaves, trunks, roots, spores, and pollen, plus half-rotted and unrecognizable fragments. The coalfields of the Northern Hemisphere, particularly those in the Late Carboniferous (Pennsylvanian), formed the energy basis for the Industrial Revolution in Western Europe and North America and still provide large quantities of fuel for many industrial countries (Figure 10.10). Carboniferous coalfields have been studied so intensively that we can reconstruct their plant communities very well; we can tell, for example, that plants had spread away from the rivers and lakes into so-called uplands—probably not very high above sea level but with distinctly drier air and soil than the lowland swamps.

The rich floras provided a food base for insects at first, but large terrestrial herbivores—amphibians as well as pelycosaurs—appeared in the latest Carboniferous and Early Permian of Euramerica. The advanced amphibian *Diadectes*, for example, ranged up to as much as 4 meters (13 feet) long as an adult (Figure 10.11). Large herbivores appeared at the same time as a major change in plants, when upland plants replaced the coal swamp forests.

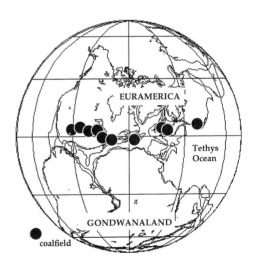

Figure 10.10 The paleogeography of the Carboniferous coalfields, which formed in swampy tropical forests. Each black dot represents a major coalfield that has survived to be exploited today.

Figure 10.11   The amphibian *Diadectes* was one of the first vertebrates (and one of the few amphibians) to become a vegetarian. (Negative 35634. Photograph by A. E. Anderson. Courtesy of the Department of Library Services, American Museum of Natural History.)

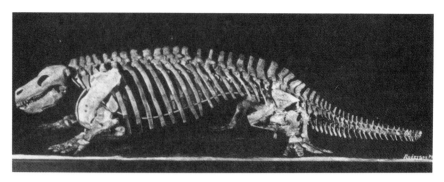

Why were vertebrates so slow to evolve herbivory? First, because the wet tropical forest in which land vertebrates evolved is a poor habitat for ground-dwelling herbivores. As in today's tropical forests, most leaves were in the canopy, and leaf litter was broken down quickly by fungi and arthropods. The first vertebrate herbivores could not have found much green material on the floor of the coal forests (Figure 9.9). Vertebrates could not have evolved herbivory until they could survive well on the forest margin, away from the watery habitats most likely to be preserved.

Second, any large-bodied vegetarian eats large volumes of low-calorie plant material and needs gut bacteria to help digest the cellulose. Gut bacteria work well only in a fairly narrow range of temperature, so an additional requirement for the first successful large-bodied vegetarians was some kind of thermoregulation.

Thermoregulation in Reptiles

Body functions are run by enzymes, which are sensitive to temperature. Other things being equal, enzymes work best at some optimum temperature; any other body temperature implies a loss of efficiency—in digestion, in locomotion, in reaction time, and so on. Birds and mammals have a sharp peak of efficiency that drops off radically with a small rise or fall in body temperature. Reptiles are called cold-blooded, but in fact they take on the temperature of their surroundings, so can be warm or cold. Their bodies can function over quite a range of internal temperature, but they also have an optimal temperature, and reptiles try to control it at that level by behavioral thermoregulation.

Generally, reptiles try to maintain their body temperatures at the highest level that is consistent with safety and cost. Although it takes energy to stay warm, the higher activity levels that are possible at higher temperatures give greater hunting or foraging efficiency, greater food intake, faster digestion, and faster growth. As long as the climate is warm and food supply is abundant enough to fuel a reptile, thermoregulation that produces or maintains warm body temperature gives a net gain in reproductive rate and so is selectively advantageous. The same principles should apply to all cold-blooded vertebrates.

Body size is a vital factor in thermoregulation. Small bodies have a low mass with a relatively large surface area. Small reptiles bask in the sun, sit in the shade, hide in burrows or in leaf litter, or exercise violently (often with push-ups) to change their body temperatures. Their small mass allows them to respond quickly to temperature changes by behavioral means, giving them sensitive control over their body processes. Large reptiles have a natural resistance to temperature change because of their mass: it takes a long time to heat them up or cool them down (just as it takes a long time to boil a full kettle of water). Behavioral thermoregulation is more energy-consuming and much less responsive for larger reptiles than for smaller ones. So large reptiles today live in naturally mild tropical climates with even temperatures day and night and season to season (like the large monitor lizards of Indonesia, Australia, and Africa), or they live near or in water, which buffers any changes in air temperature (like crocodiles and alligators, which even then are never found far outside the tropics). There are no large lizards at high latitudes.

Thermoregulation in Pelycosaurs

The spectacular pelycosaurs *Dimetrodon* and *Edaphosaurus* were not closely related, but they were both large (over 3 meters long). They both independently evolved very long neural spines on some of their vertebrae, forming a row of long vertical spines along the backs of these creatures (Figures 10.7 and 10.9). In life, the bones were covered with tissue to form a huge vertical sail that was probably used for thermoregulation.

Here is the simplest version of the story. *Dimetrodon* and *Edaphosaurus* were too large to hide from temperature fluctuations. Each probably used its sail to bask in the early morning and the late afternoon, turning its body so that the sail intercepted the sunshine. By pumping blood through the skin of the sail, it could transfer solar heat quickly and efficiently to the central body mass. Once warm and active, it would face no further problem unless it overheated. It could shed heat from the sail by the reverse process, turning the sail end-on to the sun. At night, heat would be conserved inside the body by shutting off the blood supply to the sail (Figures 10.7, 10.9, and 10.12).

The sail, as an add-on piece of solar equipment, allowed rapid and sensitive control over body temperature. Enzyme systems could have been fine-tuned to work at high biochemical efficiency within narrow temperature limits, and the animal could have foraged even in environments where air temperatures fluctuated widely. The activity levels, locomotion, and digestive systems of *Dimetrodon* and *Edaphosaurus* would all have improved. Smaller reptiles that lived alongside them would have been able to heat up quickly in the morning, simply because they were smaller, and it would have been important for the large animals to be equally active at that time—*Dimetrodon* for effective hunting and *Edaphosaurus* for effective escape or defense.

This solar scenario is usually cast in terms of an arms race between the two giants, with *Dimetrodon* warming up in the morning in order to chase *Edaphosaurus*, and *Edaphosaurus* warming up to escape. But that isn't necessarily the whole story. *Edaphosaurus* would have been an easy

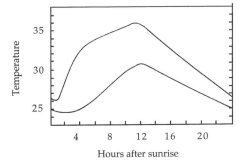

Figure 10.12 Calculations of the thermoregulatory capacity of a large *Dimetrodon* with (upper curve) and without (lower curve) its sail help to confirm why the sail evolved. It allowed the animal to warm up quickly in the morning and to reach a high body temperature close to ours. But I mistrust this particular model when it suggests that *Dimetrodon* would cool faster at night with a sail than without, and that its body temperature would fluctuate more with a sail than without. (Data from Haack, 1987.)

prey for a group of midsized predators if it were paralyzed by cold, and *Dimetrodon* no doubt took other prey. Furthermore, *Edaphosaurus* needed thermoregulation if it had gut bacteria to help it digest plant cellulose.

Some pelycosaurs did not have a sail, and the small pelycosaur *Ianthasaurus* had a small sail. Young *Dimetrodon* and *Edaphosaurus* had small versions too. The area of the sail was clearly related to body size, which makes sense only if pelycosaurs were ectothermic. Solar input and heat shedding were governed behaviorally, just as lizards thermoregulate by hiding in the shade, basking, burrowing, hibernating, and so on. The heat inertia of a large body gives some buffering against large changes in body temperature, but more energy is needed for a high activity level. If that energy is not to come in the style of birds and mammals, by burning large amounts of food in a built-in, high, internal metabolism, then it must come from outside. As in almost all living reptiles, solar input, mediated behaviorally, is the secret. So some pelycosaurs developed add-on technology for thermoregulation, using an adaptation that made them super-reptiles rather than birds or mammals.

In 1996 Christopher Bennett made two very astute observations which complicate the simple story. First, *Edaphosaurus* has knobs on the bony spines on its sail. By testing a model in a wind tunnel, Bennett found that the knobs would have generated eddies in breezes blowing past the sail. This would have no effect on solar collection by the sail, which is a radiation effect, or on its cooling by radiation, but it would make it a better cooling device by increasing convection over the skin. (Moving air cools bodies better than it heats them, as we all know from personal experience.)

Bennett also found that *Edaphosaurus* had a smaller sail than a *Dimetrodon* of the same body size, implying that the *Edaphosaurus* sail was more efficient. The only way in which it looks more efficient is in its cooling effect, and this implies even more strongly that the major function of the sail *in both animals* was cooling.

Bennett also thinks that early morning basking was probably inefficient in both animals (though it was better than nothing!). Early in the morning, their circulation systems would have been sluggish. The deep body was cool, the heart would have pumped slowly, and the blood would have been thick. They may not have been able to pump blood round the sail fast enough to transfer all the solar energy that was shining on the sail. However, once they were hot, the heart would have pumped strongly, and the blood would have flowed more freely, allowing them to cool efficiently by pumping hot blood from their overheated bodies round the sail to be cooled in the breeze. Altogether, the major problem of these big pelycosaurs was probably overheating in the heat of the day.

If pelycosaurs with sails thermoregulated, then other pelycosaurs (caseids, for example) probably thermoregulated too, in behavioral ways that left no traces on the skeleton. It is easy to imagine that the fine-tuning of the internal enzyme system that accompanied the first attempts at solar thermoregulation encouraged the evolution of internal control systems, such as those that varied blood flow to the surface. After the Permian, as advanced pelycosaurs evolved into therapsids and the rest disappeared, we see little sign of thermoregulatory devices preserved in the skeleton. There is indirect evidence, however, that therapsids had limited thermoregulation; that evidence is presented later in this chapter.

Permian Changes

Shifting continental geography resulted in major biogeographic changes in the Permian (Chapter 5). The large southern continent Gondwanaland moved north to collide with Euramerica, and by the Middle Permian these blocks formed a continuous land mass. A little later, Asia crashed into Euramerica from the east, buckling up the Ural Mountains to complete the assembly of the continents into the global supercontinent Pangea (Figure 5.9).

These tectonic events put an end to the wet climates that had fostered the system of large lakes, swampy deltas, and shorelines along the south coast of Euramerica, where the Carboniferous coal forests had flourished (Figure 10.10).

World-wide, Permian floras were dominated by gymnosperms, mostly gingkos, conifers, and cycads. Conifers had evolved in Carboniferous times. Compared with other Late Paleozoic plants, they were better adapted for drought resistance, and they probably evolved in much drier uplands, because they are rare on lowland floodplains. Tree-sized lycopods disappeared from coal swamps in the Late Carboniferous as climates became drier. As the drying trend continued into the Permian, conifers extended into lowlands at the expense of the swamp plants that had dominated Carboniferous floras.

Total plant diversity dropped in this turnover. The drop was spread over a long time, however. It was not catastrophic or even abrupt. It was a response to changing climate, perhaps based on geographic changes, and it can be explained by normal evolutionary processes.

The Invasion of Gondwanaland

Geological evidence from Gondwanaland shows that a huge ice sheet was centered on the South Pole (Figure 10.13) in Late Carboniferous and Early Permian times. Ice sheets moving northward scoured rock surfaces and deposited stretches of glacial debris on a continental scale.

The continental collisions that formed Pangea allowed land animals to walk into Gondwanaland. But pelycosaurs, which had always had a narrow tropical distribution, remained in the tropical areas, much reduced in diversity; Late Permian pelycosaurs are found only in North America and Russia. Instead, Gondwanaland was invaded by their descendants, who evolved into the synapsid reptiles called **therapsids**. Therapsids had larger skull openings than pelycosaurs did, indicating that they had more powerful jaws. The whole skull was strengthened and thickened, and the jaw always had prominent canine teeth. Therapsids also had much better locomotion than pelycosaurs.

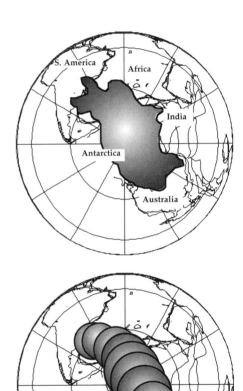

Figure 10.13 Traces of an ice age are widespread over the continents that formed Gondwanaland in the Late Paleozoic. The edges of the ice can be marked with confidence. The simplest explanation is that an ice cap formed on the South Pole at the time. Gondwanaland slowly drifted over the Pole, and the edges of the migrating ice cap left an irregular trace. The traces ended as the last little Permian ice cap melted.

Thermoregulation in Therapsids

Dimetrodon, with its sail and its great skin surface area, must have been able to maintain thermoregulation over a reasonable range of external environmental variation, and must have lived in an environment where such thermoregulation was both required and possible. It must also have lost very little water through its skin. These adaptations of a sphenacodont may have been preadaptive for the therapsid invasion of more challenging habitats. *Dimetrodon* itself survived a climatic change in the Permian of Texas that eliminated most other pelycosaurs from the region, presumably because it was more tolerant of greater temperature ranges and drier conditions than they were. But in the end *Dimetrodon* and the other pelycosaurs became extinct, to be replaced in tropical latitudes by diapsid reptiles rather than synapsids (Chapter 11).

Therapsids evolved from sphenacodont pelycosaurs, and lived mainly at middle or high latitudes rather than in tropical regions. It's not clear whether they were outcompeted in the tropics by other reptile groups. The restriction, or adaptation, of therapsids to drier and more seasonal habitats may have encouraged their success in southern Gondwanaland, away from the tropics and toward higher latitudes. Thousands of specimens of therapsids have been collected from Late Permian rocks in South Africa, and someone with time on his hands estimated that these beds contain about 800 billion fossil therapsids altogether! There are literally dozens of species, and we have a good deal of evidence about their environment. The glaciations were over and vegetation was abundant, with mosses, tree ferns, horsetails, true ferns, conifers, and a famous leaf fossil, *Glossopteris*. The climate may well have been mild considering that South Africa was at 60° S latitude. But it must have been seasonal, so the supply of plant food would have been seasonal too.

When we find large extinct reptiles at such high latitudes, we can be reasonably sure that they were unlike living reptiles in their metabolism. Their thermoregulation must have been more sophisticated than simple behavioral reactions.

We know that mammals evolved from late therapsids in the Late Triassic. Did Permian therapsids already have a mammalian style of thermoregulation, with automatic internal control, a furry skin, and a high metabolic rate? We have too little evidence to say, and the scrappy evidence available suggests that the answer is no.

Therapsids had sprawling forelimbs and did not move very efficiently compared with later reptiles and mammals. Many of them lacked the secondary palate that would have allowed them to chew and breathe at the same time. Unlike other reptiles, many therapsids had short, compact, stocky bodies, with short tails—good adaptations for conserving body heat if not generating it (Figure 10.14). They may also have had hair or thick hides for conserving heat, but there is no way of detecting that from their fossils. All this suggests that therapsids did not have a large energy budget. They may have had some moderate form of internal temperature control, but nowhere near as good as that in living mammals.

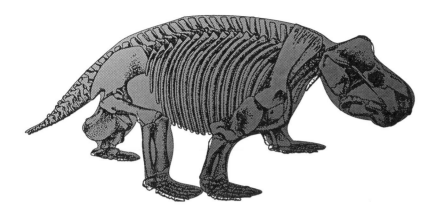

Figure 10.14 The body plan of therapsids looks as if it was good for conserving heat. This animal is *Kannemeyeria*, a rhino-sized dicynodont from Gondwanaland. (Skeletal reconstruction after Pearson.)

Therapsid Evolution

Therapsids evolved quickly in the Late Permian into several different ways of life. Their evolutionary history has not yet been properly worked out, and the classification is still changing rapidly. Figure 10.15 shows one hypothesis about the evolution of therapsids.

Theriodonts are the earliest abundant therapsids. They include the most successful Late Permian carnivores, the **gorgonopsians**, and the **cynodonts**, which evolved into mammals in the latest Triassic. **Dinocephalians** were also large-bodied and included herbivores as well as some impressive carnivores. **Dicynodonts** were the dominant herbivores of the Late Permian. The phylogram of Figure 10.15 suggests that therapsids evolved from carnivorous ancestors, and that herbivory evolved at least twice among therapsids, in dicynodonts and dinocephalians.

Theriodonts

Early theriodonts dominated the early Late Permian of Russia. All synapsids tended to have long canines, but some of these large, early Russian theriodonts approached a sabertooth condition (Figures 10.16 and 10.17) that was also evolved by later theriodonts, the gorgonopsians.

Gorgonopsians were specialist carnivores on large prey and had some exciting adaptations. They were the dominant carnivores of Late Permian times, but they became extinct abruptly at the end of the period, leaving no descendants. Their sabertoothed killing action clearly involved a very wide gape of the jaw and a slamming action that drove the canine teeth deep into the prey. The incisors were strong, but the back teeth were small and must have been practically useless. The snout was rather short, but deep enough to hold the roots of the canines. The limbs were long and fairly slender, and gorgonopsians were perhaps comparatively agile. The skull is only 50 cm (20 inches) long in the largest known gorgonopsian, and even in adults the limbs were not required to be purely load-bearing as they were in earlier, larger therapsids. The joints could therefore be more lightly built, and the whole locomotion was improved. The hind limb could be swung into an erect position,

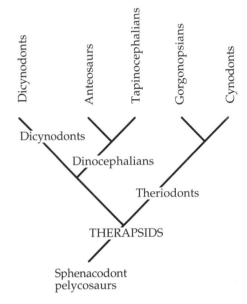

Figure 10.15 One possible hypothesis (phylogram) of the evolution of the major groups of therapsids. A group of advanced cynodonts evolved into mammals in the latest Triassic.

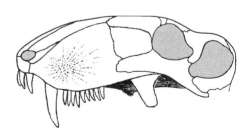

Figure 10.16 Late Permian therapsids from Russia that evolved a sabertooth arrangement included early theriodonts and their gorgonopsian descendants. *Eotitanosuchus* is an early theriodont. (After Chudinov.)

Figure 10.17 *Ivantosaurus* is a Late Permian theriodont with saberteeth about 10 cm (4 inches) long. (After Chudinov.)

stride length was greater, and the foot was lighter, altogether indicating greater speed.

Cynodonts

Cynodonts are the therapsids from which mammals evolved. They were important small- and medium-sized animals in Triassic times throughout Gondwanaland; they are discussed in Chapter 15.

Dinocephalians

Dinocephalians also moved better than pelycosaurs. Their spine was quite stiff, and limb length and stride length were longer than in pelycosaurs. The forelimbs were still sprawling, but the hind limbs were set somewhat closer to the vertical, accentuating the wheelbarrow mode of walking we have already described for pelycosaurs. Dinocephalians became very large. They appeared first in Russia, but they soon radiated in South Africa too. They had large skulls and, like all therapsids, they had strong canines. They also had well-developed incisors that seem to have been both efficient and important in feeding. Dinocephalians had unusual front teeth: their upper and lower incisors, and sometimes the canines too, interlocked along a line when the mouth closed, forming a formidable zigzagged array that would bite off pieces of food as well as piercing and tearing (Figure 10.18).

The earliest dinocephalians, or **anteosaurs**, were carnivores with skulls up to a meter long. Like the sphenacodonts, they killed prey mainly by slamming the long, sharp front teeth into them, then tearing and piercing. Apparently the back teeth were not used very much; they were fewer and smaller than in sphenacodonts.

Most other dinocephalians, or **tapinocephalians**, look carnivorous at first sight (Figure 10.18), with large canines and incisors in the front of the jaw. But they had a broad, hippo-like muzzle, a large array of flattened back teeth, and massive bodies with a barrel-like rib cage that must have contained a capacious gut. These animals may have been omnivorous, but more likely the incisors were cropping, cutting teeth used on vegetation, and the canines were fighting tusks, not carnivorous weapons. (Look inside the mouth of a hippo sometime.) The jaw exerted most pressure when closed, for efficient chewing rather than slamming.

Some tapinocephalians were particularly bizarre, with horns; some of them, probably males, had great bony flanges on the cheeks (Figure 10.18). All tapinocephalians had thick skull bones, sometimes up to 11 mm (half an inch) thick (Figure 10.19). Herbert Barghusen suggested that individuals butted heads, presumably to establish dominance within a group. I have shown a tapinocephalian in a head-butting posture in Figure 10.20. Large vegetarians today tend to fight by head-butting or pushing, while carnivores today are quick and agile and tend to use claws and teeth as they fight. Early therapsids, even the carnivores, were heavy and

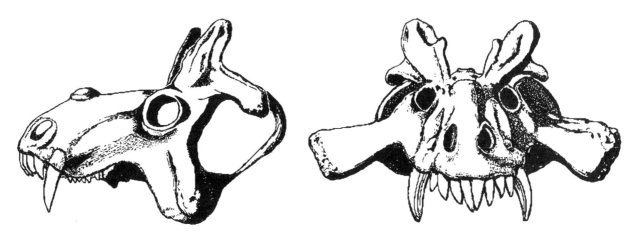

clumsy; they had sprawling limbs that were so committed to supporting their weight that they could not have used claws as weapons.

Dicynodonts

Dicynodonts were the first truly abundant worldwide herbivores. They provide 90% of therapsid specimens and much of the therapsid diversity preserved in Late Permian rocks. They are also the earliest therapsids known from South Africa. The earliest dicynodonts were already so specialized as herbivores at their first appearance that they show no close resemblance to other therapsid groups and are difficult to classify (Figure 10.15). Early dicynodonts already had a secondary palate, so they could breathe and chew at the same time.

At their peak in the Late Permian, dozens of species of dicynodonts were living in Gondwanaland, and they survived long into the Triassic. They differed from other therapsids in having very short snouts, and they had lost practically all their teeth except for the tusklike upper canines, which were probably used for display and fighting more than for eating (Figure 10.21). The jaw joint was weak, and moved forward and back in

Figure 10.18 Tapinocephalians had large canines, probably for display and fighting rather than feeding. *Estemmenosuchus*, from the Permian of Russia, also had bizarre flanges on the skull that may have distinguished males. (After Chudinov.)

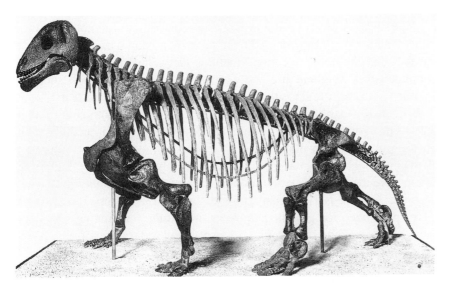

Figure 10.19 The skeleton of the massively built tapinocephalian *Moschops*, which was 2.5 meters (8 feet) long. (Negative 39137. Photograph by A. E. Anderson. Courtesy of the Department of Library Services, American Museum of Natural History.)

Figure 10.20 In 1975 Herbert Barghusen suggested that some tapinocephalians, including *Moschops*, butted heads. This reconstruction shows two more primitive therapsids, males of *Estemmenosuchus*, in head-butting position. The bony knobs on the nose and the bones above the eye all probably had sharp horns mounted on them. I have left room for them but have not shown them on this figure.

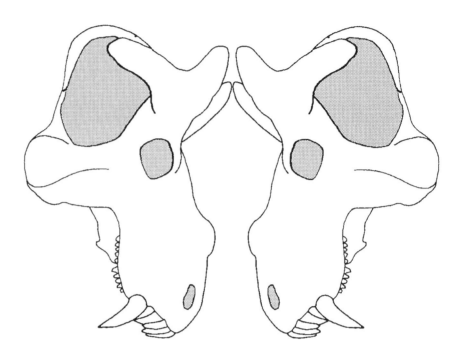

a shearing action instead of sideways or up and down. As part of this system, the jaw musculature was unusual, set far forward on the jaw, and took up a good deal of space on the top and back of the skull. These unusual jaw characters had their effect on the whole shape of the skull, which was short yet high and broad, almost boxlike. The extensive muscle attachments resulted in the eyes being set relatively far forward on a short face. Because there were practically no teeth, the jaws must have had some sort of horny beak like that of turtles for shearing off pieces of vegetation at the front and grinding them on a horny secondary palate while the mouth was closed. Dicynodonts look as if they cropped relatively tough vegetation with their beaks, and then ground it up in a rolling motion in the mouth. As in other herbivores, the body was usually bulky, with short, strong limbs (Figure 10.14).

The success of dicynodonts is astonishing. Most dicynodonts were rather small, though they ranged from rat-sized to cow-sized. Presumably the fact that the horny feeding structures of dicynodonts were replaced continuously throughout life had a great deal to do with their success. Reptiles with teeth replace them throughout life, but intermittently, so it is difficult for them to achieve continuously effective tooth rows. Other therapsids evolved effective cutting and grinding teeth, but teeth do wear out with severe and prolonged use.

Dicynodont jaws varied a lot, presumably because of their diet. There were dicynodonts with cropping jaws and with crushing jaws (perhaps for large seeds), and many browsers and grazers. Some dicynodonts were specialized for grubbing up roots, and some for digging holes, although they remained vegetarian. In a spectacular discovery in South Africa, several skeletons of a little dicynodont were found at the bottom of sophisticated spiral burrows (Figure 10.22).

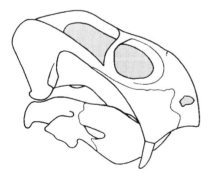

Figure 10.21 Dicynodonts had strange-looking faces because of their style of jaw construction. This is *Dicynodon*, with a skull about 15 cm (6 inches) long. (After Cluver and Hotton.)

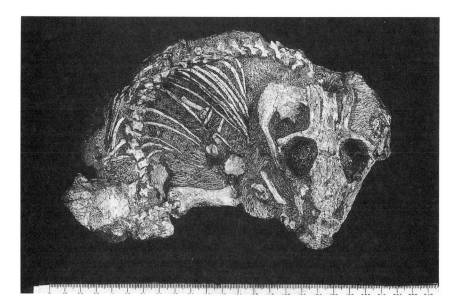

Figure 10.22 Several specimens of the little Permian dicynodont *Diictodon* have been found fossilized inside their burrows. The markings that look suspiciously like hair on this specimen are actually scratches from the needles used to clean the rock away from the bones. The scratches add an eerie realism to the fossil! (Photograph courtesy of Dr. R. M. H. Smith of the South African Museum.)

The extent of specialization among dicynodonts suggests that the climate was reasonably mild and food supply reasonably reliable at the time, in spite of the high latitude and inevitable seasonal changes. Most Permian dicynodonts were small, with skulls about 20 cm (8 inches) long. Possibly many of them were small so that they could burrow to avoid seasonal changes in temperature and food supply.

Dicynodonts declined abruptly at the end of the Permian, but a few lineages persisted, often in great numbers. The best-known dicynodont of all is a very specialized Early Triassic form, *Lystrosaurus*. It has been found in India, Antarctica, South Africa, and South China; its distribution helped and is still helping to identify fragments of Gondwanaland.

Other Triassic dicynodonts became unusually massive, with short legs and barrel-like trunks. Some grew to the size of rhinos (Figure 10.14) and must have been very slow-moving. Perhaps they were ecological analogues of giant pandas, gorillas, or ground sloths. The last of the dicynodonts even lost their tusks, which may once have been used for visual display. Possibly, therefore, the last clumsy dicynodonts were nocturnal or hid in thick undergrowth in the forest, driven there by competition with more advanced reptiles. In a gloomy scene reconstructed by Gordon Gow,

> "Thus died out the Dicynodontia, lurking in semi-obscurity in the depths of the bushes and only venturing out to feed at night. The archosaurs had come into their own."

Synapsids and Diapsid Replacements

With their generally large bodies, their radiation into herbivores and carnivores of varying sizes, and their experimentation with horns, fangs, and fighting, a Late Permian therapsid community would not seem totally strange to a modern ecologist, especially one familiar with the large mammals of the African savanna. The comparison would not stand close examination. The Permian carnivores were larger and much more numerous than modern African carnivores, and the locomotion of all Permian therapsids was very clumsy. But a Late Permian fauna would not have looked as foreign as a Triassic fauna. Faunas did not evolve to look more mammal-like. Instead, therapsids were replaced by archosaurian diapsid reptiles, which had evolved from quite different Permian ancestors.

Review Questions

The "sail-backed" pelycosaurs like *Dimetrodon* lived only in the tropics. Why didn't they live close to the poles?

You are looking at a large fossil reptile. How would you tell whether it was carnivorous or herbivorous?

Describe carefully how *Dimetrodon* or *Edaphosaurus* controlled its body temperature over 24 hours.

A South African geologist has recently found a fossil mammal-like reptile preserved in Permian rocks in the bottom of a fossil burrow. Explain how this fits with our understanding of life in Permian times in southern Africa.

There are two reasons for thinking that mammal-like reptiles might have been hairy. One is direct evidence from the fossils—what is it? The other piece of evidence is a geographic and climatic argument—what is it?

Look at the drawing of one skull of *Estemmenosuchus* (Figure 10.18). One can speculate about the social life of *Estemmenosuchus*, based on this one specimen. But what other information from other specimens would be useful in testing the speculation?

Further Reading

More Technical Reading

Barghusen, H. R. 1975. A review of fighting adaptations in dinocephalians (Reptilia, Therapsida). *Paleobiology* 1: 295–311.

Bennett, S. C. 1996. Aerodynamics and the thermoregulatory function of the dorsal sail of *Edaphosaurus*. *Paleobiology* 22: 496–506.

Cruickshank, A. R. I. 1978. Feeding adaptations in Triassic dicynodonts. *Palaeontologia Africana* 21: 121–132.

Haack, S. C. 1987. A thermal model of the sailback pelycosaur. *Paleobiology* 12: 450–458.

Hotton, N., et al. (eds.). 1986. *The Ecology and Biology of Mammal-like Reptiles*. Washington, D. C.: Smithsonian Institution Press. A set of essays on therapsids.

Kemp, T. S. 1982. *Mammal-like Reptiles and the Origin of Mammals*. London: Academic Press.

Laurin, M., and R. R. Reisz. 1990. *Tetraceratops* is the oldest known therapsid. *Nature* 345: 249–250.

Reisz, R. R. 1981. A diapsid reptile from the Pennsylvanian of Kansas. *Special Publications of the Museum of Natural History of the University of Kansas* 7: 1–74. *Petrolacosaurus*.

Ruben, J. A., et al. 1987. Selective factors in the origin of the mammalian diaphragm. *Paleobiology* 13: 54–59.

Schultze, H.-P., and L. Trueb. 1991. *Origins of the Higher Groups of Tetrapods: Controversy and Concensus*. Ithaca, New York: Cornell University Press. Chapters 17 and 18 deal with synapsids.

Smith, R. M. H. 1987. Helical burrow casts of therapsid origin from the Beaufort Group (Permian) of South Africa. *Palaeogeography, Palaeoclimatology, Palaeoecology* 60: 155–169.

Sues, H.-D., and R. R. Reisz. 1998. Origins and early evolution of herbivory in tetrapods. *Trends in Ecology & Evolution* 13: 141–145. Review.

The Triassic Takeover

Chapter 10 may give the impression that the only significant evolution among Permian and Triassic reptiles took place among synapsids. This, of course, is not true. Permian reptile faunas were dominated ecologically and numerically by synapsids, first by pelycosaurs and then by therapsids. But a good deal of evolution was going on among diapsid reptiles (Figure 10.1), and in the Triassic the diapsids came to replace synapsids as the dominant land vertebrates. The replacement was so dramatic that it has come to be a debating ground for the general question of replacement of one vertebrate group by another. Questions and some possible answers relating to the diapsid takeover form the major themes of this chapter.

The Diapsid Reptiles

Diapsids are probably descended from *Petrolacosaurus*. There are three major groups of diapsids: **turtles**, **lepidosauromorphs** (lepidosaurs and a few others) and **archosauromorphs** (archosaurs and some others). The latter two groups each contain several orders.

   Turtles are difficult to classify. As we saw in Chapter 10, they are diapsids that lost the two holes in the skull, part of a set of specialized adaptations that includes the shell. Molecular evidence places them a little closer to crocodiles than to iguanas. At the moment, it's probably best to leave them in a diapsid group of their own.

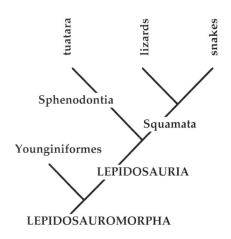

Figure 11.1 Phylogram of the lepidosauromorphs.

Lepidosauromorphs

These reptiles were originally small-bodied, lizard-like, ectothermic, and terrestrial. Some later lepidosaurs evolved into aquatic ways of life, with remarkable results, as we will see in Chapter 15.

On land, the lepidosauromorphs have been the dominant group of small-bodied reptiles since the Mesozoic. They can be divided into two major clades (Figure 11.1). **Squamata** are the numerous and diverse smaller living reptiles, including lizards and snakes. **Sphenodontia** are a sister group of the Squamata but include only one living form, the **tuatara**, *Sphenodon* (Figure 11.2). The tuatara is an outwardly lizardlike animal that survives today only on a few islands off the coast of New Zealand (Chapters 19 and 22), but its skull characters show that it is not a true lizard. Sphenodonts are known as far back as the Triassic.

Some Late Paleozoic reptiles, the **Younginiformes**, are a small but varied group of early diapsids best classified as the sister group of lepidosaurs within the lepidosauromorphs (Figure 11.1). Early Younginiformes were terrestrial, amphibious, or aquatic, but the major group of later Younginiformes, the **sauropterygians**, were beautifully adapted to life in the water (Chapter 15).

Archosauromorphs

Archosauromorphs include various diapsid reptiles that came to dominate the land in the Triassic (Figure 11.3). They were much more prone to evolve to large body size than the lepidosauromorphs, and they include the largest aerial and terrestrial animals that have ever lived.

Archosauria are conveniently divided into four groups (Figure 11.3). **Thecodonts** are the stem or ancestral archosaurs; they gave rise to the other groups in the later Triassic, although they are not strictly a clade. Several groups of thecodonts were medium-sized or large aquatic, terrestrial, or amphibious carnivores. Some highly successful thecodont descendants include the clades **Crocodylia** (crocodiles and alligators), **Dinosauria** (prosauropods, sauropods, ornithischians, and theropods),

Figure 11.2 The tuatara, *Sphenodon*, from New Zealand. (From Lydekker.)

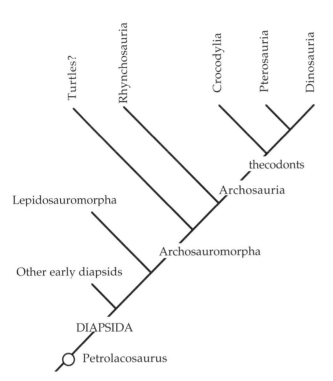

Figure 11.3 Phylogram of the archosauromorphs.

and **Pterosauria**. Dinosaurs were also the ancestors of birds, which are therefore derived diapsid, archosauromorph, archosaurian, dinosaurian reptiles. Pterosaurs evolved true flapping flight much earlier than birds did, and they dominated the skies throughout the Mesozoic (Chapter 13).

Rhynchosauria are classified among the archosauromorphs with the archosaurs proper (Figure 11.3). They were briefly the dominant Triassic herbivores.

Finally, there are some early diapsids whose relationships are not properly understood. I have placed them in an artificial clade, "other early diapsids" in Figure 11.3. They include **weigeltisaurs** (Permian gliding reptiles [Chapter 14]), and some aquatic groups, including *Mesosaurus*, a strange little Permian fish-eating reptile that was the earliest amniote to reach the southern continents (Figure 11.4). **Ichthyosaurs** (Chapter 16) are Triassic or later in age.

All this radiation among the diapsids took place in the Triassic, and the diapsid takeover from synapsids during that time was an astonishing series of events.

Figure 11.4 *Mesosaurus*, a small Permian reptile that was the first amniote to reach Gondwanaland. Its discovery in freshwater Permian sediments in South Africa and Brazil helped to confirm that Africa and South America had once been joined in Gondwanaland. *Mesosaurus* looks like an aquatic fish-eater. About 1 meter (3 feet) long. (After MacGregor.)

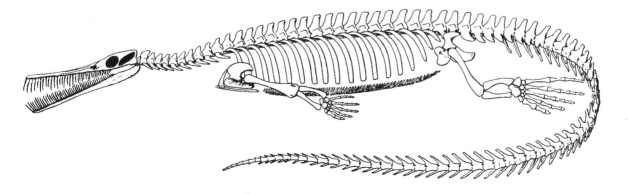

Late Permian Diapsids

The few Late Permian diapsids had already evolved into a very wide range of ways of life. In Africa, for example, *Heleosaurus* was terrestrial, *Hovasaurus* was aquatic, and *Coelurosauravus* (Chapter 14) was a glider.

Heleosaurus is a small diapsid, perhaps 30 cm (1 foot) from snout to anus, known from only one specimen with poorly preserved limbs. Robert Carroll analyzed the shoulders and pelvis, and suggested that although *Heleosaurus* sprawled most of the time, the hind limbs could be raised to an erect position for rapid bipedal running. Sharp bladelike teeth suggest a carnivorous way of life. *Heleosaurus* is an important fossil because it shares some characters with the earliest thecodonts; it may be ancestral to them and perhaps to the lepidosauromorphs too (Figure 11.3). It is advanced over earlier diapsids such as *Petrolacosaurus* in running ability.

Three hundred specimens of *Hovasaurus* make it one of the best-known Permian diapsids. Overall, it was lizardlike, perhaps only 30 cm (1 foot) long from snout to vent. But the tail was exceptionally long, strong, and deep (Figure 11.5), so the whole animal was close to a meter (3 feet) in length. The tail contained at least 70 vertebrae and looked like a swimming appendage. Inside the fossils, the abdominal cavity consistently contains a mass of small quartz pebbles. The pebbles often have a characteristic shape, tapering at both ends. Presumably they were swallowed by the animal during life. They are too small to be food-grinding pebbles and too far back in the abdomen to have occupied the stomach in life. Probably they were contained in a specially adapted abdominal sac. *Hovasaurus* almost certainly ingested the stones as ballast for diving. Living Nile crocodiles do the same thing, and the extinct plesiosaurs may have done so too.

Figure 11.5 *Hovasaurus* was an aquatic diapsid from the Late Permian of Madagascar. The tail was very long and strong, and its abdominal cavity was fossilized with pebbles inside; they can be interpreted as ballast. Total length about a meter (3 feet). (From Currie.)

The Triassic Diapsid Takeover: The Pattern

Over 600 Permo-Triassic reptiles have been described. Their history is best preserved for those that lived in lowland faunas. Upland faunas are not often preserved, and aquatic reptiles are not as numerous as terrestrial ones.

Large pelycosaurs dominated the tropical forests of Euramerica in the Early Permian. The only tetrapod outside this area was aquatic, the little reptile *Mesosaurus* (Figure 11.4), which lived in and around the African and Brazilian parts of Gondwanaland. By the Middle Permian, reptiles could walk into Gondwanaland. There was still an oceanic barrier where Siberia had not yet joined Euramerica, and mountains and/or climatic barriers along the southern end of Africa prevented reptiles from reaching the Indian and Australasian parts of Gondwanaland.

Therapsids had replaced pelycosaurs as the dominant land reptiles. New advanced therapsids dominated the Late Permian, particularly dicynodonts. But in some areas, diapsids were radiating into new ways of life. There were terrestrial, aquatic, and gliding diapsids in Madagascar; more important, the success of *Heleosaurus* in South Africa suggests that terrestrial diapsids were invading cooler regions for the first time.

Gondwanaland had rich Triassic faunas; and by this time, apparently, land animals were free to disperse throughout Pangea. Therapsid diversity dropped sharply in the Early Triassic, although the species that survived were widespread and numerous. Dicynodonts were extraordinarily abundant at larger sizes, and cynodonts were medium-sized herbivores. There were few therapsid predators: most of them were small- and medium-sized cynodonts such as *Cynognathus*. The first archosaurs were small carnivorous thecodonts, although therapsids outnumbered them 65 to 1 at first. But thecodonts were larger by the end of the Early Triassic, some of them 5 meters (16 feet) long, with massive skulls a meter long. In South Africa, *Euparkeria* was a fast, lightly built thecodont carnivore.

Therapsids were the dominant herbivores well into the Late Triassic, with huge numbers of specimens in the Middle Triassic of Brazil. But Middle Triassic thecodonts showed marked improvements in running ability over earlier forms. By the end of the Middle Triassic, rhynchosaurs became abundant vegetarians alongside the dicynodonts. There were even greater changes among the carnivores. Thecodonts of various sizes became abundant in the Middle Triassic, ranging from large quadrupedal rauisuchians to small bipedal ornithosuchids, with crocodile-like forms also present. Cynodont carnivores were at most medium-sized but remained abundant and diverse.

Therapsids and rhynchosaurs declined distinctly in the Late Triassic, though they were still important ecologically. By the latest Triassic most of the therapsids had disappeared, along with rhynchosaurs and the early thecodonts. The vegetarians of the latest Triassic were almost all prosauropod dinosaurs; thecodont carnivores were larger, more diverse, and more mobile than before, and they were joined by the first theropod and ornithischian dinosaurs. The first mammals were few and small.

The major players in the diapsid takeover in the Late Triassic were the archosauromorphs. Rhynchosaurs and thecodonts were the major new terrestrial groups, while nothosaurs and ichthyosaurs became fully adapted marine reptiles, and pterosaurs evolved powered flight.

Finally, at the end of the Triassic, dinosaurs quickly overwhelmed terrestrial ecosystems throughout the world, replacing thecodonts and therapsids alike in every medium- and large-bodied way of life, to form a land fauna dominated by dinosaurs that lasted throughout the Jurassic and Cretaceous Periods.

The replacement of therapsids by archosaurs took place worldwide. Therapsid carnivores were replaced first, gradually during the Middle Triassic. Carnivorous archosaurs became gradually larger, more diverse, and more abundant with time. They also came to have a more erect limb structure, which indicates better locomotion, including bipedal running in many cases. The herbivore replacement was much more rapid, taking place in the early part of the Late Triassic.

There are arguments about the relative suddenness of some of the replacements. Detailed studies in Triassic rocks which extend from North Carolina to Nova Scotia have shown that here at least there was a sudden, major extinction of Late Triassic land vertebrates. Some people have wondered whether a meteorite impact was involved, pointing to the huge crater of Manicouagan in Québec, Canada (70 km [40 miles] across), which is dated with some uncertainty as Late Triassic. Others deny that the replacements were sudden at all, or, if they were, that there were at least two separate episodes of extinction and replacement. The arguments mostly reflect the difficulty of judging relative time in terrestrial deposits; they should be resolved by careful research.

There are more fundamental arguments about the relationship between archosaurs and therapsids. Some people argue that the therapsids largely died out for environmental reasons (climatic change, for example, or an asteroid impact) and were replaced by archosaurs that radiated after the therapsids had largely gone. Others suggest more direct competition, in which carnivorous archosaurs first outcompeted therapsid carnivores and then hunted out the remaining therapsid vegetarians. Vegetarian archosaurs then evolved to take advantage of the abundant plant life. In this model, the archosaur success resulted from competitive superiority.

When it comes to real evidence, we can show clearly that archosaurs were much superior to other contemporary land vertebrates in respiration and locomotion. The difference between Triassic archosaurs and synapsids is so clear and so profound that it provides all the reason we need to account for the decline of the synapsids, except for their shifty, vicious, nocturnal survivors, the mammals.

Respiration, Metabolism, and Locomotion

David Carrier put together some simple but powerful ideas about the links between respiration, locomotion, and physiology.

Fishes have no problem maintaining high levels of exercise. Many sharks swim all their lives without rest, for example. Gill respiration gives all the oxygen exchange needed for such exercise levels, and with good hunting skills, the necessary food supply is readily available.

The evolution of lungs did not change that relationship between anatomy and physiology. Air has to be pumped in and out of an internal body cavity, however, and living lungfishes may have rather low exercise levels because their oxygen exchange is not geared for high performance.

Tetrapods moving about on land face a much more serious problem. The shoulder girdle and the forelimbs in particular, powered in part by the muscles of the trunk, are largely devoted to supporting and moving the body over the ground. In the sprawling gait of amphibians and living reptiles, the trunk is twisted first to one side and then the other in walking and running. As the animal steps forward with its left front foot, the right side of the chest and the lung inside it are compressed while the left side expands (Figure 11.6). Then the cycle reverses with the next step. This distortion of the chest interferes with and essentially prevents normal breathing, in which the chest cavity and both lungs expand uniformly and then contract. If the animal is walking, it may be able to breathe between steps, but *sprawling vertebrates cannot run and breathe at the same time*. I shall call this problem **Carrier's Constraint**.

Animals can run for a while without breathing: for example, Olympic sprinters usually don't breathe during a 100-meter race. Animals can generate temporary energy by anaerobic glycolysis, breaking down food molecules in the blood supply without using oxygen. But this process soon builds up an oxygen debt and a dangerously high level of lactic acid in the blood. Mammalian runners (cheetahs and humans, for example) often use anaerobic glycolysis even though they can breathe while they run; it's a useful but essentially short-term emergency boost, like an afterburner in a jet fighter.

Living amphibians and reptiles, then, can hop or run fast for a short time, first using up the oxygen stored in their lungs and blood, then switching to anaerobic glycolysis. They cannot sprint for long, however. If lizards want to breathe, they have to stand still with feet symmetrical. Lizards run in short rushes, with frequent stops. By attaching recorders to the body, Carrier showed that the stops are for breathing, and that lizards don't breathe as they run. Therefore, all living amphibian and reptilian carnivores use ambush tactics to capture agile prey: chameleons and toads flip their tongues at passing insects, for example.

The giant varanid lizard, the ora or Komodo dragon, which eats deer, pigs, and tourists (most notably, Baron Rudolf von Reding on 18 July 1974), goes a little way toward solving Carrier's Constraint by pumping air into its lungs from a throat pouch; but that only gives it a small improvement in performance. The Komodo dragon has a short sprinting range, but it prefers to ambush prey from 1 meter away.

Amphibians and most living reptiles have a three-chambered heart, which has usually been regarded as inferior to the four-chambered heart of living mammals and birds. But the three-chambered heart is useful to a lizard. Lizards run to catch food or to get away from danger, so they must use their resources most efficiently at this time. In a run, it is useless and perhaps dangerous for the lizard to waste energy pumping blood to lungs that cannot work. The lizard thus uses all the heart and blood capacity it has to circulate its store of oxygen around the whole body. The price it pays is a longer recharging time when it has to resupply oxygen to the blood, but it is usually able to do this at a less critical moment.

Early tetrapods all had sprawling gaits and faced a great problem. Their respiration and locomotion used much the same sets of muscles, and both systems could not operate at the same time. Imagine the laborious journey of *Ichthyostega* from the water to its breeding pools, with a few steps and a few gasps repeated for the whole journey. One can understand why so many early amphibians remained adapted to life spent

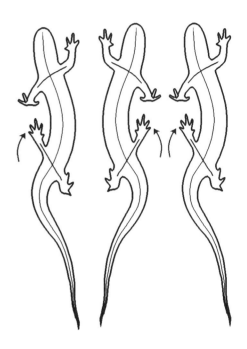

Figure 11.6 David Carrier pointed out that the sprawling locomotion of a lizard or salamander forces it to compress each lung alternately as it moves. Therefore, it can't breathe if it runs. Some salamanders have evolved to live without lungs, but probably for other reasons.

The reptilian idea of fun

Is to bask all day in the sun.

A physiological barrier,

Discovered by Carrier,

Says they can't breathe, if they run.

largely in water, and why many early reptiles often looked amphibious. *Eryops*, for example, swam with its tail (Figure 9.4) and would have had no major difficulty in devoting its rib-cage muscles to taking deep breaths at the surface. Horned nectrideans evolved ways to cut down undulation of their bodies as they swam, and perhaps they did this for air breathing rather than feeding (compare the discussion in Chapter 9).

When we see land animals such as pelycosaurs, with stiffened backbones and teeth designed for carnivorous and vegetarian diets rather than fish eating, we have to conclude that the problem had at least partially been solved. It's no good, for example, to raise metabolic rate by solar thermoregulation if there is no reliable oxygen supply to tissues.

I suggest that the secret of the pelycosaurs was the stiffening of the backbone. They simply did not twist the body much as they moved. They had long bodies and relatively short limbs compared with lizards, and in any case a short step would not have rotated the trunk very much or distorted the lungs. The stiffening of the body also meant that most of the forelimb rotation was taken up at the shoulder joint, rather than being transmitted to the trunk. Furthermore, pelycosaurs had wheelbarrow locomotion, and the front limbs were mainly reactive support props, so the muscles operating them did not exert forces on the chest wall except to support the shoulder joint. On the other hand, the driving muscles of the pelvic girdle attached far from the chest wall.

The pelycosaurs thus had a special synapsid *mitigation* of Carrier's Constraint: they evolved adaptations that went some way toward reducing its consequences. (If you understand that point, look back at the skeleton of *Ichthyostega* [Figure 8.15]. Perhaps there is an analogous reason for its peculiar rib structure?) But pelycosaurs could not *solve* Carrier's Constraint. There is no way that they were running freely, or breathing while they ran.

Fishes can swim in water with sustained energy because Carrier's Constraint does not apply to gill breathing. The same is probably true for the lung breathing of turtles, because their shell does not allow the lungs to be distorted as they swim and come to the surface to breathe.

Many living land vertebrates have evolved a beautiful answer to Carrier's Constraint. They have freed the mechanics of respiration from the mechanics of locomotion by evolving an **erect stance**. The body is suspended more freely from the shoulders, allowing the thorax to make its breathing movements with hardly any twisting.

The evolutionary solution to Carrier's Constraint that resulted from erect stance is shown best today in mammals. Mammals evolved the **diaphragm**, a set of muscles to pump air in and out of the chest cavity. Air is sucked in as the diaphragm contracts, and forced out by the reaction of the elastic tissues of the lung. At the same time, the locomotion in most mammals has evolved to encourage breathing on the run. The backbone flexes and straightens in an up-and-down direction with each stride, alternately expanding and compressing the rib cage evenly (Figure 11.7). This rhythmic pumping of the chest cavity in the running action can be synchronized with the action of the diaphragm to move air in and out of the lungs with little effort. Thus quadrupeds running at full speed—

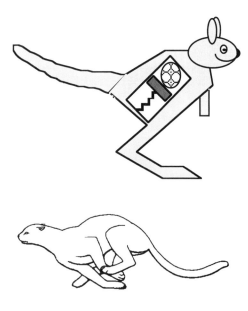

Figure 11.7 An animal with erect limbs does not rotate its chest as it runs as much as a sprawling animal does. It is easy to see that the mechanical wallaby can force air symmetrically in and out of its lungs as it hops along. But the cat, other mammals, and certain other groups of tetrapods can also breathe as they run, usually in synchrony with their steps, so they can reach a high exercise level.

gerbils, jackrabbits, dogs, horses, and rhinoceroses—take one breath per stride, and wallabies take one breath per hop (Figure 11.7). Trotting is far more complex, but that doesn't harm the line of argument presented here. Human runners usually take a breath every other stride. It is such a natural action that we don't notice it: runners should try to breathe out of phase to get some idea of the mechanism.

Animal locomotion often involves cyclic movements such as the strides and strokes of running or swimming limbs, or wingbeats in flight. Breathing may be made more efficient if it is synchronized with certain phases of limb movement. This is particularly important in human swimming, but it is a general principle. Flying insects synchronize their respiration with their wingbeats: the same muscular actions that raise and lower the wings also act to expand and compress the body, forcing air in and out of the spiracles. Birds do much the same thing (Chapter 14).

These principles are pieces of basic animal physiology, and they should be as true for extinct animals as they are for modern ones. Therefore, erect stance might be necessary for sustained running in any land animal, and its evolution should represent a great breakthrough in any tetrapod lineage, giving the basis for greatly improved running speed and stamina. Living reptiles are successful, but they are limited in the ecological roles they can perform because they have a sprawling gait and cannot sustain fast movement for very long.

Diapsids living today, such as lizards, don't have erect stance or sustained energy output, but we must not be fooled into thinking that all diapsids have always lacked those capabilities. David Carrier suggests that the Triassic diapsids—the archosauromorphs in particular—were the first tetrapods to make the breakthrough to erect gait and rapid, sustained locomotion. That breakthrough is preserved in the fossil record in the structure of the limbs and shoulder girdles of the early archosauromorphs. Erect gait and sustained locomotion was most likely the key innovation that made possible the diapsid, and in particular the archosaur, takeover of the Late Triassic.

I suggest here that an accident of history played an important role in forming the differences between Triassic diapsids and synapsids. Therapsids evolved largely in cool climates of the late Permian, in northern Laurasia and southern Gondwanaland, while Permian diapsids evolved in warmer climates. Part of a heat-retaining syndrome in cool climates is to have stocky, compact bodies and short limbs and appendages, and therapsids are characteristically built that way (for example, Figure 10.14). Diapsids, on the other hand, had long, strong tails, and much of their body weight was on the hind limbs. Many diapsids became partly or totally bipedal, and it may have been comparatively easy for them to evolve erect limbs from a bipedal stance. Therapsids, with short tails, did not have that option, and all of them were quadrupeds with a good deal of weight on the front feet. It may have been difficult to escape from the wheelbarrow locomotion that the therapsids inherited from the pelycosaurs, especially at larger body size. Truly erect gait—the solution to Carrier's Constraint—did not evolve among synapsids until the mammals of the Early Jurassic.

Figure 11.8 One of several possible phylograms to show the relationships of the various groups of Archosauria.

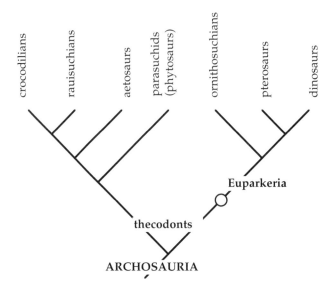

Thecodonts

Thecodonts are the stem group of the archosaurs (Figures 11.3 and 11.8). Early thecodonts were sprawling reptiles, but there was a repeated evolution of advanced locomotion throughout the history of thecodonts and their immediate descendants.

For most of the Middle and Late Triassic, the largest carnivorous thecodonts were rauisuchians and ornithosuchians (Figure 11.9). They included bipeds and quadrupeds, but they were large animals and geographically widespread. *Saurosuchus* from the Late Triassic of Argentina is 6 to 7 meters long (about 20 feet), with a massive skull and fairly long legs. *Postosuchus*, from the Late Triassic of Texas, is about 4 meters long including the tail, and stood 2 meters high. It was lightly built and walked (and ran) bipedally. But it was a hunter, with a heavy killing head, impressive wide-opening jaws, and serrated stabbing and cutting teeth. The eyes were large, set for forward stereoscopic vision, with bony eyebrows to shade them. *Postosuchus* is uncannily like a small version of the much larger and later carnivorous dinosaurs *Tyrannosaurus* and *Allosaurus* in structure, adaptation, and presumably in ecology. These advanced thecodonts had erect limbs, allowing them long stride length and efficient, sustained running.

Advanced thecodonts were perhaps close in ecology to living monitor lizards, which have a semi-erect gait and are active predators with a preferred body temperature close to 37°C (98°F). The Komodo dragon of Indonesia is the top predator in its ecosystem, weighing over 100 kg (200 pounds). Advanced Triassic thecodonts are about the same size and, if anything, were more active, because many had fully erect gait and could probably run faster and further.

Parasuchids (sometimes called **phytosaurs**) were large, long-snouted carnivorous thecodonts from the tropical belt of the Late Triassic. They evolved toward a crocodilian appearance and ecology (Figure

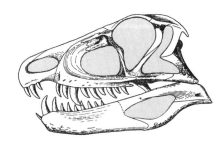

Figure 11.9 The advanced thecodont *Ornithosuchus* evolved many adaptations in parallel not only with rauisuchians such as *Postosuchus* but also with later carnivorous dinosaurs. Skull about 25 cm (a foot) long. (After Walker.)

11.10). The major visible external difference is that parasuchid nostrils were at the base of the skull rather than the end of the snout. Parasuchid limbs permitted three different walking styles: upright, with the body well off the ground; slow, sinuous sprawling and crawling; and tobogganing over mud into the water. At least some parasuchids were effective predators on terrestrial animals. Two specimens from India more than 2 meters (7 feet) long had stomach contents that included small bipedal archosaurs, and one had eaten a rhynchosaur. Parasuchids disappeared at the end of the Triassic, with many other thecodont groups. They were not replaced by dinosaurs but by true crocodilians, which had been small and terrestrial through the Triassic but increased in size into the Jurassic as they occupied aquatic habitats.

Thecodonts are crocodile ancestors, and living crocodiles may well be some guide to the physiology, locomotion, and ecology of parasuchid thecodonts. Crocodiles have a good circulatory system, with more advanced heart and lung modifications than other living reptiles. Although they normally walk slowly on land, in a sprawling stance, they are also capable of a faster run in which the limbs are nearly vertical. The little freshwater crocodile of Australia can gallop (briefly) at 16 kph (10 mph), but some parasuchids could probably have done even better. Crocodiles show that it's artificial to categorize animals as having only one possible gait and stance: even so, many animals specialize in one gait or another.

Early crocodilians that lived alongside and competed with thecodonts were small, long-legged, terrestrial predators. Some, from the Late Triassic of western Europe and eastern North America, look like lightly built running carnivores (Figure 11.11); early crocodilians may have evolved a close-to-erect stance independently of other archosaurs. Later crocodilians became adapted to water, replacing the parasuchids, and only then did they become much larger. They evolved a secondary palate so that they could bite and chew under water without flooding their nostrils, and they also lost some of the features of their terrestrial gait, becoming secondarily sprawling.

Figure 11.10 The Triassic parasuchid *Rutiodon*. (Negative 319167. Photograph by Edward Bailey. Courtesy of the Library Services Department, American Museum of Natural History.)

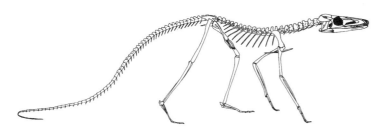

Figure 11.11 A reconstruction by P. J. Crush of a fast-running early terrestrial crocodile from the Late Triassic of Britain. It was very lightly built and only about 50 cm (18 inches) long.

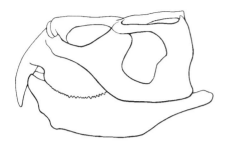

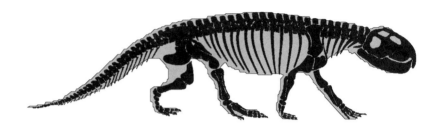

Figure 11.12 Left: The skull of a Late Triassic rhynchosaur, *Scaphonyx*. (After Sill.) Right: The general body plan of *Scaphonyx*, which was about a meter long (3 feet).) (Skeleton after von Huene.)

Rhynchosaurs

Rhynchosaurs evolved in the Middle and Late Triassic with the decline of most large vegetarian therapsids and the disappearance of some. They were all herbivores, pig-sized animals with hooked snouts bearing a powerful cutting beak and hind limbs that look as if they might have been used for digging (Figure 11.12). Strong jaws bore batteries of slicing teeth, which are unusual among reptiles in that they were fused to bone at the base, not set into normal sockets. The teeth were ever-growing and were not replaced during life. As rhynchosaurs grew, they simply added more bone and more teeth at the back of the growing jaw as the teeth at the front became worn out. This style of tooth addition allowed rhynchosaurs great precision in tooth emplacement, so their bite was very effective for slicing vegetation with a scissor-like action.

Rhynchosaurs have been difficult to classify because of their peculiar features. They are probably a sister group of the other archosaurs. They were abundant and widespread in the Middle and Late Triassic and may have replaced therapsid groups because they too evolved an erect gait. However, rhynchosaurs rapidly became extinct at the end of the Triassic.

Dinosaur Ancestors

The thecodont *Euparkeria*, also from South Africa, was a larger, even more agile predator than *Heleosaurus*. It was about a meter long, very lightly built, and had a long, strong tail to give balance in running. Its skull was long and light, with many long, sharp stabbing teeth (Figure 11.13). *Euparkeria* was evidently fast-running, probably a bipedal runner but walking on four feet. Its speed and agility may have promoted its success in comparison to contemporary therapsids. The limbs were set directly under the body, reducing the mechanical stresses of fast running and allowing sustained running. *Euparkeria* is an advanced thecodont that could be the ancestor of both pterosaurs and dinosaurs (Figure 11.8).

The first dinosaurs appeared over much of Gondwanaland in the Late Triassic, around 225 Ma. They were small, agile, bipedal carnivores, and their characteristically large to enormous sizes evolved later. *Herrerasaurus* and *Eoraptor*, the best-known early dinosaurs, lived in Argentina alongside a fauna dominated by rhynchosaurs, with synapsids present

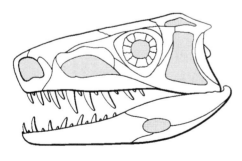

Figure 11.13 The advanced thecodont *Euparkeria* is a reasonable candidate for a generalized archosaur and dinosaur ancestor.

too. The community seems to have been stable for at least 10 m.y. or so, with the dinosaurs forming perhaps one-third of the carnivores. After that, the ecology seems to have changed rapidly, and dinosaurs became dominant, as we shall see in Chapter 12.

Review Questions

Look at *Mesosaurus*. You have two pictures: a drawing of the whole animal, and a close-up inside-the-mouth view of its jaw. Infer what you can about its habitat and its diet from the features shown on the picture. If you recognize the name *Mesosaurus*, then you can earn bonus points by telling me what else is interesting about *Mesosaurus*.

The archosaurs (the group that included dinosaurs) took over most of the ways of life of animals on land in the late Triassic. What feature of their anatomy was most important in this?

Explain why lizards can't run for very long without stopping.

Further Reading

More Accessible Reading

Benton, M. J. 1993. Late Triassic extinctions and the origin of the dinosaurs. *Science* 260: 769–770. Benton tries to argue that dinosaurs replaced Triassic reptiles, but didn't outcompete them.

Owerkowicz, T., et al. 1999. Contribution of gular pumping to lung ventilation in monitor lizards. *Science* 284: 1661–1663. Monitor lizards avoid Carrier's Constraint (a little) by pumping air from a throat pouch.

Rieppel, O. 1999. Turtle origins. *Science* 283: 945–946.

More Technical Reading

Carrier, D. R. 1987. The evolution of locomotor stamina in tetrapods: circumventing a mechanical constraint. *Paleobiology* 13: 326–341. In my view, one of the real breakthrough papers of the 1980s. I think that Carrier is wrong only in linking endothermy so tightly to athletic ability, and as we shall see, this point becomes very important in considering dinosaurs.

Gatesy, S. M. 1991. Hind limb movements of the American alligator (*Alligator mississippiensis*) and postural grades. *Journal of Zoology, London* 224: 577–588.

Gower, D. J., and E. Weber. 1998. The braincase of *Euparkeria*, and the evolutionary relationships of birds and crocodilians. *Biological Reviews* 73: 367–411. Euparkeria is basal to advanced archosaurs.

Olsen, P. E., et al. 1987. New Early Jurassic tetrapod assemblages constrain Triassic-Jurassic tetrapod extinction event. *Science* 237: 1025–1029; see also discussion, v. 241, pp. 1358–1360.

Padian, K. (ed.). 1987. *The Beginning of the Age of Dinosaurs.* Cambridge, England: Cambridge University Press.

Parrish, J. M. 1987. The origin of crocodilian locomotion. *Paleobiology* 13: 395–414.

Dinosaurs

We are familiar with dinosaurs in many ways: they have been with us since kindergarten or before, in comic strips, toys, stories, movies, nature books, TV cartoons, and advertising. Yet it's still not easy to understand them as animals. The largest dinosaurs were more than ten times the weight of elephants, the largest land animals alive today. Dinosaurs dominated land communities for 100 million years, and it was only after they disappeared that mammals became dominant. It's difficult to avoid the suspicion that dinosaurs were in some way competitively superior to mammals and confined them to small body size and ecological insignificance. We would dearly love to know the basis for that superiority.

Dinosaurs were descended from small, bipedal thecodonts, and they appeared in the Late Triassic of Gondwanaland at about the same time as the first mammals. The earliest dinosaurs were small bipedal carnivores. All the spectacular variations on the dinosaur theme came later. Even so, all four major dinosaur groups (Figure 12.1) had evolved by the end of the Triassic.

Theropods

Theropods were all bipedal carnivores, retaining the body plan and the ecological character of ancestral dinosaurs. Yet some of the most advanced dinosaurs were theropods, and the theropod clade also includes living birds. The earliest known dinosaurs were theropods: *Eoraptor* and *Herrerasaurus*, both from the Late Triassic of Argentina. *Eoraptor* was a tiny animal with a skull only about 2.5 cm (an inch) long. Even so, it was

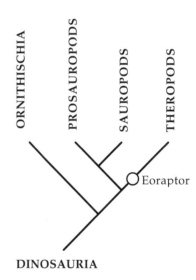

Figure 12.1 A phylogram showing the major groups of dinosaurs. The earliest known dinosaur is the little theropod *Eoraptor*, and prosauropods are also known from Late Triassic rocks. Therefore, the cladogram predicts that all four groups of dinosaurs had diverged by that time, and that there are "ghost ornithischians" and "ghost sauropods" still to be discovered somewhere in Late Triassic rocks.

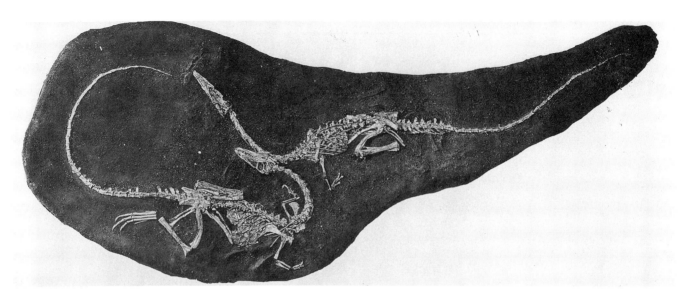

Figure 12.2 *Coelophysis*, an early theropod dinosaur (a ceratosaur) from the Late Triassic of New Mexico. (Negative 329319. Photograph by Boltin. Courtesy of the Library Services Department, American Museum of Natural History.)

well adapted as a fast-running carnivore, with sharp teeth and grasping claws on its forelimbs. *Herrerasaurus* was very like *Eoraptor* but much larger, between 3 and 6 meters (10–20 feet) long. *Coelophysis* from the Late Triassic of North America was 2.5 meters (8 feet) long, lightly built (perhaps only 20 kg or 45 pounds), and was clearly adapted for fast running. The bones of its skeleton were more extensively fused into stronger units than in the earliest theropods, so *Coelophysis* is placed into the first of the derived theropod groups, the **ceratosaurs** (Figure 12.2).

More advanced theropods (Figure 12.3) separated early into the **allosaurs** (massive, powerful predators) and the lighter and more agile **coelurosaurs** (Figure 12.4). Within coelurosaurs, one lineage of small, agile carnivores includes the **birds** and the **dromaeosaurs**, which include *Velociraptor* and *Deinonychus*. As we collect more theropods from this lineage, it becomes increasingly difficult to draw the line between dinosaurs

Figure 12.3 A phylogram of the theropods (simplified from Padian et al., 1999).

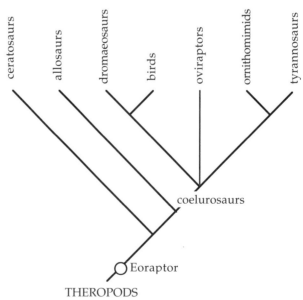

Figure 12.4 The coelurosaur *Stenony-chosaurus*, a lightly built dinosaur with a relatively large brain. (Reconstruction by Bob Giuliani. © Dover Publications Inc., New York. Reproduced by permission.)

and birds. For example, *Unenlagia*, from Argentina, had arms that were very bird-like in the way they folded, and like birds but unlike other theropods, the shoulder joint faced sideways. *Mononykus*, from the Late Cretaceous of Mongolia, is built much like a small theropod, including a rather long tail, but it apparently has a breastbone like a bird. Its wings/arms were much modified, so the hand had only one strong, blunt, clawed finger. Obviously, *Mononykus* and others like it (the **alvarezsaurids**) are very close to birds: rumors in early 1999 suggest that they lie within theropods as the immediate sister group to *Archaeopteryx* and other birds.

The scientists who described *Mononykus* wondered whether it dug with these strange hands, but they recognized that burrowing would not suit a rather tall, long-legged bird/theropod (see Norell et al., 1993). I suggest that it used the hands for digging out its prey (small mammals?) or for molding its nest. Ecologically, I would compare it with the big, flightless, megapod birds of Australia and New Guinea, which build a huge nest from piles of leaves that they collect and shape by kicking backwards with their enlarged feet. This is a truly comical sight, especially as the bird has to keep looking over its shoulder to see what it is doing. *Mononykus*, I suspect, had a much easier time making its nest. The arms of *Mononykus* seem especially well designed for adduction—moving together under a load—so I envisage *Mononykus* digging a shallow nest, then sweeping together vegetation or sand to cover its eggs.

The other two lineages of coelurosaurs are distinct groups. **Oviraptors** are the nest-building dinosaurs from Mongolia, to be discussed later. **Ornithomimids** are the so-called ostrich dinosaurs. They may have been the most intelligent of all dinosaurs and evolved a body plan much like a living ostrich, except that they had long arms and slim, dexterous fingers instead of wings. They had feet that were especially well-suited for running. Tyrannosaurs are related to ornithomimids.

A small, agile, bipedal body form was successful throughout theropod history. Theropods included the smallest dinosaurs, such as *Compsognathus*, which weighed only about 1 kg (2 pounds), but despite their light build most theropods were larger than the mammalian predators we

Figure 12.5 *Deinonychus*. Above: A skeletal reconstruction by John Ostrom, showing the animal as a powerful, fast-running predator. The tail was so tied by tendons that it acted as a wonderful counterweight for agile turning. Below: Flesh reconstruction of *Deinonychus* by Robert Bakker. Even apart from the physical weaponry of this dinosaur, Bakker's reconstruction of muscles and ligaments gives a vivid image of its power. (Courtesy of John Ostrom, Yale University.)

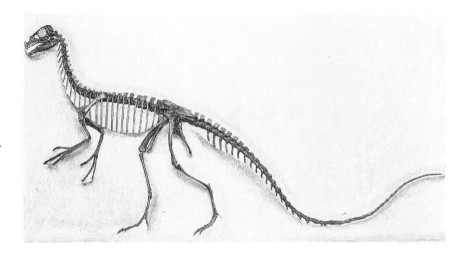

are familiar with today. Ornithomimids had long legs and necks, large eyes, and rather large brains, but no teeth. They could have been formidable carnivores, of course, but perhaps they specialized on smaller prey animals. Ornithomimids often lacked the large slashing claws and had

Figure 12.6 The skull of *Deinonychus*, an impressive Cretaceous predator from Montana, described by John Ostrom. The skull is long, light, and strong, and the teeth are designed for seizing, slashing, and cutting. The "velociraptors" of the movie Jurassic Park owed more to *Deinonychus* than they did to the real *Velociraptor*. (Courtesy of John Ostrom, Yale University.)

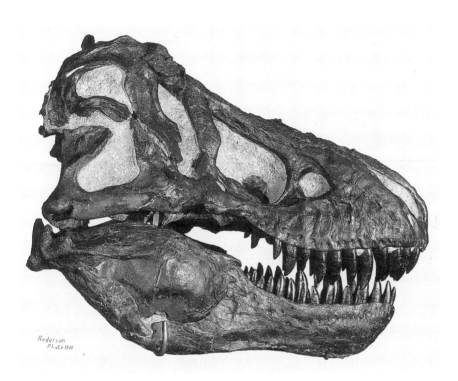

Figure 12.7 *Tyrannosaurus rex* had a massive skull. These jaws must have been the primary killing weapons. (Negative 35491. Photograph by A. E. Anderson. Courtesy of the Library Services Department, American Museum of Natural History.)

long fingers that could have been used to manipulate objects.

Deinonychus, from the Early Cretaceous, is one of the most impressive carnivores ever evolved, though it was only about 2 meters long (Figures 12.5, 12.6). It was fast and agile and had murderous slashing claws on both hands and feet, and a most impressive set of teeth.

Several lineages of theropods evolved to giant size. The giants are known informally as "carnosaurs" though they are not a clade. Giant size evolved at least four times—among Jurassic allosaurs and in three Cretaceous groups: in North American tyrannosaurs (Figure 12.7), in *Giganotosaurus* from Argentina., and in *Carcharodontosaurus* from North Africa. These theropods were the largest land carnivores of all time, each weighing about 6 or 7 tons (more than an elephant), and standing about 6 meters (20 feet) high, with a total length around 12 meters (40 feet). At present, *Giganotosaurus* seems to be the largest, with an estimated length close to 14 m (40+ feet), and a weight of 6–8 tons. They all must have relied on massive impact from the head for killing, aided by huge stabbing teeth that would have caused severe bleeding, usually lethal, in a prey animal.

Deinocheirus is an enormous coelurosaur from the Late Cretaceous rocks of Mongolia. We don't know much about it, because all we have is a set of giant shoulder bones and forelimbs with slashing claws, with no trace of the rest of the skeleton. The forelimb alone was 2.4 meters (7 feet) long, so this animal was well outside the range of most coelurosaurs in size and power.

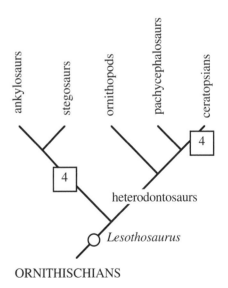

Figure 12.8 One of several alternative phylograms for ornithischian dinosaurs. The boxes marked "4" indicate separate evolution of a four-footed stance.

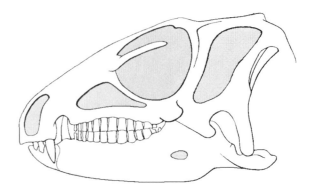

Figure 12.9 Two early ornithischians, both from southern Gondwanaland. Like all basal dinosaurs, they were erect, bipedal, and small. Left, *Lesothosaurus*. (Reconstruction by Bob Giuliani. © Dover Publications Inc., New York. Reproduced by permission.) Right, *Heterodontosaurus*, from southern Gondwanaland, was one of the earliest ornithischians. It had teeth that varied greatly along the jaw in size, shape, and presumed function. Skull about 10 cm (4 inches) long. (After Charig and Crompton.)

Ornithischians

The earliest ornithischians were the first herbivorous dinosaurs, and they gave rise to a spectacular radiation of dinosaurs, all of which were also herbivorous. Judging by their teeth, most ornithischians ate rather coarse, low-calorie vegetation, so many of them tended to be at least medium-sized. They were the most varied and successful herbivorous animals of the Mesozoic era, and they were abundant in terrestrial ecosystems right to the end of the Cretaceous. Small bipedal ornithischians, such as *Lesothosaurus* and the heterodontosaurids, gave rise at various times to derived groups that were much heavier, with some animals weighing 5 tons or more (Figure 12.8). The armored dinosaurs form one derived clade that includes **stegosaurs** and **ankylosaurs**; the horned dinosaurs or **ceratopsians** form another; but most of the larger ornithischians were **ornithopods**, which include **iguanodonts** and the so-called "duckbill" dinosaurs, the **hadrosaurs**.

The best-known early ornithischians are small dinosaurs from the Early Jurassic of Gondwanaland. The earliest one, *Lesothosaurus* (Figure 12.9, left), was small, agile, and fast-running, but it clearly had vegetarian teeth. *Heterodontosaurus* was very much like it, but its teeth were even more specialized for a vegetarian diet (Figure 12.9, right). Small teeth at the front of the upper jaw bit off vegetation against a horny pad on the lower jaw. The back teeth evolved into shearing blades for cutting vegetation. (The sharp incisors were for display or fighting.) The cheek teeth were set far inward, with large pouches outside them to hold half-chewed food for efficient processing.

The earliest ornithopods were less than a meter long, but they soon increased significantly in size. The jaw enlarged to contain a large number of grinding teeth. Many ornithopods continued to be moderate in size and retained their bipedal stance: *Hypsilophodon* is a good example. A general theme of ornithopod evolution was the successive appearance of groups that in different ways evolved toward the 5- to 6-ton size that seems to have been a weight limit for most terrestrial herbivores. Even at this size, many ornithopods remained bipedal. Others probably walked most of the time on all fours but raised themselves up on two limbs for running, or browsing on high vegetation (like goats and gerenuks today (Figure 12.10).

Figure 12.10. The hadrosaur *Corythosaurus*, an advanced ornithopod, showing its ability to rear on its hind legs to reach vegetation. (Reconstruction by Bob Giuliani. © Dover Publications Inc., New York. Reproduced by permission.)

Figure 12.11 The skeleton of a hadrosaur, the duckbilled dinosaur *Edmontosaurus*. Ossified tendons helped to hold the tail rigid over the pelvis when necessary. (From Marsh.)

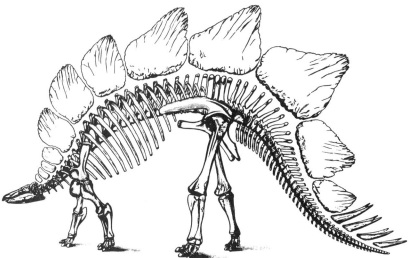

Figure 12.12 The skeleton of a stegosaur. (From Marsh.)

Ornithopods evolved large batteries of teeth, and newly evolved modifications of the jaws and jaw supports allowed complex chewing motions. **Iguanodonts** were particularly abundant in the Early Cretaceous: they reached 9 meters (30 feet) in length and stood perhaps 5 meters (16 feet) high. They cropped off vegetation with powerful beaks before grinding it in powerful batteries of teeth. Most iguanodonts were replaced ecologically by a variety of **hadrosaurs** (duckbilled dinosaurs) in the Middle and Late Cretaceous (Figures 12.10, 12.11). Hadrosaurs were about the same in size and body plan as iguanodonts, but had tremendous tooth batteries, with several hundred teeth in use at any time.

The other ornithischians were dominantly quadrupeds, but they betrayed their bipedal ornithopod ancestry with hindlimbs that were usually

Figure 12.13 *Ankylosaurus.* Modern reconstructions like this one show ankylosaurs as relatively active dinosaurs. Older reconstructions showed them more like giant tortoises. (Reconstruction by Bob Giuliani. © Dover Publications Inc., New York. Reproduced by permission.)

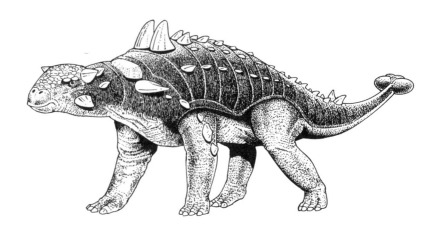

longer and stronger than the forelimbs. **Stegosaurs** (Figure 12.12), with their characteristic plates set along the spine, were the major quadrupedal ornithischians in the Jurassic, but they were replaced in the Early and Middle Cretaceous by the armored **ankylosaurs** (Figure 12.13). Later in the Cretaceous, the ornithischians were particularly abundant and varied in their body styles. Many quadrupedal forms lived alongside the hadrosaurs, and **ceratopsians**, or horned dinosaurs, became perhaps the most successful of all ornithischians (Figure 12.14).

Figure 12.14 *Triceratops* is the most famous of the horned dinosaurs or ceratopsians. (Negative 310437. Courtesy of the Library Services Department, American Museum of Natural History.)

Sauropodomorphs

Some early dinosaurs evolved to become very large, heavy quadrupedal vegetarians with broad feet and strong pillar-like limbs. The **sauropodomorphs** had an early radiation as **prosauropods** and a later radiation as the famous **sauropods** with which we are all familiar. Prosauropods were abundant, medium to large dinosaurs of the Late Triassic (Figure 12.15). They were typically about 6 meters (20 feet) long, but *Riojasaurus* was unusually large at 10 meters (over 30 feet). Prosauropods lived on all continents except Antarctica, with rich faunas known from Europe, Africa, South America, and Asia. They ranged into the Early Jurassic, when they were replaced by sauropods.

Prosauropods were all browsing herbivores. The teeth were generally good for cutting vegetation but not for pulping it. Opposing teeth did not contact one another, and all the grinding must have been done in a gizzard. (Masses of small stones have been found inside the skeletons of several prosauropods.) Prosauropods have particularly long, lightly built necks and heads, and light forequarters. They were clearly adapted to browse high in vegetation, perhaps reaching up from the tripod formed by the hind limbs and heavy tail. Only *Riojasaurus*, the largest, was always quadrupedal because of its weight, but it had a very long neck to compensate. Prosauropods were the first animals to browse on vegetation high above the ground, and they represent a completely new ecological group of herbivores exploiting an important new resource in the zone up to perhaps 4 meters (13 feet) above ground. The same adaptation was re-evolved later in sauropods, and again in mammals such as the giraffe.

Prosauropods were the largest and heaviest members of their communities, and they were abundant. *Plateosaurus* (Figure 12.15) accounts for 75% of the total individuals in a well-collected site in Germany. Coupled with their size, this means that they sometimes made up over 90% of the animal mass in their communities. Some prosauropods had unusual characters: *Yunnanosaurus* from China had teeth that met to provide self-sharpening surfaces as they wore against one another.

Sauropods are the largest land animals that ever evolved. Remember that there is a natural human ambition to discover the largest or the oldest of anything; we must be cautious in assessing claims about the size and weight of dinosaurs without also surveying the evidence. But even a cautious person must admit that well-documented sauropod body weights are at least 50 tons; *Supersaurus* from Colorado may have approached 80 tons, and *Argentinosaurus*, from Argentina of course, may have been close to 100 tons. Famous names and enormous numbers are associated with sauropod anatomy: *Apatosaurus* (or "Brontosaurus"), for example, 20 meters (65 feet) in length and weighing 30 tons; and *Diplodocus*, once thought to be the longest land animal, at over 25 meters (80 feet) long. *Brachiosaurus* has long forelimbs carrying it over 12 meters (40 feet) high, as tall as a four-story building, and with a weight estimated at 50 to 80 tons. But *Supersaurus* may have been 40 meters (130+ feet) long, and *Seismosaurus* from New Mexico, was at least 28 meters (90 feet), and more likely 40-50 meters (130–170 feet) long. Figure 12.16 gives some idea of the relative size of some large sauropods.

Sauropods were all herbivores, of course; no land animals that size could have been carnivorous. They had curiously small heads and very long necks that allowed them to browse on anything within 10 meters (33 feet) of ground level. The tails were long also, but the body was massive, with powerful load-bearing limb bones and pelvis. All sauropods were quadrupedal, though perhaps the earliest forms could briefly have been bipedal. The major body mass was centered close to the pelvis, which was accordingly more massive than the shoulder girdle.

Figure 12.15 The prosauropod *Plateosaurus*, about 7 meters (23 feet) long. One can imagine this animal browsing high in vegetation. (Reconstruction by Bob Giuliani. © Dover Publications Inc., New York. Reproduced by permission.)

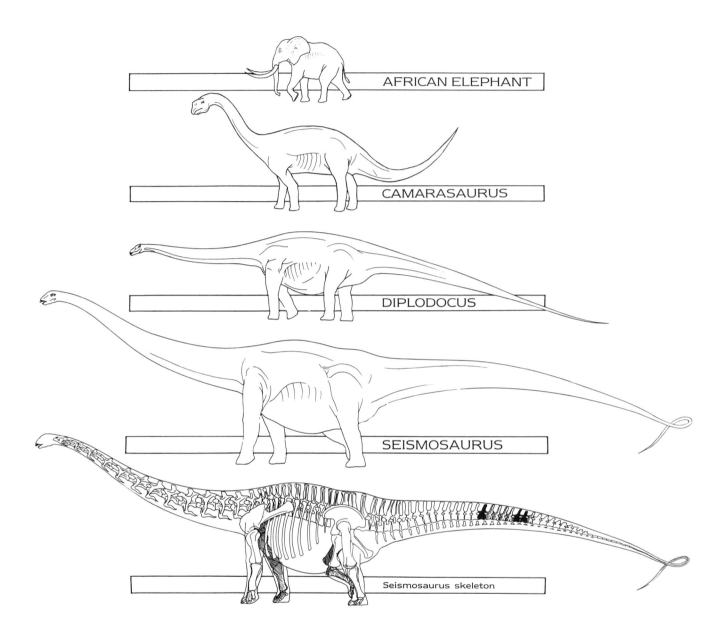

Figure 12.16 All sauropods are much alike in general body plan, differing mostly in scale. The diagram compares sizes and body plans of some sauropods with the largest living land mammal, the African elephant. (Courtesy of David D. Gillette and the New Mexico Museum of Natural History, Albuquerque.)

Dinosaur Paleobiology: Life at Large Size

There are small dinosaurs: *Compsognathus* was only the size of a chicken, for example. But the dominant feature of dinosaurs, and the dominant aspect of their paleobiology, has to be the enormous size of the largest ones (Figure 12.16). Ornithischian dinosaurs are perhaps easier to understand than the others because they were vegetarians in the 5-ton range, comparable with elephants or rhinos, perhaps. On the other hand, there are no 5-ton carnivores alive today on land that we can compare with carnosaurs such as *Tyrannosaurus*, and there are no 80-ton vegetarians that we can compare with the sauropods. Despite this, we can make some reasonable inferences about dinosaur biology.

Vegetarian Dinosaurs

Animals on purely vegetarian diets almost always have bacteria in their guts to help them break down cellulose. Large animals have slower metabolic rates than small ones, and for vegetarians this means a slower passage of food through the gut and more time for fermentation. Alternatively, a large vegetarian can digest a smaller percentage of its food and live on much poorer quality forage. Vegetarians usually grind their food well so that it can be digested faster, and this was accomplished in two different ways in the two major groups of vegetarian dinosaurs, the ornithischians and the sauropods.

Sauropods had very small heads for their size, and small, weak teeth (Figure 12.17). This has sometimes been thought to indicate a soft diet. Instead, sauropods probably had a small head for the same mechanical reason that giraffes do: the head sits on the end of a very long neck. In both sets of animals the food is gathered by the mouth and teeth, then swallowed and macerated later. Giraffes are ruminants, and boluses of food are regurgitated and chewed at leisure with powerful batteries of molar teeth.

Sauropods probably used a different system for grinding food. They probably had enormous, powerful gizzards in which food was ground up between stones that the dinosaurs swallowed. Wild birds seek and swallow grit to help them grind food, and poultry farmers can increase egg production by feeding grit to their chickens. A fossil moa of New Zealand was preserved with 2.5 kg (over 5 pounds) of stones in its stomach area. There are literally millions of dinosaur gizzard stones or gastroliths in terrestrial Cretaceous rocks all over the western interior of the United States. A high proportion of these are made of very hard rocks, often colored cherts. William Stokes made the irresistible suggestion that dinosaurs specifically sought brightly colored, rounded pebbles in stream beds for swallowing, providing themselves with perfectly shaped and very hard grinding stones. (Dinosaurs as the first rockhounds!)

Ornithischians, on the other hand, generally had massive batteries of grinding teeth, especially in the hadrosaurs and ceratopsians, and they

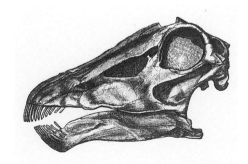

Figure 12.17 The head of a large sauropod, *Diplodocus*, showing the comparatively weak teeth. *Diplodocus* is shown also in Figure 12.16, to give some idea of the relative size of head and body. (From Marsh.)

Figure 12.18 An early reconstruction of dinosaur paleobiology: a carnivore leaping on a rival. The reconstruction is lively and attractive until one realizes the scale involved: each of these animals weighs two tons. No creature of this mass could leap in this way without injury. This reconstruction was made by Charles R. Knight in 1897. (Negative 335199. Photograph by R. E. Logan. Courtesy of the Department of Library Services, American Museum of Natural History.)

would have chewed up their food thoroughly, as living mammals do. Only a few ornithischians had weaker teeth, and there are gizzard stones associated with fossils of the little dinosaur *Psittacosaurus*.

Food gathering and processing do not seem to have posed problems difficult enough to prevent vegetarian dinosaurs from reaching enormous size. Dinosaurs evolved to be much larger than other land vertebrates for reasons not connected with diet.

Posture and Habitat

We can use our knowledge of the mechanics of bone, muscle, and ligament to interpret dinosaur skeletons. Early workers had difficulty coming to grips with the size of dinosaurs, mostly because they did not understand biomechanics. Thus, early reconstructions show dinosaurs sitting in trees or jumping high in the air (Figure 12.18). In fact, they would have broken bones on impact after even a small fall. Older reconstructions showing large dinosaurs with bent knees and bent elbows, usually near water, are equally inappropriate, because they are based on comparison with lizards. A 5-ton stegosaur and especially a 50-ton sauropod must have had erect limbs to support their weight, like elephants or rhinos rather than lizards.

Early workers, who suspected that sauropods would have found it difficult to support their weight with bent limbs, suggested that instead they spent much of their life in swamps, buoyed up by water. The long necks were interpreted as snorkels (Figure 12.19). Even today, many drawings of sauropods show water nearby. The snorkel idea is very unlikely: in any depth of water, the pressure would have been great enough to prevent the rib cage from expanding, and the sauropod would have suffocated. All sauropods and most other dinosaurs should be envisioned as having lived in drier plains country. Duckbilled dinosaurs did not live

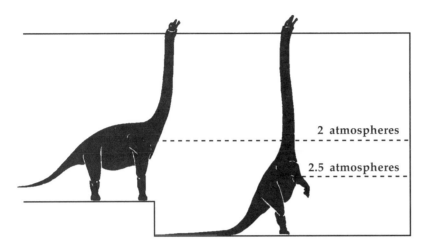

Figure 12.19 Sauropods could not have used their necks as snorkels. Water pressure on the rib cage would have prevented it from expanding, and the animal would have suffocated. (Sauropod postures after Bakker.)

2 atmospheres

2.5 atmospheres

in water either; their jaws had enormous batteries of teeth with large grinding surfaces, and they ate coarse vegetation in dry country.

Dinosaur Behavior

Dinosaur behavior can be judged by footprints; for example, a dinosaur stampede has been discovered (Figure 12.20). Rocks laid down about 90 Ma as sediments near a Cretaceous lakeshore in Queensland, Australia, bear the track of a large carnosaur heading down toward the lake with a 2-meter (6-foot) stride. Superimposed on this track are thousands of small footprints made by small, bipedal, lightly built dinosaurs, running back up the creek bed away from the water. More than 3000 footprints have now been uncovered, showing all the signs of a panicked stampede. At least 200 animals belonging to two species were stampeding. One of the species, probably a coelurosaur, ranged up to about 40 kg (90 pounds), and the other, probably an ornithopod, ranged up to twice as

Figure 12.20 A dinosaur stampede from Cretaceous rocks in Queensland. A huge carnosaur walked or stalked from left to right along an ancient creek bed toward a shallow lake. Thousands of little footprints were made by small dinosaurs all stampeding the other way, across and over the tracks the carnosaur made on its way down to the lake. (Courtesy of the Queensland Museum, Australia.)

Figure 12.21 Left, skull of the dome-headed dinosaur *Pachycephalosaurus*, arranged in head-butting position. (Negative 319456. Photo by C. H. Coles. Courtesy of the Library Services Department, American Museum of Natural History.) Right, *Pachycephalosaurus*. (Reconstruction by Bob Giuliani. © Dover Publications Inc., New York. Reproduced by permission.)

Figure 12.22 Large (probably male) *Parasaurolophus* with enormous crest on the head. (Reconstruction by Bob Giuliani. © Dover Publications Inc., New York. Reproduced by permission.)

large. Juveniles and adults of both species were digging in their toes as they tried to accelerate: 99% of the footprints lack heelmarks. The footprints show slipping, scrabbling, and sliding, and the smaller species usually avoided the tracks of the larger one. They must have been hemmed against the lakeshore, breaking away in a terrified group.

The stampede sheds light on ecology as well as behavior, telling us that at least some dinosaurs gathered in herds and behaved just as African plains animals do today at waterholes on the savanna, responding immediately and instinctively to the approach of dangerous predators.

Other indicators of behavior are preserved in dinosaur skeletons. The ornithischian dinosaur *Pachycephalosaurus* had a dome-shaped area of bone on the top of its skull (Figure 12.21). It is not solid bone but air-filled and porous, with an internal structure very much like that found in the skulls of sheep and goats that fight by ramming their heads together. Did *Pachycephalosaurus* do that? It weighed eleven times as much as a bighorn sheep, so the most recent interpretation of this dinosaur is that it probably butted opponents from the side rather than head-to-head. *Triceratops* and other ceratopsians, however, also had air spaces between the horns and the brain case, which may indicate that they competed by direct head-to-head impact.

Some hadrosaurs had huge crests on the head (Figure 12.22). The crests were not solid but contained tubes running upward from the nostrils and back down into the roof of the mouth (Figure 12.23). Only large males had large crests; females had smaller ones, and juveniles had none at all. The tubes are unlikely to have evolved for additional respiration or thermoregulation. (If so, adults would have needed large tubes whether they were female or male.) When the tubes are reconstructed they look like medieval horns and can be blown to give a note (Figure 12.23). The varying sizes of crests allow us to infer differences between the sounds produced by young, by adult females, and by adult males, to go with the different visual signals provided by the crests. These hadrosaurs may have had a sophisticated social system, perhaps as complex as those we take for granted in mammals and birds.

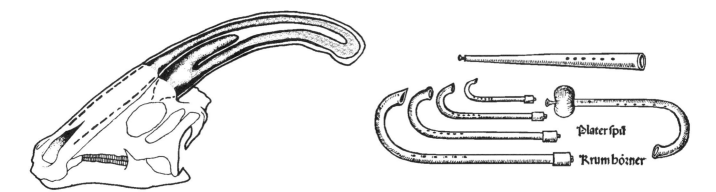

Dinosaur Eggs and Nests

Major finds of fossilized dinosaur eggs and nests (Figure 12.24) have been found in Cretaceous rocks in Montana, Alberta, Mongolia, Spain, and Argentina. We now have eggs from all three major dinosaur groups.

The Montana nests are carefully constructed bowls of mud or sand lined with vegetation. In each nest the eggs were laid or arranged in a neat pattern so they would not roll around. Embryonic dinosaurs are preserved in some nests. Many of the nests are clustered together at regular close intervals, suggesting that they were in communal breeding grounds —nesting colonies, if you like. The nests had sometimes been remodeled and reused, perhaps in successive seasons. Many large (long-lived) birds do this today, including the red-tailed hawks in my pine tree.

There is good evidence for long-term parental care by dinosaurs. One nest from Montana contained 15 baby duckbilled dinosaurs. We know that these baby dinosaurs were not new hatchlings because they are about twice too large for the eggshells found nearby, and because their teeth had been used long enough to have wear marks. But they were together in the nest when they died and were buried and fossilized, with an adult close by—*Maiasaura*, the "good mother."

This interpretation has been challenged on the grounds that hatchling dinosaurs had well-developed limbs and pelvis, showing that they were capable of moving about as soon as they hatched. Therefore, they did not receive parental care. Of course, this argument does not work. At a very different body size, partridges hatch a large number of chicks, which are taken by a parent on trips from the nest to forage very soon after hatching. Ostriches run a crèche system for the care of foraging young.

Dinosaur eggs and nests were found in Mongolia in the 1920s (Figure 12.24). They were naturally associated with the most abundant dinosaur in the area, *Protoceratops*. To everyone's surprise, when an embryo was finally discovered inside one of the eggs, it was well enough preserved to be identified not as *Protoceratops*, but as the little theropod *Oviraptor* (Figure 12.25). The irony here is that an adult *Oviraptor* had originally been identified (and named) as a nest-robbing, egg-eater, preying on an innocent *Protoceratops*!

A block of rock collected in Mongolia in 1993 turned out to contain an adult *Oviraptor* that had been buried in a sandstorm while it was

Figure 12.23 Left, large (probably male) *Parasaurolophus* had enormous crests on the head. A cross-section shows that the crest carried hollow tubes that were part of the nostril system. Left and right tubes from each nostril met in a central chamber above the roof of the mouth. Females and juveniles had smaller crests or none at all. (After Weishampel and Norman.) Right, medieval German musical instruments that are very much like the air passages of an adult male *Parasaurolophus*. David Weishampel made this striking analogy.

Makela and Horner have stressed

That maiasaur parents were best

The chicks' teeth are worn

So they're fairly well grown,

But they're still in a group in the nest!

Figure 12.24   Dinosaur eggs (now known to be from the dinosaur *Oviraptor*) photographed at their discovery site during the American Museum of Natural History expedition to Mongolia in 1923. The eggs were shaped and arranged so that they would not roll around. Some eggs never hatched. (Negative 410765. Photo by Shackelford. Courtesy of the Department of Library Services, American Museum of Natural History.)

crouched over a nest of *Oviraptor* eggs. The only reasonable explanation of this find is that the dinosaur was brooding its eggs, just as most living birds do. By 1996, three of the seven known *Oviraptor* adults had been discovered on or near nests.

So, in terms of their posture and reconstructed behavior, including parental care, complex social structure, and intelligence (needed to run a complex society), dinosaurs should be compared not with living reptiles, but with living mammals and birds.

Mammals and birds are warm-blooded, with high metabolic rates, and they are much more intelligent than other living vertebrates. But do these characters also apply to dinosaurs? The question is usually worded, "Were dinosaurs warm-blooded?" This question deserves its own chapter, not just for the question itself, but because it illustrates how scientific debate works.

Figure 12.25   The small theropod dinosaur *Oviraptor*, misnamed as an egg thief when actually it was a builder and guarder of its own nests. (Reconstruction by Bob Giuliani. © Dover Publications Inc., New York. Reproduced by permission.)

Review Questions

Give a typical adult body weight for a duckbill or horned dinosaur or a stegosaur.

An older idea suggested that brontosaurs lived in swamps and used their long necks as snorkels. What is wrong with this idea?

Give some indication of the noise that dinosaurs might have made. What's your evidence?

What's the evidence that some dinosaurs had parental care well after the young had hatched?

What did duckbill dinosaurs eat? And how do we know?

Describe the heart of a brontosaur, or any other sauropod.

THOUGHT QUESTION—Most mammals, birds, reptiles, and amphibians are and were small. Why were so few dinosaurs small?

Discuss the science involved in reaching the interpretation expressed in the limerick below. What further evidence would be useful, and how would you go about getting it?

> Impelled by their dinosaur need
>
> They went down by the lake for a feed
>
> But a carnosaur near
>
> Made them scamper with fear
>
> In the first recorded stampede.

Further Reading

Easy Access Reading

Ackerman, J. 1998. Dinosaurs take wing: new fossil finds from China provide clues to the origin of birds. *National Geographic* 194 (1): 74–99.

Alexander, R. McN. 1991. How dinosaurs ran. *Scientific American* 264 (4): 130–136.

Bakker, R. T. 1986. *The Dinosaur Heresies*. New York: William Morrow.

Barreto, C., et al. 1993. Evidence of the growth plate and the growth of long bones in juvenile dinosaurs. *Science* 262: 2020–2023.

Chen, P.-J., et al. 1998. An exceptionally well-preserved theropod dinosaur from the Yixian Formation of China. *Nature* 391, 147–152. *Sinosauropteryx*.

Chiappe, L. 1998. Dinosaur embryos. *National Geographic* 194 (6): 34–41.

Chiappe, L. M., et al. 1998. Cranial morphology of the avian Alvarezsauridae: evidence from a new relative of *Mononykus*. *Nature* 392: 275–278. *Shuvuuia*, a bird-like theropod?

Chiappe, L. M., et al. 1998. Sauropod dinosaur embryos from the Late Cretaceous of Patagonia. *Nature* 396: 258–261.

Chin, K., et al. 1998. A king-sized theropod coprolite. *Nature* 393: 680–682.

Currie, P. J. 1989. Long-distance dinosaurs. *Natural History* June 1989: 60–65.

Fastovsky, D. E., and D. B. Weishampel. 1996. *The Evolution and Extinction of the Dinosaurs*. New York: Cambridge University Press. The best text book.

Forster, C., et al. 1998. The theropod ancestry of birds: new evidence from the Late Cretaceous of Madagascar. *Science* 279: 1915–1919, and comment, pp. 1851–1852. *Rahonavis*, a Cretaceous bird with the slashing claw of a dromaeosaur.

Galton, P. M. 1970. Pachycephalosaurids: dinosaurian battering rams. *Discovery* 6: 23–32. Suggested head-to-head impact.

Geist, N. R., and T. D. Jones. 1996. Juvenile skeletal structure and the reproductive habits of dinosaurs. *Science* 272: 712–714, and comment, p. 651.

Gillette, D. G. and M. Hallett. 1994. *Seismosaurus, the Earth Shaker*. Princeton University Press. A good read; explains very well how paleontologists work.

Hammer, W. R., and W. J. Hickerson. 1994. A crested theropod dinosaur from Antarctica. *Science* 264: 828–830.

Hargens, A. R., et al. 1987. Gravitational haemodynamics and oedema prevention in the giraffe. *Nature* 329: 59–60, and comment, pp. 13–14.

Horner, J. R., and J. Gorman. 1988. *Digging Dinosaurs*. New York: Workman. An excellent read and a persuasive insight into the scientific investigation of dinosaurs. Published in paperback, 1990.

Horner, J. R., and D. Lessem. 1994. *The Complete* T. rex. New York: Simon and Schuster. Paperback. Light, bright and breezy.

Ji, Q. et al., 1998. Two feathered dinosaurs from northeastern China. *Nature* 393: 753–761, and comment, pp. 729–730.

Lessem, D. 1992. *Kings of Creation*. New York: Simon and Schuster. A journalist's book, not a scientist's, but a very good read.

Morell, V. 1996. A cold, hard look at dinosaurs. *Discover* 17 (12): 98–108. Includes John Ruben's work.

Norell, M., et al. 1993. New limb on the avian family tree. *Natural History* 102 (9): 38–42. *Mononykus*.

Norell, M. A., et al. 1994. A theropod dinosaur embryo and the affinities of the Flaming Cliffs dinosaur eggs. *Science* 266: 779–782, and comment, p. 731. See comment also in *Nature*, v. 372, p. 130. *Oviraptor* embryo.

Norell, M. A., et al. 1995. A nesting dinosaur. *Nature* 378: 774–776, and comment, pp. 764–765.

Norman, D. B. 1985. *The Illustrated Encyclopedia of Dinosaurs*. New York: Crescent Books. The best of the "coffee-table" books, with good science.

Novacek, M. J., et al. 1994. Fossils of the Flaming Cliffs. *Scientific American* 271 (6): 60–69. Some of the new discoveries in Mongolia.

Novacek, M. J. 1996. *Dinosaurs of the Flaming Cliffs*. New York: Anchor Books. Entertaining account of the Mongolian expeditions of the 1990s.

Novas, F. E., and P. F. Puerta. 1997. New evidence concerning avian origins from the Late Cretaceous of Patagonia. *Nature* 387: 390–392, and comment, pp. 349–350. *Unenlagia*.

Paladino, F. V., et al. 1990. Metabolism of leatherback turtles, gigantothermy, and thermoregulation of dinosaurs. *Nature* 344: 858–860.

Ruben, J. A., et al. 1996. The metabolic status of some Late Cretaceous dinosaurs. *Science* 273: 1204–1207.

Ruben, J. A., et al. 1997. Lung structure and ventilation in theropod dinosaurs and early birds. *Science* 278: 1267–1270, and comment, p. 1229–1230.

Ruben, J. A., et al. 1999. Pulmonary function and metabolic physiology of theropod dinosaurs. *Science* 283: 514–516, and comment, p. 468. Interpretation based on *Scipionyx*.

Sanz, J. L., et al. 1995. Dinosaur nests at the sea shore. *Nature* 376: 731–732.

Sereno, P. C., and F. E. Novas. 1992. The complete skull and skeleton of an early dinosaur. *Science* 258: 1137–1140. *Herrerasaurus*.

Sereno, P. C., et al. 1993. Primitive dinosaur skeleton from Argentina and the early evolution of Dinosauria. *Nature* 361: 64–66. *Eoraptor*.

Sereno, P. C., et al. 1996. Predatory dinosaurs from the Sahara and Late Cretaceous faunal differentiation. *Science* 272: 986–991, and comment, pp. 971–972. *Carcharodontosaurus*.

Sereno, P. C., et al. 1998. A long-snouted predatory dinosaur from Africa and the evolution of spinosaurids. *Science* 282: 1298–1302, and comment, pp. 1276–1277. *Suchomimus*.

Sereno, P. C. 1999. The evolution of dinosaurs. *Science* 284: 2137–2147.

Seymour, R. S. 1976. Dinosaurs, endothermy and blood pressure. *Nature* 264: 207–208.

Shreeve, J. 1997. Uncovering Patagonia's lost world. *National Geographic* 192 (6): 120–137.

Thomas, D. A., and J. O. Farlow. 1997. Tracking a dinosaur attack. *Scientific American* 277 (6): 74–79.

Varricchio, D. J., et al. 1997. Nest and egg clutches of the dinosaur *Troodon formosus* and the evolution of avian reproductive traits. *Nature* 385: 247–249.

More Technical Reading

Alexander, R. McN. 1989. *Dynamics of Dinosaurs and Other Extinct Giants*. New York: Columbia University Press.

Dodson, P. 1996. *The Horned Dinosaurs*. Princeton, N.J.: Princeton University Press.

Farlow, J. O., and M. K. Brett-Surman (eds.). 1997. *The Complete Dinosaur*. Bloomington: Indiana University Press. Many fine essays on all aspects. See especially Part 4: Biology of the Dinosaurs.

Forster, C. A. 1996. New information on the skull of *Triceratops*. *Journal of Vertebrate Paleontology* 16: 246–258. Head-butting?

Galton, P. M. 1985. Diet of prosauropod dinosaurs from the late Triassic and early Jurassic. *Lethaia* 18: 105–123.

Gillette, D. D., and M. G. Lockley (eds.). 1989. *Dinosaur Tracks and Traces*. Cambridge: Cambridge University Press. Paperback, 1991.

Padian, K., et al. 1999. Phylogenetic definitions and nomenclature of the major taxonomic categories of the carnivorous Dinosauria (Theropoda). *Journal of Vertebrate Paleontology* 19: 69–80.

Stokes, W. L. 1987. Dinosaur gastroliths revisited. *Journal of Paleontology* 61: 1242–1246.

Sues, H.-D. 1978. Functional morphology of the dome in pachycephalosaurid dinosaurs. *Neues Jahrbuch für Geologie und Paläontologie Monatshefte* 1978 (8): 459–472. Suggests side-butting.

Thulborn, R. A. 1990. *Dinosaur Tracks*. London: Chapman and Hall.

Thulborn, R. A., and M. Wade. 1979. Dinosaur stampede in the Cretaceous of Queensland. *Lethaia* 12: 275–279.

Weishampel, D. B. 1981. Acoustic analysis of potential vocalization in lambeosaurine dinosaurs (Reptilia: Ornithischia). *Paleobiology* 7: 252–261.

Weishampel, D. B., et al. (eds.). 1990. *The Dinosauria*. Berkeley: University of California Press. The paleontologists' encyclopedia of dinosaurs.

Warm-Blooded Dinosaurs?

There have been tremendous arguments about the body temperatures and metabolic rates of dinosaurs. The arguments are intense and interesting, and they are not yet resolved.

There are several aspects of this question. First, what does "warm-blooded" mean? Did dinosaurs control their body temperature precisely —that is, were they **homeotherms**? Or were they **heterotherms**, allowing their body temerature to vary quite widely, depending on their needs? Were they **endotherms**?—did they produce a high body temperature by generating a lot of metabolic heat, raising their overall energy budget? Or were they ectotherms, controlling their body temperature dominantly by behavior, by clever use of sunlight, shade, shelter, activity level, and so on? If they were homeotherms, did they control their body temperature at a much higher level than their average environment, say 35° to 40°C (95° to 104°F), like most mammals and birds? Was it lower, say 30°C (86°F), more like a platypus or a hedgehog? Or was it lower still, hardly any warmer than the average temperature of an equatorial habitat, say 27°C (81° F)? Along with these answers come different estimates of the resting metabolic level, the energy output, and the activity level of dinosaurs. Were they comparable with elephants, ostriches, or lions, or more like living reptiles, or like none of the above? Were all dinosaurs alike in their physiology?

We are still in the process of learning the differences between living reptiles and mammals. Any vertebrate <u>cells</u> are capable of high-energy output if they are kept fueled with oxygen and food. Thus, the secret of evolving thermoregulation at high levels and at high resting metabolic rates lies in the engineering around the cells rather than in their biochemistry. Respiration and circulation systems, which transport oxygen and food, are the crucial factors. For example, the hearts of living mammals and reptiles are very different, and David Carrier has shown how and why their respiratory systems and locomotory systems are different too (Chapter 11). To understand dinosaurs, we need to be able to reconstruct the circulation and respiration of these fossil vertebrates, as well as their locomotion. It is certainly not easy, but it's a better prospect than trying to infer the biochemistry of their cells!

Then there is an ecological factor. Whatever the advantages of high resting metabolic rate, it has a cost—more food must be eaten. The higher the metabolic rate, the greater the cost.

For nearly thirty years, professional and amateur paleontologists alike have been increasingly persuaded that dinosaurs were warm-blooded, with connotations of high levels of activity. Perhaps the movie *Jurassic Park* represented the acme of this wave of thought. The author Michael Crichton was very much influenced by the work of John Ostrom and Robert Bakker, and the immense success of the movie in turn set for a while the popular image of dinosaurs.

But we need to look at evidence rather than a movie screen. Evidence has been proposed in favor of warm-blooded dinosaurs, but much of it is fuzzy or ambiguous. There are arguments on both sides.

Dinosaurs With Feathers

Feathers have always been regarded as structures unique to birds: in fact, they have been used for 200 years as one of the most important characters that define birds. Recently discoveries have shown that some dinosaurs had feathers too.

Four theropods from Early Cretaceous beds in China are of great interest. *Sinosauropteryx* and *Beipiaosaurus* are preserved with a halo of very fine structures on the body surface that look like down. *Protarchaeopteryx* is very much like a small *Velociraptor* except that it has down feathers on its body, tail, and legs, and a fan-shaped bunch of long feathers, several inches long, at the very end of its tail. *Caudipteryx* also has down and strong tail feathers, but it has feathers on its arms as well. They are shorter toward the fingers, and longest toward the elbow, in contrast to the feathers on the wings of flying birds.

This does not mean that all dinosaurs were feathered: but the scanty evidence suggests that a number of theropods were. We have direct evidence from skin impressions that (some) ornithischians and sauropods were leathery. But the finds do confirm that birds evolved from small ground-running predatory theropods very much like *Velociraptor*, and that feathers evolved before birds and before flight. Of the four Chinese theropods, *Protarchaeopteryx* is the closest relative yet discovered to birds, especially to the most primitive bird, *Archaeopteryx*.

So how do we now distinguish a bird from a small theropod? With difficulty, and certainly not by its feathers! (Technically, a minor change in a skull bone called the quadratojugal is the best indicator of the evolution from one group to the other, but this definition may change with the next discovery.)

This is perhaps a good opportunity to illustrate how insignificant the transition can be from one group to another. The first bird hatched out of an egg laid by a theropod dinosaur, but unless there were many other hatchlings at almost exactly the same evolutionary stage to form a breeding population, the lineage would have gone extinct. Even if you had been there, watching the chicks hatch and grow up, you would not have been aware that you were seeing the transition from theropod to bird in that generation of that evolving population. (And, of course, that is true of any other evolutionary transition that has ever occurred.) That means that if the fossil record is relatively complete, the change that defines the transition will necessarily be one so trivial that it will look artificial—and, of course, it is trivial and it is artificial!

The Origin of Feathers

The proteins that make feathers in living birds are completely unlike the proteins that make reptilian scales today. Feathers originate in a skin layer deep under the outer layer that forms scales. It is very unlikely that feathers evolved from reptilian scales, even though that thought is deeply embedded in the minds of too many paleontologists. Feathers probably arose as new structures under and between reptile scales, not as modified scales. Many birds have scales on their lower legs and feet where feathers are not developed, and penguins have such short feathers on parts of their wings that the skin there is scaly for all practical purposes. So there is no real anatomical problem in imagining the evolution of feathers on a scaly reptilian skin. But feathers evolved in theropods as completely new structures, and any reasonable explanation of their origin has to take this into account.

Obviously, feathers did not evolve for flight. They evolved for some other function and were later modified for flight.

Feathers may have evolved to aid thermoregulation. The feathered Chinese theropods all have down, probably as insulation to keep their bodies at an even temperature. It would not matter whether they used their feathers to conserve heat in cold periods, or to keep heat out in hot periods, or both. In either case, insulation would have been useful.

The thermoregulatory theory for the origin of feathers is probably the most widely accepted one today, but it does have problems. Why feathers? Feathers are more complex to grow, more difficult to maintain in good condition, more liable to damage, and more difficult to replace than fur. Every other creature that has evolved a thermoregulatory coat—from bats to bees and from caterpillars to pterosaurs—has some kind of furry cover. There is no apparent reason for evolving feathers rather than fur even for heat shielding.

Our store of good data is slight

We don't know which answer is right:

A cutaneous feather

May be better than leather

But for warmth? or for show? or for flight?

Figure 13.1 Forearm display would have drawn attention to the powerful claws of the first bird, and derived theropod, *Archaeopteryx*. (From Heilmann, 1926.)

Furthermore, thermoregulation cannot account for the length or the distribution of the long feathers on *Protarchaeopteryx* or *Caudipteryx*. Short feathers (down) can provide good thermoregulation, but thermoregulation does not require long feathers, and it would not help thermoregulation very much to evolve long strong feathers on the arms and tail. So it is difficult to suggest that feathers evolved for thermoregulation alone. It would be better to think of another equally simple explanation.

I naturally prefer an idea that I developed years ago, with my colleague Jere Lipps. In living birds, feathers are for flying, for insulation, but also for camouflage and/or display. Lipps and I suggested that feathers evolved for display. The display may have been between females or between males for dominance in mating systems (sexual selection), or between individuals for territory or food (social selection), or directed toward members of other species in defense.

Living reptiles and birds often display for one or all of these reasons, using color, motion, and posture as visual signals to an opponent. Display is often used to increase apparent body size; the smaller the animal, the more effectively a slight addition to its outline would increase its apparent size. Lipps and I therefore proposed that erectile, colored feathers would give such a selective advantage to a small displaying theropod that it would encourage a rapid transition from a scaly skin to a feathery coat. Display would have been advantageous as soon as *any* short feathers appeared, and it would have been most effective on movable appendages, such as forearms and tail. (Display on the legs would not be so visible or effective.) Forearm display by a small theropod would also have drawn particular attention to the powerful weapons the theropod carried there, its front claws (Figure 13.1). The Chinese theropod *Caudipteryx* carried long feathers on its middle finger, between the two outside claws, and it could fold that middle finger away, with the feathers out of harm's way, during a strike.

The display hypothesis explains more features of the feathered theropods and the first bird *Archaeopteryx* than other hypotheses, with fewer assumptions. It explains completely the feather pattern—the evolution of long strong feathers on arms and tail.

Once they evolved, feathers could quickly have been co-opted for thermoregulation, and the down coat on the Chinese theropods may show that process. Down can only be for thermoregulation. Although down is not proof of warm blood, it is very strong evidence in favor of it. In living birds, down feathers are associated with the problem of heat loss for hatchlings.

Egg Brooding

In early 1999, several dinosaur specialists informally reported that the skeleton of the Chinese theropod *Caudipteryx* was most like that of oviraptors. Then feather-like structures were reported on *Beipiaosaurus*, which is a therizinosaur, an Asian dinosaur group related to oviraptors (Figure 12.3). These new data have very important consequences for the number of theropods that had feathers. Look at the cladogram of Figure 12.3. If oviraptors and birds have feathers, and if feathers only evolved once (which seems almost inevitable, given their complexity), then many other theropods could have had feathers too, including the dromaeosaurs and *Velociraptor*. There are still other implications. We know already that *Oviraptor* brooded its eggs, just as living birds do (p. 200). A few desert-dwelling birds protect their eggs from the sun by sitting on them, but most birds brood their eggs to keep them warm: even the desert-dwellers keep their eggs warm at night. If *Oviraptor* had feathers, then it was even more bird-like than we had imagined, and it is very unlikely that it was cold-blooded! It now becomes a matter of judgment how far to extend *Oviraptor*'s body temperature to other dinosaurs. We do not know whether other dinosaurs brooded their eggs, but we do know that all dinosaur groups built nests. Among the whole array of dinosaurs, it is clear that we can make the best case for homeothermy, perhaps even endothermy, among lineages of theropods. A completely different line of evidence can be applied to other dinosaur groups.

High-Latitude Dinosaurs

It is fairly easy to imagine a large adult dinosaur having a fairly even temperature during small temperature swings from day to night, especially in mild tropical climates. But there are rich plant floras and rich dinosaur faunas containing hadrosaurs and tyrannosaurs in Late Cretaceous rocks of the north slope of Alaska, which at the time was at high northern latitude—at least 70° N and possibly as much as 80° N. Equally rich cool-climate floras and Early Cretaceous dinosaurs flourished in southern Australia, which was then at about 80° S, and dinosaurs have been found in Antarctica too. In these high latitudes, vegetarian dinosaurs had to survive strong seasonal changes in light, temperature, and plant food supply.

Cretaceous polar temperatures were rather mild: luxuriant floras grew at high latitudes in both hemispheres. But luxuriant growth in polar latitudes can only occur in summer. Winter darkness prevents plant growth, and winter forage would have been scarce even if average Cretaceous polar winter temperatures were above freezing. (Summer growth in Arctic plants today is often limited by low temperatures even though there is a lot of light.) Late Cretaceous floras of North America were dominated by deciduous plants everywhere north of Montana, presumably because light levels were too low to make it worthwhile for plants to maintain leaves all through the winter. There must have been at least occasional frosts in polar latitudes, and if it's dark, there's no way of reheating a region once a winter storm cools it down.

There are no large *ectotherms* in high latitudes today: crocodiles and turtles live mostly in the tropics and only occasionally range into temperate climates. There are no reptiles at all in Alaska. Yet large *endotherms* survive successfully even near today's ice caps. Carnivores such as polar bears, and large swimming mammals such as walruses, penguins, seals, and whales live year-round in polar regions, in spite of the large amount of food they need for body heating. Terrestrial omnivores such as brown bears and grizzly bears avoid the worst period of low food supply and cold by hibernating. But large herbivores dare not hibernate: they would be vulnerable to attack by insomniac carnivores. They migrate instead; caribou herds cover hundreds of kilometers between seasons.

If living animals are reliable ecological guides, the big vegetarian dinosaurs of Alaska were endotherms and migrated with the seasons. Today the great migrants are fliers (birds), swimmers (whales), or walkers (large mammals with large, precocious young that can walk and run very quickly after birth—caribou, or African plains animals). The Alaskan dinosaurs included many juvenile hadrosaurs, which implies that hatchlings were reasonably precocious and grew quickly. Robert Spicer has pointed out that all herbivores 5 tons or larger must be able to forage over a lot of ground. Probably the foraging of Cretaceous hadrosaurs involved significant northward and southward movements with the seasons. Some hadrosaurs may have migrated annually to nesting grounds in the north, just as pregnant caribou migrate northward today. Young hadrosaurs would then have had the advantage of feeding on the growth of the Arctic summer. Meanwhile, tyrannosaurs and smaller predators followed the migration, just as wolves follow the caribou herds today.

The same argument for migration does not apply in southern Australia, where the land mass inhabited by the dinosaurs was separated from the rest of the continent. There is no escaping the conclusion that the dinosaurs lived there year-round. And many of them were small: the little ornithischian *Leaellynasaura* was only chicken-sized, another was dog-sized, and even an *Allosaurus* was much smaller than its North American counterparts. *Leaellynasaura* had very big eye-sockets for its size, suggesting large eyes, perhaps to see during the dark polar winter. It is difficult to avoid the inference that these dinosaurs were endothermic. Tim Flannery has suggested that some feathers found in lake beds in the region may have been shed from the dinosaurs.

Doubts about Endothermy

David Carrier showed us a link between erect posture and endothermy (Chapter 11). Erect posture allows birds and mammals to breathe enough oxygen to run endothermy, but it does not *require* them to do so. If dinosaurs were able to migrate, they could have avoided the polar winters that require endothermy to survive. *Oviraptor* could have successfully brooded its eggs as long as it was *homeothermic*: it did not have to be *endothermic*. Down feathers might have been used to regulate temperature in an ectothermic theropod, not necessarily in an endothermic one. The little Australian dinosaurs may have been able to run endothermy at a cheap rate (say heating to what we might call cool temperatures, around 70°F or so). All these lines of evidence suggest endothermy, sometimes strongly, but they do not *require* it.

Research on dinosaur bone structure has yielded ambiguous evidence on metabolic levels, and therefore on body temperature and behavior. For example, a hypsilophodont (small ornithischian herbivore) from southern Australia had continuous bone deposition, while an ornithomimosaur (small theropod "ostrich dinosaur") from the same beds had cyclic bone deposition. Normally, one would interpret continuous bone deposition as signifying year-round activity and/or homeothermy, and cyclic bone deposition as signifying strongly seasonal bone deposition and/or heterothermy. But these dinosaurs were too small to migrate, and the continuous deposition in the hypsilophodont suggests that it at least did not hibernate. We are left with the uncomfortable feeling that dinosaur bone structure is not easy to interpret in any simple way.

Dinosaur Noses: Turbinates

The most powerful argument against endothermic dinosaurs comes from evidence associated with their noses. In living mammals and birds, there are complex nasal passages called **turbinates**, formed by thin bones. All air breathed through the nose must pass through these passsages, over the enormous skin area that covers the turbinates. Some of the passages are lined with chemical detectors for the sense of smell, but I shall discuss only the passages that are used for another function. Special turbinate passages are lined with mucus, which acts to warm and humidify air as it is breathed in, and to cool and dry air on its way out. This function conserves body moisture that would otherwise be lost from the lungs with each breath. Turbinates do not occur in living reptiles, and have been independently evolved in birds and mammals. As you might imagine, camels have spectacular turbinates.

John Ruben and his colleagues argue that turbinates are required if air-breathing vertebrates are to have a consistently high resting metabolic rate and therefore be endothermic. Typically, in the wild, a living bird or mammal eats roughly 20 times the food consumed by a reptile of the same weight, and breathes at least 20 times the oxygen in order to oxidize that food. A bird or mammal would lose a great deal of moisture in that process if it breathed out damp air. Turbinates that recover moisture from air on the way out are therefore foolproof indicators of endothermy in living vertebrates, and they are potentially foolproof indicators in fossil vertebrates.

Ruben and his colleagues looked at the nasal passages of three theropods and one ornithischian dinosaur, and found that there was no room to fit turbinates in any of them. In fact, the passages through the nasal bones were as small or smaller than those of living reptiles, while living mammals and birds have much more space in their nasal passages to make room for turbinates. This is strong evidence that dinosaurs did not have as high a resting metabolic rate as living mammals and birds do: a dinosaur would have had a resting metabolic rate comparable with that of living reptiles.

However, I should point out that the absence of turbinates is strong evidence that the *resting* metabolic rate of dinosaurs was comparable with that of living reptiles. Turbinates say nothing about the metabolic rate that can be reached *during bursts of exercise*. The only constraint here is Carrier's Constraint: that dinosaurs should have been able to breathe while they ran. Evidence from two dinosaurs preserved with their soft parts suggested to Ruben that they had structures comparable with our mammalian diaphragms, allowing them to breathe with a pump-like action, and to reach high activity levels. In any case, evidence from the Australian stampede footprints <u>proves</u> that dinosaurs were able to run fast.

Behavioral and Passive Thermoregulation

Temperature can be controlled to some extent by behavior (basking, bathing, seeking shade) as in living lizards and crocodiles. Many of the advantages of thermoregulation are gained even if only the extremes of environmental temperature fluctuation are avoided. And **behavioral thermoregulation** is cheap: cooling does not involve sweating away valuable water, and solar heating does not burn food.

Most living reptiles are comparatively small, and they gain and lose heat quickly. This may be good or bad for them, but it follows the laws of physics. However, most dinosaurs were large enough that they would have had **passive thermoregulation** (thermoregulation that needs no action). Think how long it takes to defrost a frozen chicken, and even more, a frozen turkey. It would have taken most of the day to heat up even a one-ton dinosaur, and most of the night to cool it down. The largest dinosaurs would have had practically uniform temperature whether they chose to or not: they would have warmed and cooled so slowly relative to day and night that their inner temperature would not have changed much at all. Passive thermoregulation is also called *inertial thermoregulation* or *gigantothermy*. It would have maintained medium to large dinosaurs at a body temperature close to that of the daily average temperature of their surroundings—homeothermy, perhaps, but not endothermy.

For example, a large passively-thermoregulating dinosaur in a mild tropical climate like that of modern-day Amazonia would have had a body temperature around 27°C (80°F), day and night, season after seaon. So in some habitats it is easy to believe that many large dinosaurs—say most ornithischians and sauropods—could function well with this sort of thermoregulation, at a body temperature somewhat lower than ours. As we have seen, one could possibly argue that polar dinosaurs are evidence for thermoregulation at comparatively cool temperatures.

The Answer?

So (finally!), what's the answer? At least small theropod dinosaurs had thermoregulation (down feathers). Dinosaurs did not have a high resting metabolic rate (turbinates). Dinosaurs could breathe while they ran (erect gait), so had an ability for fast sustained running (confirmed by footprints). There are no living animals with this combination of features.

Living mammals and birds are endothermic athletes, living reptiles and amphibians are ectothermic sprawlers. Most likely, then, dinosaurs had a uniquely "dinosaurian" physiology.

Insight came with new research on Australian crocodiles. Crocodiles can bask in the sun to gain heat (Chapter 8). They can also walk into the water to cool down: overall, they can use behavioral thermoregulation to maintain fairly even body temperatures. In 1999 Frank Seebacher and his colleagues found, as expected, that 1-ton crocodiles could regulate their body temperatures more easily than smaller crocodiles (because of the effects of passive thermoregulation). But they also found, to their surprise, that large crocodiles were on average <u>warmer</u> than small ones. There is no explanation for this (yet), but it means that if dinosaurs worked the same way, that large dinosaurs could have maintained high body temperatures (say 38° C, 100° F) simply by passive and behavioral thermoregulation.

Crocodiles are archosaurs, and are the nearest living relatives of dinosaurs (except for birds). I shall summarize the work of Seebacher's team as the **archosaurian model** for dinosaur thermoregulation. In the archosaurian model, environmental heating and cooling costs only time, not energy. Dinosaurs, then, would have had a resting metabolic level about the same as crocodiles, much lower than living mammals.

All vertebrates have a maximum exercise metabolism about seven times their resting value, so the archosaurian model would allow an impressive performance by dinosaurs, though not up to the level of mammals and birds. Remember that dinosaurs, with erect posture, could have maintained fast running. I have seen an Australian crocodile lunging for prey, and I have no illusions about the speed of movement of living reptiles. Add the ability to sustain that speed (dinosaurs had solved Carrier's Constraint), and dinosaur performance becomes truly formidable.

The archosaurian model fits the evidence from turbinates. If dinosaurs had low resting metabolic rates, they had low oxygen consumption, they did not lose much water as they breathed, and they did not need turbinates. Turbinates, of course, are irrelevant to the question of dinosaur running. Animals that are running (for food, or for their lives) use up water whether they have turbinates or not, because they mouth-breathe. We humans do not run along breathing through our turbinate noses.

The archosaurian model also accounts for dinosaurs in high latitudes. Life in cool climates could have been possible, given that the animals would need food only at reptilian levels.

The archosaurian model accounts for the fact that many dinosaurs are large: size alone would give relatively high body temperature and its advantages, at little cost.

The archosaurian model also accounts for some of the cooling adaptations of dinosaurs that seem puzzling at first sight. Larger dinosaurs probably had more problems cooling down than they did heating up.

Some astute paleontologists have been arguing for years that dinosaurs had a different, dinosaurian, thermal biology. And they turn out to be right. James Farlow suggested that dinosaurs had the ability to turn their metabolic rate up and down, possibly with the seasons. This would be almost an automatic result of the archosaurian model, especially in seasonal climates outside the tropics.

Size and Metabolic Rates

Metabolic rates among living vertebrates fall with increasing size in any group of similar animals; it's cheaper to feed an elephant than the same weight of horses. To put it simply, an animal ten times the size of another will require only six times the food and oxygen. Exceptions usually make sense. For example, hummingbirds have a higher metabolic rate than one would expect because of the (high-energy) way in which they hover to collect nectar, their (high-energy) food. One can predict, therefore, that even if sauropods operated at the metabolic rates of mammals, they would have needed perhaps six times the food of an elephant, not ten times as much. But if sauropods operated on the archosaurian model, they would have needed even less food. The problems of sauropod life at very large size thus may not be as awesome as they seem.

Cooling Large Herbivorous Dinosaurs

The macerated plant material inside the gut of a herbivorous dinosaur must have produced an enormous amount of heat as it fermented. Anyone familiar with garden compost will realize that this could have raised body temperature considerably, especially in a 50-ton sauropod. This is more "free" heat to add to solar energy. However, the high body temperature may have been difficult to control and would have responded to how often, how much, and what the dinosaur ate. Without any other control, rapid fermentation would have raised body temperature, which would have increased fermentation rate, and so on, in a potentially dangerous runaway reaction. So herbivorous dinosaurs may have had more problems in controlling their temperatures than carnivorous dinosaurs. The larger the dinosaur, the easier the problem may have been, because a larger body size would have provided better passive temperature regulation. Other things being equal, this may help to explain why the sauropod dinosaurs evolved to such extraordinary size.

Stegosaurus may provide an example of a large (5-ton) herbivorous dinosaur with a fermentation problem. *Stegosaurus* is famous for the plates on its back (Figure 12.12). They are not spines, and they are offset from the spinal column. The plates are staggered right and left along the back, and their surfaces are corrugated, as if in life they were covered in skin and had large blood vessels running up and down them. Engineering experiments with aluminum and steel models of stegosaurs, heated by electricity and put into wind tunnels, have shown that the plates are practically ideal for shedding excess heat to a breeze. They are not designed for basking. This implies that *Stegosaurus* had heat-stress problems, and needed to shed heat, at least at times. Of course, when it did not need cooling, the blood supply to the plates would have been shut off.

A complicated system like this implies that *Stegosaurus* needed an efficient cooling system but did not need warming. *Stegosaurus* is also famous for having a large cavity in its spinal column that was once interpreted as the site of a large posterior "brain." This cavity is much better explained by comparing it with the air cavities found in bird skeletons, which are used to pass cooling air through the body before it reaches the lungs. Other dinosaurs also had these cavities, and therefore they probably had this accessory air-cooling system too. In the enormous sauropods, blood-vessels passing up and down the long neck and tail could have been used to cool the body.

Triceratops (Figure 12.14) has a very large bony frill on its skull, and large horns. Recent research by Reese Barrick and colleagues suggests that the frill, largely covered by skin, was an effective heat-shedding surface for the whole body. The horns may also have been used to shed heat to keep the brain cool!

Small Carnivorous Dinosaurs

In the archosaurian model, thermoregulation is much easier for large animals than small ones. Yet all dinosaurs were small as babies, and a number of dinosaurs were small as adults, especially theropods. The evolution of egg brooding by adults can perhaps be seen as an important dinosaur innovation that crocodiles and other reptiles did not achieve. And perhaps it would seem reasonable that small theropods would also have evolved a thermoregulatory coat (of feathers) that was not evolved in other dinosaur lineages. Those two innovations (feathers and brooding) were critical theropod innovations (to make them better theropods) that led to the evolution of a completely new group of dinosaurs: the birds.

Review Questions

There is uncertainty about the relative warm-bloodedness of dinosaurs. Give two arguments in favor of dinosaurs being warm-blooded and one in favor of them being cold-blooded.

The sails on the backs of *Dimetrodon* and *Stegosaurus* probably did different things for the two animals. Please explain.

Further Reading

Easy Access Reading

Ackerman, J. 1998. Dinosaurs take wing: new fossil finds from China provide clues to the origin of birds. *National Geographic* 194 (1): 74–99.

Barreto, C., et al. 1993. Evidence of the growth plate and the growth of long bones in juvenile dinosaurs. *Science* 262: 2020–2023.

Chen, P.-J., et al. 1998. An exceptionally well-preserved theropod dinosaur from the Yixian Formation of China. *Nature* 391, 147–152. *Sinosauropteryx.*

Currie, P. J. 1989. Long-distance dinosaurs. *Natural History* 6/1989: 60–65.

Ji, Q. et al., 1998. Two feathered dinosaurs from northeastern China. *Nature* 393: 753–761, and comment, pp. 729–730.

Morell, V. 1996. A cold, hard look at dinosaurs. *Discover* 17 (12): 98–108. Includes John Ruben's work.

Norell, M. A., et al. 1995. A nesting dinosaur. *Nature* 378: 774–776, and comment, pp. 764–765.

Paladino, F. V., et al. 1990. Metabolism of leatherback turtles, gigantothermy, and thermoregulation of dinosaurs. *Nature* 344: 858–860.

Ruben, J. A., et al. 1996. The metabolic status of some Late Cretaceous dinosaurs. *Science* 273: 1204–1207.

Ruben, J. A., et al. 1997. Lung structure and ventilation in theropod dinosaurs and early birds. *Science* 278: 1267–1270, and comment, pp. 1229–1230.

Ruben, J. A., et al. 1999. Pulmonary function and metabolic physiology of theropod dinosaurs. *Science* 283: 514–516, and comment, p. 468. Interpretation based on *Scipionyx*.

Seymour, R. S. 1976. Dinosaurs, endothermy and blood pressure. *Nature* 264: 207–208.

Swisher, C. C., et al. 1999. Cretaceous age for the feathered dinosaurs of Liaoning, China. *Nature* 400: 58–61, and comment, pp. 23–24.

Vickers-Rich, P., and T. H. Rich. 1993. Australia's polar dinosaurs. *Scientific American* 269 (1): 50–55.

Xu, X., et al. 1999. A therizinosaurid dinosaur with integumentary structures from China. *Nature* 399: 350–354. *Beipiaosaurus*.

More Technical Reading

Barrick, R. E., et al. 1998. The thermoregulatory function of the *Triceratops* frill and horns: heat flow measured with oxygen isotopes. *Journal of Vertebrate Paleontology* 18: 746–750. Cooling devices.

de Buffrénil, V., et al. 1987. Growth and function of *Stegosaurus* plates: evidence from bone histology. *Paleobiology* 12: 459–473.

Chinsamy, A., et al. 1998. Polar dinosaur bone histology. *Journal of Vertebrate Paleontology* 18: 385–390. Study of the polar dinosaurs from Australia, with ambiguous results.

Desmond, A. J. 1975. *The Hot-Blooded Dinosaurs*. London: Blond and Briggs; issued in paperback by Hutchison, 1990. A lively historical account.

Farlow, J. O., and M. K. Brett-Surman (eds.). 1997. *The Complete Dinosaur*. Bloomington: Indiana University Press. Many fine essays on all aspects, especially in Part 4: Biology of the Dinosaurs.

Paul, G. S. 1988. Physiological, migratorial, climatological, geophysical, survival, and evolutionary implications of Cretaceous polar dinosaurs. *Journal of Paleontology* 62: 640–652.

Seebacher, F., et al. 1999. Crocodiles as dinosaurs: behavioural thermoregulation in very large ectotherms leads to high and stable body temperatures. *Journal of Experimental Biology* 202: 77–86.

CHAPTER FOURTEEN

The Evolution of Flight

There are four kinds of flight: passive flight, parachuting, soaring, and powered flight. **Passive flight** can be used only by very tiny organisms light enough to be lifted and carried by natural winds and air currents, and light enough to suffer no damage on landing. Tiny insects, baby spiders, and many kinds of pollen, spores, and seeds can be transported this way. But the only control available to them is the timing of takeoff; after that they are entirely at the mercy of chance events.

The first land organisms and the first aerial organisms were microscopic. As reproduction adjusted to the problems of life in air, spores were evolved by fungi, plants, and other small organisms—their soft reproductive cells were protected by a dry, watertight coating, rather than the damp slime that is sufficient in water. Dry spores could then be spread as passive floaters on the wind—Earth's first fliers. Plant spores occur in Ordovician rocks (Chapter 8), and they are numerous and widespread enough in Devonian rocks to be used as guide fossils in relative age dating. Apart from anything else, this suggests that some plant species had such a large area of tolerable habitats open to them that a long-range dispersal method had become worthwhile.

Gliding flight includes **parachuting**, in which the flight structures slow a fall, and **soaring**, in which the flight structures allow an organism to gain height by exploiting natural air currents. Parachuting and soaring may seem to grade into one another, but their biology is very different, and the two flight modes probably have distinctly different origins. Parachuting organisms have simpler flight structures and much less control over the direction, speed, and height of flight than soarers. They seek short-range travel from one point to another, and their landing point is reasonably predictable because they do not seek external air currents for lift. Parachuting is used in habitats where external air currents are minimal, especially in forests. Wind gusts and air currents are potentially disastrous to animal parachutists, just as they are to human paratroops.

Powered flight is usually accomplished by some sort of flapping motion with special structures (wings). It requires considerable energy expenditure but gives much greater independence from variations in air currents, and it is usually accompanied by a high level of control over flight movements. Because powered flight is achieved by controllable appendages, almost all powered fliers can glide to some extent, some very poorly (no better than parachutists) and some very well indeed. Raptors and soaring seabirds are examples of powered fliers that glide well.

Soaring is used by flying organisms that range widely over a broad habitat. It is a low-energy flight style because the lift comes from external air currents rather than muscular expenditure by the flier. Energy costs are mainly related to the maintenance and adjustment of gliding surfaces in the air flow. Soarers may need occasional bursts of flapping flight if there are no up-currents, or in transferring from one up-current cell to another. Flapping is sometimes needed for takeoff, until airspeed exceeds stalling speed, or for final adjustments of attitude and speed in landing. Because flapping flight is needed occasionally by all soarers today (especially in emergencies), soaring probably cannot evolve from parachuting but only from powered flight.

Flight of all kinds demands a light, strong body. Soaring especially emphasizes lightness in muscle mass as well as overall structure. Powered flight has more requirements, including a significant output of energy and strength. Even the best soarers among living birds, albatrosses and condors, cannot flap for long before they are exhausted, because their flight muscles are small relative to their size and total weight. It might be difficult for a specialized soarer to re-evolve the ability to sustain flapping flight.

One would think that powered flight could evolve easily from gliding. Any evolving wing should be a fail-safe device, allowing a gliding fall during flight training. But an animal that has already evolved efficient gliding would not easily improve its flight by flapping in mid-glide, because that would disturb the smooth airflow over the gliding surfaces. Aerodynamic analysis shows that an evolutionary transition is possible from gliding to flapping, but only in very special circumstances. The glider must add fairly large, rapid wing beats, not little flutters, and because wing beats require considerable expenditure of energy, there must be a corresponding payoff in energy saved (for example, the animal must save some walking or climbing, or must reach a larger food supply).

The evolution of flight demands lightening and strengthening of the whole body structure, and the evolution of a flight organ from a pre-existing structure (a limb, for example) that could otherwise perform some other function. Flight may have strong advantages in locomotion—for food-gathering, escape, rapid travel between base and food supply, or migration—but it also has costs, not just in energy but also in constraints on body form and function that may have accompanying drawbacks. Flight has evolved many times in spite of all these problems.

Flight usually involves relatively large lifting structures; in almost every case a small lifting structure is no better than none at all. Lifting structures must already be present before flight can evolve, and they must therefore have evolved for some different function. This theme dominates the discussion of flight origins in this chapter.

Flight in Insects

Primitive insects are known from Devonian rocks, but flying insects may not have evolved until the Late Carboniferous, as insects radiated in the Carboniferous forest canopy. Many insects of all sizes are known from the coal beds of the Carboniferous, and half of all known Paleozoic insects had piercing and sucking mouthparts for eating plant juices. In turn, these smaller insects were a food source for giant predatory dragonflies (Figure 9.9) and for early reptiles (Chapter 9).

In living insects (except mayflies), only the last molt stage, the adult, has wings, and there is a drastic metamorphosis between the last juvenile stage (the nymph) and the flying adult. Wings have to be as light and strong as possible; in living insects this is achieved by withdrawing as much live tissue as possible. Most of the wing is left as a light mass of dead tissue that cannot be repaired. This gives great flying efficiency, though it usually means a short adult lifespan. The automatically short life expectancy of flying insects has played a strong part in the evolution of social behavior among some insects, in which the genes of a comparatively few breeding but nonflying adults are passed on with the aid of a great number of cheap, throwaway, sterile flying individuals (worker bees, for example). Some insects shed their wings. In many ants, for example, the wings are functional only for a brief but vital period during the mating flight. Insects do not have a long enough life expectancy to have the luxury of learning, so they carry with them a "read-only memory" that seems to govern their behavior entirely by "instinct."

But these are characters of living insects, and they would not necessarily apply to early insects that had not yet evolved flight. However, there is now good evidence from Carboniferous insects on the evolution of flight in this group, the first animals to take to the air under control. Jarmila Kukalová-Peck has pointed out that insects (and angels) are the only flying creatures that evolved flapping flight without sacrificing limbs to form the wings. Insects have thus lost little of their ability to move on the ground.

Many living insects are good gliders. Dragonflies, which were among the earliest insects to evolve flight, have wings arranged so that they are very stable in a gliding attitude, but dragonflies use flapping flight to chase rapidly and expertly after their prey. We still do not understand dragonfly flight. Somehow, complex eddies are produced between the two sets of wings, which beat out of phase. At some phases in the wing cycle, dragonfly wings produce lift forces that are 15 to 20 times the body weight.

Other insects have complex locking devices to hold their wings in a gliding position without energy expenditure. They need these locking devices for gliding because in powered flight their wings flap freely during complex movements. This line of reasoning suggests that insects evolved flight as flappers and later adjusted in a complex way to gliding.

The critical fact about the evolution of insect wings is that arthropod limbs consisted originally of two branches: a walking leg and another jointed unit—the **exite**—that was used as a filtering device or a gill. These structures are still found in most marine arthropods, but at first sight they seem to have been lost in insects, which have only walking legs. They were not lost: exites disappeared because they evolved into wings. We can see some of the stages in this evolution. In the young water-dwelling stages—**nymphs**—of living mayflies, the exites along the abdomen are shaped into platelike gills. The same structure is found on the thoracic exites of larval dragonflies, some beetles, and several other groups of insects.

Swimming as a Preadaptation

The earliest known insects lived in swampy forests. Their larvae probably developed from eggs laid in water, and, like dragonflies today, they spent most of their lives as water-dwelling nymphs, with only a short, late stage of life in air. There are fossilized nymphs of Late Carboniferous insects, and many of them had exites modified into platelike gills (Figure 14.1). The plates were probably also used for swimming. Living mayfly nymphs use their plated gills for swimming in the same way.

At first the plates probably acted as flaps over the spiracles that opened and closed as the nymph or adult left and entered the water. They could also have waved in the water to improve respiration, particularly in low oxygen conditions, by increasing fluid flow over the gills. In either case, the plates would have strengthened their muscular flapping action. They could then easily have evolved into aids for locomotion and then into *the* locomotory system under water. Such a system can be adapted to air, if body size is small. At small size, insects perceive the viscosity of air as very high. Viscous effects are greater than the inertial drag imposed by body mass, and the differences between hydrodynamics and aerodynamics, between motion in water and motion in air, are small at small size. Some living waterbugs use their wings to fly in water and in air, using the same wing motions and the same muscles.

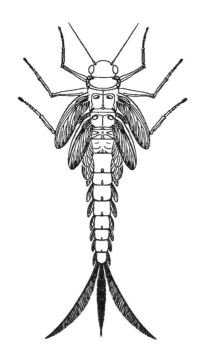

Figure 14.1   A mayfly nymph from the Lower Permian of Oklahoma. The wings on the thorax are big and look functional, but they are for underwater rowing rather than flight. There are also smaller winglets on all the segments of the abdomen. (Redrawn after a reconstruction by Jarmila Kukalová-Peck.)

So when early adult insects emerged into air by crawling up plant stems, they carried with them hinged plates that had evolved to become locomotory devices. As insects evolved air breathing through spiracles, the plates became available as protowings without the constraint of being used for any other function.

In this scenario, young adult insects emerged to feed on plant juices above the water, or on other plant-eating insects. They used the wing movements they had already evolved under water to climb, to keep steady, and to break falls as they moved around on waving plant surfaces. As movement in air became more important to their feeding, mating, escape from predators, and dispersal, the adults evolved larger, better-designed wings. It became more and more difficult for them to have a normal arthropod growth pattern with many molts, because it is diffcult to molt and regrow ever larger and lighter wings.

Larval nymphs came to swim more and more with their limbs and tail, or with water jets, while the developing wings were folded back against the body. Wings were not developed and stretched out as delicate membranes until the nymph left the water permanently. Then it became important to breed quickly as an adult, soon after emergence, while the new wings were in good condition. Successive molts during adult life apparently did not pay off. A wing could only have been molted and re-grown by retaining a good deal of living tissue in it, making it heavy and clumsy, and forcing an adult to be flightless during the molt. Adults that molted would therefore have been poor fliers even after their new wings grew, so natural selection would have worked against them. The anatomical and ecological differences between the aquatic larvae and the aerial adults became so extreme that insect metamorphosis at the final emergence also became more dramatic.

Since adult insects do not molt, they do not grow. The suppression of molting in adults had the additional effect of limiting the average size of insects, and it is probably significant that the largest insects that ever evolved are the huge dragonflies of the Late Carboniferous.

As wings were de-emphasized in all stages except the adult, they became much smaller, more streamlined organs in nymphs, and the wing mounting became adapted for use only in air. The characteristic wings of advanced insects evolved, with their ability to fold back over the body.

An intermediate stage in the evolution of insect flight may still exist in some primitive living insects, mayflies and stoneflies. James Marden and Melissa Kramer showed that these primitive insects "skim" across water surfaces, using a wing action that is exactly like flying, but they also receive some lift from the legs, which remain in contact with the water surface.

Basking as a Preadaptation

An alternative idea suggests that early insect wings evolved to a size useful for flying because they were used in thermoregulation. Some living butterflies use their wings as solar collectors, in an essential component

Figure 14.2 Some butterflies use their wings as solar collectors in pre-flight warm-up. (From Barnes et al., *The Invertebrates: A New Synthesis.* © 1993 Blackwell Scientific Publications.)

of preflight warm-up (Figure 14.2). Even a very small wing is useful for thermoregulation, not just for absorbing heat by acting as a solar panel, but for trapping warm air under its translucent surface as glass does in a greenhouse. In this theory, wings were preadapted for light structure and large size well before they were used seriously for flight.

But I suspect that basking alone as a preadaptation would not lead directly to flight, because the necessary muscular equipment and wing motions for flight would not have evolved from simple static solar panels. I think it's more likely that basking evolved in butterflies because they had already evolved wings, rather than the reverse. It's certainly possible, however, that basking behavior helped to perfect insect wings.

Parachuting Vertebrates

Several living forest-dwelling vertebrates have evolved parachuting flight, using skin flaps as flight surfaces. They include flying squirrels, three different lines of Australian gliding marsupials (greater gliders, squirrel gliders, and feathertail gliders), *Draco* the flying lizard, flying geckos, flying frogs (Figure 14.3), and even a flying snake. This suggests that parachuting adaptations evolve in animals of the forest canopy that habitually jump from branch to branch, from tree to tree, or from trees to the ground. Any method of breaking the landing impact or of leaping longer distances would be advantageous and might evolve rapidly.

None of these parachuting animals has powered flight, however. The energy for gliding flight is gravitational, generated as the animal climbs in the tree and released as it parachutes off the branch. Parachuting can evolve in animals with rather low metabolic rates. It does not require the

Figure 14.3 A parachuting vertebrate, the flying frog. (From Lydekker.)

high metabolic rate of birds and bats, which have powered flight. Characteristically, parachuting animals have short limbs, long trunks, flexible spines, and quadrupedal stance.

Early Gliding Vertebrates

The earliest known gliding vertebrate is the Late Permian reptile *Coelurosauravus*. It has been found in Germany, Britain, and Madagascar, so it was widespread and presumably successful. At the time, all these areas were near tropical shores. *Coelurosauravus* was an ordinary, rather primitive, small diapsid reptile in the structure of its skull and body, about the size of a small squirrel. A few normal ribs made up a cage that surrounded the heart and lungs. However, the trunk is dominated by 20 or so long, curving, lightly-constructed rod-shaped bones that extended outward and sideways from the body. They supported a skin membrane that was close to an ideal airfoil in shape, 30 cm (1 foot) across, and could only have been used for gliding (Figure 14.4, left). More impressive still, the bones are jointed so that the airfoil could have been folded back along the body when it was not in use. Extra-long vertebrae allowed space for this folding along the spinal column. These bones are not ribs, but must have evolved specifically as a gliding structure. Because of this unique character, *Coelurosauravus* is placed in a major group of its own, the Weigeltisauria.

We can judge how well *Coelurosauravus* was adapted for gliding flight by comparing it with Triassic and living reptilian gliders.

Kuehneosaurids are a family consisting of two gliding reptiles from the Late Triassic, *Kuehneosaurus* from Britain and *Icarosaurus* from New Jersey (Figure 14.4, right). They too had effective airfoils, but since they were stretched out on elongated ribs, gliding must have evolved independently in kuehneosaurids and *Coelurosauravus*. No gliding lepidosaur is known from the fossil record after the Triassic, so the living lizard *Draco*, which also uses elongated ribs to support an airfoil, must represent yet another independent evolution of gliding. Ligaments and muscles between the ribs of *Draco* give precise control of the gliding surface, and this was probably true also in the fossil gliders. In Permian, Triassic, and living gliders, all four limbs remain free for walking, grasping, and climbing. All three groups can fold up the airfoil when it is not in use. But there are interesting differences in the way the airfoil is organized.

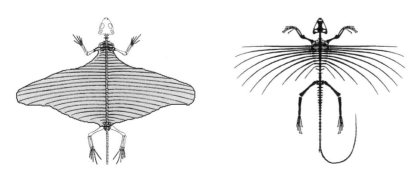

Figure 14.4 Left: *Coelurosauravus*, a gliding reptile from the Permian. (Airfoil reconstruction based on a skeletal diagram by Robert Carroll.) Right: *Icarosaurus*, a gliding reptile from the Triassic of New Jersey, represents another independent evolution of gliding flight among vertebrates. (After Colbert.)

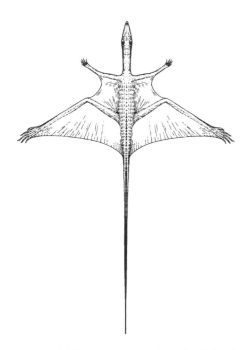

Figure 14.5 Reconstruction of a Triassic reptile from Kazakhstan, *Sharovipteryx*, as suggested by A. G. Sharov from the single specimen discovered. (After Sharov.)

Figure 14.6 Reconstruction of *Sharovipteryx* suggested by Carl Gans and his Russian colleagues. They were able to fly a successful model by adding a canard to the leading edge of the flight surface. (After Gans et al., 1987.)

In *Draco* and in the Triassic gliders, the ribs are single, unjointed bones. When *Draco* folds its airfoil, the spinal ends of the ribs have to be moved between the back muscles, which means that the ribs cannot be very big or very strong. The Triassic gliders had long levers mounted on the spinal ends of their ribs to get around this problem, and of course *Coelurosauravus* avoided the problem altogether by having a separate jointed airfoil that was not made from ribs. In some ways, then, *Coelurosauravus* was better designed than its later counterparts.

Sharovipteryx was discovered by accident in a search for fossil insects in Late Triassic rocks in Central Asia. It was a small reptile, and preserved skin clearly indicates that it had a gliding membrane. *Sharovipteryx*, however, was unique in that the membrane was stretched between very long, strong hind limbs and a long tail, so a large, broad wing surface was set well behind the trunk and head (Figure 14.5), rather like some modern aircraft—the Concorde, for example. Carl Gans and his Russian colleagues suggested that *Sharovipteryx* glided very well, perhaps with the aid of a canard, or accessory membrane, associated with the short, normal-looking forelimbs (Figure 14.6). Sensitive control over flight could have been maintained by slight backward-and-forward motion of the hind limbs, as in many gliding birds today. A similar but cruder system is used in swing-wing aircraft.

Longisquama is a strange reptile from the same Triassic rocks as *Sharovipteryx*. Its remains include a series of long, flattened bones with flared, curved tips. Susan Evans suggested that these were ribs from a gliding airfoil. The lightness and flattening of the bones and the curvature of their tips would all make sense if that were true. *Longisquama* was probably a glider, very much like the kuehneosaurids.

An airfoil does not appear by magic, especially a folding one. Robert Carroll points out that there may have been other good reasons for evolving a folding, extended rib structure. The great area of exposed skin could have been used in thermoregulation, for example. If the extinct reptiles behaved like *Draco*, they may have used their airfoils for display as well as for flight, and may have evolved them first for display.

We recognize these fossil reptiles as gliders because they had specialized skeletons. By comparison with small, insect-eating vertebrates in forests today, there were probably many other jumping and gliding reptiles in Permian and Triassic forest canopies, with skin flaps rather than specialized ribs. The forest canopy was probably rich in many species of small insectivorous amphibians and reptiles.

Flying Fishes

Flying fishes are a spectacular component of warm-water seas today. Their flight is apparently a means of escape from predators. Swimming rapidly upward through the water surface and sculling furiously with the tail fin as it finally leaves the water, the fish gains enough speed to glide for some distance through the air. Once airborne, the fish extends its pectoral fins, and sometimes its pelvic fins too, into airfoils that allow it to

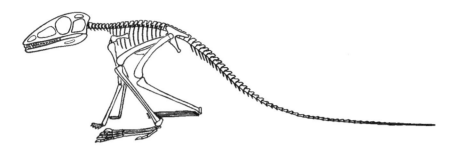

Figure 14.7 The small, lightly built Triassic archosaur *Scleromochlus*, which may be the closest relative of the early pterosaurs. (After von Huene.)

extend its glide path and give it some degree of control over its angle of re-entry. Because the fins are shaped like gliding surfaces, they have a characteristic outline that can be recognized in the fossil record. *Thoracopterus*, from Triassic warm-water seas, was a flying fish with fins laid out at two levels as it glided, like an old-fashioned biplane.

Pterosaurs

Pterosaurs are the most famous flying reptiles. They too evolved in the Triassic period, but they are archosaurs, quite unrelated to lizards. They are most likely closely related to the dinosaurs. The earliest pterosaur so far discovered was already fully adapted for flight, so its ancestors are not yet known. The most likely candidate is a small, Late Triassic archosaur called *Scleromochlus* (Figure 14.7), a lightly built, long-legged terrestrial runner that shares many characters with early pterosaurs.

Pterosaurs have very lightly built skeletons, with air spaces in many of the bones. Their forelimbs were extended into long struts that supported a wing membrane, as in birds and bats. Pterosaurs were unique, however, in that most of the wing membrane was supported on one extraordinarily long finger, while three other fingers were normal and bore claws. The fourth finger was about 3 meters (10 feet) long in the largest pterosaurs. In contrast, birds support the wing with the whole arm, and bats use all their fingers as bony supports through their wing membranes (Figure 14.8). Pterosaurs thus have a unique wing anatomy, but as the largest flying creatures ever to evolve and as a group that flourished for more than 140 m.y., they can't be dismissed as primitive or poorly adapted.

Most pterosaurs had large eyes sighting right along the length of long, narrow, lightly-built jaws. The teeth were usually long, thin, and pointed, often projecting slightly outward and forward (Figure 14.9). Where stomach contents have been preserved with pterosaur skeletons, they always contain fish remains such as spines and scales. And almost all pterosaur fossils are preserved in sediments laid down on shallow seafloors. Some pterosaurs may have fished on the wing, like living birds such as gadfly petrels or skimmers, which fly along just above the water surface and dip in their beaks to scoop up fish or crustaceans. Other pterosaurs may have fed like terns, which dive slowly so that only the head, neck, and front of the thorax reach under the water, while the wings remain above the surface. Pterosaurs also may have fished standing in the

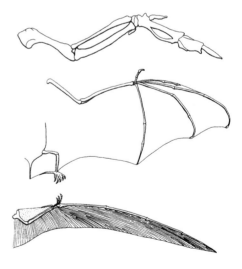

Figure 14.8 The skeletal differences between the wings of a bird (top), a bat (center), and a pterosaur (bottom). The pterosaur is constructed with a long narrow wing, in keeping with today's majority opinion.

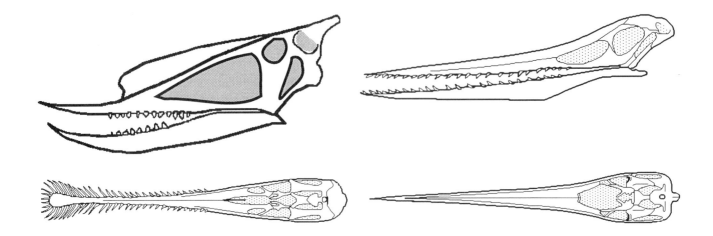

Figure 14.9 The jaws and teeth of some pterosaurs. Above: Two typical fishing pterosaurs, *Dzungaripterus* (left) and *Ornithocheirus* (right) in side view, each with sharp pointed teeth that protruded slightly to impale a fish in a strike. Their skull mechanics were obviously rather different. Below: The skulls of *Gnathosaurus* (left), and *Nyctosaurus* (right). *Gnathosaurus* probably scooped water and filtered small prey from it. *Nyctosaurus* presumably caught prey by stabbing at it like a heron. (After Wellnhofer.)

water like herons, or sitting on the water. It seems unlikely that they crash-dived into water like pelicans or gannets, or swam underwater like penguins: pterosaur wings were too long and too fragile. At least one pterosaur, *Pterodaustro* from Argentina, had teeth that were so fine, long, and numerous that it must have been a filter feeder, perhaps like a flamingo. Some short-jawed pterosaurs may have eaten shore crustaceans or insects.

There are two main groups of pterosaurs. **Rhamphorhynchoids** (Late Triassic to Late Jurassic) are the stem group of early pterosaurs, rather than a clade. Most of them had wingspans under 2 meters (6 feet), and some were as small as sparrows. They had long, thin, stiff tails that carried a vertical vane on the tip, perhaps as a dynamic stabilizer in flight. **Pterodactyloids** are a clade of advanced pterosaurs that replaced rhamphorhynchoids in the Late Jurassic and flourished until the end of the Cretaceous. Pterodactyloids had no tails, and many were much larger than rhamphorhynchoids. The large forms were adapted for soaring rather than continuous flapping flight, although they all flapped for takeoff. *Pterodactylus* itself was sparrow-sized, but *Pteranodon*, from the Cretaceous of North America, had a wingspan of about 7 meters (22 feet); and the gigantic pterosaur from Texas, *Quetzalcoatlus*, was at least 11 meters (35 feet) in wingspan, the largest flying creature ever to evolve.

Although pterosaur bones were light and fragile, several examples of outstanding preservation have shown us many details of their structure. Black shales in Lower Jurassic rocks of Germany have shown details of rhamphorhynchoids; Late Jurassic members of both pterosaur groups have been found exquisitely preserved in the Solnhofen Limestone of Germany and in lake deposits in Kazakhstan in Central Asia; from the Lower Cretaceous of Brazil we have partial skeletons preserved without crushing; and the Upper Cretaceous chalk beds of Kansas have yielded huge specimens of *Pteranodon*. Discoveries of skin, wing membranes, and stomach contents allow reliable biological interpretation of these exciting animals. These discoveries are important because pterosaurs have no living descendants that we can study, and we have not found their ancestors. It is clear now that all pterosaurs, including the giant forms, were capable of powered, flapping flight. All had a large bony plate on the front of the rib cage to which powerful flight muscles were attached. At the other end, the flight muscles attached to a wide flange at the top of the arm (Figure 14.10).

Almost all reconstructions of pterosaurs show the wing attached to the ankle (Figure 14.11, top). Some people, such as Peter Wellnhofer, support this reconstruction. Most people, however, including Kevin Padian, argue instead that the wing was attached only to the arm and body, and that pterosaurs flew much like birds. The disagreement continues because the wing is often preserved lying across the skeleton after death, and it is difficult to tell whether the contact between wing and leg is anatomical or simply accidental. Padian argues that the idea of attachment to the ankle is based on analogy with bats, rather than evidence from fossils. There is no evidence that any pterosaur had a wing attached any further down the leg than mid-thigh, and even that is controversial.

If Padian is correct, pterosaurs had long, narrow wings, much like a falcon or an albatross (Figure 14.11, bottom). All the small rhamphorhynchoids and many of the pterodactyloids had active, flapping flight. Naturally, the gigantic pterosaurs could not have flapped for long, and they probably spent most of their time soaring, as does the living albatross. Aerodynamic analysis shows that pterosaurs were the best slow-speed soaring fliers ever to evolve.

The wing itself was not simply a giant skin membrane: that would have been too weak to power flapping flight. Furthermore, with bones, joints, and ligaments only on the leading edge of the wing, a pterosaur needed a way to control the aerodynamic surface of the wing. Beautifully preserved specimens show that the wing had special adaptations. It was

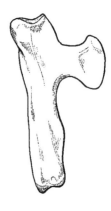

Figure 14.10 The humerus of all pterosaurs has a special flange where powerful flight muscles attached. This bone is from *Nyctosaurus*. (After Williston.)

Figure 14.11 Above: A pterosaur reconstructed in an older fashion, with its wing membrane attached at the ankle. Below: Padian's reconstruction of the pterosaur wing as long and falconlike.

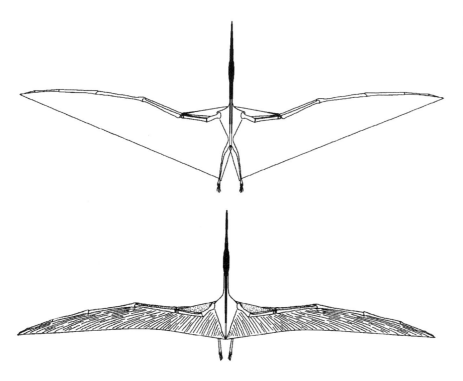

stiffened by many small, cylindrical fibers, which were probably tied together by small muscles. The combination of structural stiffeners and muscles allowed fine control over the surface, and at the same time made the wing reasonably strong, not easily damaged or warped, and not likely to billow in flight like the fabric of a hang glider.

Padian and Jeremy Rayner have suggested that the major lifting forces acted on the inner third of the wing, and that the stress was transmitted through the arm bones to the shoulder and body. The long wing finger did not have to resist much lifting force; it bent downward and backward and acted to prevent the wing from twisting in flight.

Overall, there is a great deal of similarity between the flight of birds and pterosaurs, despite the structural differences between them.

The second unresolved question is about the leg structure of pterosaurs and their walking ability. Most reconstructions of pterosaurs show them as clumsy and batlike on the ground, reasonable if the wings tied up the hind limbs. But if pterosaur wings were birdlike, they did not constrain the hind limbs in any way.

Pterosaur pelvises had the hip joint facing sideways and upward, but then so do birds. The femur was horizontal in life, and the hind limbs of pterosaurs were perfectly good for walking and running, as in birds and small dinosaurs. The ankle joints were hinges, moving only forward and backward, unlike the ankles of lizards or bats.

Padian thinks that small pterosaurs could have been good ground runners (Figure 14.12). In his view, rhamphorhynchoids, with their long tails tipped with rudders would have been really agile. Large pterosaurs would have had no more difficulty than vultures or albatrosses in takeoff from the ground—that is, difficult, but possible with some effort. However, a beautifully preserved foot of *Dimorphodon* shows that it could not have flexed in the way Padian shows in Figure 14.12, so *Dimorphodon* could not have run in the way he shows. Instead it probably waddled flat-footed. Was *Dimorphodon* a typical pterosaur, or an unusual one?

This question is not likely to be resolved any time soon. Pterosaurs were not exactly like birds or exactly like bats. They had a distinct biology and ecology. Perhaps different pterosaurs had different flying and walking systems.

Flapping flight involves very high energy expenditure. Birds are warm-blooded, as are bats and many large insects when they are in flight —moths, dragonflies, and bees are examples. Thus one might guess that pterosaurs too were warm-blooded, or at least had the athletic ability of

Figure 14.12   Kevin Padian's reconstruction of an early little pterosaur, *Dimorphodon*, suggesting that it could have been a fast and agile runner.

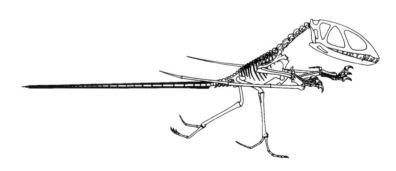

other early archosaurs such as dinosaurs (Chapter 13). Pterosaur bones had air spaces running through them in the same way that living bird bones do. In birds, this system helps to provide air cooling, and it is reasonable to interpret pterosaur bone structure in the same way. Several Jurassic pterosaurs are now known that have fur preserved on the skin: if pterosaurs had fur, they were probably warm-blooded. This fits with their active, flapping flight, with their brain size (large for reptiles), and with the idea that they were closely related to dinosaurs. Flapping flight has evolved only three times among vertebrates (in pterosaurs, in birds, and in bats) and in each case the animal was apparently warm-blooded before or just as it achieved flight.

If most pterosaurs ranged widely over the ocean searching for fish, it would have been impossible for pterosaur nestlings to feed themselves until they had reached a fairly advanced stage of growth and flight capability. Nesting behavior and care of the young would therefore have been mandatory. The social behavior of pterosaurs may have been complex. *Pteranodon* at least was dimorphic. Males were larger, with long crests on the back of the head and with relatively narrow pelvic openings. Females were smaller, with smaller crests but larger pelvic openings.

The largest pterosaur, *Quetzalcoatlus* (Figure 14.13), lived right at the end of the Cretaceous Period. Its bones are found in nonmarine beds, deposited perhaps 400 km (250 miles) inland. Perhaps it was the ecological equivalent of a vulture, soaring above the Cretaceous plains and scavenging on carcasses of dinosaurs rather than wildebeeste. *Quetzalcoatlus*

Figure 14.13 The wingspans of various fliers. *Pteranodon*, a large pterosaur, conservatively estimated to have a wingspan of 7.5 meters (25 feet); a Sopwith Camel F1, wingspan 8.5 meters (28 feet); the pterosaur *Quetzalcoatlus*, the largest flying animal ever to evolve, conservatively estimated with a wingspan of 13.5 meters (45 feet); a Supermarine Spitfire Mark 21, wingspan 11 meters (37 feet); and a General Dynamics F16A Fighting Falcon, wingspan 10 meters (33 feet). The size of the pilots in the airplanes will serve as an extra reminder of the size of the pterosaurs. Pterosaur wings are reconstructed after Padian's interpretation (see Figure 14.12).

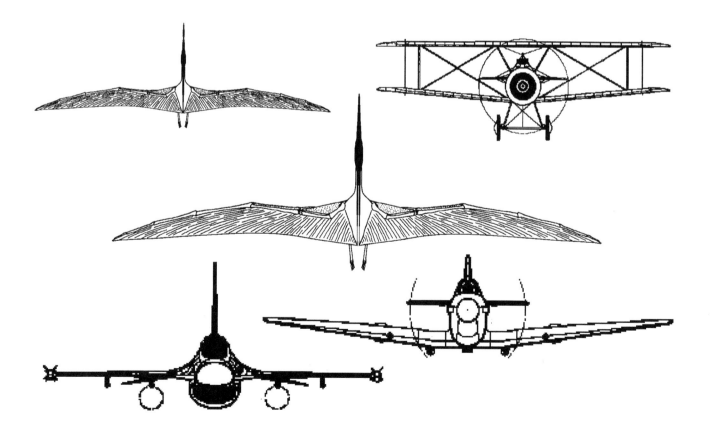

did have a strangely long, strong neck, but its beak seems too lightly built for this method of feeding. Perhaps it was more like a gigantic heron, standing 3 meters (10 feet) tall and weighing 100 kg (250 pounds) and fishing in inland lakes or picking up frogs, turtles, or arthropods such as crayfish from shallow water.

There are two great unsolved questions about pterosaurs. The first is the origin of flight in pterosaurs. They were the first vertebrates to achieve flapping flight, and it's unlikely that they did so in the same way that insects did. Padian has argued that pterosaurs could run fast, so they were terrestrial bipeds as well as good fliers. If the earliest pterosaurs were closely related to *Scleromochlus*, as Padian suggests, then pterosaur flight evolved among ground-running animals, not among tree dwellers. If *Dimorphodon* could not run, then the question is wide open.

We do not know why pterosaurs became extinct. As we have seen, they were most likely active, warm-blooded animals with flapping flight much like that of birds. Yet pterosaurs became extinct at the end of the Cretaceous, at the same time as the dinosaurs disappeared, while birds did not. We shall return to that question in Chapter 18.

Birds

Living birds are warm-blooded, with efficient thermoregulation that maintains body temperatures higher than our own. Birds breathe more efficiently than mammals, pumping air through their lungs rather than in and out. They have better vision than any other animals. Birds build extraordinarily sophisticated nests: bowerbirds are second only to humans in their ability to create art objects. And above all, birds can fly better, farther, and faster than any other animals, an ability that demands complex energy supply systems, sensing devices, and control systems.

Birds include ostriches and penguins, which cannot fly, and hummingbirds, which can hardly walk. But birds share enough characters for us to be sure that they form a single clade, descended from archosaurian reptiles. The skull, pelvis, feet, and eggs of birds and reptiles are so similar that Darwin's friend T. H. Huxley called birds "glorified reptiles."

The earliest known bird is *Archaeopteryx* from Upper Jurassic rocks in Germany, perhaps the most famous fossil in the world (Figure 14.14). Only seven specimens, plus a single feather, have been found (one of those is currently missing, probably stolen). All the specimens come from the Solnhofen Limestone, which formed as a lime mud on the floor of a tropical lagoon. In addition to *Archaeopteryx*, the exquisite fossils occasionally found in these rocks include pterosaurs, a few land animals, small pieces of plant fragments, and marine creatures that swam or were carried into the Solnhofen lagoon. The bodies were hardly disturbed at all as they were buried in the mud and preserved in extraordinary detail. Paleoclimatic evidence suggests an environment rather like that of the Persian Gulf today—a hot, dry desert shore with only sparse vegetation.

Figure 14.14 There are only a few known skeletons of *Archaeopteryx*, the first bird. This is the "Berlin" specimen, which belongs to the Humboldt Museum in Berlin. (From an early engraving.)

The first complete *Archaeopteryx* was immediately recognized as a fossil bird, because it had feathers on its wings and tail. But without feathers, it looks very much like a small theropod dinosaur. Two of the six known specimens lay unrecognized for a long time, labeled as small theropods.

Archaeopteryx has a theropod pelvis, not the tight, boxlike structure of living birds. It has a long, bony tail, clawed fingers, and a jaw full of savage little teeth (Figure 14.14), all theropod features. It lacks many features of living birds: there is no breastbone, and there is no hole through the shoulder joint through which to pass the large tendon that gives the rapid, powerful, twisting wing upstroke in living birds (Figure 14.15). The only birdlike features on the entire bony skeleton of *Archaeopteryx* are a few characters of the skull. Even the wishbone or furcula is also found in theropod dinosaurs.

Compsognathus is a small theropod dinosaur that was also preserved in the Solnhofen Limestone (Figure 14.16). It was a little larger than *Archaeopteryx*, and was a fast-running predator. One specimen has the skeleton of a long, slim, fast-running lizard neatly folded up inside it. Both *Compsognathus* and *Archaeopteryx* are always preserved in an unusual body attitude, with the neck severely ricked back over the body. We know why this happens. If an animal dies today on or near the beach or on a desert salt pan, it may be mummified by wind and salt spray before it rots or is eaten by predators. The muscles slacken and the tendons dry out. The long tendons that support the head contract severely, dragging the skull backwards over the spine. At the same time, any body feathers on a bird usually drop off, but the strong feathers set into the wings and tail stay fixed in position.

Occasionally, mummified birds may be washed out to sea on a high tide, or blown into the water by a gale. They may float for several weeks before becoming waterlogged, and even when they finally sink, they retain their peculiar body attitude. *Compsognathus* did not have feathers, and it was without question a terrestrial biped. *Archaeopteryx* and *Compsognathus* were fossilized in the same way, as mummified bodies that floated out from a shoreline some distance away. There is no need to suggest that *Archaeopteryx* could fly because it sank and was buried at sea.

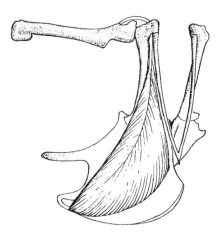

Figure 14.15 *Archaeopteryx* lacked many features of living birds. All living birds have a powerful muscle, the supracoracoideus, shown here passing from the breastbone up through a special hole in the shoulder joint to the top side of the humerus. It acts to fold up (and streamline) the wing as it is raised. Meanwhile the flexible wishbone acts as a spring, to prevent the shoulder girdle from distorting too much as the wing muscles pull on the shoulder joints. *Archaeopteryx* had not evolved this system. There is no hole in the shoulder joint, no muscle attachment for the supracoracoideus on the humerus, and its wishbone was big, solid, and rigid rather than flexible and springlike.

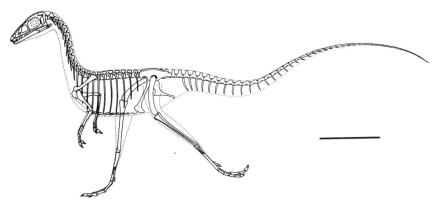

Figure 14.16 *Compsognathus* is a small terrestrial theropod dinosaur preserved in the Solnhofen Limestone along with *Archaeopteryx*, and in the same style. (Courtesy of John Ostrom, Yale University.)

The Origin of Birds

The dinosaurs fly past in herds

Singing their song without words

    They're small ones, it's true,

    Warm and feathery too,

But they're here—and we call them birds.

There has always been resistance to the idea that birds, and specifically *Archaeopteryx*, evolved from small theropod dinosaurs. The German scientist who described the first specimen of *Archaeopteryx*, two years after Darwin published *The Origin of Species*, stressed that the appearance of the fossil, transitional between reptiles and birds, was certainly not to be regarded as evidence that Mr. Darwin was correct.

The arguments continue: Alan Feduccia recently claimed in a comprehensive book on the evolution of birds that birds evolved from Triassic reptiles that we have not yet discovered. One major problem is the hand, where the three fingers of birds develop from the middle three "bone buds", while the theropod fingers develop from the first three. However, almost all paleontologists are now convinced that birds evolved from theropods, and the finger discrepancy arose from a shift in developmental genes. Furthermore, there are now many transitional fossils between theropods and the first birds. In particular, feathers have been found on several theropods that were certainly ground-running animals (Chapter 13).

The Origin of Powered Flight in Birds

Since ground-running theropods had feathers, the question of the origin of flight in birds has nothing to do with the appearance of feathers. (Flight evolved in bats and pterosaurs without feathers.) There have been three important hypotheses for the origin of bird flight, and I shall add a fourth.

The Arboreal Hypothesis

The arboreal hypothesis suggests that ancestral birds evolved flight by jumping out of trees. The arboreal theory was the most favored until recently, and it still has supporters. But it must be abandoned in the face of the new theropods from China. With long, erect limbs, a comparatively short trunk, and bipedal locomotion, *Archaeopteryx* and the feathered theropods are exactly the opposite in body plan of all living mammals and reptiles that jump and glide from tree to tree. There is nothing in the ancestry of birds as we now know it to suggest any arboreal adaptations at all.

Flapping arms or proto-wings—in fact, any feathers at all on wings or tail—increase drag. Aerodynamically, the transition from gliding to flapping is difficult: there is only a narrow theoretical window through which the transition could have been made (page 216). Such a transition would have been especially difficult for *Archaeopteryx* because it had such a long, bony tail with long feathers on it. This kind of tail adds much more drag than it adds lift.

The Cursorial Hypothesis

Perhaps some adaptations in a ground-dwelling theropod could provide some of the anatomy and behavior necessary for flight, such as lengthening the forearms, especially the hands, placing long, strong feathers in those areas, and evolving powerful arm movements. An early version of the cursorial hypothesis suggested that a fast-running reptile might evolve long scales on the arms. In this theory, the scales generated lift as the arms were actively flapped on the run (Figure 14.17). The animal could now take long leaps, perhaps encouraging the scales to evolve into feathers and the leaps to evolve into powered flapping flight.

Feathers did not evolve from scales, but in any case the idea does not work mechanically. Any lift generated by a flapping arm decreases the ground traction given by the feet, and acceleration is lost. A racing car is held *down* on the track by its airfoils for good traction, and an aircraft cannot be driven through its wheels on the takeoff run. A running theropod that flapped its arms would increase drag: the faster the run, the greater the drag. Only a very small amount of thrust would have been generated by the arms in the early stages of the process. A new version of this theory was published in 1999, but I suspect that the assumptions in that paper were not correct.

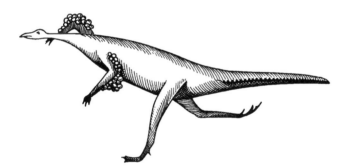

Figure 14.17 The original version of the cursorial theory suggested that long scales on the forearms of a protobird could have given enough lift for takeoff if the arms were flapped vigorously on the run. (After Nopsca.)

The Running Raptor

More recent versions of the cursorial hypothesis are much better: they are mechanically sounder and include behavior that involved strong, synchronized arm strokes and the evolution of strong pectoral muscles.

Gerald Caple and his colleagues suggested that a proto-bird hunted by running fast and leaping after flying or jumping insects it disturbed. To catch an elusive dodging prey while its feet were off the ground, a protobird would have to be able to adjust its body attitude in the air and then regain a stable position for landing. Such adjustments could be made aerodynamically by generating a small amount of lift or drag, and that would be best added at the tips of the arms. If the right arm movements were added, the effect would be greater still. The proto-bird would now be well on the way to flapping takeoff, and the flights would be gradually prolonged until complete control had been reached.

But this proposed activity would consume a lot of energy. No predator today, bird or otherwise, runs at high speed to flush out insects it can leap after. Furthermore, effective attitude control for a leaping animal requires a critical airspeed that is high in the early stages of the process, when the proto-wings are just beginning to generate lift. The required speed might have been 10 meters per second, over 25 mph, far too much for any reasonable proto-bird.

The Display and Fighting Hypothesis

Jere Lipps and I suggested that display was involved in the evolution of flight as well as feathers. Theropods had long, strong display feathers on arms and tail (Chapter 13). Successful display was increased by lengthening the arms, especially the hand, and by actively waving them, perhaps flapping them rapidly and vigorously. Flapping in display would have encouraged the evolution of powerful pectoral muscles, and the supracoracoideus system.

Display can be very effective, and not just for sexual ends. Frigate birds and bald eagles often try to rob other birds of food instead of catching prey themselves. Because the penalty for wing injury is high, many birds can be intimidated by display into giving up their catch rather than fighting to defend it.

But a threat display must not be exposed as an empty bluff. Fighting is the last resort. Living birds often fight on the ground, even those that fly well. The wings no longer have claws but are still used as weapons in forward and downward smashes (steamer ducks are particularly deadly at this). Beaks and feet can be used as weapons too, and are most effective when used in a downward or forward strike.

A strong wing flap, directed forward and downward, is also the power stroke that gives lift to a bird in takeoff. Lipps and I suggest that strong wing flapping is a simple extension of display flapping, encouraged by fighting behavior. Powerful flapping used to deliver forearm smashes could have lifted the bird off the ground, allowing it also to rake its opponent from above with its hind claws. The more rapidly the wings could be lifted for another blow, the more effective the fighting. This would rapidly encourage an effective wing-lifting motion that minimized air resistance, so the wing action would then be almost identical to a takeoff stroke.

A variant of our idea has also been proposed by Kevin Padian, who prefers to think of the wing stroke evolving from the arm strike used by a theropod in predation. It is not clear how this could have led easily to whole-body takeoff, however. *Caudipteryx* was able to fold its feathers away while making its fighting stroke.

Archaeopteryx fits our display-and-fighting hypothesis well. It was well adapted for display. Like any small theropod, it was well equipped for fighting with its teeth and the strong claws on hands and feet. *Archaeopteryx* did not have long primary feathers on its fingers (Figure 13.1), probably because they would have hidden the claws in display and would most likely have broken in a fight.

Archaeopteryx could not fly well; I suspect that it hardly flew at all. It may have been able to glide a short distance, but it could not have sustained flapping flight. It did not have the supracoracoideus system (Figure 14.15). This muscle passes through the shoulder joint, and as well as raising the wing, it twists it. On the upstroke, the twist arranges the wing and feathers so that they slip easily through the air, with little drag. At the top of the upstroke, the wing is in exactly the right position to give a powerful downbeat. Without the supracoracoideus (which is easily identified because it leaves a strong trace on the shoulder joint), a bird cannot fly by wing flapping. In fact, it cannot even take off and land, because the greatest power from the wings is required during slow flight.

In small flying birds today, the wishbone acts as a spring that repositions the shoulder joints after the stresses of each wing stroke. It is needed to give the rapid flaps necessary for flight (a starling flies with 14 complete wing beats *per second*). The wishbone also helps to pump air in and out of the lungs, and to recover some of the muscular energy put into the downstroke. In *Archaeopteryx* and in theropods with wishbones, the bone is U-shaped and strong and solid; it could not have acted as an effective spring. Furthermore, *Archaeopteryx* did not have the long primary feathers on the wing tips, or the breastbone anchoring the muscles that are needed for routine takeoff and landing. It could not have raised its arms high above its body for an effective downstroke. In fact, *Archaeopteryx* evolved structures that were active deterrents to flight. Its tail was long and bony, with long feathers. Among living birds with display feathers, this sort of tail is aerodynamically the worst of all possible tail styles, adding a lot of drag and little lift.

Archaeopteryx, then, was a fierce little fast-running, displaying bird, which probably spent its life scurrying around the Solnhofen shore, hunting for small prey such as crustaceans, reptiles, and mammals. In hunting style, *Archaeopteryx* was probably much like the roadrunner of the dry country of the American Southwest, but its ecological setting was like that of a steamer duck: on a shoreline with year-round food supply. *Archaeopteryx* did not compete in the air with the pterosaurs that are also found in the Solnhofen Limestone.

From Fight to Flight

Display and fighting in birds takes a lot of energy, whether it is for territory, dominance, or food, but it provides an enormous payoff in survival and selection. Sexual display in most living birds must be done correctly, or no mating takes place. New behaviors can be evolved rapidly, and they are often evolutionarily cheap, because they usually don't require important morphological changes in their early stages.

The display hypothesis suggests that the earliest birds gained flight behavior, anatomy, and experience at low ground speed and low height: ideal preflight training. The selective payoff for successful mastery of the flight motions gave significant advantages, even before flight itself was possible. From that point, the many advantages of flight were added to those of social or sexual competition. I do not think it is a coincidence that the males of the Early Cretaceous Chinese birds *Confuciusornis* and *Changchengornis* had extravagantly long (display) feathers on the tail!

Once liftoff was achieved, flapping flight quickly followed. In more advanced birds than *Archaeopteryx*, the supracoracoideus tendon system evolved in the shoulder, while the wishbone evolved into a spring. The breastbone evolved as the anchor for the flight muscles. The forearms became longer, lighter, and more fragile in bone structure, becoming specialized as wings, and losing the finger claws. Meanwhile, the feet and beak became the dominant fighting weapons, as in most living birds.

Earlier Bird Evolution?

Sankar Chatterjee found a small fossil in Late Triassic rocks in Texas that he claimed was a bird, so called it *Protoavis*. He identified small bumps on the forelimbs as feather attachments, and he interpreted *Protoavis* as the earliest bird, 75 m.y. before *Archaeopteryx*. Others disagree!

New fossils turn up all the time. Sometimes there is enthusiastic and immediate acceptance of new claims; sometimes there is deep and immediate suspicion about them until they are finally proven beyond question; sometimes opinion is strongly divided. This is partly a question of individual attitude toward anything new, but we can apply scientific judgment as well. Certain questions must be asked about a new find such as *Protoavis*, especially a fossil that could be an important evolutionary link between major lineages. The same questions should be asked about *Tulerpeton* (Chapter 8), "Lizzie" (Chapter 9), and *Ardipithecus* (Chapter 22). Is the age accurately known? What makes this particular fossil an early member of a new clade, rather than an advanced member of an ancestral group? So what makes *Protoavis* a bird rather than a theropod?

A major and convincing criticism of *Protoavis* is that its "birdlike" characters are not exclusively birdlike but are characters that were shared by all little early theropods, and later inherited by birds as well as many advanced theropods. There is no particular reason to identify *Protoavis* as anything more than a small, rather poorly preserved, early theropod.

Cretaceous Birds

After *Archaeopteryx*, the evolution of birds was very rapid. Early Cretaceous rocks have yielded bird remains in all the northern continents and in Australia. *Sinornis*, a sparrow-sized bird from the Early Cretaceous of China, had many features directly related to much better flight and perching than was possible in *Archaeopteryx*. The body and tail were shorter, and the tail had fused vertebrae at its end that provided a firm but light base for strong tail feathers. The center of mass of the body was much farther forward, closer to the wings. *Sinornis* had a breastbone, a shoulder joint that allowed it to raise its wings well above the horizontal, and fingers that were adapted to support feathers rather than grasping and tearing claws. The wrist could fold much more tightly forward against the arm than the 90° seen in *Archaeopteryx*, so the wing could be folded away cleanly in the upstroke or on the ground, reducing drag. The foot was much better adapted for perching. Even so, *Sinornis* still had some very primitive features: the skull and pelvis were much like those of *Archaeopteryx*.

New Early Cretaceous fossils from Spain and from China tell the same story. Rapid evolution among Early Cretaceous birds dramatically improved their flying and perching ability; perhaps this is why most of them were small and light. *Confuciusornis* and *Changchengornis*, from China, had lighter bones than *Archaeopteryx*, and had genuine beaks, rather than jaws with teeth. *Eoalulavis* from Spain had evolved the alula, the arrangement of feathers near the wing tip that allows slow flight without stalling.

Clearly, there was a radiation of Cretaceous birds, but we have such a scanty fossil record that we can't tell how fast or how global it was. Most Cretaceous bird fossils are from shoreline habitats, but that may reflect preservation bias rather than ecological reality. We have good fossils of Late Cretaceous diving birds such as *Hesperornis* (Figure 14.18), and *Ichthyornis* was ternlike in its adaptations (Figure 14.19).

Cenozoic Birds

When the dinosaurs died out at the end of the Cretaceous, there must have been a very interesting opportunity for surviving creatures to invade the ecological niches associated with larger body size on the ground. The two leading contenders were birds and mammals, and although mammals quickly became large herbivores, it was birds that became the dominant land predators in some regions in the Paleocene. These birds evolved to become flightless terrestrial bipeds once more. (This is the only way a bird can become heavy and powerful.)

Large, flightless birds called diatrymas (after *Diatryma*, one of the largest) lived across the Northern Hemisphere in the Paleocene and Eocene. They stood about 2 meters (6 feet) tall, and they had massive legs with vicious claws and huge, powerful beaks that look like killing instruments as efficient as the heads of tyrannosaurs. A full-grown *Diatryma* could have killed many of the mammals in its community. Carnivorous birds with very similar adaptations, the phorusrhacids, dominated the plains ecosystem of South America somewhat later (Figure 14.20). Diatrymas became extinct at the end of the Eocene, but the phorusrhacids survived until the late Pliocene. Some phorusrhacids were 2.5 meters (7 feet) tall, and a spectacular late phorusrhacid, *Titanis*, crossed to Florida from South America less than 3 m.y. ago. It was larger than an ostrich and no doubt caused at least temporary consternation among the Floridian mammals of the time.

The southern continents have a number of large flightless birds. Living forms such as the ostrich, cassowary, rhea, and emu are familiar enough, but even more interesting forms are now extinct. All these birds

Figure 14.18 A reconstruction of a flightless, diving Cretaceous bird, *Hesperornis*. It looks as though it had a flying ancestor, because its shoulder and breastbone have a structure associated with rapid and powerful upward wingbeats. It had teeth, however, a primitive character for birds. (From Marsh.)

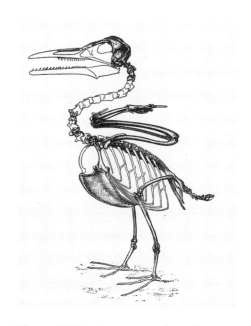

Figure 14.19 A reconstruction of a Cretaceous bird, *Ichthyornis*. It appears to have been an agile flier that ate fish, with a way of life rather like a living tern. Note that *Ichthyornis* had teeth. (From Marsh.)

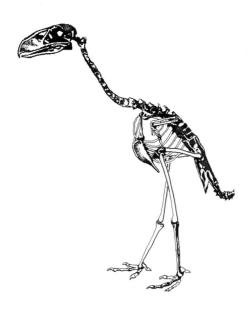

Figure 14.20   One of the large flightless carnivorous birds of the Cenozoic of South America, a phorusrhacid from Argentina. The original reconstruction by Sinclair was shown with a greatly curved neck so that it would fit onto the size of the page he was allowed; this is my attempt to straighten out the neck into a normal posture. (After Sinclair.)

Figure 14.21   Left, an ostrich is shown for scale next to *Aepyornis*, the elephant bird of Madagascar that became extinct only a few hundred years ago. On the right is the roc Smizurgh of Arab and Persian legend carrying off three elephants at once. The legend of the roc may have been based on ostriches, but ostriches must have been familiar to Arab traders visiting East Africa. Most likely the much larger and rarer elephant bird was the real inspiration for the stories. (From a nineteenth-century story book.)

are loosely grouped together as **ratites**, although they may be a mixture of birds that evolved in parallel from different ancestors. The moas of New Zealand reached well over 3 meters (10 feet) in height. *Aepyornis*, the "elephant bird" of Madagascar (Figure 14.21, left), was living so recently that its eggshells are still found lying loose on the ground. The eggs are unmistakable because they had a volume of 11 liters (2 gallons). Early Muslim traders along the African coast certainly saw these eggs, and they may even have seen living elephant birds in Madagascar, giving rise to folktales about the fearsome roc that preyed on elephants and carried Sinbad the Sailor on its back (Figure 14.21, right). *Aepyornis* and *Dromornis*, a giant extinct Australian bird (Chapters 20 and 23), are close competitors for Heaviest Bird Ever to Evolve. The Guinness Book of World Records currently favors *Dromornis*, which was powerfully built and weighed perhaps 500 kg (1100 pounds).

Ratites live now on widely separated southern continents (Figure 14.22), but this is probably not a relic of Gondwanaland geography. A good ratite is known from the Middle Eocene of Europe.

The Largest Flying Birds

The largest flying birds so far discovered are teratorns, immense birds from South America, now extinct, who reached North America during the Pleistocene. Hundreds of specimens have been found in the tar pits of La Brea in Los Angeles, California (Figure 14.23), and from Florida and Mexico. But the largest teratorn was *Argentavis* from the Late Miocene of Argentina: it had a wingspan of 7.5 meters (24 feet) and an estimated body weight close to 75 kg (170 pounds). By contrast, the largest living bird is the royal albatross, just over 3 meters (10 feet) in wingspan.

The beak of *Argentavis* suggests that it was a predator, not a scavenger. It probably stalked prey on the ground. With a skull 55 cm (2 feet) long and 15 cm (6 inches) wide, it could have swallowed prey animals 15 cm across. Its bones are associated with other vertebrate fossils, but 64% of those are from *Paedotherium*, a little mammal about the size of a jackrabbit (in other words, an easy swallow for *Argentavis*).

In the same size range as teratorns were pelagornithids, gigantic marine birds that must have spent most of their time soaring over water.

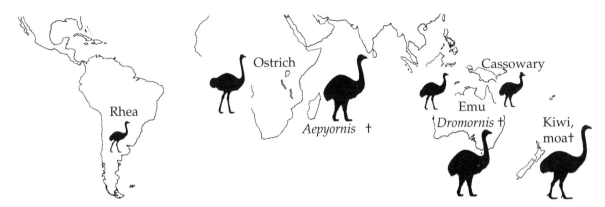

Figure 14.22 The distribution of some living and fossil ratites.

They ranged worldwide from the Eocene to the Late Miocene. They were lightly built, but the wingspan was close to 6 meters (nearly 20 feet) in the largest specimens. They weighed perhaps 40 kg (90 pounds). Their beaks were very long, with toothlike projections built into their edges, presumably to help them hold squirming prey. More than any other known birds, pelagornithids were the ecological equivalents of the pterosaurs, and it will be fascinating when further research allows us to reconstruct their mode of life accurately.

Figure 14.23 A Pleistocene teratorn, *Teratornis*, from the La Brea tar pits of Los Angeles. The preservation is so good that the wingspan can be measured accurately at nearly 4 meters (12 feet). This is enormous for a bird, although not for a pterosaur (compare Figure 14.13).

Bats

The latest evolution of flapping flight among vertebrates took place among bats. In all bats, the wing is stretched between arm, body, and leg, with the fingers of the hand splayed out in a fan toward the wingtip (Figure 14.8). The wing membrane has little strength of its own, but it is elastic, and tension has to be maintained in it by muscles and tendons. The hind leg is used as an anchor for the trailing edge of the membrane, which means that the limb is not free for effective walking and running. Bats therefore are forced into unusual habits, which include roosting in inaccessible places where they hang upside down. Because bats are placental mammals, they have evolved special adaptations to maintain flight during pregnancy and nursing. For example, the pelvis has features that allow the body to be streamlined yet still have a rather large birth canal.

Figure 14.24 The earliest known fossil bat, *Icaronycteris* from the Early Eocene of Wyoming. The fossil is shown here with the shadow outline of the wings reconstructed. (From the cover of *Science*, December 9, 1966. © 1966 The American Association for the Advancement of Science.)

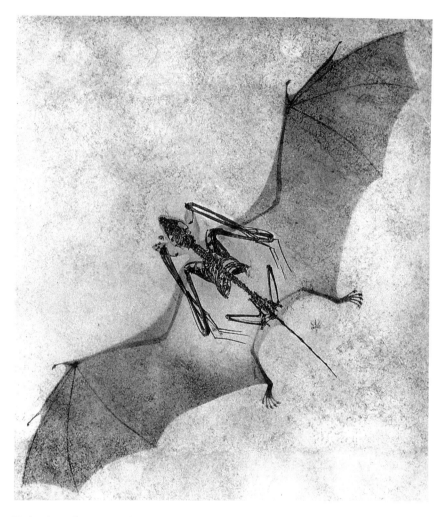

Baby bats have needle-sharp milk teeth that allow them to hold tightly to the mother's fur in flight (ouch at feeding time!).

The earliest known bat, *Icaronycteris*, is an extremely well-preserved fossil from Early Eocene lake beds in Wyoming (Figure 14.24). The Messel Oil Shale, in Middle Eocene rocks of Germany, has yielded dozens of bat skeletons. Some of them still contain the bats' last meals (primitive moths). Even the smallest ear bones are preserved, and they tell us that these bats were already equipped with the echo-locating sonar that all insect-hunting, fishing, and frog-eating bats have today. The baby bats at Messel already had sharp milk teeth.

Current opinion is that bats evolved flight in trees, through a parachuting stage, though there is no evidence at all from bat ancestors, because we have not identified them. Bat sonar presumably evolved from the acute hearing of little, nocturnal, insect-hunting mammals in the forest canopy of the Cretaceous. Obviously, bats must already have had an eventful evolutionary history before the Eocene, but we still have to find these fossils.

Review Questions

Some Permian and Triassic lizards have been reconstructed as gliders. What evidence is there for such a reconstruction?

What were the ancestors of birds? Be as precise as you can.

Which pair of these are the closest relatives? Birds, crocodiles, dinosaurs.

Summarize arguments which suggest that *Archaeopteryx* could fly.

Summarize arguments which suggest that *Archaeopteryx* could not fly.

What did most pterosaurs eat, and how do we know?

How would you tell the difference between a well-preserved bird skeleton and a well-preserved pterosaur skeleton?

What evidence suggests that pterosaurs were warm-blooded?

Over time, three groups of vertebrates evolved powered flight, that is, flight that is not gliding but involves flapping flight that uses the forelimbs. What are they?

If we believe in evolution, we would predict that if dinosaurs evolved into birds, we should find "intermediates," or examples of parallel evolution. Give an example of: a) a dinosaur that evolved at least one bird-like character, and say what the character was; and b) a bird that retained at least one dinosaur-like character, and say what the character is.

Further Reading

More Accessible Reading

Ackerman, J. 1998. Dinosaurs take wing: new fossil finds from China provide clues to the origin of birds. *National Geographic* 194 (1): 74–99.

Averof, M., and S. M. Cohen 1997. Evolutionary origin of insect wings from ancestral gills. *Nature* 385: 627. Genetic evidence from living arthropods.

Balmford, A., et al. 1993. Aerodynamics and the evolution of long tails in birds. *Nature* 361: 628–631. *Archaeopteryx*-type tails make flight worse, not better.

Burgers, P., and L. M. Chiappe. 1999. The wing of *Archaeopteryx* as a primary thrust generator. *Nature* 399: 60–62.

Clark, J. M., et al. 1998. Foot posture in a primitive pterosaur. *Nature* 391: 886–889. *Dimorphodon* could not have run.

Dial, K. P., et al. 1997. Mechanical power output of bird flight. *Nature* 390: 67–70.

Feduccia, A. 1993. Evidence from claw geometry indicating arboreal habits of *Archaeopteryx*. *Science* 259: 790–793. But see comment, pp. 764–765.

Frey, E., et al. 1997. Gliding mechanism in the Late Permian reptile *Coelurosauravus*. *Science* 275: 1450–1452, and comment, p. 1419.

Houde, P. 1986. Ostrich ancestors found in the Northern Hemisphere suggest new hypothesis of ratite origins. *Nature* 324: 563–565, and comment, p. 516.

Kramer, M. G., and J. H. Marden. 1996. Almost airborne. *Nature* 385: 403–404.

Marden, J. H., and M. G. Kramer. 1994. Surface-skimming stoneflies: a possible intermediate stage in insect flight evolution. *Science* 266: 427–430, and further comment, v. 270, p. 1685.

Marden, J. H. 1995. Flying lessons from a flightless insect. *Natural History* 104 (2): 4–8.

Marshall, L. G. 1994. The terror birds of South America. *Scientific American* 270 (2): 90–95.

Padian, K., and L. M. Chiappe. 1998. The origin of birds and their flight. *Scientific American* 278 (2): 28–37, and correspondence, v. 278 (6): pp. 8–8A.

Poore, S. O., et al. 1997. Wing upstroke and the evolution of flapping flight. *Nature* 387: 799–802.

Sanz, J. L., et al. 1996. An Early Cretaceous bird from Spain and its implications for the evolution of avian flight. *Nature* 382: 442–445.

Sanz, J. L., et al. 1997. A nestling bird from the Lower Cretaceous of Spain: implications for avian skull and neck evolution. *Science* 276: 1543–1546, and comment, p. 1501.

Sereno, P. C., and C. Rao. 1992. Early evolution of avian flight and perching: new evidence from the Lower Cretaceous of China. *Science* 255: 845–848. *Sinornis*.

Shipman, P. 1998. *Taking Wing:* Archaeopteryx *and the Evolution of Bird Flight*. New York: Simon and Schuster.

Wellnhofer, P. 1991. *The Illustrated Encyclopedia of Pterosaurs*. New York: Crescent Books. Beautifully illustrated; Wellnhofer's ideas on pterosaurs.

Zimmer, C. 1992. Ruffled feathers. *Discover* 13 (5): 44–54. Sankar Chatterjee and the *Protoavis* affair.

Zimmer, C. 1997. Terror, take two. *Discover* 18 (6): 68–74. *Titanis* and friends.

Zimmer, C. 1998. Into the night. *Discover* 19 (11): 110–115. Bat evolution and echolocation.

More Technical Reading

Brush, A. H. 1993. The origin of feathers: a novel approach. Chapter 2 in Farner, D. S., et al. (eds.) *Avian Biology*, vol. 9. London: Academic Press.

Campbell, K. E., and E. P. Tonni. 1981. Preliminary observations on the paleobiology and evolution of teratorns (Aves, Teratornithidae). *Journal of Vertebrate Paleontology* 1: 265–272.

Caple, G., et al. 1983. The physics of leaping animals and the evolution of pre-flight. *American Naturalist* 121: 455–476.

Chatterjee, S. 1991. Cranial anatomy and relationships of a new Triassic bird from Texas. *Philosophical Transactions of the Royal Society of London B* 332: 277–342. This is the alleged Triassic bird, but see a comment by John Ostrom, *Nature* 353: 212.

Chatterjee, S. 1997. *The Rise of Birds*. Baltimore: Johns Hopkins University Press.

Colbert, E. C. 1970. The Triassic gliding lizard *Icarosaurus*. *Bulletin of the American Museum of Natural History* 143: 85–142.

Cowen, R., and J. H. Lipps. 1982. An adaptive scenario for the origin of birds and of flight in birds. *Proceedings of the 3rd North American Paleontological Convention, Montréal*, 109–112.

Feduccia, A. 1996. *The Origin and Evolution of Birds*. New Haven: Yale University Press. Contains a lot of data and a lot of opinion, and it's often difficult to tell them apart.

Gans, C., et al. 1987. *Sharovipteryx*, a reptilian glider? *Paleobiology* 13: 415–426.

Gatesy, S. M., and K. P. Dial. 1996. Locomotor modules and the evolution of avian flight. *Evolution* 50: 331–340. Theropod tails allowed bird evolution.

Haubold, H., and E. Buffetaut. 1987. A new interpretation of *Longisquama insignis*, an enigmatic reptile from the Upper Triassic of Central Asia. *Comptes rendus de l'Académie des Sciences de Paris, Série II* 305: 65–70.

Jenkins, F. A. 1993. The evolution of the avian shoulder joint. *American Journal of Science* 293-A: 253–267.

Ji, Q., et al. 1999. A new Late Mesozoic confuciusornithid bird from China. *Journal of Vertebrate Paleontology* 19: 1–7. *Changchengornis*.

Kingsolver, J. G., and M. A. R. Koehl. 1994. Selective factors in the evolution of insect wings. *Annual Review of Entomology* 39:425–451.

Kukalová-Peck, J. 1987. New Carboniferous Diplura, Monura, and Thysanura, the hexapod ground plan, and the role of thoracic side lobes in the origin of wings (Insecta). *Canadian Journal of Zoology* 65: 2327–2345.

Novacek, M. J. 1987. Auditory features and affinities of the Eocene bats *Icaronycteris* and *Palaeochiropteryx* (Microchiroptera, incertae sedis). *American Museum Novitates* 2877.

Ostrom, J. H. 1997. How bird flight might have come about. In Wolberg, D. L., et al. (eds.), *Dinofest International*, 301–310. Philadelphia: The Academy of Natural Sciences.

Padian, K., and L. M. Chiappe. 1998. The origin and early evolution of birds. *Biological Reviews* 73: 1–42. Latest major survey, with extensive reference list.

Rayner, J. M. V., and K. A. Padian. 1993. The wings of pterosaurs. *American Journal of Science* 293-A: 91–166.

Rich, P. V. 1980. The Australian Dromornithidae: a group of extinct large ratites. *Contributions in Science of the Los Angeles County Natural History Museum* 330: 93–104.

Sanz, J. L., et al. 1995. The osteology of *Concornis lacustris* (Aves: Enantiornithes) from the Lower Cretaceous of Spain and a re-examination of its phylogenetic relationships. *American Museum Novitates* 3133: 1–33.

Sigé, B., et al. 1998. The deciduous dentition and dental replacement in the Eocene bat *Palaeochiropteryx tupaiodon* from Messel: the primitive condition and beginning of specialization of milk teeth among Chiroptera. *Lethaia* 31: 349–358.

Tintori, A., and D. Sassi. 1992. *Thoracopterus* Bronn (Osteichthyes: Actinopterygii): a gliding fish from the Upper Triassic of Europe. *Journal of Vertebrate Paleontology* 12: 265–283.

Wagner, G. P., and J. A. Gauthier. 1999. 1,2,3 = 2,3,4: a solution to the problem of the homology of the digits in the avian hand. *Proceedings of the National Academy of Sciences* 96: 5111-5116, and comment by Feduccia, pp. 4740–4742.

Witmer, L. M., and K. D. Rose. 1991. Biomechanics of the jaw apparatus of the gigantic Eocene bird *Diatryma*: implications for diet and mode of life. *Paleobiology* 17: 95–120.

CHAPTER FIFTEEN

The Origin of Mammals

Living reptiles and living mammals are very different, with no surviving intermediates, and this requires us to make some mental adjustments as we try to understand how their ancestral counterparts, the diapsids and synapsids, evolved in such divergent ways in the Triassic.

Living mammals suckle their young, and they are warm-blooded—endothermic and homeothermic. They have hair, not scales. They have only one bone along their lower jaw, instead of the reptilian four bones, and their teeth are not replaced continuously during life. Instead, milk teeth are replaced only once, and other teeth, such as the big molars or wisdom teeth, are formed only once. Mammalian teeth meet very accurately and work very efficiently, at the cost of severe problems if teeth are damaged, lost, or worn out.

The three bones that are "missing" from the mammal jaw evolved into the middle ear of mammals, giving mammals particularly acute hearing at high frequency (squeaks). Perhaps more important than anything else, the mammal brain is enlarged and specialized. The forebrain has huge lobes that wrap around older parts of the brain and contain a completely new structure, the **neocortex**, found only in mammals. The parts of the brain that are greatly increased in volume provide improved sensitivity to hearing, smell, and touch, and they are divided into the left and right lobes that psychologists talk about so much.

It's impossible to imagine all these differences arising overnight, but we can see some of them evolving gradually within the therapsids that were the ancestors of mammals. Most of the evidence comes from Gondwanaland, especially from South America and South Africa. The record is richest in jaws and teeth. The dentary bone in the therapsid jaw, originally the small section at the front, came to dominate the jawbone until the rearmost three bones on each side were only little nubbins near the hinge. The teeth became even more differentiated, and, in particular, the teeth behind the canines became larger and more complex in their shape and structure. This may suggest that tooth replacement during life became slower, but that is difficult to judge from the fossil record. Later therapsids evolved the secondary palate, the division between the mouth and the nasal passages that allows mammals, including humans, to breathe and chew at the same time.

In terms of soft parts and thermoregulation, we have no direct evidence and must make indirect deductions. We've already seen evidence for seasonal climates in southern Gondwanaland, suggesting that temperature control of some sort probably evolved among therapsids long before their bony characters became mammalian, and at some point in evolution there must have been warm-blooded hairy reptiles. (Hairy reptiles are not a contradiction to a paleontologist!)

Therapsids were abundant and diverse at the beginning of the Triassic (Chapter 10). But the larger therapsids gradually disappeared, and by the end of the Triassic the survivors were rather small. Cynodonts need special attention because they evolved into mammals.

Cynodonts

Cynodonts (Figure 15.1) were the last major therapsid group to appear in the geological record, in the Late Permian. They are best known from Gondwanaland. At least six groups of carnivorous and herbivorous cynodonts evolved some mammalian characters during the Triassic, and a small-bodied carnivorous cynodont group probably evolved into the first true mammals in the very late Triassic. I give only a general account of the evolution of mammalian characters in cynodonts, to show that the changes were gradual ones that produced more efficient cynodonts.

Evolving Mammalian Characters

There's a paradox about the evolutionary transition from therapsid to mammal: it is too well known and complete. Everyone agrees that the therapsids are a clade, that cynodonts are a clade within therapsids, and that mammals are a clade within cynodonts (Figure 15.1). But there is

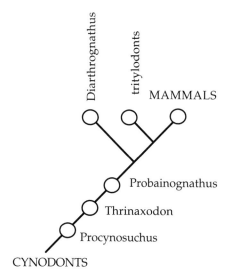

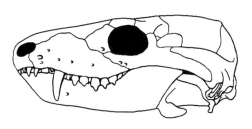

Figure 15.1 Above: Phylogram to show the probable relationship between various cynodonts. Below: The skull of a representative cynodont, *Thrinaxodon* from the Lower Triassic of Gondwanaland. Skull about 8 cm (3 inches) long. The small size and close packing of the structures around the jaw joint would allow substantial changes without major shifts in bony components. (Skull after Hopson and Jenkins.)

controversy over which therapsid was actually the first mammal (and which animals the clade Mammalia includes). Everyone agrees that the clade Mammalia should be marked by the acquisition of some novel, derived, "mammalian" character, but which one? Mammalian characters did not all evolve at once.

Everyone has their own favorite branch point on the cladogram. I accept as "mammals" those Late Triassic cynodonts and their descendants in which a projection on the dentary bone fits into a socket on the squamosal bone, to form a hinged jaw. These animals have been identified as mammals by most paleontologists for several decades. Now we should look at the evolution of the characters that define them as mammals.

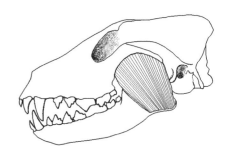

Figure 15.2 The masseter muscle is set into the angle of the jaw in mammals.

Jaws

The secondary palate, which allows chewing and breathing at the same time, evolved in other therapsids as well as cynodonts. However, cynodonts evolved a key innovation involving the rearrangement of the jaw: the **masseter**, a large muscle that runs from the skull under the cheekbone to the outer side of the lower jaw (Figure 15.2). In living mammals it is the most powerful muscle that closes the jaw. (Put your fingers on the angle of your own jaw, clench your teeth and relax again, and you will feel the masseter at work.) The evolution of the masseter had several important effects.

First, jaw movements were easier to control and could become more precise and complex. There was much more accurate lateral and back-and-forward movement of the lower jaw in chewing. Second, biting became more powerful. Third, the force of the bite was transmitted more directly through the teeth rather than indirectly by leverage around the jaw hinge. The lower jaw was slung in a cradle of muscles, and stresses acting on the jaw joint during chewing were much reduced.

In reptiles, the lower jaw is made of several bones, but as chewing efficiency improved, the dentary bone took up more and more of the lower jaw. The other bones became smaller and were crowded back towards the jaw joint. Stresses on the jaw joint itself were reduced as the masseter evolved, and the three bones behind the dentary on each side became specialized for transmitting vibrations to the stapes rather than strengthening the back of the jawbone. Eventually the dentary became the only bone in the lower jaw and the others became part of the ear.

As this happened, the jaw joint was gradually remodeled. In reptiles the jaw is hinged between the articular and quadrate bones, but in typical mammals the jaw hinges between the dentary on the lower jaw and the squamosal bone of the upper jaw. Many people have worried about the apparent jump of the jaw joint from one pair of bones to another, since evolution is a gradual process. All the relevant bones in cynodont skulls were small and close together, however, allowing a major structural shift without major displacement of the jaw hinge (Figure 15.3).

Probainognathus, from the Middle Triassic of South America, is very close to the cynodont ancestor of mammals (Figure 15.3). Later

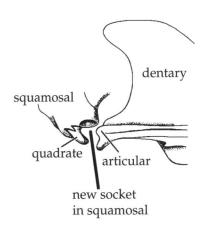

Figure 15.3 The structure of the back of the jaw in the advanced cynodont *Probainognathus*, showing how small a transition would be needed to change the hinge from the articular and quadrate, as normal in reptiles, to the dentary and squamosal, as normal in mammals. (After Romer.)

changes still needed to complete the transition included smaller size; completion of the change in jaw structure to hinge only on the squamosal and dentary; completion of the middle ear from the three "excess" bones on each side of the lower jaw; enlargement of the brain; formation of definite premolars and molars in the jaw and reduction of tooth replacement to only two sets; better sculpture of the molars, with the mammalian jaw movements that go with it; and changes in the backbone that made it more flexible in curling up during mammalian springing and hopping. None of these changes would have been difficult or unlikely.

Teeth and Tooth Replacement

Cynodonts had teeth as well differentiated as those of many later mammals. They had complex, multi-cusped teeth behind the canines, which implies more complex food processing than in other therapsids. The jaw changes gave greater biting forces near the hinge and smaller errors in occlusion. The teeth themselves, meeting their counterparts accurately, came to be exquisitely sculptured to perform their functions precisely. Different cynodonts, presumably with different diets, evolved shearing, crushing, or shredding actions. *Procynosuchus*, an early cynodont (Figure 15.1), may have been the first therapsid to chew insects rather than crush them and swallow them whole. Shearing is well seen in later carnivorous cynodonts, and there may have been limited self-sharpening of the teeth. Among herbivores, the teeth were organized for crushing; even here, slightly worn (self-wearing) opposing surfaces made a better crushing surface than new tooth surfaces. Look at a newly exposed permanent tooth of a child to see how irregular an edge it has when it first erupts.

Reptiles replace their teeth often during life, and although the process has some systematic pattern to it, any adult reptile has a mixture of larger, older teeth and smaller, newer teeth along its jaw. This means that top and bottom teeth cannot be relied upon to meet precisely against one another, so that tooth functions are comparatively crude. In advanced cynodonts, however, the jaw was slung in a rearranged set of muscles so that jaw control was more precise; the teeth also show precise occlusion between top and bottom jaws. Tooth replacement must have been more controlled and less frequent among cynodonts than in normal reptiles, and the fossil record confirms that. Cynodont teeth were replaced precisely, to maintain good occlusion of different, specialized teeth along a growing jaw. Thus the molarlike teeth of young animals were replaced by canines, while new molarlike teeth were added to the back of the jaw.

Hearing

Like its ancestors, the early cynodont *Procynosuchus* had a hearing system that transmitted ground-borne vibrations through the forelimbs and shoulder girdle to the brain, by way of the bones of the lower jaw and a massive stapes. As therapsid feeding came to emphasize chewing and slicing, it was important for teeth to be arranged all the way along the jaw (actually along the dentary bone), far back toward the hinge. The three bones on each side of the jaw behind the dentary became smaller, and so did the stapes, especially as therapsid body size became smaller. The hearing system evolved to detect and transmit airborne sound, and the posterior jaw bones evolved into the bones of the middle ear.

Clearly, airborne sound was increasingly important to late cynodonts and early mammals. Perhaps they hunted insects at least partly by sound. The middle ear bones were linked to the jaw in very early mammals, but later they came to be suspended from the skull. As the hearing pathway was separated from the jaw, the mammal no longer had to listen to its own chewing so much, and would have had much better hearing. If we define mammals as animals that have a single dentary bone forming the lower jaw, then it took some time after mammals first evolved to reorganize the other three bones into the "mammalian" middle ear. As far as we can tell, only advanced mammals evolved the complex spiral inner ear, rather late in their history.

Brains

The huge increase in brain size and complexity between advanced cynodonts and early mammals occurred at the same time as the changes in the jaw and ear structure. Tim Rowe has suggested that these changes were connected. Essentially, he says, the growth clock was reset, allowing the brain to grow for much longer than the structures around it. As the skull and jaw adjusted to accommodate a bigger brain, interesting changes could occur. Rowe's suggestion is based on the fact that in the living opossum, the ear bones reach adult size after three weeks, while the brain grows for twelve weeks.

Rowe's suggestion does not explain the changes in jaw and ears, but it sets up an evolutionary situation in which the changes could happen. It provides an ecological and/or behavioral context in which a relatively large, more complex brain evolved, and it encourages us to ask why such a brain would have been important to an early mammal.

Locomotion

Cynodonts still had wheelbarrow locomotion (Chapter 10): the hind limbs provided propulsion while the forelimbs gave only passive support. Cynodont hind limbs evolved to become semi-erect, whereas the forelimbs remained sprawling (Figure 15.4). The change in the hind limbs brought the feet closer together, and the ankle changed enough to give more direct propulsion along the line of travel. Some improvement in the shoulder joints allowed better locomotion, but it was only a better wheelbarrow style. The spine shows adaptations toward greater stiffness, so that power was transmitted more efficiently from the hind limbs. Later cynodonts evolved more flexible neck vertebrae, for example, so that the

Figure 15.4 The cynodont *Thrinaxodon*, showing the limb structure. (Based on skeletal reconstruction by Farish Jenkins.)

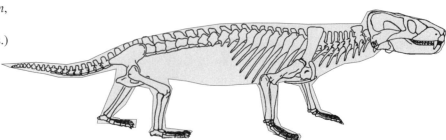

head could swivel freely on the stiffened body. Even with these changes among cynodonts, erect limbs were not evolved by the earliest mammals but came much later.

Thermoregulation and Metabolic Level

Because their jaws and teeth show such an emphasis on efficient food processing, cynodonts probably had higher metabolic rates than pelycosaurs. This does not mean that cynodonts reached the metabolic levels of modern mammals, especially as their limbs (especially the forelimbs) were semi-erect at best. The spine of therapsids still flexed laterally rather than up and down (though note the strangely widened ribs on *Thrinaxodon* in Figure 15.4, which perhaps were retrofitted devices that cut down on lateral flexing).

Several lines of evidence suggest that therapsids, and cynodonts in particular, were evolving toward endothermy. Mammals have evolved a **diaphragm** as an important part of the breathing system. This sheet of muscle forces the lungs to expand, helping respiration. A diaphragm can work only when there are no ribs around the abdomen, so its evolution can be detected in fossil vertebrates. It seems to have evolved within the cynodonts, which lost their abdominal ribs (*Thrinaxodon*, Figure 15.4).

W. J. Hillenius examined hundreds of therapsid skulls for evidence of the presence of the turbinates that indicate high resting respiration rates in living air-breathing vertebrates (Chapter 13). It is clear that pelycosaurs did not have them, but there is fragmentary evidence of turbinates in a few South African therapsid skulls from the late Permian—from four species, to be precise, one therocephalian and three cynodonts, including *Thrinaxodon*.

Primitive mammals today tend to have comparatively low metabolic levels, and they thermoregulate at temperatures far below those of most mammals. Therapsids were mostly medium-sized, with stocky bodies. Perhaps they operated at a body temperature of 28°–30°C (82°–86°F), only a little less than primitive mammals today. In other words, they could have been moderately warm-blooded, with at least some primitive kind of thermoregulation.

Remember that turbinates are nothing to do with maximal performance: they conserve water during "normal" nose breathing. When we need to run, we run, we breathe through our mouths, and we don't care about water loss because we care more than anything else about running.

We may not be running for our lives as often as our ancestors did, and we don't usually run to catch our next meal, but we temporarily forget about water conservation.

I suggest, then, that therapsids evolved some sort of endothermy but they did not evolve great locomotory performance. They improved their breathing enough to maintain a fairly high basal metabolic rate (diaphragm, turbinates, perhaps the ribs of *Thrinaxodon*), but they were not erect athletes the way that dinosaurs were, and they could not support sustained high speed because of Carrier's Constraint. If this suggestion is true, therapsids evolved limited endothermy without solving Carrier's Constraint (their descendants the mammals solved both). Dinosaurs solved Carrier's Constraint without evolving endothermy (their descendants the birds solved both). Birds and mammals have converged on endothermy and high performance from completely different pathways. Therapsid physiology differed dramatically from that of living mammals, from that of living reptiles, and from that of dinosaurs.

Other Mammalian Characters

The bones of the cynodont snout have holes and grooves that suggest important blood vessels and nerve canals. Evidence from the early cynodont *Procynosuchus* suggests tight-fitting skin on the snout except immediately around the mouth. Perhaps there were well-developed lips in *Procynosuchus* and later cynodonts, to go with the extra chewing inferred from the teeth and jaws. A well-developed snout blood supply might suggest important sensory organs such as whiskers and noses.

Procynosuchus had lower incisors arranged in a horizontal comb. A similar arrangement occurs in living lemurs, who use the incisors to groom the fur of other members of the troop. If that was true of *Procynosuchus*, it suggests strongly that all cynodonts had hair or fur.

Mammalian Reproduction

The major biological differences between living reptiles and living mammals are not in the skeleton, but in other characters. Reptiles have large eggs with a large energy store, and their young hatch as independent juveniles capable of living without parental care. Mammals have small eggs, and their young depend on parental care. Other major differences are physiological: most living mammals have high body temperatures and hair to insulate them, while reptiles lack hair and are cold-blooded.

Small, warm-blooded animals have a high ratio of body surface to volume, so in extremes of temperature they tend to heat up and cool down too quickly. This is especially true for young (tiny) individuals. If therapsids were warm-blooded, how did they deal with this problem? And how might the problem bear upon the origin of mammals, especially in view of the fact that the earliest mammals were tiny (smaller than mice)? The problem is compounded by the fact that tiny, warm-blooded animals must find and eat very large quantities of food compared with their body size.

We can find more clues from living small, warm-blooded vertebrates —the birds. Many nestling birds are helpless and cold-blooded. They depend on their parents for food and for warmth, and they have very low metabolic rates. But because they do not have to find their own food to keep warm, nestlings can devote all their food intake to growth. Helpless nestlings have very large digestive tracts for their size. Warm blood, temperature control, and the ability to make coordinated movement come later and gradually. This strategy avoids the energy problem of warm blood at small size, and nestlings are essentially cold-blooded until they have grown to considerable size. Furthermore, most birds cut down environmental temperature fluctuations in their nestlings by caring for them in nests designed to maintain a uniform temperature. But the system demands intensive care by one or both parents.

Most likely, some similar strategy was followed by late therapsids and early mammals, but in burrows rather than nests. The little therapsid *Diictodon* was digging burrows by the end of the Permian (Chapter 10). As therapsids evolved to very small size in the Late Triassic, the need for parental care would have become more and more acute. As the pelvis became smaller, eggs would necessarily have become smaller and smaller, with less and less yolk, and the young would have hatched earlier and been more helpless. Although freed from the anatomical problem of laying large eggs, the parent(s) were committed to providing a steady supply of food to the young after hatching, like birds and unlike most reptiles. On the other hand, smaller eggs and the rapid growth of helpless hatchlings gave an opportunity for very rapid reproductive rates in closely spaced litters (or clutches).

Suckling

Living monotremes still have the kind of reproduction that we infer for advanced cynodonts and early mammals. The duckbilled platypus lays and hatches tiny eggs in a nest inside a burrow. Monotremes also nourish their hatchlings by suckling, rather than collecting food for them. This behavior has advantages: the parent does not have to leave the hatchling to search for suitable food for it, because any normal adult food can be converted into milk. The hatchling digests milk easily, and its parent is never far away, providing protection and warmth.

Charles Darwin suggested how suckling might have evolved in mammals, even before Western science discovered monotremes. His theory survives with only minor modifications. Let's assume that mammalian ancestors were already caring for eggs by incubating them. A special gland may have secreted moisture to keep the eggs humid during incubation. Hatchlings that licked the incubation gland benefited by gaining water to help deal with the food brought back by the parents, and perhaps the secretions had the added advantage of being antibacterial. The adaptation was strengthened as long as the fluid helped hatchlings to survive and grow. Gradually, as the secretions came to contain mineral salts and trace elements and then nutritious organic compounds (milk) as well as water, the mother's excursions for food could be reduced and the hatchlings benefited even more by her increased attendance. Rapid evolution of full lactation from specialized nipples followed, with efficient suckling by the hatchlings.

The mammalian system is interesting because only the female parent is specialized to have milk glands, so that the male may take little or no role in caring for the young. Male mammals have nipples, of course, and there is no obvious biochemical reason why baby mammals should suckle only from the female, so the reason is probably genetic.

It turns out that the development of the milk glands in mammals is controlled by a set of the Hox genes that are universal among metazoans, typically laying out nerves, vertebrae, segments, limbs and other body systems. Almost certainly, the lactation system is switched on, under genetic control, as the female goes through pregnancy and delivery. The switching system is complex, and has components from three of the four separate Hox gene clusters that mammals carry. Even in females, there are mutations that upset this complex system. Since males do not go through pregnancy, they would not generate the appropriate signals to switch on the lactation genes.

The development of suckling can be dated indirectly. Cynodonts had a secondary palate and could chew while still breathing, but even the tiniest baby cynodont had teeth and so probably did not suckle. Perhaps the parent brought food to the nest or into the burrow. But the first mammals had very limited tooth replacement, possibly related to their small size and short lifespans, and they probably suckled in some fashion. The evolutionary transition from licking to suckling was not as simple in baby mammals as it might seem—suckling demands full and flexible cheeks. Cheeks must have evolved, along with many other mammalian characters, among Triassic cynodonts. Certainly some sort of cheek would have been needed to cover the newly evolving masseter muscle.

> Early mammals suckled their brood
>
> They breathed in and out as they chewed,
>
> Their molar tooth facets
>
> Were masticatory assets,
>
> But their locomotion was crude.

Live Birth

Suppose early mammals had reached the point of being reproductively mammalian, perhaps like monotremes: they laid eggs but suckled their hatchlings. What would cause or encourage viviparity—the evolution of live birth?

There is nothing unusual about live birth. It has evolved independently many times in fishes, amphibians, reptiles, and mammals: in fact, in every living vertebrate group but birds. It has evolved independently in at least 90 different groups of lizards and snakes, and some insects have evolved it too. But how did it evolve in the earliest mammals?

Laying eggs is a difficult proposition below a critical size of mammal. The egg must be laid through a pelvic opening, and a shelled egg with a reasonable amount of yolk must have a certain minimum size to be viable. For a very small mammal to lay one or more shelled eggs might demand such flexible pelvic bones that other functions, such as agility or fast running, might be jeopardized. (Flying birds may not share that problem.)

Constraints on the pelvis that would forbid laying a large, shelled egg may not apply to a fetus, which is structurally and physiologically more flexible than an egg. A fetus does not need a yolk or shell during its development; it can be squeezed through a birth canal more safely than can an eggshell. Inside a warm-blooded mother, the fetus develops at a more uniform temperature than in a nest, and enzymes for growth and digestion that work best at the mother's body temperature can be present throughout life. The growing fetus has an unlimited supply of water and oxygen, and an easy way of getting rid of CO_2 and other wastes, all of which are problems for an embryo inside an eggshell. There is far less chance of predation or infection. Finally, if suckling has already been evolved, the young never need be separated from the care and protection of the mother, even if they are helpless at birth.

Egg-laying monotremes survive today, proving that viviparity is not essential for mammals in spite of the list we have just compiled. But the dominance of viviparity among modern mammals suggests that it evolved early in mammalian history. The necessary steps would include the gradual evolution of ways to transport material between mother and fetus, the beginnings of the placenta. The first viviparity would have been on the marsupial pattern, with or without a pouch to contain the young, but it need not have been as specialized a process as it is in living marsupials. We do not know when mammals evolved live birth, but indirect evidence suggests that it took place in the Cretaceous.

The Invasion of the Night

If cynodonts were moderately warm-blooded, their evolution to smaller size would almost automatically have produced adaptations such as insulation, parental care, and so on. But why should they have evolved to smaller size? Given our ideas about dinosaur biology and physiology (Chapter 13), it was probably because of competition from late thecodonts and early dinosaurs, which may not have evolved endothermy but certainly had solved Carrier's Constraint. Ecologically squeezed between the first dinosaurs (fast-moving predators with sustained running) and the small, lizardlike reptiles of the Triassic (running on cheap solar energy with a low resting metabolic rate), the later cynodonts may have escaped extinction only by evolving into a habitat suitable for small, warm-blooded animals and no-one else: the night. In doing so, they underwent the radical changes in body structure, physiology, and reproductive pattern that now mark them as a separate class of vertebrates, the mammals.

Burrowing in the dark, the first mammals were invading an underground habitat that required much greater sensitivity to hearing, smell, and touch. This requirement may have selected for the relatively large, complex brain that did them no good against the dinosaurs, but in the end allowed them to emerge as the dominant large land animals after the dinosaurs were gone.

Early Mammals

The first mammals were tiny, and their fossils are rare and difficult to collect except by washing and sieving enormous volumes of sediment. But after years of effort we now have fragments of early mammals (mostly teeth) from many localities in many continents, all lying close to the Triassic-Jurassic boundary.

The family Morganucodontidae, named after *Morganucodon*, could be ancestral to many Mesozoic mammal groups that are now extinct. A family named after *Kuehneotherium* may be ancestral to all other mammals, including the three surviving groups, monotremes, marsupials, and placentals. Other fragmentary fossils are puzzling. Current research on early mammals may soon provide more details on this part of the story.

Morganucodonts

Morganucodonts are fairly well known from two nearly complete skeletons found in southern Africa (Figure 15.5). They were small, perhaps only 10 cm (4 inches) to the base of the tail, and only about 25 g, or 1 ounce, much like modern shrews. They had small but nasty jaws and were obviously little carnivores, probably eating insects, worms, and grubs. They had longer snouts and much larger brains than cynodonts. The skeletons show that they were agile climbers and jumpers. The neck was very flexible, as in living mammals, and the spine could have flexed up and down in addition to the lateral bending of therapsids.

The jaw joint was still like that of late cynodonts. But the teeth were fully differentiated into mammalian types, and the molars had double roots. As in most mammals today, the front teeth were replaced once, and there was only one set of molars.

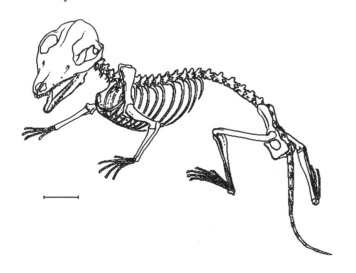

Figure 15.5 A morganucodont. The scale bar is 2 cm. (After Jenkins and Parrington, 1976, and Jenkins, 1984.)

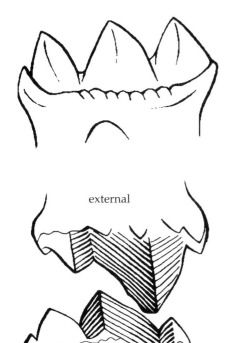

external

internal

Figure 15.6 Triconodont teeth have three cusps in a row (above, after Simpson). They had an action rather like that of pinking shears (below, after Jenkins, 1984).

The molar teeth had three cusps in a line, so the name **triconodont** has been used for the group of early mammals that contains morganucodonts. Triconodont molars worked by shearing vertical faces up and down past one another (Figure 15.6), giving a zigzag cut exactly like that of pinking shears in dressmaking. This is efficient, especially for thin or soft material, but requires precise up-and-down movement.

Some Late Jurassic mammals called **docodonts** had the cutting edges of the molars arranged at a slightly oblique angle, so their most effective cutting action required a slightly sideways jaw motion as well as up and down. This improvement gave a crushing component as well as a cutting action. Other skull features of docodonts look more primitive than those of morganucodonts, however, and they may very well have had much earlier Triassic ancestors that are still to be found.

Triconodonts and docodonts were successful into Early Cretaceous times. The docodont *Gobiconodon* evolved to be the size of a possum. *Jeholodens* is a strange, newly discovered triconodont from China, which has sprawling hind limbs but erect forelimbs. However, all living mammals probably descended from the other Late Triassic family, the kuehneotheriids.

Kuehneotheriids

Kuehneotheriids are not as well known as morganucodonts because we have found only their jaws and teeth. They were also small and shrewlike, and they retained a complex contact at the jaw hinge. But they had more complex molar teeth, so that food was trapped, squeezed, and sliced into small pieces from two directions at once as the molar cusps passed one another. The efficient operation of these teeth depended on precise engineering and on coordinated cutting and chewing jaw action.

Jurassic mammals are poorly known, but cladistic analysis suggests that several major groups evolved during that period. Two of those groups survive: the monotremes and the therians (which include marsupials and placentals). Symmetrodonts are a small extinct group of early therians, and multituberculates are a larger extinct group. The relationships among these groups are still under debate.

Monotremes

Steropodon and *Kollikodon* are the earliest known monotremes. They are each known from one partial jaw, found in an opal mine in Cretaceous rocks at Lightning Ridge in New South Wales, Australia. Each narrowly escaped being made into jewelry! Monotremes had been considered as very primitive mammals, so it is reassuring to have found such ancient fossils. However, it may be that the monotreme lineage is older still, and we can look forward to finding Jurassic monotremes some day.

Multituberculates

Multituberculates are small mammals that evolved superficially rodent-like teeth, and were successful in the Late Jurassic, Cretaceous, and early Cenozoic. They often made up more than half the mammals in Late Cretaceous faunas. Multituberculates survived the great extinction at the end of the Cretaceous and reached their greatest diversity in the Paleocene, before being replaced by more modern-looking mammals, especially the true rodents. They form a distinct clade of mammals. Multituberculates are sometimes called the "rodents of the Mesozoic," but their ecology may not be so simple to reconstruct. Their origin is controversial: some paleontologists claim that their direct ancestors were a Triassic family, the Haramiyidae, but that seems very early indeed for the separation of multituberculates from other mammals.

The incisor teeth of multituberculates were usually specialized for grasping and puncturing, rather than gnawing, but there were six in the upper jaw and only two in the lower. The very large, sharp-edged premolars were designed for holding and cutting, while the molars were grinding teeth. The system looks well suited for cropping and chewing vegetation with a back-and-forward jaw action (Figure 15.7). Although their radiation corresponds with the general rise of the flowering plants, many multituberculates were probably omnivores, like rats rather than guinea pigs. Specific forms can be interpreted more precisely. Some incisors were ever-growing and self-sharpening, well designed for gnawing. (Gnawing teeth may not have evolved for chewing nuts and seeds, but to open up wood to get at insects.) Other multituberculates had long, thin, saberlike incisors, like some modern insectivores that use them to impale insects. Still others probably used the shearing premolars and the crushing molars to eat fruits or seeds (Figure 15.7). The range in body size (mouse- to rabbit-sized) indicates a fairly wide ecological range among multituberculates. Some later forms from the Early Cenozoic were clearly tree dwellers. *Ptilodus* had a prehensile tail and squirrel-like hind feet that could rotate backwards for climbing downward (Figure 15.8).

Kryptobataar, from the Late Cretaceous of Mongolia, was an important multituberculate because we can say confidently that it had live birth. It had a narrow, rigid pelvis that was incapable of widening during birth. Thus the birth canal would have been at most only 3 to 4 mm wide. The animal could not have laid any reasonable-sized egg, but it could have borne a very small fetus (newborn marsupials weigh about 1 gram).

Monotremes and multituberculates both lack the advanced molar teeth of placentals and marsupials. The ear structure of monotremes and multituberculates is similar. Both of them have a well-developed middle ear, but neither has an ear that is good at hearing high-frequency sounds. At least some multituberculates had a pivoting clavicle (collarbone) to brace the shoulder , even though they did not have a fully erect forelimb. In contrast, monotremes still have a slight sprawl. In other words, multituberculates and symmetrodonts are probably more closely related to living marsupials and placentals than monotremes are (Figure 15.9).

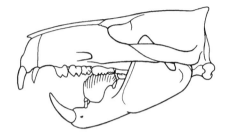

Figure 15.7 The teeth of a multituberculate, *Ptilodus* (see also Figure 15.8). (After Simpson, Krause, and Kielan-Jaworowska.)

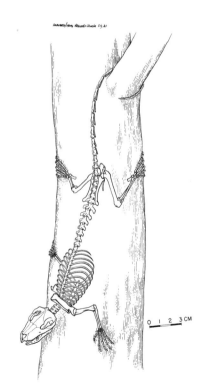

Figure 15.8 Reconstruction of *Ptilodus*, a tree-climbing multituberculate from the Early Cenozoic. (Courtesy of David Krause, SUNY Stony Brook.)

Figure 15.9 The best current hypothesis for the evolution of advanced mammal groups. <u>Zhang</u> is *Zhangheotherium*, a newly described fossil from the Jurassic-Cretaceous boundary in China; and <u>Delta</u> is *Deltatheridium*, a basal marsupial from the Cretaceous of Mongolia. Given the age of *Zhangheotherium*, we ought to find Jurassic monotremes some day.

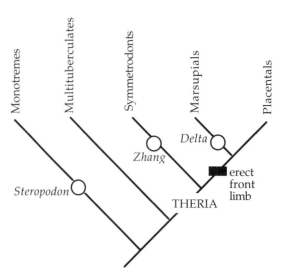

Therian Mammals

Therian mammals are more advanced, and include the common ancestor of living marsupials and placentals, with some early related forms.

Symmetrodonts

Symmetrodonts are Jurassic and Cretaceous mammals. In 1998 the first complete symmetrodont, *Zhangheotherium*, was described from the same remarkable rocks in China that also yielded feathered theropod dinosaurs (Chapters 12, 13), many early birds (Chapter 14), and the earliest flowering plant (Chapter 16). *Zhangheotherium*, like all early mammals, was small, only a few inches long. Its ear structure was primitive, and it had sprawling front limbs, though its shoulder joint was beginning to show advanced characters. All this new information allows us to understand symmetrodonts as mammals that are basic therians, without the more advanced characters of marsupials and placentals (Figure 15.9).

The Tribosphenic Molar

Jurassic mammals were probably all small and nocturnal, and carnivorous, insectivorous, or perhaps omnivorous: only a few had teeth that could chew up fibrous vegetation. With the spread of flowering plants in the Early Cretaceous, herbivorous dinosaurs, insects, and mammals all increased in diversity. The increase in food in the form of insects, seeds, nuts, and fruits provided a great ecological opportunity for small mammals, and a new type of molar tooth was invented. Tribosphenic molars are found in advanced **therians**—marsupials and placentals. They are complex in shape and can perform a large variety of functions as upper and lower teeth interact. They evolved from simpler teeth by adding new surfaces that shear past one another as the jaw moves sideways in a chewing motion. Tribosphenic molars are particularly well suited for puncturing and shearing, and especially for grinding, superbly fitting mammals for a diet of insects and high-protein seeds and nuts.

However, there is a problem. The tribosphenic molar is so efficient a shape for its function that it has probably evolved more than once. An early mammal may have molars that look tribosphenic, but in fact are subtly different. This makes for confusion and controversy. For example, an Early Cretaceous mammal from Australia was given the awesome name of *Ausktribosphenos* because of the tribosphenic shape of its molar teeth. It is early in time and very far geographically from other therian mammals. It probably does not have genuine tribosphenic molars, despite its name, but molars that evolved much the same shape. This leaves it as a curiosity in early mammal history, but not a contradiction.

Marsupials and Placentals

Early mammals probably reproduced by delivering small, helpless young once they had evolved beyond the monotreme stage of egg-laying. The divergence of mammals into separate marsupial and placental clades probably took place early in the Cretaceous. Each style of reproduction, in its own way, solves some of the problems of the mammalian way of life. Marsupial and placental styles of reproduction are now quite distinct, but they probably both evolved from a state that we would now identify as simple but largely marsupial. Even so, it's usually impossible to tell whether any given Cretaceous mammal was monotreme, marsupial, or placental in its reproduction, especially when the pelvic regions are not well preserved.

Even today more than 90% of all mammals weigh less than 5 kg (11 pounds) as adults. All small mammals give birth to tiny helpless young, probably because they do not have enough body volume to pack into their babies all the requirements for fully independent mammal life. Tiny mammal babies are cold-blooded at first and absolutely dependent on parental care. Mammals with large bodies can accommodate and give birth to larger, more competent offspring. This factor may have been the key to the success of the large placental mammals as opposed to large marsupials, but it doesn't apply to small placentals and marsupials.

Living marsupials bear fetuses surrounded by a membrane like the eggshell membrane of a bird or a monotreme. Its most important component is the **trophoblast**, a cell layer that allows very close contact between fetal and maternal tissue yet prevents the passage of substances that would cause the mother to reject the foreign body growing within her. Only a limited amount of nutrition can be passed to the growing embryo from the mother, and after a certain gestation time it is better for the fetus to be born so that it can take nutrition more efficiently by suckling. Living marsupials, therefore, have short gestation periods followed by long suckling periods, often with the young in a pouch.

Sometime early in the Cretaceous, a line of small mammals evolved a new derived character, the true **placenta**. This is a specialized structure built in the uterus jointly by the fetus and the mother. The placenta has an enormous surface area (fifty times the skin area of a newborn human), and it is used to supply the fetus with nutrition, oxygen, and hormones, and to pass waste products from the fetus to the mother for disposal. Essentially, the placenta is a large, discriminatory, two-way pump. The trophoblast of placental mammals is much more effective than it is in marsupials, allowing the placenta to support a growing fetus much longer. As a result, placental mammals can evolve a long gestation period, so they can have shorter lactation periods before the young reach a stage where they are independent of the mother.

Marsupials have never evolved a placenta or a trophoblast as efficient as that of placental mammals, so they cannot supply the fetus with all its needs past a certain stage of development. Their trophoblast separates mother from fetus but allows only a limited range of materials to pass between them. As a result, marsupial newborns are fetuses that must be agile enough to reach the nipple.

None of this means that marsupials are inferior to placentals. A marsupial mother who experiences a natural crisis can easily abandon her young while they are fetuses, because she already carries them as an external litter. She may be ready to breed again quickly. A few placentals can absorb their fetuses, but most placental mothers must carry their internal young to term for a comparatively long gestation period, even during a flood, drought, or harsh winter, often at the risk of their own lives. Marsupial females can delay fetal development after implantation, whereas placental females rarely can. The marsupial reproductive system stresses flexibility in the face of an unpredictable environment, so it may sometimes be superior to the placental system. Native marsupials and introduced placentals of the same body weight in the same environments in Australia (wallabies versus rabbits, for example) take on average about the same time to rear their young successfully. We still don't know the relative energy cost of the two methods. Placental and marsupial styles of reproduction, each in their own way, reduce the hazards of rearing young at small body size, but one is not always more efficient than the other.

Note that the flexibility of marsupials in abandoning their young is comparable with that of birds, who may abandon a nest in a crisis, even if there are eggs or young in it. Herons and storks will abandon a single chick if there is enough time left in the year for them to start another clutch of eggs that gives them a greater chance of rearing several chicks. A principle called the **Concorde Fallacy** seems to operate in human affairs—if a great deal has been invested in a project, then a great deal more will be invested in order to see it through to the bitter end, even after it is clear that the project will never repay its cost. The supersonic Concorde airliner is one case, but there are many others, such as the Vietnam War and the Space Shuttle. Animals operating under natural selection cannot afford to waste anything and must be ruthless in cutting their losses as soon as they detect eventual failure. Lions and cheetahs should abandon the chase as soon as they see they cannot catch their prey, and prospective parents should abandon their young if they cannot be reared successfully. In these terms, the allegedly superior placental reproductive system is more likely to result in wasteful expense than either the egg-laying of birds or the marsupial system. It is simply a bigger gamble than the others. In the long run, the three methods must be about equal in their results, because different animals practice them all successfully.

Other major differences between marsupials and placentals today are in thermoregulation and metabolic rate. Size for size, placentals thermoregulate at slightly higher temperatures and have slightly higher metabolic rates. They are "faster livers," as one writer has put it. This need not affect reproduction, because female marsupials increase their metabolic rate during pregnancy and lactation, up to placental levels. It is true that the brain grows faster in fetal placental mammals than in marsupials, and there is a small but significant difference in adult brain size, weight for weight, between the two groups. In turn, the metabolically active brain uses more oxygen in placentals, partly accounting for their higher energy budget. In spite of the metabolic differences, however, there is no systematic difference in at least one vitally important aspect: locomotion. Marsupials can run at about the same maximum speeds as equivalent placentals, and they have about the same stamina.

Small marsupials and small placentals were clearly separate groups by Late Cretaceous times. The little mammal *Deltatheridium* from Mongolia is the most complete early marsupial, though the lineages separated earlier in the Cretaceous. At first marsupials and placentals would not have been greatly different ecologically. Early placental mammals would still have had tiny, helpless young. The evolution of precocious young such as colts, calves, and fawns, which are large and can run soon after birth, had to wait until placental mammals reached large size; not until then did the placental mammals become more successful in their distribution and diversity. Placental mammals may well have little or no advantage over marsupials when both are small, but large precocious young are not an option for marsupials, while they are for placentals.

The Inferiority of Mammals

The first mammals were vicious, shrew-sized carnivores. They certainly did not roam about in the open and certainly did not dominate the landscape as their Permian ancestors the therapsids had. The earliest mammals probably skulked around at night eating insects and grubs.

What went wrong? If mammals are so dominant today, why weren't they in the Late Triassic? If being a mammal is so progressive and advanced and superior, why wasn't the invention of various mammalian characters one after another a Good Thing?

It may have depended on the competition. By the end of the Triassic, thecodonts had replaced and probably outcompeted the therapsids, driving them underground, deep into forests, or into nocturnal habits all over the world. And as the last few therapsids became extinct or were confined to tiny body size, some thecodonts evolved into one of the most spectacular vertebrate groups of all time, the dinosaurs.

Mammals increased in diversity through the Cretaceous, but not in a spectacular way. Their moderate increase in diversity is correlated with the evolution of the flowering plants and with the insects that diversified with them. Forest canopy ecosystems had flourished since Carboniferous times (Chapter 9). Mesozoic mammals, small-bodied and insectivorous, were clear candidates to have invaded canopy ecosystems. By the end of the Cretaceous, with angiosperms well established, it is easy to envisage a diverse set of mammals occupying many small-bodied ways of life in the forest, particularly at night. The ancestors of primates and bats most likely evolved their special characters in the canopy. Small mammals are very important in the canopy even today: the equatorial forest has many species of birds active by day and mammals at night, each occupying a small-bodied way of life, eating insects, seeds, nuts, and fruit.

Immediately after the end of the Cretaceous and the disappearance of the dinosaurs, mammals began a tremendous radiation into all body sizes and many different ways of life. The inverse relationship between the success of Mesozoic archosaurs, especially dinosaurs, and Mesozoic therapsids and mammals is probably not a coincidence. It reflects some real inability of the mammalian line to compete successfully in open terrestrial environments at the time. The extinction that finally seems to have "released" the evolutionary potential of mammals must be seen in the context of the rest of the world's life, and we shall look at that in the next three chapters.

Review Questions

List the characters that were acquired during the evolutionary transition from therapsids to true mammals.

If we believe in evolution, we would predict that as reptiles evolved into mammals, there would be "intermediates". Name a reptile that evolved at least one mammal-like character, and say what the character is.

If we believe in evolution, we would predict that as reptiles evolved into mammals, there would be "intermediates". Name a mammal that has at least one reptile-like character, and say what the character is.

The mammal-like reptiles almost went extinct. Why? and when?

Describe an early mammal: size, ecology, diet, reproduction.

I suggested that late therapsids or early mammals "invaded the night". *Why* would they have done that? and what evidence can you put together that suggests they did.

Further Reading

Easy Access Reading

Cifelli, R. L., et al. 1996. Fossil evidence for the origin of the marsupial pattern of tooth replacement. *Nature* 379: 715–718.

Flannery, T. F., et al. 1995. A new family of monotremes from the Cretaceous of Australia. *Nature* 377: 418–420.

Hu, Y., et al. 1997. A new symmetrodont mammal from China and its implications for mammalian evolution. *Nature* 390: 137-142. *Zhangheotherium*

Kielan-Jaworowska, Z., et al. 1987. The origin of egg-laying mammals. *Nature* 326: 871–873.

Meng, J., and A. R. Wyss. 1995. Monotreme affinities and low-frequency hearing suggested by multituberculate ear. *Nature* 377: 141–144, and comment, pp. 104–105.

Rougier, G. W., et al. 1998. Implication of *Deltatheridium* specimens for early marsupial history. *Nature* 396: 459–463.

Rowe, T. 1996. Coevolution of the mammalian middle ear and neocortex. *Science* 273: 651–654.

Sereno, P. C., and M. C. McKenna. 1995. Cretaceous multituberculate skeleton and the early evolution of the mammalian shoulder girdle. *Nature* 377: 144–147; and comment, pp. 104–105.

More Technical Reading

Blackburn, D. G. et al. 1989. The origins of lactation and the evolution of milk: a review with new hypotheses. *Mammal Review* 19: 1–26.

Duboule, D. 1999. No milk today (my Hox have gone away). *Proceedings of the National Academy of Sciences* 96: 322–323. Short comment on an accompanying detailed paper.

Hillenius, W. J. 1992. The evolution of nasal turbinates and mammalian endothermy. *Paleobiology* 18: 17–29.

Hillenius, W. J. 1994. Turbinates in therapsids: evidence for Late Permian origins of mammalian endothermy. *Evolution* 48: 207–229.

Jenkins, F. A. 1984. A survey of mammalian origins. In P. D. Gingerich and C. A. Badgley (organizers). *Mammals: Notes for a Short Course. University of Tennessee Studies in Geology* 8: 32–47.

Kemp, T. S. 1982. *Mammal-like Reptiles and the Origin of Mammals*. London: Academic Press.

Kielan-Jaworowska, Z., et al. 1992. Interrelationships of Mesozoic mammals. *Historical Biology* 6: 185–202.

Krause, D. W., and F. A. Jenkins. 1983. The post-cranial skeleton of North American multituberculates. *Bulletin of the Museum of Comparative Zoology, Harvard University* 150: 199–246.

Rosowski, J. J., and A. Graybeal. 1991. What did *Morganucodon* hear? *Zoological Journal of the Linnean Society* 101: 131–168. We don't know yet.

Rowe, T., and J. Gauthier. 1992. Ancestry, paleontology, and definition of the name Mammalia. *Systematic Biology* 41: 372–378. One side of the argument!

Savage, R. J. G., and M. R. Long. 1987. *Mammal Evolution: An Illustrated Guide*. London: British Museum (Natural History).

Schmidt-Nielsen, K., et al. (eds.). 1975. *Comparative Physiology: Primitive Mammals*. Cambridge: Cambridge University Press.

Sidor, C. A., and J. A. Hopson. 1998. Ghost lineages and "mammalness": assessing the temporal pattern of character acquisition in the Synapsida. *Paleobiology* 24: 254–273.

Szalay, F. S., et al. (eds.). 1993. *Mammal Phylogeny: Mesozoic Differentiation, Multituberculates, Monotremes, Early Therians, and Marsupials*. New York: Springer-Verlag.

Marine Reptiles

Large-bodied animals on land (dinosaurs) and in the air (pterosaurs) had their counterparts in large marine reptiles in Mesozoic seas. Most of these evolved in Triassic times but reached their greatest abundance in the Jurassic and Cretaceous. Several different clades of reptiles evolved spectacular adaptations to life at sea.

Turtles

We have already seen that turtles are very specialized diapsids. The first well-known turtle is from the Late Triassic of Europe, and it already had bony plates on its surface, though it had not yet accomplished the turtle trick of having the shoulder blades inside the ribs.

Turtles were widespread and successful in Jurassic and Cretaceous seas and estuaries. Perhaps the most famous is the giant Cretaceous turtle *Archelon*, which was 3 meters (10 feet) long and nearly 4 meters (13 feet) in flipper span. It was so large that it couldn't have swum with a complete solid carapace, and it had only a bony framework (Figure 16.1). Large marine turtles are anything but primitive in their biology. Their limbs are modified into hydrofoils, and they "fly" underwater. Marine turtles can navigate precisely over thousands of kilometers and they are warm-blooded, maintaining their body temperatures at levels significantly higher than the water around them.

Figure 16.1 The giant Cretaceous marine turtle, *Archelon*, evolved a carapace that was lightened so that it could maintain buoyancy in the water. (Courtesy of the Peabody Museum of Natural History, Yale University.)

Crocodiles

Crocodiles are archosaurs like thecodonts and dinosaurs, and their ancestry is clearly terrestrial, probably among thecodonts of some sort. All crocodiles were terrestrial predators in the Late Triassic. There were large, powerful crocodile-like aquatic thecodonts such as parasuchids (Chapter 11) in the Triassic, and true crocodiles did not become aquatic until after the others became extinct. *Terrestrisuchus*, a small crocodile from the Late Triassic of Britain, had long, slim, erect limbs, and with a length of less than 1 meter it probably ran quite fast on land (Figure 11.11). Terrestrial crocodiles lived on well into the Jurassic, and in the end may have been outcompeted on land by bipedal theropod dinosaurs.

Ever since the Early Jurassic, most crocodiles have been amphibious. Many of them are predators at or near the water's edge. Some became almost entirely aquatic, and others returned to land in the Cenozoic to become powerful terrestrial predators. Crocodiles that became amphibious or aquatic evolved to large size and were reasonably common in Mesozoic seas and rivers. *Deinosuchus* from the Late Cretaceous of Texas is the largest crocodile that ever lived. It had a skull 2 meters (6 feet) long. It was about 10 meters (33 feet) long, and weighed about 5 tons. It may have taken duckbilled dinosaurs as prey (they are found in the same rock formations) in the same way that living Nile crocodiles take hippos.

Figure 16.2 *Deinosuchus*, a gigantic crocodile from the Late Cretaceous of Texas that may have preyed on dinosaurs at the water's edge. Scene reconstructed by Walter Ferguson for the American Museum of Natural History. (Negative 337265. Courtesy of the Department of Library Services, American Museum of Natural History.)

Crocodiles today are not equipped to kill large prey quickly. They usually kill large prey by holding them under the water until they drown. There's no reason to suppose that *Deinosuchus* did anything more sophisticated as it hunted large dinosaurs (Figure 16.2). *Stomatosuchus* was a gigantic duckbilled crocodile from the Middle Cretaceous of Africa. There are no living duckbilled crocodiles, so it is difficult to reconstruct its ecology.

Ichthyosaurs

Ichthyosaurs are not easily related to any other reptile groups, but the best guess is that they were highly derived diapsids. They were shaped much like dolphins, except that the tail flukes are horizontal in dolphins and vertical in ichthyosaurs. The most advanced ichthyosaurs had a continuation of the spine running into the lower tail fin (Figure 16.3). The main propulsion would then have been a side-to-side body motion, like a fish rather than a dolphin. The limbs were modified into small, stiff fins for steering and attitude control, again like dolphins, so that ichthyosaurs would have been very maneuverable up and down in the water as well as sideways. The tail fin was usually very deep, which is characteristic of swimmers that use fast acceleration in hunting prey. Ichthyosaurs were

Figure 16.3 An ichthyosaur, beautifully preserved with skin, from the Jurassic of Germany. (Courtesy of the Library Services Department, American Museum of Natural History.)

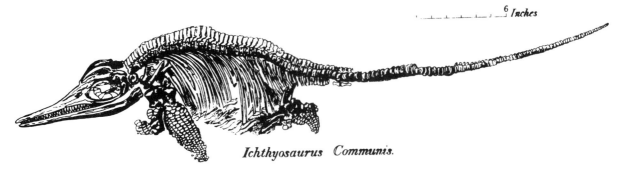

Ichthyosaurus Communis.

6 *Inches*

Figure 16.4 Beautiful fossils of ichthyosaurs were discovered early in the nineteenth century and caused much discussion about evolution. (From Buckland.)

beautifully streamlined, but would have been unable to move on land.

Beautiful ichthyosaur fossils have been known for 200 years, and they figured in many early discussions of evolutionary theory because everyone could recognize their exquisite adaptations for life in water (Figure 16.4). Ichthyosaurs all had good vision, with large eyes looking right along the line of the jaw. In advanced ichthyosaurs the jaw was long and thin, with many piercing conical teeth that were well designed for catching fish. Preserved stomach contents include fish scales and hooklets from the arms of cephalopods, possibly soft-bodied squids. One spectacular Jurassic ichthyosaur, *Eurhinosaurus*, had a swordlike upper jaw projecting far beyond the lower, with teeth all along its length. *Eurhinosaurus* was probably an ecological equivalent of the swordfish, using its upper jaw to slash its way through a school of fish and spinning around to catch its crippled victims.

Most early ichthyosaurs had blunt, shell-crushing teeth and may have hunted and crushed ammonites and other shelled cephalopods in a way of life that did not demand high levels of hydrodynamic performance. The earliest ichthyosaurs found to date, from the Early Triassic of the Northern Hemisphere, were small, about 1 meter (3 feet) long, but they were already specialized for marine life. *Mixosaurus* was a typical small, early ichthyosaur, known from Middle Triassic rocks ranging from the Arctic to Nevada to Indonesia; but the best-preserved specimens come from the Alps. The spine had not yet turned down to form the lower tail fin, but almost all the other features show excellent adaptation to swimming, with the limbs totally modified into effective fins.

Shonisaurus from the Late Triassic of Nevada, at 15 meters (50 feet) long, is one of the largest ichthyosaurs known. It was robust too, with a huge, strong deep body and long, powerful fins (Figure 16.5). Thirty-seven specimens of *Shonisaurus* were fossilized together, most of them

Figure 16.5 A huge, powerful, deep-bodied early ichthyosaur, *Shonisaurus*, up to 15 meters (50 feet) long, from the Triassic of Nevada. (After Merriam.)

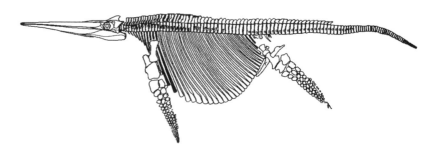

facing in the same direction—probably the result of a mass stranding, like the calamity that happens occasionally to whales and dolphins today.

Jurassic ichthyosaurs were abundant and varied, but there was only one Cretaceous ichthyosaur, *Platypterygius*. It had lost the large tail for fast acceleration, and instead its limbs were modified into large fins. This suggests that it had more of a cruising style of hunting than most ichthyosaurs did. It may have used the limb fins as underwater wings for propulsion rather than steering, in the style used by sea turtles and penguins and reconstructed for some plesiosaurs.

Sauropterygians

Sauropterygians (Figure 16.6) are a clade of reptiles whose ancestry can be traced back to the Younginiformes of the Permian (Chapter 11). Their nearest known relative is the aquatic diapsid reptile *Claudiosaurus*. Sauropterygians had unusually large limbs for land animals that had evolved toward life in water (compare crocodiles and whales). Most had small heads and comparatively long necks for their body size, so their prey (presumably fishes) must have been relatively small.

Placodonts are early but specialized sauropterygians known only from the Triassic of Europe. They had their own set of adaptations that reflected a specialized ecology like the living walrus, which dives down to shallow seafloors to dig and crush clams. *Placodus* itself had unusual teeth that suited it for this way of life. The large comblike teeth at the front of the jaw (Figure 16.7, left) were probably used to dig into the seafloor to scoop up clams, and sediment could be washed off them by shaking the head with the mouth open. The clean clams were then crushed between flat molar teeth in the lower jaw and flat plates on the roof of the mouth (Figure 16.7, right). Placodonts did not need great maneuverability or speed, and many had heavy plated carapaces that covered them dorsally and ventrally, rather like a turtle.

Pachypleurosaurs were the simplest early sauropterygians in structure, though we have good specimens only from the Middle Triassic. The intermediate forms between them and their Permian ancestors remain to be discovered. Pachypleurosaurs are small marine reptiles that are best known from Middle Triassic rocks of the Alps, but they also occur in China. Hundreds of specimens are known, but only a few are well preserved. Their limbs were not very strong, and they were modified for life in water in the sense that they could be folded back against the body for

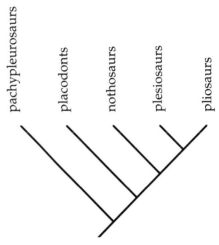

Figure 16.6 A cladogram of the sauropterygians, which included some of the more spectacular marine reptiles of the Mesozoic. (Simplified from Storrs.)

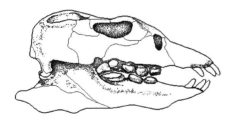 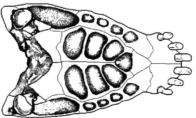

Figure 16.7 Left: The skull of *Placodus*, a marine reptile from the Triassic of Germany that may have had an ecology much like the living walrus. Right: The lower jaw of *Placodus*, showing the clam-crushing teeth of the palate. (What did it do with its tongue during this process?) (After Broili and Romer.)

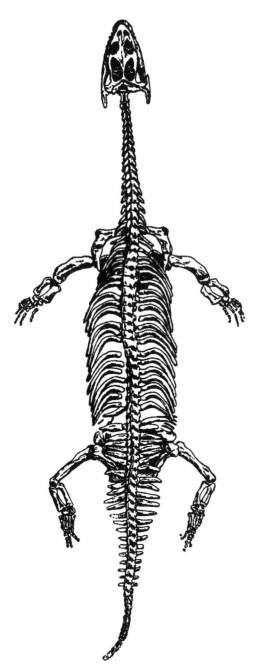

Figure 16.8 *Lariosaurus*, a nothosaur from the Triassic of Switzerland, about 60 cm (2 feet) long. (Idealized after Peyer.)

extremely low water resistance; the tail was powerful. Pachypleurosaur swimming style accentuated power and flexibility of the body and tail, though this did not include the rib cage: pachypleurosaurs had thick ribs that presumably made up a very stiff thorax. This adaptation is a solution to Carrier's Constraint (Chapter 11) in active, air-breathing swimmers. I suspect that pachypleurosaurs swam like living monitor lizards, with the front limbs tucked away against the rib cage and the hind limbs used as rudders. The forelimbs were relatively short but quite powerful, perhaps for dragging the animal out onto land for breeding and egg-laying.

The **nothosaurs** were more advanced Late Triassic sauropterygians. All nothosaurs were large compared with their pachypleurosaur ancestors. They extended the rigid thorax of pachypleurosaurs by evolving ribs far back along the body. With their bodies stiffened in this way, nothosaurs probably relied less on the tail for propulsion than pachypleurosaurs did, and they had strong forelimbs that may have contributed compensating swimming power. The nothosaur forelimb was not really winglike and may have used a rowing action to give propulsion. The hindlimbs were quite strong but not particularly well adapted for a swimming stroke (Figure 16.8).

One nothosaur group evolved in Early Jurassic times into the largest and best-known sauropterygian clade, the **plesiosaurs**. Plesiosaurs had large bodies, and limbs that were very strong, equally well developed front and back, and highly modified for swimming. They swam with all four limbs that used the stiffened body as a solid mechanical base, in a further extension of the swimming style of nothosaurs. The limbs were strengthened and further modified for efficient swimming strokes, eventually becoming much more important in swimming than the tail.

Plesiosaurs flourished worldwide in marine ecosystems from the Early Jurassic until the end of the Cretaceous. They came in two versions. Pliosaurs had short necks and long, large heads, and they looked rather like powerful, long-headed ichthyosaurs. They swam mainly with the strong limbs, however, all four of which were large, paddle-shaped structures, shaped into effective hydrofoils (Figure 16.9). Pliosaurs reached very large size. For example, *Kronosaurus* from Australia was 12 meters (40 feet) long and had a skull close to 4 meters (12 feet) long.

True plesiosaurs had the same limb structure but had very long necks and small heads (Figure 16.10). An average adult was about 3 meters (10 feet) long, with a neck that had 40 vertebrae. But some plesiosaurs were giants too. *Elasmosaurus* from the Cretaceous of Kansas was 12 meters (40 feet) long, with 76 neck vertebrae!

Figure 16.9 *Leiopleurodon*, a pliosaur. (After Newman and Tarlo.)

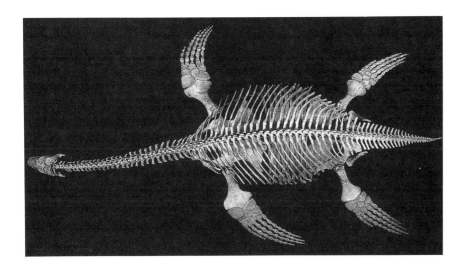

Figure 16.10 The plesiosaur *Cryptocleidus*. The limbs must have dominated the swimming—but how? (Negative 35605. Photograph by A. E. Anderson. Courtesy of the Library Services Department, American Museum of Natural History.)

Plesiosaur limbs were jointed to massive pectoral and pelvic girdles, presumably by very strong muscles and ligaments. Jane Robinson has suggested that these structures could be explained if all four limbs were used in an up-and-down power stroke, in underwater "flying" like that of penguins—except, of course, that four limbs were involved instead of two. She realized that the plesiosaur body had to be tightly strung with powerful ligaments to transmit the propulsion generated by the limbs to the body that they pulled through the water; and she found grooves in the skeleton where the ligaments had run.

But plesiosaur limbs were not jointed well enough to the shoulder and pelvic girdles to allow strictly "flight" power strokes, and they could not have been lifted above the horizontal. They also show no sign of powerful muscle attachments. Steven Godfrey has suggested instead that the propulsion stroke was downward and backward in a combination of "flying" and rowing; living sea lions swim this way.

However it worked, plesiosaur swimming required precise coordination between the limb strokes. But here is an unresolved question: how did the limb strokes coordinate? Did all four limbs work in synchrony? Did the power stroke of both front limbs alternate with the power stroke of both back limbs? Or did right front and left back limb strokes coincide with left front and right back? Most people favor the first technique of synchronous strokes, which is also used by sea lions. The second option would involve too much stress on the trunk, which would be alternately extended and compressed if power strokes alternated between front and back limbs. The third option, however, would require only resistance to trunk twisting, which could easily be accomplished by the ligaments along the spine and those connecting the large bony masses along the underside (Figures 16.9, 16.10): this could potentially make a plesiosaur much more maneuverable than the other techniques.

It is difficult to envisage how plesiosaurs hunted. Perhaps, with their large heads, pliosaurs hunted large fishes at fairly high speed. But true plesiosaurs are different. They have large bodies but small heads and long necks. Perhaps they stalked smaller prey and used sustained underwater "flight" mainly for migration or for cruising to feeding grounds.

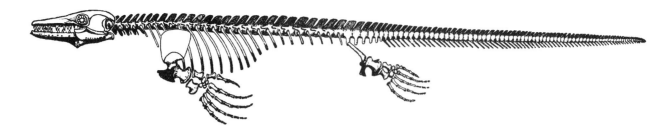

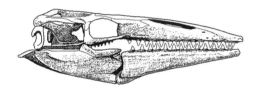

Figure 16.11 The Cretaceous mosasaur *Platecarpus*, essentially a gigantic lizard adapted for carnivorous life in the sea. About 4 meters (13 feet) long. (After Merriam.)

Figure 16.12 The jaws and teeth of the mosasaur *Clidastes*. (After Russell.)

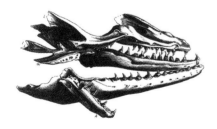

Figure 16.13 *Mosasaurus*, from Maastricht, the original mosasaur. The tremendous jaws of this creature were the talk of Europe in the late 1700s (along with the American Revolution, the French Revolution, and the Industrial Revolution). (From Buckland.)

Mosasaurs

Mosasaurs were essentially very large Late Cretaceous monitor lizards, up to 10 meters (30 feet) in length, the largest lizards that have ever evolved. Their evolution of aquatic adaptations in parallel with ichthyosaurs and plesiosaurs is astonishing.

Mosasaur bodies were long and powerful, with tails and limbs adapted for swimming. The main propulsion came from flexing the body and sculling with the tail, which was flat and deep, as it is in living crocodiles. But, in addition, the limbs were modified into beautiful hydrofoils. The elbow joint was rigid, and the shoulder joint was designed for up-and-down movement. Although the forelimbs could have given some lift, most mosasaurs probably used them as steering surfaces, as dolphins do. Some forms like *Platecarpus*, however, had well-developed forelimbs (Figure 16.11), and may have used them in a kind of underwater flying, like penguins. The hind limbs were like the forelimbs, though smaller, with the major muscle attachments also giving up-and-down movement. Because the pelvis was not fixed to the backbone, the hind limb strokes cannot have delivered much power. The hind limbs could rotate, and probably worked like aircraft elevators to adjust pitch and roll.

Mosasaurs had long heads set on a flexible but powerful neck. The large jaws often had a hinge halfway along the lower jaw, which may have served as a shock absorber as the mosasaur hit a large fish at speed. This hinge and the powerful stabbing teeth suggest that most mosasaurs ate large fishes (Figure 16.12, 16.13). Other mosasaurs had large, rounded, blunt teeth like those of *Placodus* (Figure 16.7), and they probably crushed mollusc shells to reach the flesh inside.

Air Breathers at Sea

All these Mesozoic reptiles were air breathers and therefore faced special problems for life in the sea. Precisely the same problems are faced today by marine mammals. The major one, of course, is the fact that air breathers must visit the surface for air, but there are also problems in introducing young to a complex and dangerous world where they must be prepared to use sophisticated skills immediately after birth. Many marine reptiles, mammals, and birds return to the shore for reproduction. Turtles simply lay large clutches of eggs and leave them buried in the sand, a method that results in horrific mortality but has obviously worked successfully for 200 million years. Seals, sea lions, and penguins have their young on shore in safe nurseries, so that they can breathe air, be fed, and grow for a while before they take to a swimming and foraging life at sea.

But living dolphins and whales never come ashore. They have special adaptations for air breathing, breeding, giving birth and caring for the young at sea. The young are born tail first, and mothers and other related adults will push them to the surface until they learn to breathe properly. The young must be able to dive immediately to suckle, and whales feed their babies milk under high pressure.

There is evidence that Mesozoic marine reptiles solved the same kinds of problems in spectacular fashion. Several fossils of ichthyosaurs have been found with young preserved inside the rib cage of adults, evidence that ichthyosaurs at least had evolved live birth. The preserved fetuses have long, pointed jaws, showing that they would have been able to feed for themselves immediately after birth, and they were born tail first as whales are. All this implies that ichthyosaurs had special mechanisms for training the young to swim and feed, and it also suggests parental care on a scale comparable with that of dinosaurs (Chapter 12).

Mosasaurs do not look as if they would readily have come ashore to lay eggs, and no fetuses or even juveniles have been found associated with adults. There is some indirect evidence that they had live birth at sea: the pelvis is very unusual. This may have resulted simply from adaptation to swimming, but perhaps the normal pelvis was expanded to give birth to live offspring much bigger than any normal egg.

The early sauropterygians had limbs that would have allowed them to haul themselves out onto a beach to lay eggs or to give birth, rather like sea lions. The fairly strong limbs of placodonts and the small body size of pachypleurosaurs make them particularly easy to imagine on the shore. Nothosaurs and plesiosaurs are usually much larger, and would have had to work much harder to drag themselves up a beach. Plesiosaurs, with limbs modified into long hydrofoils, may have been totally sea-going, with live birth at sea as in ichthyosaurs, whales, and dolphins.

Air-breathers at sea are subject to Carrier's Constraint (Chapter 11) —they cannot swim fast if they flex the body side-to-side. As in their terrestrial counterparts, marine mammals and birds do not have a problem: their bodies flex up and down as they swim. Mosasaurs, as lizards, certainly could not have swum at speed for long. As nothosaurs evolved into plesiosaurs, they also evolved stiffened trunks that avoided Carrier's Constraint (compare Figure 16.8 with Figures 16.9 and 16.10), and their underwater flight is a reflection of that evolutionary breakthrough. What about ichthyosaurs? They certainly look fast, yet their tail fin flexes sideways, and the body does not look stiff (Figure 16.3). I suspect that the ichthyosaurs had only one way to avoid Carrier's Constraint: by leaping out of the water as they swam, as penguins and dolphins do at high speed. In the leap, the body can be straight as the animal takes a deep breath. In my view, only those ichthyosaurs that evolved leaping could have swum fast for any length of time.

Fast swimming air breathers are rare,

Some ichthyosaurs did it with flair:

They swam up with a leap

(It's energetically cheap)

And they took a big breath in mid-air.

Many large and powerful swimming creatures today are warm-blooded to some extent: many sharks, tuna, and several turtles, as well as dolphins. Thus, one could guess that ichthyosaurs and plesiosaurs were warm-blooded, and that ichthyosaurs at least had live birth as well. This does not make them mammals, but it does suggest that they were most impressive creatures.

Almost all these magnificent marine reptiles became extinct at the end of the Cretaceous, along with dinosaurs, pterosaurs, and a significant number of marine invertebrates. Only crocodiles and turtles have survived to give us some clues about the mode of life of large reptiles. Unfortunately, these survivors are far from being typical Mesozoic reptiles!

Review Questions

Whales give birth at sea. What evidence is there that some Mesozoic reptiles did too?

Name an important group of marine reptiles that lived in the Mesozoic. If your chosen group is rather like a living group, name that living group: if your chosen group is not like any living group, draw it.

Over time, some groups of land-dwelling vertebrates have returned to life in the sea, as air-breathing swimmers. Name one group that is extinct, and one that is alive.

THOUGHT QUESTION.—Here is a picture of a plesiosaur. You remember that they probably "fly" under water with their limbs, but it is not clear whether they fly with left front/right back then right front/left back alternating strokes, or front together-then-back together beats. Now think about Carrier's Constraint, and tell me whether you can suggest a style that's better than the other. Give your reasoning.

Further Reading

Easy Access Reading

Erickson, G. M., and C. A. Brochu. 1999. How the 'terror crocodile' grew so big. *Nature* 398: 205–206.

More Technical Reading

Alexander, R. McN. 1989. *Dynamics of Dinosaurs and Other Extinct Giants.* New York: Columbia University Press. Chapter 9 discusses swimming in extinct marine reptiles.

Carroll, R. L. 1981. Plesiosaur ancestors from the Upper Permian of Madagascar. *Philosophical Transactions of the Royal Society of London B* 293: 315–383.

Carroll, R. L., and P. Gaskill. 1985. The nothosaur *Pachypleurosaurus* and the origin of plesiosaurs. *Philosophical Transactions, Royal Society of London B* 309: 343–393.

Cowen, R. 1996. Locomotion and respiration in aquatic air-breathing vertebrates. In D. Jablonski et al. (eds.), *Evolutionary Paleobiology*, pp. 337–352. Chicago: University of Chicago Press.

Dobie, J. L., et al. 1986. A unique sacroiliac contact in mosasaurs (Sauria, Varanoidea, Mosasauridae). *Journal of Vertebrate Paleontology* 6: 197–199.

Godfrey, S. 1984. Plesiosaur subaqueous flight: a reappraisal. *Neues Jahrbuch für Geologie und Paläontologie Monatshefte* 11: 661–672.

Lingham-Soliar, T. 1992. A new mode of locomotion in mosasaurs: subaqueous flight in *Plioplatecarpus. Journal of Vertebrate Paleontology* 12: 405–421.

McGowan, C. 1988. Differential development of the rostrum and mandible of the swordfish (*Xiphias gladius*) during ontogeny and its possible functional significance. *Canadian Journal of Zoology* 66: 496–503. Compares with ichthyos.

McGowan, C. 1991. *Dinosaurs, Spitfires, and Sea Dragons.* Cambridge, Massachusetts: Harvard University Press. Chapters 8–10 deal with marine reptiles.

Robinson, J. A. 1975. The locomotion of plesiosaurs. *Neues Jahrbuch für Geologie und Paläontologie Abhandlungen* 149: 286–332.

Robinson, J. A. 1976. Intracorporal force transmission in plesiosaurs. *Neues Jahrbuch für Geologie und Paläontologie Abhandlungen* 153: 86–128.

Storrs, G. W. 1993. Function and phylogeny in sauropterygian (Diapsida) evolution. *American Journal of Science* 293A: 63–90.

Taylor, M. A. 1987. A reinterpretation of ichthyosaur swimming and buoyancy. *Palaeontology* 30: 531–535.

CHAPTER SEVENTEEN

Why Flowers Are Beautiful

As plants invaded drier habitats from Devonian times onward, they evolved ways to retain water and protect their reproductive stages from drying out. The major advance was the perfection of seeds, which are fertilized embryos packed in a reasonably watertight container filled with food. The embryo can survive in suspended animation within the seed until the parent plant arranges for its dispersal. Germination can be delayed until after successful transport to a favorable location. The seedling then bursts its seed coat and grows, using the nutrition in the seed until its roots and leaves have grown large and strong enough to support and maintain the growing plant.

Seeds had evolved in Late Devonian times, and seed ferns were a successful component of Late Paleozoic floras, including the coal forests; they flourished into the Triassic. But Mesozoic gymnosperms perfected the seed system, making up 60% of Triassic and 80% of Jurassic species. Gymnosperms include conifers, cycads, and gingkos. Mesozoic forests had trees up to 60 meters (200 feet) high, forming famous fossil beds such as the Petrified Forest of Arizona. Conifers were the dominant land plants during the Jurassic and Early Cretaceous, and they are still by far the most successful of the gymnosperms. Finally, around the Jurassic-Cretaceous boundary, the flowering plants or **angiosperms** evolved and eventually came to dominate land floras.

Seed plant reproduction has two phases, fertilization and seed dispersal. The plant must be pollinated, and after the seed has formed it must be transported to a favorable site for germination. A major factor in the evolution of angiosperms is their manipulation of animals to do these two jobs for them.

Mesozoic Plants and Pollination

Conifers and many other plants are pollinated by wind. They produce enormous numbers of pollen grains, which are then released to blow in the wind in the hope that a grain will reach the pollen receptor of a female plant of the same species. The pollen emitters and receptors are often packed closely together in cones.

Wind pollination works, just as scattering sperm and eggs into the ocean works for many marine invertebrates. But the process looks very expensive. The pollen receptor in conifers is only about one square millimeter in area, so to achieve a reasonable probability of fertilization, the female cone must be saturated with pollen grains at a density close to one million grains per square meter. Parent plants do some things to cut the costs of wind pollination. Male cones release pollen in dry weather in just the right wind conditions, for example, and female cones are aerodynamically shaped to act as efficient pollen collectors. But for practical purposes, wind pollination can be consistently successful only if many individuals of the same plant species live in closely packed groups: conifers in temperate forests or grasses in prairies and savannas. An ecological setting like a modern jungle, where many species have well-scattered individuals, is not the place for wind pollination.

We can imagine Jurassic floras dependent on wind pollination, with plants ready to release large supplies of pollen. Insects then, as now, probably foraged for the food offered by plentiful pollen and soft, unripe female organs waiting for fertilization. We know that there were large, clumsy beetles in the Jurassic, and they probably visited plants for food. As they moved from plant to plant, they may have visited the same species frequently, collecting and transferring pollen by accident. Insects could help even by visiting one plant or one sex: in some living cycad gymnosperms, wind carries pollen to the surface of the female cone, but insects clustering around the cone carry it into the pollen receptors.

Over time, the plant structure may have evolved toward cooperation with insects in certain ways. Perhaps delicate structures were protected, but pollen was made easier to gather, and female pollen collectors were moved closer to the male pollen emitters. Such changes would have made pollen transfer by insects more likely, and less costly to the plant. Devices to attract insects—strong scents at first, then brightly colored flowers—perhaps evolved side by side with rewards such as nectar. Those plants that successfully attracted insects would have benefited by increasing their chances of fertilizing and being fertilized. Insects deliver pollen much more efficiently than wind.

An ideal pollinator should be able to exist largely on pollen and nectar, so that it can gather all its food requirements by visiting plants. It should visit as many (similar) plants as possible, so it should be small, fast-moving, and agile. A nocturnal pollinator should have a good sense of smell, and a daytime pollinator should have good vision or a good sense of smell, or both.

The only Jurassic candidates to fit this job description were insects. Birds and bats had not yet evolved, and small mammals were probably too sluggish. Pollinators had an increasing incentive to learn and remember certain smells and sights, and the insects that evolved rapid and error-free recognition of pollen sources, and clever search patterns to find them, would have become superior food gatherers and probably superior reproducers. Today, insects discriminate strongly between species, even between color varieties of particular species. Some insects congregate for mating around certain plant species.

Magnolias and Moths, Cycads and Beetles

It has long been known that living magnolias are primitive flowering plants. The recent discovery of the earliest flowering plant has confirmed that early angiosperms were very much like magnolias (Figure 17.1). Early Cretaceous flowers, though small, had relatively large petals, and the flowers could have produced many small seeds.

The Winteraceae are a family of primitive, medium-sized trees related to magnolias. Their fossils date back to Early Cretaceous times. About fifty or so species of Winteraceae are found today, in moist, tropical forests. The Winteraceae have an intriguing pollination system.

The trees bloom all summer, but there are never more than a few flowers open at once (usually only one per tree). Each flower lasts for two days. On the first day it displays female organs, and on the second morning it extends male organs, thus avoiding inbreeding. On the second day the male phase produces pollen in a sticky, nutritious oil. A primitive moth is attracted to the flower and feeds on the oily pollen, getting much of it entangled and dried on its body. The moths arrive in dozens and are strong fliers. Presumably they go off to search for another tree when they have stripped off all the male pollen.

The female phase extrudes a strong scent, but it provides no food for the moths. Instead, the scent seems to act as a mating stimulant. Both female and male moths are attracted in large numbers to the flowers, and in the display and rapid movement involved, pollen previously collected from another tree is delivered to the female stigma.

The moths involved in the pollination are among the most primitive known. Instead of sucking mouth parts for collecting nectar, they have grinding jaws that they use to chew pollen and spores. Their fossil record also dates back to the Early Cretaceous, so this style of pollination may be very ancient indeed, perhaps a good clue to the success of the early angiosperms. Many other magnolialike angiosperms have large, fragrant flowers, where insects congregate to feed and mate (and pollinate).

Primitive angiosperms were not the only plants to evolve insect pollination. Living cycad gymnosperms are often pollinated by insects, and they have evolved adaptations that are surprisingly like those of magnolias. *Zamia*, a Florida cycad, attracts a particular species of beetle to its male cones. The cones make perfect egg-laying sites for the beetle, because they are surrounded by thick, starchy leaves in which the beetle

Figure 17.1 Beautiful flowers had already evolved in the Early Cretaceous. This figure is a reconstruction of a flower discovered in the Early Cretaceous of North America. (After Basinger and Dilcher.)

larvae can burrow and feed. Female beetles are attracted to the cones by a minty scent. (Other cycads have special biochemical reactions that heat up the cone, evaporating scented volatile chemicals very efficiently.) Male beetles have already staked out territories on the cones, fighting off other males with special long, strong forelimbs. After all the fighting, mating, and egg-laying activity, the beetles are coated with pollen. It's not yet clear why beetles then visit female cones, but it's probably by mistake because male and female cones are similar. The female cone is sticky, but probably to help pollen transfer rather than to provide scent or nectar. Visits to the female cone are short, because there's no incentive to stay; in fact, many cycad seeds are toxic to animals, including humans. But brief insect visits to the female cone are perfect for the plant, which needs fertilization without insect damage.

Study of the magnolia system has led to the suggestion that the success of angiosperms resulted from their selection as insect mating sites. Mating is concentrated in and around the flower, encouraged originally by scents.

Animal pollination can give another advantage to plants because a large mass of pollen is dumped on the stigma, rather than single wind-blown grains. Competition between individual pollen grains to fertilize the ovule allows the female plant more mate choice than in other plants (remember Chapter 3). Pollen grains are haploid, so they cannot carry hidden recessive genes (as we do). A female plant could in theory select certain pollen grains over others by placing chemical or physical barriers between the stigma and the ovule; the first pollen grain to breach the barrier is selected over others for fertilization. Experimentally, plants that are allowed to exercise pollen choice in this way have stronger offspring than others. This aspect of angiosperm reproduction may have been one of the most important factors in their success.

Of course, pollination encouraged tremendous diversity among the pollinators, as they came increasingly to specialize on particular plants. The astounding rise in diversity of beetles and bees began in Cretaceous times, so that there are now tens of thousands of species of each. Those beetles and bees associated with angiosperms are many times more diverse than those associated with gymnosperms.

Pollination cannot be the whole story, however, because insects help to pollinate cycads too, yet angiosperms are enormously successful while cycads have always been a relatively small group of plants. Several other Mesozoic plants experimented with ways of persuading organisms to transport pollen, and flowerlike structures evolved more than once. Furthermore, if pollination were the key to angiosperm success, flowers could have evolved as soon as flying insects became abundant in the Late Carboniferous. There are some signs that insect pollination began then as something of a rarity. But angiosperms appeared much later and rather suddenly in the Early Cretaceous (Figure 17.2). Therefore, angiosperm success is not related simply to their evolution of flowers. In fact, all Mesozoic plants that had insect pollination must have paid a significant price for fertilization, in insect damage.

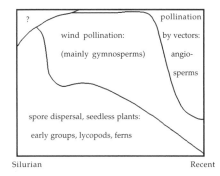

Figure 17.2 The percentage of plants pollinated by vectors increased sharply in the Cretaceous with the rise of angiosperms. But because many of those vectors (insects, for example) had been present since the Late Carboniferous, there must have been more to the rise of angiosperms than pollination by vectors. (Data simplified from Niklas, Tiffney, and Knoll.)

Mesozoic Plants and Seed Dispersal

If seeds fall close under the parent plant, they may be shaded so that they cannot grow, or they may be eaten by animals that have learned that tasty seeds are often found under trees. Many plants rely on wind to disperse their seeds. Sometimes seeds are provided with little parachutes or air-foils to help them travel far away from the parent; winged seeds evolved almost as soon as seeds themselves, in the Late Devonian. But seeds dispersed far away by wind will often fall into places that are disastrous for them. Although wind dispersal works, it seems very wasteful. Only plants that produce great numbers of seeds will consistently survive using this method, and if they are to produce great numbers of seeds that can be carried by wind, they cannot pack much food for the embryo into each seed. Wind-dispersed seeds must be light, so they cannot carry much energy for seedling growth. They have to germinate in relatively well-lit areas, where the seedling will be able to photosynthesize soon after emerging above ground.

An alternative strategy is for a plant to have its seeds carried away by an animal and dropped into a good place for growth. Many animals can carry larger seeds than wind can, and larger seeds can successfully germinate in darker places. As with pollination, animals must be persuaded, tricked, or bribed to help in seed dispersal.

Some animals visit plants to feed on pollen or nectar, and others browse on parts of the plants. Others simply walk by the plant, brushing it as they pass. Small seeds may be picked up accidentally during such visits, especially if the seed has special hooks, burrs, or glues to help to attach it to the visitor, and especially if the visitor has a coat of hair or feathers. Seeds may be carried some distance before they fall off. Small seeds may be eaten by a visiting herbivore, but some may pass unharmed through the battery of gnawing or grinding teeth, through the gut and its digestive juices, to be automatically deposited in a pile of fertilizer.

Plants face two different problems in persuading animals to disperse seeds and in persuading them to pollinate. In pollination there is often a payment on delivery: the pollinator collects nectar or another reward as it picks up and again as it delivers the pollen. There is no such payment on delivery of a seed. Any payment is made by the plant in advance, so that seed dispersers have no built-in payment for actual delivery of the seed. It would be better for them to cheat and to eat every seed. Thus plants often rely on tricks (burrs, for example) to fix seeds to dispersers. Velcro was evolved by plants long before the idea was copied by an astute human. Alternatively, plants may pack many small seeds into a fruit so that the disperser will concentrate on the fruit and swallow the seeds without crushing them (in strawberries, for example).

Some plants actually invite seed swallowing. They have evolved a tasty covering around the seed (a berry or fruit), and if the animal or bird eats the seed along with the fruit, every surviving seed is planted in fertilizer. Some seeds are tiny compared with the fruit. Although they are likely to be swallowed without being chewed, tiny seeds can carry little food for the developing embryo. Large seeds loaded with nutrition are often protected by a strong seed coat or packed inside a nut.

Seed dispersal by animals is not a cost-free service. Many seed dispersers eat seeds, passing only a few unscathed through their gut. So there is a significant wastage of seeds that depends on a delicate balance between the seed coat and the teeth and stomach of the disperser. Too strong a seed coat, and the disperser will turn to easier food or germination will be too difficult; too weak a seed coat, and too many seeds will be destroyed. Some plants species are so delicately adjusted to a particular disperser species that the seeds germinate well only if they are eaten by that disperser.

Angiosperms evolved carpels as a new and unique protection for their ovules, and eventually for the developing seeds. Carpels probably evolved to protect against large, hungry insects that visited the plant looking for pollen and nectar but were happy to settle for developing seeds. Very soon, however, the angiosperm seed coat began to protect seeds as they passed through vertebrate guts. A seed with a strong coat was proof against many possible predators, but perhaps at the same time came to be sought by one or a few animals that could break the seed coat. A plant could evolve to a stable relationship with a few successful seed predators: the predators would receive enough food from the seeds to keep them visiting the plant regularly, but would pass enough seeds unscathed through the gut that the plant benefited too.

The Jurassic was a time of insects and reptiles of all kinds. Insects may be good pollinators, but they are too small to be large-scale seed transporters, with all respect to the insect that moves Mexican jumping beans. Reptiles are large enough, but they often have low metabolic rates, so any seeds they swallow are exposed to digestive juices for a long time. Reptiles do not even have fur in which seeds can be entangled (though feathered theropod dinosaurs might have done!).

In the later Mesozoic, dinosaurs coexisted first with gymnosperms and then with early angiosperms. Seeds were undoubtedly dispersed by dinosaurs to some extent, since the giant vegetarian ornithischians and sauropods consumed great quantities of vegetation. But in spite of the size of the deposit of fertilizer that must have surrounded seeds passing through a dinosaur, it's likely that browsing dinosaurs damaged plants more than they helped them. It's unlikely that any Mesozoic plant would have encouraged dinosaur browsing.

Transport over long distances can take a seed beyond the range of its normal predators and diseases, and it can allow a plant to become very widespread provided that there are pollinators in its new habitat. As angiosperms adapted to seed dispersal by animals, they probably became capable of dispersal into new habitats much faster than other plants. Thus, other things being equal, we might expect a dramatic increase in the angiosperm fossil record as they adapted toward seed dispersal by animals rather than wind. (Some living angiosperms are pollinated by wind but have their seeds dispersed by animals. These include grasses, which did not evolve until well into the Cenozoic.)

We have seen that there were few effective animal seed-transporters in the Jurassic, and that dinosaurs were unlikely candidates in the Cretaceous. Philip Regal has suggested that birds and mammals triggered the radiation of angiosperms by aiding them in seed dispersal. Birds and mammals have feathers and fur in which seeds easily become entangled; seeds pass quickly through their small bodies with their high metabolic rates and are likely to be unharmed unless they have been deliberately chewed. Birds commonly range over long distances. Although other plants have seeds too, angiosperm seeds would have been especially suited to vertebrate transport because of their extra protective coating. Conifer seeds are usually small and light, designed to blow in the wind, and conifers depend on close clusters for pollination. Isolated conifers are likely to be unsuccessful reproducers, and additional transport would make little difference to their long-term success. Angiosperms are more likely to succeed in the long term when planted as isolated units.

However, the great angiosperm radiation took place in the Early and Middle Cretaceous, when mammals and birds were still minor members of the ecosystem. This early success of angiosperms may be explained better by the "fast seedling" hypothesis. This idea is based on the fact that angiosperm seeds germinate sooner, and the seedlings grow faster and photosynthesize better than those of gymnosperms. Angiosperms may simply have outcompeted gymnosperms in the race for open spaces.

Regal's idea applies better to the later radiation of mammals and land birds in the Cenozoic, when there was a further large increase in angiosperm diversity, abundance, and dominance. Angiosperms evolved to become forest trees as well as smaller weeds. Bruce Tiffney showed that Cenozoic angiosperm seeds were much larger than Cretaceous ones. The ability of angiosperms to become dominant forest trees in ecosystems, and their successful evolution of large-seed dispersal aided by animals and birds, were Early Cenozoic events.

The rise to dominance of the angiosperms in the Early Cenozoic provided a food bonanza for seed dispersers. Birds and small mammals, especially early primates and bats, all joined the seed- and fruit-eating guilds in Early Cenozoic times. Some tropical flowers today still rely for pollination on bats, small marsupials, or lemurs.

We can imagine a whole set of pollinators and seed dispersers evolving together with the plants on which they specialized. For most plants, it would be best not only to be conspicuous, but also to be different from other plant species, to encourage pollinators and seed seekers to be faithful visitors. Suppose that particular techniques are needed to extract seeds or pollen from a plant. A visitor that learns the secret has an advantage over others and will tend to visit that species rather than foraging at random, which might require learning several collecting techniques. Fewer strangers are likely to visit the plant and rob its regular visitors of their rewards. The plant is much more likely to be fertilized or dispersed by faithful visitors than random browsers. An insect, which has only a short adult life, a limited memory, and a limited learning capacity, is more likely to be a faithful pollinator to the first plant it learns to forage from, or to the species for which it is genetically programmed. It's easy to imagine the evolution of a great variety of bright and highly scented flowers and fruits, together with a great variety of their specialized pollinators and seed seekers. Again, the evolution of the faithful visitor may have been much later than the evolution of angiosperms themselves. Only in the Early Cenozoic do angiosperm flowers show evidence of pollination by faithful visitors such as bees, wasps, bats, and other small animals, and seed dispersal by birds, mammals, and large insects.

Living angiosperms have developed extraordinary devices for pollination as well as seed dispersal. One Arctic flower provides its insect pollinators with a bowl of petals that forms a perfect parabolic sunbathing enclosure. Orchids have petals shaped and colored like female insects, and they are pollinated by undiscriminating and optimistic males. It's quite by accident that we happen to sense and appreciate the scents and colors of the flowers around us, because most of them were selected for the eyes and senses of insects. But we can gain a scientific as well as an aesthetic kick from looking at flowers if we admire their efficiency as well as their beauty.

Angiosperms and Mesozoic Ecology

The first angiosperms appeared around the Jurassic-Cretaceous boundary and were abundant by the Middle Cretaceous, especially in disturbed environments such as riverbanks. Since then there has been a steady expansion of angiosperms, so that today ferns are characteristic only of damp environments, conifers dominate mainly in temperate forests, and other ancient plants such as cycads and gingkos are rare.

All big herbivores were short-limbed, short-necked, low browsers until Late Triassic times (possibly because of constraints on heart structure). Examples are *Edaphosaurus* and the caseids in the Permian, and the rhynchosaurs and the therapsid herbivores in the Triassic. But in the Late Triassic the first prosauropods appeared, adapted by long limbs and necks to browse on high vegetation. In Gondwanaland, an important area for prosauropods, the seed ferns typical of Triassic floras became almost extinct and were replaced by a wave of conifers. By the Middle Jurassic, the great expansion of sauropod dinosaurs had added to the number, mass, and reach of high browsers. Perhaps the aromatic compounds typical of conifers evolved as a chemical defense against dinosaur browsing. (These compounds deter mammals and insects today.)

At the end of the Jurassic, we see a reduction of the sauropods that probably had been higher browsers, and the rise of the low-browsing ornithischians. More seedlings would now have been cropped off before reaching maturity, and any plant that could reproduce and grow quickly would have been favored.

Conifers reproduce slowly. It takes two years from fertilization until the seed is released from the cone, and wind dispersal typically does not take the seed very far. The whole reproductive system of conifers depends on wind and works best in a group situation such as a forest.

On the other hand, most angiosperms are designed for pollination by animals, especially insects; for rapid germination and growth; and for rapid release of seeds (within the year). An angiosperm is much more likely to succeed as a weed, rapidly colonizing any open space, and is more likely to be widely distributed because of its dispersal method. The earliest angiosperms were small, weedy shrubs, exactly the kind of plant that could survive heavy dinosaur browsing. A conifer forest, once broken up by dinosaur browsing or natural accident, would most likely have been recolonized by shrubs and weeds that could invade and grow rapidly (look at the results of clear-cutting in a conifer forest today). The weeds themselves would have reproduced quickly, so would have been more resistant to browsing than were young conifer seedlings.

Even without dinosaur browsing, angiosperms would have found habitats where they would have been very successful. In Middle Cretaceous rocks, for example, angiosperm leaves dominate sediments laid down in river levees and channels. Shifting and changing riverbank areas favor weeds because large trees are felled by storms and frequent floods. Most Middle Cretaceous pollen, on the other hand, comes from sediments laid down in lakes and near-shore marine environments. This is the windblown pollen from stable forests on the shores and on lowland plains away from violent floods, and it is dominantly conifer pollen.

Angiosperms did not take over the entire Cretaceous world, however. They were very slow to colonize high latitudes. (I suspect this is a reflection of their greater dependence on insect pollinators, which drop off in both number and diversity in higher latitudes.)

Furthermore, Late Cretaceous fossil floras preserved in place under a volcanic ash fall in Wyoming show that even if angiosperms dominate a local flora in diversity of species, they may make up only a small percentage of the biomass. In the Big Cedar Ridge flora, angiosperms made up 61% of the species, but covered only 12% of the ground. We have to be careful in distinguishing between the diversity, the abundance, and the ecological importance of angiosperms. They cannot really be said to have dominated the *ecology* of any Cretaceous area.

Nevertheless, one can argue, as Bruce Tiffney and others have done, that the angiosperm radiation provided the basis for the radiations of the 5-ton ornithischians of the later Cretaceous. They seem to have lived in much larger herds than Jurassic dinosaurs, up to several thousand in the case of *Maiasaura*, and later Cretaceous dinosaurs were much more diverse as well as more abundant than their predecessors.

Ants and Termites

The success of angiosperms benefited pollinators and seed dispersers, and the evolution of angiosperms was related to the ecology of large animal browsers. But today, some of the most effective tropical herbivores are leaf-cutting ants, and most terrestrial vegetation litter is broken down by termites. One-third of the animal biomass in Amazonia is made up of ants and termites. In the savannas of West Africa there are more like

2000 ants per square meter! There may be 20 million individuals in a single colony of driver ants, but the world record is held by a supercolony of ants in northern Japan, which has 300 million individuals, including a million queens, in 45,000 interconnected nests spread over 2.7 square kilometers (one square mile).

The higher social insects (bees, ants, termites, and wasps) began a major evolutionary radiation in the Late Cretaceous, as angiosperms became dominant in terrestrial ecosystems. The earliest known bee, found in Cretaceous amber from New Jersey, is a female worker bee adapted for pollen gathering. Bee society already had a sophisticated structure.

Angiosperm Chemistry

As we have seen, many angiosperms attract animals to themselves for pollination and seed dispersal. The plant usually pays a price in the production cost of substances such as nectar and in the cost of seeds eaten. Browsing animals and plant- and sap-eating insects often eat more plant material than they return in the form of services to the plant, and attracting such creatures results in a net loss of energy.

Angiosperms have therefore evolved an amazing variety of structures and chemicals that act to repel herbivores. These can be as simple and as effective as spines and stings, they can be contact irritants as in poison ivy and poison oak, or they can be severe or subtle internal poisons. Cyanide is produced by a grass on the African savanna when it is grazed too savagely. Many of our official and unofficial pharmacological agents were originally designed not for human therapy but as plant defenses. More than 2000 species of plants are insecticidal to one degree or another. Caffeine, strychnine, nicotine, cocaine, morphine, mescaline, atropine, quinine, ephedrine, digitalis, codeine, and curare are all powerful plant-derived chemicals, and it is not a coincidence that many of them are important insecticides or act strongly on the nervous, reproductive, or circulatory systems of mammals (some are even contraceptive and would act directly to decrease browsing pressure). One hundred and fifty million pyrethrum flowers are harvested every day, to fill a demand for 25,000 tons of "natural" insecticide per year. A million tons of nicotine per year were once used for insect control, until it was found that the substance was extremely toxic to mammals (self-destructive humans still smoke it, however!). Other plant chemicals are powerful but can be used to flavor foods in low doses. All our kitchen flavorings and spices are in this category. Garlic keeps away insects as well as vampires and friends.

For paleobiologists, the problem of angiosperm chemistry is its failure to be preserved in the fossil record. Clearly, the increasing success of angiosperms in the Late Cretaceous and Early Cenozoic occurred in the face of intense herbivory by the radiating mammals and insects of that time. The chemical defenses of angiosperms probably evolved very early in their history.

Review Questions

List the qualities that make a good pollinator for small flowering plants.

List the qualities that make a good seed disperser for small flowering plants.

What candidates were available to pollinate early flowering plants?

What candidates were available to disperse the seeds of early flowering plants?

Describe how dinosaurs might have affected the ecology and biology of Jurassic plants.

You are an early angiosperm. Make a list of the trade-offs involved in persuading insects to transport (or deliver) your pollen.

You are an early angiosperm. Make a list of the trade-offs involved in persuading animals to disperse your seeds.

Further Reading

Easy Access Reading

Balandrin, M. F., et al. 1985. Natural plant chemicals: sources of industrial and medicinal materials. *Science* 228: 1154–1160.

Buchmann, S. L., and G. P. Nabhan. 1996. *The Forgotten Pollinators*. Washington, D.C.: Island Press. Read this book! You will look at the world around you in a different way.

Crane, P. R., et al. 1986. Lower Cretaceous angiosperm flowers: fossil evidence on early radiation of dicotyledons. *Science* 232: 852–854.

Labandeira, C., and J. J. Sepkoski. 1993. Insect diversity in the fossil record. *Science* 261: 310–315.

Lewis, A. C. 1986. Memory constraints and flower choice in *Pieris rapae*. *Science* 232: 863–865.

Lidgard, S., and P. R. Crane. 1988. Quantitative analyses of the early angiosperm radiation. *Nature* 331: 344–346, and comment, pp. 304–305.

Mulcahy, D. L. 1979. The rise of the angiosperms: a genecological factor. *Science* 206: 20–23.

Mulcahy, D. L., and G. B. Mulcahy. 1987. The effects of pollen competition. *American Scientist* 75: 44–50.

Regal, P. J. 1977. Ecology and evolution of flowering plant dominance. *Science* 196: 622–629.

Sun, G., et al. 1998. In search of the first flower: a Jurassic angiosperm, *Archaefructus*, from Northeast China. *Science* 282: 1692–1695, and comment, pp. 1653–1654.

More Technical Reading

Bond, W. J. 1989. The tortoise and the hare: ecology of angiosperm dominance and gymnosperm persistence. *Biological Journal of the Linnean Society* 36: 227–249.

Crane, P. R. 1985. Phylogenetic analysis of seed plants and the origin of angiosperms. *Annals of the Missouri Botanical Garden* 72: 716–793.

Crepet, W. L. 1984. Advanced (constant) insect pollination mechanisms: pattern of evolution and implications vis-a-vis angiosperm diversity. *Annals of the Missouri Botanical Garden* 71: 607–630.

Dilcher, D. L., and P. R. Crane. 1984. *Archaeanthus*: an early angiosperm from the Cenomanian of the Western Interior of North America. *Annals of the Missouri Botanical Garden* 71: 351–383.

Friis, E. M., et al. (eds.). 1987. *The Origins of Angiosperms and Their Biological Consequences*. Cambridge: Cambridge University Press.

Futuyma, D. J., and M. Slatkin (eds.). 1983. *Coevolution*. Sunderland, Massachusetts: Sinauer. Excellent chapters by Futuyma, Feinsinger, and Janzen.

Michener, C. D., and D. A. Grimaldi. 1988. The oldest fossil bee: apoid history, evolutionary stasis, and antiquity of social behavior. *Proceedings of the National Academy of Sciences* 85: 6424–6426.

Pellmyr, O., and L. B. Thien. 1986. Insect reproduction and floral fragrances: keys to the evolution of the angiosperms? *Taxon* 35: 76–85.

Tang, W. 1982. Heat and odor production in cycad cones. *Fairchild Tropical Garden Bulletin* 42 (3): 12–14.

Tang, W. 1987. Insect pollination in the cycad *Zamia pumila* (Zamiaceae). *American Journal of Botany* 74: 90–99.

Tiffney, B. H. 1984. Seed size, dispersal syndromes, and the rise of the angiosperms: evidence and hypotheses. *Annals of the Missouri Botanical Garden* 71: 551–576.

Tiffney, B. H. 1998. Land plants as food and habitat in the age of dinosaurs. In Farlow, J. O., and M. K. Brett-Surman (eds.), *The Complete Dinosaur*, pp. 352–370. Bloomington: Indiana University Press.

Waser, N. M. 1986. Flower constancy: definition, cause, and measurement. *American Naturalist* 127: 593–603.

Wing, S. L., et al. 1993. Implications of an exceptional fossil flora for Late Cretaceous vegetation. *Nature* 363: 342–344.

The End of the Dinosaurs

Almost all the large vertebrates on Earth, on land, at sea, and in the air—all dinosaurs, plesiosaurs, mosasaurs, and pterosaurs—suddenly became extinct about 65 Ma, at the end of the Cretaceous Period. At the same time, most plankton and many tropical invertebrates, especially reef-dwellers, became extinct, and many land plants were severely affected. This extinction event marks a major boundary in Earth's history, the K–T or Cretaceous–Tertiary boundary, and the end of the Mesozoic Era. The K–T extinctions were worldwide, affecting all the major continents and oceans. There are still arguments about just how short the event was. It was certainly sudden in geological terms and may have been catastrophic by anyone's standards.

Despite the scale of the extinctions, however, we must not be trapped into thinking that the K–T boundary marked a disaster for all living things. Most groups of organisms survived. Insects, mammals, birds, and flowering plants on land, and fishes, corals, and molluscs in the ocean went on to diversify tremendously soon after the end of the Cretaceous. The K–T casualties included most of the large creatures of the time, but also some of the smallest, in particular the plankton that generate most of the primary production in the oceans.

There have been many bad theories to explain dinosaur extinctions. More bad science is described in this chapter than in all the rest of the book. For example, even in the 1980s a new book on dinosaur extinctions suggested that they spent too much time in the sun, got cataracts, and because they couldn't see very well, fell over cliffs to their doom. But no matter how convincing or how silly they are, any of the theories that try to explain only the extinction of the dinosaurs ignore the fact that

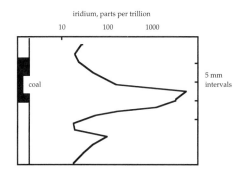

iridium, parts per trillion

coal

5 mm
intervals

Figure 18.1 A typical iridium spike at the K–T boundary. This data set is from New Mexico, where the K–T boundary lies in a coal bed. The scale for iridium is logarithmic. (Data from C. J. Orth et al., 1982.)

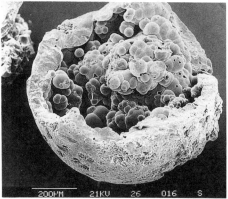

Figure 18.2 Top: A fragment of quartz from the K–T boundary in Montana, showing two sets of intersecting lamellae characteristic of quartz crystals that have undergone severe shock. Bottom: A tiny spherule from the K–T boundary clay in Wyoming. (Photographs courtesy of Bruce Bohor, U. S. Geological Survey, Denver.)

extinctions took place in land, sea, and aerial faunas, and were truly worldwide. The K–T extinctions were a global event, so we should examine globally effective agents: geographic change, oceanographic change, climatic change, or an extraterrestrial event. The most recent work on the K–T extinction has centered on two hypotheses that suggest a violent end to the Cretaceous: a large asteroid impact and a giant volcanic eruption.

An Asteroid or Cometary Impact?

A meteorite big enough to be called a small asteroid hit Earth precisely at the time of the K–T extinction. The evidence for the impact was first discovered by Walter Alvarez and colleagues. They found that rocks laid down precisely at the K–T boundary contain extraordinary amounts of the metal **iridium** (Figure 18.1). It doesn't seem to matter whether the boundary rocks were laid down on land or under the sea. In the Pacific Ocean and the Caribbean the iridium-bearing clay forms a layer in ocean floor sediments; it is found in continental shelf deposits in Europe; and in North America, from Canada to New Mexico, it occurs in coal-bearing rock sequences laid down on floodplains and deltas. The dating is precise, and the iridium layer has been identified in more than 100 places around the Earth. Where the boundary is in marine sediments, the iridium occurs in a layer just above the last Cretaceous microfossils, and the sediments above it contain Paleocene microfossils from the earliest part of the Cenozoic.

The iridium is present only in the boundary rocks and therefore was deposited in a single large **spike**—a very short event. Iridium occurs in normal seafloor sediments in microscopic quantities, but the iridium spike at the K–T boundary is very large. Iridium is rare on Earth, and although it can be concentrated by chemical processes in a sediment, an iridium spike of this magnitude must have arisen in some unusual way. Iridium is much rarer than gold on Earth, yet in the K–T boundary clay iridium is usually twice as abundant as gold, sometimes more than that. The same high ratio is found in meteorites. The Alvarez group therefore suggested that iridium was scattered worldwide from a cloud of debris that formed as an asteroid struck somewhere on Earth.

An asteroid big enough to scatter the estimated amount of iridium in the worldwide spike at the K–T boundary may have been about 10 km (6 miles) across. Computer models suggest that if such an asteroid collided with Earth, it would pass through the atmosphere and ocean almost as if they were not there and blast a crater in the crust about 100 km across. The iridium and the smallest pieces of debris would be spread worldwide by the impact blast as the asteroid vaporized into a fireball.

If indeed the spike was formed by a large impact, what other evidence should we hope to find in the rock record? Well-known meteorite impact structures often have fragments of **shocked quartz** and **spherules** (tiny glass spheres) associated with them (Figure 18.2). The glass is formed as the target rock is melted in the impact, blasted into the air as a spray of droplets, and almost immediately frozen. Over geological time, the glass spherules may decay to clay. Shocked quartz is formed when quartz crystals undergo a sudden pulse of great pressure. If they are not heated enough to melt, they may carry peculiar and unmistakable microstructures (Figure 18.2, top).

All over North America, the K–T boundary clay contains glass spherules (Figure 18.2, bottom), and just above the clay is a thinner layer that contains iridium along with fragments of shocked quartz. It is only a few millimeters thick, but in total it contains more than a cubic kilometer of shocked quartz in North America alone. The zone of shocked quartz extends west onto the Pacific Ocean floor, but shocked quartz is rare in K–T boundary rocks elsewhere: some very tiny fragments occur in European sites. All this evidence implies that the K–T impact occurred on or near North America, with the iridium coming from the vaporized asteroid and the shocked quartz coming from the continental rocks it hit.

The K–T impact crater has now been found. It is a roughly egg-shaped geological structure called **Chicxulub**, deeply buried under the sediments of the Yucatán peninsula of Mexico (Figure 18.3). The structure is about 180 km across, one of the largest impact structures so far identified with confidence on Earth. A borehole drilled into the Chicxulub structure hit 380 meters (more than 1000 feet) of igneous rock with a strange chemistry. That chemistry could have been generated by melting together a mixture of the sedimentary rocks in the region. The igneous rock under Chicxulub contains high levels of iridium, and its age is 65 Ma, exactly coinciding with the K–T boundary.

On top of the igneous rock lies a mass of broken rock, probably the largest surviving debris particles that fell back on to the crater without melting, and on top of that are normal sediments that formed slowly to fill the crater in the shallow tropical seas that covered the impact area.

Well-known impact craters often have **tektites** associated with them as well as shocked quartz and tiny glass spherules. Tektites are larger glass beads with unusual shapes and surface textures. They are formed when rocks are instantaneously melted and splashed out of impact sites in the form of big gobbets of molten glass, then cooled while spinning through the air.

Haiti was about 800 km from Chicxulub at the end of the Cretaceous (Figure 18.3). At Beloc and other localities in Haiti, the K–T boundary is marked by a normal but thick (30 cm) clay boundary layer that consists mainly of glass spherules (Figure 18.2). The clay is overlain by a layer of turbidite, submarine landslide material that contains large rock fragments. Some of the fragments look like shattered ocean crust, but there are also spherical pieces of yellow and black glass up to 8 mm across that are unmistakably tektites. The Beloc tektites apparently formed at about 1300°C from two different kinds of rock; and they are dated precisely at 65 Ma. The black tektites formed from continental volcanic rocks and the yellow ones from evaporite sediments with a high content of sulfate and carbonate. The rocks of the Yucatán around Chicxulub are formed dominantly of exactly this mixture of rocks, and the igneous rocks under

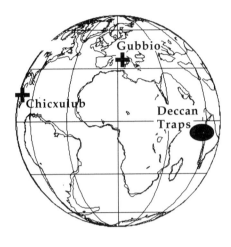

Figure 18.3 Important localities for discussing the K–T boundary, shown in the positions they occupied at the time, 65 Ma. Gubbio is the town in Italy where Walter Alvarez collected the sample that yielded the first evidence of the iridium spike.

Chicxulub have a chemistry of a once-molten mixture of the two. Above the turbidite comes a thin red clay layer only about 5–10 mm thick that contains iridium and shocked quartz.

One can explain much of this evidence as follows: an asteroid struck at Chicxulub, hitting a pile of thick sediments in a shallow sea. The impact melted much of the local crust and blasted molten material outward from as deep as 14 km under the surface. Small spherules of molten glass were blasted into the air at a shallow angle, and fell out over a giant area that extended northeast as far as Haiti, several hundred kilometers away, and to the northwest as far as Colorado. Next followed the finer material that had been blasted higher into the atmosphere or out into space and fell more slowly on top of the coarser fragments.

The egg-shape of the Chicxulub crater shows that the asteroid hit at a shallow angle, about 20°-30°, splattering more debris to the northwest than in other directions. This accounts in particular for the tremendous damage to the North American continent, and the skewed distribution of shocked quartz far out into the Pacific.

Other sites in the western Caribbean suggest that normally quiet, deep-water sediments were drastically disturbed right at the end of the Cretaceous, and the disturbed sediments have the iridium-bearing layer right on top of them. At many sites from northern Mexico and Texas, and at two sites drilled on the floor of the Gulf of Mexico, there are signs of a great disturbance in the ocean at the K–T boundary. In some places, the disturbed seafloor sediments contain fossils of fresh leaves and wood from land plants, along with tektites dated at 65 Ma (Figure 18.4). Around the Caribbean and at sites up the Eastern Atlantic coast of the United States, existing Cretaceous sediments were torn up and settled out again in a messy pile that also contains glass spherules of different chemistries, shocked quartz fragments, and an iridium spike. All this implies that a great tsunami or tidal wave affected the ocean margin of the time, washing fresh land plants well out to sea and tearing up seafloor sediments that had lain undisturbed for millions of years. The resulting bizarre mixture of rocks has been called "the Cretaceous-Tertiary cocktail."

Once Chicxulub was identified, it became possible to calculate that shocked quartz had been launched into a high-angle spray from the impact. This first hot fireball blew vaporized and molten debris—including glass spherules and iridium—high above the atmosphere to be deposited last and globally as it slowly drifted downward. The larger fragments, solid and molten, were blasted outward at lower angles, but not very far, and were deposited first and locally (about 15 minutes travel time to Colorado!). At the same time, smaller fragments, including shocked quartz, were blown upward between the hot fireball and the larger fragments, and were deposited second and regionally (about 30 minutes to reach Colorado). The impact energy, for comparison with hydrogen bomb blasts, was around **100 million megatons**.

Figure 18.4 Two tektites from the K–T boundary at Mimbral, Mexico, including one referred to affectionately as the "Mimbral yoyo." (Courtesy of Phillippe Claeys of the University of California, Davis.)

A Giant Volcanic Eruption?

Exactly at the K–T boundary, a new plume (Chapter 6) was burning its way through the crust close to the plate boundary between India and Africa. Enormous quantities of basalt flooded out over what is now the Deccan Plateau of western India to form huge lava beds called the **Deccan Traps**. A huge extension of that lava flow on the other side of the plate boundary now lies underwater in the Indian Ocean (Figures 18.3 and 18.5). The Deccan Traps cover 500,000 km^2 now (about 200,000 square miles), but they may have covered four times as much before erosion removed them from some areas. They have a surviving volume of 1 million km^3 (240,000 cubic miles) and are over 2 km thick in places. The entire volcanic volume that erupted, including the underwater lavas, was much larger than this (Figure 18.5).

Furthermore, the Deccan eruptions began suddenly just before the K–T boundary. The peak eruptions may have lasted only about one million years ($\pm$ 50%), but that short time straddled the K–T boundary. The rate of eruption was at least 30 times the rate of Hawaiian eruptions today, even assuming it was continuous over as much as a million years; if the eruption was shorter or spasmodic, eruption rates would have been much higher. The Deccan Traps probably erupted as lava flows and fountains like those of Kilauea, rather than in giant explosive eruptions like that of Krakatau. But estimates of the fire fountains generated by eruptions on the scale of the Deccan Traps suggest that aerosols and ash would easily have been carried into the stratosphere. The Deccan plume is still active; its hot spot now lies under the volcanic island of Réunion in the Indian Ocean.

Thus there is strong evidence for short-lived but gigantic volcanic eruptions at the K–T boundary. Some people have tried to explain all the features of the K–T boundary rocks as the result of these eruptions. But the evidence for an extraterrestrial impact is so strong that it's a waste of time to try to explain away that evidence as volcanic effects. We should concentrate instead on the fact that the K–T boundary coincided with two very dramatic events. The Deccan Traps lie across the K–T boundary and were formed in what was obviously a major event in Earth history. The asteroid impact was exactly at the K–T boundary. Certainly something dramatic happened to life on Earth, because geologists have defined the

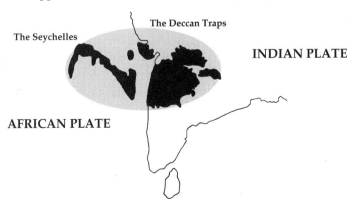

Figure 18.5 The Deccan Traps of India erupted as the Indian and African plates tore apart about 65 Ma. An enormous volume of lava was emplaced above a mantle hot spot (stippled), not only on land on the Indian plate, but offshore to form an enormous underwater mass on the African plate around what is now the Seychelles Islands and part of the Indian Ocean floor. The black areas mark surviving volumes of lava, shown as they were formed in the geography of the K–T boundary. Much more lava must have been erupted than has survived.

K–T boundary and the end of the Mesozoic Era on the basis of a large extinction of creatures on land and in the sea. An asteroid impact, or a series of gigantic eruptions, *or both*, would have had major global effects on atmosphere and weather.

There is a feeling, particularly among physical scientists, that if we can show that a physical catastrophe occurred at the K–T boundary, we have an automatic explanation for the K–T extinctions. But this connection has to be demonstrated, not just assumed. We still have to ask which catastrophe, if either, caused the K–T extinctions, and if so, how?

Did a Catastrophe Cause the Extinctions?

Almost all the scientists directly involved in trying to explain the K–T extinctions are emotionally committed to one catastrophic hypothesis or the other, or are emotionally against both. This has resulted in claims that seem to overinterpret the evidence. One must be prepared to make one's own decision, and certainly all claims must be subject to close scrutiny.

Some Impact Scenarios for Extinction

We think we understand impacts and explosions rather well, after direct study of the Moon's surface, photographic surveys of cratered surfaces on planets and satellites, and our experience with nuclear blasts. We also know that asteroids do strike the Earth. Meteor Crater in Arizona, Manicouagan Crater in Canada, and scores of others can be seen from air photographs; indeed, about 20% of the world's nickel is mined from the Sudbury impact site in Canada, where an asteroid struck about 2000 Ma. Over geological time scales, an asteroid impact is not an unusual event.

Some general predictions of the asteroid impact theory are clear and can be used as indirect tests of its plausibility. The impact of a 10 km asteroid would blow a mass of vaporized rock and steam high above the atmosphere, forming an immense dust cloud that would slowly settle out through the atmosphere over a period of weeks, perhaps several months, perhaps several years. The blast and the cloud would spread material worldwide (Figure 18.6). The scenario has been discussed extensively because similar consequences—**nuclear winter** or at least **nuclear fall**—could result from a thermonuclear war. But realistic models are still not available, and at least some of the discussion is biased one way or another because the topic is so important politically. Nuclear fall models and K–T impact models have been so intertwined in people's minds that results from one tend to be automatically applied to the other in spite of the differences between the two.

Here is one possible impact scenario. An impact at Chicxulub, where the target rocks contain high quantities of sulfur, produces enormous amounts of **sulfate aerosols** in the atmosphere that act as nucleation sites for **acid rains** much more intense and devastating than anything we have generated from industrial pollution. One model suggests rain with the strength of battery acid! The direct effect is enough to suffocate some air breathers, to destroy plant foliage, and to dissolve the shells of marine creatures living along shores and in the surface waters of the ocean. The balance of CO_2 between air and ocean is upset, and a chain of climatic events makes ocean surface waters barren for perhaps 20 years. Among other effects of the impact, dust, smoke, and aerosols cut down the sun's rays for weeks or months, so that land plants and algal plankton in the ocean cannot photosynthesize. The dust also causes freezing air temperatures within days after the impact, and maintains them below freezing for weeks or even months. This may not be an unusual situation at a pole, and may not be a problem for an organism living deep in the ocean, but it is a catastrophe for organisms on continental land masses.

Later, once the dust and aerosols have settled out, the enormous amount of CO_2 released into the atmosphere by the impact generates a greenhouse effect that elevates temperatures on Earth for a thousand years or more.

The most extreme impact scenario could be called the **microwave summer** because it contrasts so much with nuclear winter. It was put together by Jay Melosh and colleagues. In this scenario, some of the material produced in a very large asteroid impact was blasted upward at a velocity greater than Earth's escape velocity, although most of it eventually fell back into the atmosphere on ballistic trajectories after a travel time of about one hour. An asteroid of mass 10^{15}–10^{16} kg would have supplied the observed iridium and spherules, in a depositional layer averaging 10 kg/sq m (about 20 pounds per square yard of Earth's surface).

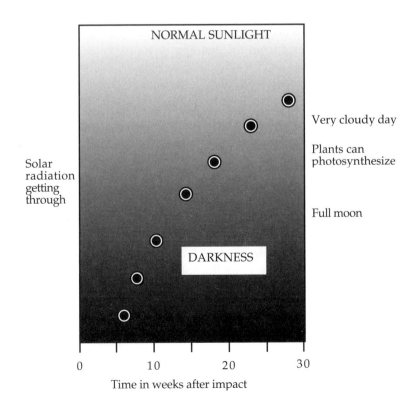

Figure 18.6 An early version of the "impact winter" scenario. The dots show the predictions of a computer program that calculates light levels at the Earth's surface as dust in the atmosphere blocks the sun's rays after an impact. This model is more drastic than some. In this one, a lot of dust falls out slowly. Phytoplankton in the ocean or land plants cannot photosynthesize under any circumstances for 20 weeks (five months) after the impact. (From data in *Geological Society of America Special Paper* 190.)

One can calculate how much thermal radiation the mass of ballistic debris would have emitted as it re-entered the atmosphere. Data on nuclear weapons suggest that the radiation pulse from infalling dust would have been 1000 times more than enough to ignite dry forests.

Ejecta radiation arrives spread over time, however, not in the single radiant pulse generated by an H-bomb. Even so, when we calculate this effect, the rates of worldwide radiation were somewhere between 30 and 100 times that of full sunshine, predominantly in the form of heat.

Of course, half of the radiation was directed upward into space, and some was absorbed by atmospheric water vapor and CO_2. Nevertheless, one-third reached the Earth's surface. It would have taken most of the radiation to evaporate dense cloud, which would therefore largely have protected the surface beneath. Light cloud or no cloud would of course have given little or no shielding. Therefore, Melosh and colleagues estimate surface heating of perhaps 10 kilowatts per square meter for several hours, comparable with the heating in a domestic oven set at Broil.

This radiant heat then generated global wildfires that allegedly left soot in the K–T boundary sections. In general a surface temperature of 545°C is needed for wood to ignite spontaneously, and the radiation could not have produced this on a worldwide basis. But the volatile gases given off by hot wood will burst into flame after 20 minutes at 380°C, which is attained in the scenario. Even local variations in received radiation would have been sufficient to begin fires.

In perhaps the most bizarre of the "What if?" scenarios, if the tropical ocean surface were to reach 50° C, hypercanes (gigantic hurricanes) might have sucked up ice and dust and blown them into the stratosphere, blocking sunlight even more and destroying the ozone layer!

What do we do with these impact scenarios? Naturally, we compare them with the evidence from the geological record. Birds, tortoises, and mammals live on land and breathe air: the evidence from the K–T boundary shows that they survived the K–T boundary event. Therefore they and the air they breathed weren't set on broil for several hours. To put it simply, these scenarios did not happen.

Volcanic Scenarios for Extinction

We also think we know rather a lot about volcanic eruptions. Gigantic eruptions could produce results similar to those of an impact. Volcanic eruptions produce ash, but, even more important, they produce vast amounts of aerosols in the form of sulfuric acid droplets, which stay suspended longer than ash and produce long-lasting effects on climate. Eruptions can sometimes blow material into the stratosphere, where it can be carried over great areas. The eruption of Tambora in 1815 blew out 30 cubic km (7 cubic miles) of ash and dust, which caused spectacular sunsets worldwide and inspired Turner's finest paintings. The darker side of the eruption was that the dust and ash blocked off enough sunlight to cause "The Year Without a Summer" in 1816. Crops failed all over the Northern Hemisphere, resulting in widespread hunger, and even starvation in some areas. The summer was so gloomy in Europe that it depressed Mary Shelley enough to write the famous novel *Frankenstein*.

The eruption of Toba, 75,000 years ago, is the largest documented eruption on Earth, perhaps 100 times the scale of Tambora. Yet the Toba ash is nowhere near the scale of the Deccan Traps. The possible results of the Deccan Traps eruptions include acid rain, ozone depletion, a greenhouse effect, a cooling effect, or any combination of the above: in other words, many of the same effects cited for an asteroid impact.

The Ecology of a Catastrophe

It's easy to imagine that a giant eruption or impact might have caused some kind of catastrophe at the K–T boundary. But it is not certain that it would. The problem with discussing impacts, nuclear war, and eruptions is that we don't know how much dust, smoke, and aerosols would be produced, even though it's absolutely critical to calculations of darkening and temperature change that we know those factors rather precisely. We don't know how far aerosols and stratospheric dust would be carried over the Earth, or in detail what effects they would have. Dust in the air can help absorb solar heat rather than reflect it, for example, and some models of nuclear war suggest that parts of the Earth would warm, parts would cool, and parts would stay at about the same temperature.

In some ways, some volcanic and impact scenarios are similar. For example, some calculations suggest that a Chicxulub impact could have produced hundreds of times more sulfate aerosol than the Tambora eruption in 1815, with its dramatic climatic effect.

The most persuasive scenarios of catastrophic extinction are quickly summarized. Regionally, there is little doubt that the North American continent would have been absolutely devastated. Globally, even a short-lived catastrophe among land plants and surface plankton at sea would drastically affect normal food chains. Pterosaurs, dinosaurs, and large marine reptiles would have been vulnerable to food shortage, and their extinction after a catastrophe seems plausible. Lizards and primitive mammals, which survived, are small and often burrow and hibernate; they would have found plenty of nuts, seeds, insect larvae, and invertebrates buried or lying around in the dark. In the oceans, invertebrates living in shallow water would have suffered greatly from cold or frost, or perhaps from CO_2-induced heating. But deeper-water forms are insulated from heat or cold shock and have low metabolic rates; they therefore would be able to survive even months of starvation. High-latitude faunas in particular were already adapted to winter darkness, though perhaps not to extreme cold. Thus, tropical reef communities could have been decimated, but deep-water and high-latitude communities could have survived much better. All these patterns are observed at the K–T boundary.

Doubts about Catastrophes

The problem with catastrophic hypotheses for the K–T extinctions is that the catastrophes must have been severe but not too severe, because so many creatures survived. Dust and soot must have fallen quickly (within a year) to satisfy some scenarios, but had to remain suspended longer in the atmosphere to produce other effects.

Some specific evidence shows that impacts and eruptions do not necessarily cause catastrophes. For example, a major impact formed the Ries crater in Germany at 15 Ma, throwing huge masses of boulders more than 100 km (60 miles) into Switzerland and the Czech Republic, and tektites several hundred kilometers. The Ries impact did not affect even the local mammal fauna. A major impact at 51 Ma formed the Montagnais crater in the North Atlantic, 45 km (28 miles) across, and an impact hit Chesapeake Bay at 35 Ma, causing a crater 90 km (56 miles) across, but neither of them caused an extinction.

One should beware, however, of dismissing catastrophic explanations because small events do not trigger catastrophes. There may be a threshold effect: if the event is not big enough it will do nothing, but if it is big enough it will do everything. Perhaps there has been only one asteroid impact in the last 500 m.y. large enough to cause a mass extinction (at the K-T boundary); perhaps there have been only two eruptions large enough, at the K–T boundary and/or the P–Tr boundary.

Despite the model predictions and despite reasonable evidence about the physical effects, we don't yet know whether an impact and/or an eruption would have catastrophic, severe, or only mild biological and ecological effects, or whether those effects would be local, regional, or global. In each scenario, however, the killing agent is transient: it would have operated for only a short time geologically. Clearly, if such events occur, they are rare. That does not make them impossible, only unlikely. And that means they have to be very persuasive before we accept them!

Paleontological Evidence from the K–T Boundary

The paleontological evidence from the K–T boundary is ambiguous. While many phenomena are well explained by an impact or a volcanic hypothesis, others are not. The fossils do provide us with real evidence about the K–T extinction events, instead of inferences from analogy or from computer models.

The best-studied terrestrial sections across the K–T boundary are in North America. Immediately this is a problem, because we know that the effects of the asteroid impact were greater here than in most parts of the world. Perhaps this has given us a more catastrophic view of the boundary event that we would gather from, say, comparable careful research in New Zealand. Even so, it is obvious that life, even in North America, was not wiped out: many plants and animals survived the K–T event.

Land Plants

North American land plants were devastated from Alberta to New Mexico at the K–T boundary. The sediments below the boundary are dominated by angiosperm pollen, but the boundary itself has little or no angiosperm pollen and instead is dominated by fern spores in a spore spike analogous to the iridium spike (Figure 18.7). Normal pollen counts occur immediately after the boundary layer. The spore spike therefore coincides precisely with the iridium spike in time and is equally intense and short-lived.

The spore spike could be explained by a short but severe crisis for land plants, generated by an impact or an eruption, in which all adult leaves died off for lack of light, or in a prolonged frost, or in acid rain. Perhaps ferns were the first plants to recolonize the debris, and higher plants returned later. This happened after the eruption of Krakatau in 1883. Ferns quickly grew on the devastated island surfaces, presumably from windblown spores, but they in turn were replaced within a few decades by flowering plants as a full flora was reestablished.

Evidence from leaves confirms the data from spores and pollen. Land plants in North America recovered from the crisis, but many Late Cretaceous plant species were killed off. The survivors probably remained safe during the crisis as seeds and spores in the soil, or even as roots and rhizomes.

Angiosperms were in the middle of a great expansion in the Late Cretaceous (Chapter 17), and the expansion continued into the Paleocene and Eocene. Yet there were important and abrupt changes in North American floras at the K–T boundary. In the Late Cretaceous, for example, an evergreen woodland grew from Montana to New Mexico in a seasonally dry, subtropical climate. Changing leaf patterns indicate that the climate was slowly warming during the latest Cretaceous. At the boundary the dominantly evergreen Late Cretaceous woodland changed to a

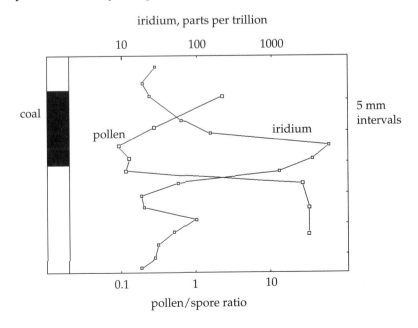

Figure 18.7 In many K–T boundary sites the iridium spike coincides with a spore spike. For some unknown period of time, hardly any angiosperm pollen was deposited, though fern spores were deposited in abundance. These data are from the same locality as those of Figure 18.1.

largely deciduous Early Cenozoic swamp woodland growing in a wetter climate. The fern spike represents a period of swampy mire at the boundary itself. Deciduous trees survived the K–T boundary events much better than evergreens did; in particular, species that had been more northerly spread southward. These changes could be explained in two different catastrophic scenarios: a regional catastrophe that wiped out all vegetation locally, with recolonization from survivors from the north; or a catastrophe that selectively destroyed evergreen plants.

Plants in Japan were affected less than North American ones, and Southern Hemisphere plants were hardly affected at all. Most likely, this reflects the fact that North America was hit far harder than any other continent by the Chicxulub impact.

Freshwater Communities

Some ecological anomalies at the K–T boundary are not easily explained by a catastrophic scenario. Freshwater communities were less affected than terrestrial ones. For example, turtles and a more primitive group of aquatic reptiles, the champsosaurids, survived in North Dakota while dinosaurs were totally wiped out. Freshwater communities are fueled largely by stream detritus, which includes the nutrients running off from land vegetation. It has been suggested that animals in food chains that begin with detritus rather than with primary productivity would survive a catastrophe better than others. That may be true generally and seems to be true for freshwater communities at the K–T boundary, but such communities would survive any ecological crises better, catastrophic or not.

Environmental Sex Determination

Most catastrophic scenarios are so severe that it's difficult to see how some groups of animals survived. Many living reptiles have environmental sex determination (ESD). The sex of an individual with ESD is not determined genetically, but by the environmental temperatures experienced by the embryo during a critical stage in development. Often, but not universally, the sex that is larger as an adult develops in warmer temperatures. This pattern probably evolved because, other things being equal, warmer temperatures promote faster growth and therefore larger final size (at least for ectotherms). Female turtles are larger than males because they carry huge numbers of large eggs, so baby turtles tend to hatch out as females if the eggs develop in warm places and as males in cooler places. (This makes turtle farming difficult.) Crocodiles and lizards are just the reverse. Males are larger than females because there is strong competition between males, so eggs laid in warmer places tend to hatch out as males. ESD is not found in warm-blooded, egg-laying vertebrates (birds and monotreme mammals), and it didn't occur in dinosaurs if they too were warm-blooded.

ESD is found in such a wide variety of ectothermic reptiles today that it probably occurred also in their ancestors. If so, a very sudden change in global temperature should have caused a catastrophe among ectothermic reptiles at the K–T boundary. But it did not. Crocodilians and turtles were hardly affected at all by the K–T boundary events, and lizards were affected only mildly.

High-Latitude Dinosaurs

Late Cretaceous dinosaurs lived in very high latitudes north and south, in Alaska and in South Australia and Antarctica. These dinosaurs would have been well adapted to strong seasonal variation, including periods of darkness and very cool temperatures. An impact scenario would not easily account for the extinction of such animals *at both poles.*

Birds

The survival of birds is the strangest of all the K–T boundary events, if we are to accept the catastrophic scenarios. Smaller dinosaurs overlapped with larger birds in size and in ecological roles as terrestrial bipeds. How did birds survive while dinosaurs did not? Birds seek food in the open, by sight; they are small and warm-blooded, with high metabolic rates and small energy stores. Even a sudden storm or a slightly severe winter can cause high mortality among bird populations. Yet an impact scenario, according to its enthusiasts, includes "a nightmare of environmental disasters, including storms, tsunamis, cold and darkness, greenhouse warming, acid rains and global fires." There must be some explanation for the survival of birds, turtles, and crocodiles through any catastrophe of this scale, or else the catastrophe models are wrong.

Where Are We?

It is clear that at least the extreme "impact winter" models are wrong. It's not clear that impact hypotheses or volcanic hypotheses can explain satisfactorily the extinction patterns we see in the fossil record. There are nagging fears that we are overstating the effects of the impact because the results are so clear in North America, close to the impact site.

Oceanographic Change

An impact or a gigantic eruption that might otherwise have caused only a regional extinction might have caused the global K–T extinction by inducing longer-term climatic changes. These changes would be best recorded in ocean sediments and marine fossils. Tropical reef communities were drastically affected in the K–T extinctions, as were microplankton in the surface waters of the ocean. The pattern of marine K–T extinctions is consistent with a massive breakdown in normal marine ecology.

Oxygen isotope measurements across the K–T boundary suggest that oceanic temperatures fluctuated markedly in Late Cretaceous times and through the boundary events. Furthermore, carbon isotope measurements across the K–T boundary suggest that there were severe, rapid, and repeated fluctuations in oceanic productivity in the 3 m.y. before the final extinction, and that productivity and ocean circulation were suppressed for at least several tens of thousands of years just after the boundary, and perhaps for 1 or 2 m.y. afterward. These changes could have devastated terrestrial ecosystems as well as marine ones. Steve D'Hondt has suggested that climatic change is the connection between the impact and the extinction: the impact upset normal climate, with long-term effects that lasted much longer than the immediate and direct consequences of the impact.

There were survivors: hardly any major groups of organisms became entirely extinct. Even the dinosaurs survived in one sense (as birds). In particular, planktonic diatoms survived well, possibly because they have resting stages as part of their life cycle. They recovered as quickly as the land plants emerged from spores, seeds, roots, and rhizomes. The sudden interruption of the food chains on land and in the sea may well have been quite short, even if full recovery of the climate and full marine ecosystems took much longer. D'Hondt et al. suspect that normal surface productivity was re-established in the oceans after a few thousand years at most. However, it took about three million years for the full marine ecosystem to recover, probably because so many marine predators (crustaceans, molluscs, fishes, and marine reptiles) had disappeared, and had to be replaced by evolution among surviving relatives.

We still do not have an explanation for the demise of the victims of the K–T extinction, while so many other groups survived. We do not know whether it was the impact alone, or the *combination* of the impact and the plume volcanism, that caused the extinction, and we do not know the linkages between the physical events and the biological and ecological effects. It would be astonishing if the impact played no role, and it would be astonishing if the volcanism played no role.

The unusual severity of the K–T extinction, its global scope, and the sudden and dramatic biological features such as the fern-spore spike may have happened because an asteroid impact **and** a gigantic eruption occurred when global ecosystems were particularly vulnerable to a disturbance of oceanic stability. We will probably gain a better perspective on the K–T boundary as we gather more information about the Late Permian and Late Devonian extinctions. It looks increasingly probable that the Permian extinction was linked with a massive plume eruption, and this may mean that mass extinctions need <u>either</u> an external (impact) trigger <u>or</u> an internal (volcanic) one, and in addition they also require a tectonic or geographic setting that made the global ecosystem vulnerable.

Cretaceous conditions weren't stressful,

Dinosaurs were very successful;

Their sudden demise

Is quite a surprise,

But our theories are woefully guessful!

Review Questions

Name three major groups of vertebrates that survived the K–T extinction.

Name three major groups of animals that became extinct at the K–T boundary.

What evidence is there that an asteroid hit Earth at the K–T boundary?

Describe one major result of an asteroid impact on Earth. How might that result affect life on Earth?

Argue the case that an asteroid impact caused the K–T extinctions.

Argue the case that an asteroid impact did not cause the K–T extinctions.

Further Reading

Easy Access References

Alvarez, L. W., et al. 1980. Extraterrestrial cause for the Cretaceous-Tertiary extinction. *Science* 208: 1095–1108. The paper that started it all.

Alvarez, W., et al. 1995. Emplacement of Cretaceous-Tertiary boundary shocked quartz from Chicxulub crater. *Science* 269: 930–935. Why the shocked quartz overlies the rest of the impact layer.

Alvarez, W. 1997. *T. rex and the Crater of Doom*. Princeton University Press. The best of many books on the extinction. Try to read it first.

Bourgeois, J., et al. 1988. A tsunami deposit at the Cretaceous–Tertiary boundary in Texas. *Science* 241: 567–570.

Chapman, C. R., and D. Morrison. 1994. Impacts on the Earth by asteroids and comets: assessing the hazard. *Nature* 367: 33–40.

D'Hondt, S., et al. 1998. Organic carbon fluxes and ecological recovery from the Cretaceous-Tertiary mass extinction. *Science* 282: 276–279.

Duncan, R. A., and D. G. Pyle. 1988. Rapid eruption of the Deccan flood basalts at the Cretaceous/Tertiary boundary. *Nature* 334: 841–843.

Head, G., et al. 1987. Environmental determination of sex in the reptiles. *Nature* 329: 198–199.

Hildebrand, A. R., et al. 1995. Size and structure of the Chicxulub crater revealed by horizontal gravity gradients and cenotes. *Nature* 376: 415–417, and comment, pp. 386–387.

Koeberl, C., et al. 1996. Impact origin of the Chesapeake Bay structure and the source of the North American tektites. *Science* 271: 1263–1266. An 90-km crater from 35 Ma under Chesapeake Bay.

MacDougall, J. D. 1988. Seawater strontium isotopes, acid rain, and the Cretaceous-Tertiary boundary. *Science* 239: 485–487.

Melosh, H. J., et al. 1990. Ignition of global wildfires at the Cretaceous/Tertiary boundary. *Nature* 343: 251–254. Microwave summer.

Morgan, J., et al. 1997. Size and morphology of the Chicxulub impact crater. *Nature* 390: 472–476.

O'Keefe, J. A., and T. J. Ahrens. 1989. Impact production of CO_2 by the Cretaceous/Tertiary impact bolide and the resultant heating of the Earth. *Nature* 338: 247–248.

Rabinowitz, D. L., et al. 1993. Evidence for a near-Earth asteroid belt. *Nature* 363: 704–706. These are not killers, but they are important.

Rampino, M. R., and T. Volk. 1988. Mass extinctions, atmospheric sulphur and climatic warming at the K/T boundary. *Nature* 332: 63–65.

Raup, D. M., and J. J. Sepkoski. 1986. Periodic extinction of families and genera. *Science* 231: 833–836.

Schuraytz, B. C., et al. 1996. Iridium metal in Chicxulub impact melt: forensic chemistry on the K-T smoking gun. *Science* 271: 1573-1576.

Sharpton, V. L., et al. 1992. New links between the Chicxulub impact structure and the Cretaceous/Tertiary boundary. *Nature* 359: 819–821.

Stothers, R. B. 1984. The great Tambora eruption of 1815 and its aftermath. *Science* 234: 1191–1198.

Turco, R. P. et al. 1990. Climate and smoke: an appraisal of nuclear winter. *Science* 247: 166–176. Revised version of their original 1983 suggestion.

White, R. S., and D. P. McKenzie. 1989. Volcanism at rifts. *Scientific American* 261(1): 62–71. Why the Deccan Traps are so enormous.

More Technical Reading

D'Hondt, S., et al. 1996. Oscillatory marine response to the Cretaceous-Tertiary impact. *Geology* 24: 611–614.

Emanuel, K. A., et al. 1995. Hypercanes: a possible link in global extinction scenarios. *Journal of Geophysical Research* 100: 13755–13765.

Finnegan, D. L., et al. 1986. Iridium emissions from Kilauea volcano. *Journal of Geophysical Research* 91: 653–663.

Heissig, K. 1986. No effect of the Ries impact event on the local mammal fauna. *Modern Geology* 10: 171–191.

Li, L., and G. Keller. 1998. Abrupt deep-sea warming at the end of the Cretaceous. *Geology* 26: 995–998.

Pope, K. O., et al. 1997. Energy, volatile production, and climatic effects of the Chicxulub Cretaceous/Tertiary impact. *Journal of Geophysical Research* 102: 21645–21664.

Rampino, M. R., et al. 1988. Volcanic winters. *Annual Reviews of Earth and Planetary Science* 16: 73–99.

Ryder, G., et al. (eds.) 1996. *The Cretaceous-Tertiary Event and Other Catastrophes in Earth History. Geological Society of America Special Paper* 307. Papers presented at the third major international conference on the K/T event. Two previous conferences were published in the same series, volumes 190 and 247 (1990).

Schultz, P. H., and S. D'Hondt. 1996. Cretaceous-Tertiary (Chicxulub) impact angle and its consequences. *Geology* 24: 963–967.

Sheehan, P. M., and T. A. Hansen. 1986. Detritus feeding as a buffer to extinction at the end of the Cretaceous. *Geology* 14: 868–870.

Sigurdsson, H., et al. 1992. The impact of the Cretaceous/Tertiary bolide on evaporite terrane and generation of major sulfuric acid aerosol. *Earth and Planetary Science Letters* 109: 543–559.

Warme, J. E., and Sandberg, C. A. 1996. Alamo megabreccia: record of a Late Devonian impact in southern Nevada. *GSA Today* 6: 1–7.

Wolbach, W. S., et al. 1990. Fires at the K/T boundary: carbon at the Sumbar, Turkmenia, site. *Geochimica et Cosmochimica Acta* 54: 1133–1146. The assumptions in their previous papers have now become facts!

Wolfe, J. A. 1987. Late Cretaceous-Cenozoic history of deciduousness and the terminal Cretaceous event. *Paleobiology* 13: 215–226.

Wolfe, J. A., and G. R. Upchurch. 1986. Vegetation, climatic, and floral changes at the Cretaceous-Tertiary boundary. *Nature* 324: 148–152.

CHAPTER NINETEEN

Cenozoic Mammals: Origins, Guilds and Trends

The end of the Cretaceous Period was marked by so many changes in life on the land, in the sea, and in the air that it also marks the end of the Mesozoic Era and the beginning of the Cenozoic Era (in which we live). Survivors of the Cretaceous extinctions built up into a very impressive and varied set of organisms, beginning in the Paleocene epoch, the first 10 m.y. of the Cenozoic. In the marine fossil record, the Cenozoic era is dominated by molluscs, especially by bivalves and gastropods, the clams and snails of beach shell collections.

On land, the Cenozoic is marked by the dominance of flowering plants, insects, and birds, and in particular by the radiation of the mammals from insignificant little insectivores into dominant large animals in almost all terrestrial ecosystems. Cenozoic mammals have a very good fossil record. There are thousands of well-preserved skeletons, and we understand their evolutionary history very well. I shall not try to give anything close to an overall survey of mammalian evolution. Instead, I shall use the mammal record to illustrate the ways in which evolution has acted on animals, because the same effects can be seen (more dimly) throughout the rest of the fossil record.

Evolution is the overall result of environmental factors acting on organisms through natural selection. But it is easier to understand evolutionary processes if we can isolate some of the different aspects involved. In this chapter and the next, I shall look at four major aspects of evolution as I survey Cenozoic life, and in each case I shall try to identify the various opportunities that allowed or encouraged evolutionary change to occur:

- The ecological setting of evolution

- Improving or changing well-defined adaptations

- Geographical influences on evolution

- Climatic influences on evolution

In this chapter we see how successive groups of mammals evolved to replace dinosaurs, and we discuss some of the major evolutionary events of the Cenozoic era. Much of the turnover in the fossil record consists of the **ecological replacement** of one group of animals by another. An older group may disappear, for various reasons, offering an ecological opportunity for a new set of species that evolves and replaces the older set. Sometimes the new group outcompetes the older group, so that we see not just ecological replacement but **ecological displacement**. There are many examples of **parallel evolution**, in which certain body patterns that are apparently well suited for a particular way of life evolve again and again in different continents at different times. Understanding these processes helps us to sort through the complexity of catalogs of fossils.

Then we look at a smaller-scale phenomenon, **evolution by improvement**. Given that a particular body plan is well suited for executing a particular way of life, we often see changing morphology through time within a single group of organisms. These evolutionary changes can often be interpreted as a series of increasingly good adaptations to the characteristic way of life, or as a set of alternative adaptations within the general way of life. **Coevolution**, as in the arms race between predator and prey, in the relationship between plant and herbivore, or between plant and pollinator, can lead to increasingly efficient adaptation. In a successful, long-lived group that survives, one can trace the various adaptations that eventually led to the characters of the survivors. One can then put together a narrative story of a lineage—The Evolution of the Horse, or Elephant, or Giraffe, or whatever the living representative of the group happens to be. Obviously, one must first have a good idea of the evolutionary relationships within the group (a reliable phylogram, in other words). In almost all cases, the evolutionary and adaptive pattern of a group is not a straight line but a winding pathway through time. But the attempt to trace a lineage through the complexity of evolution can be instructive, showing how the adaptations respond to environmental opportunities.

The Evolution of Cenozoic Mammals

The surviving major groups of terrestrial creatures after the Cretaceous extinction were mammals and birds. Crocodilians were amphibious rather than terrestrial. Most Mesozoic mammals had been small insectivores, probably nocturnal and probably with limbs adapted for agile scurrying rather than fast running. Flying birds must be small, but there is not the same constraint on terrestrial birds. There was probably intense competition between ground-dwelling birds and mammals in a kind of ecological race for large-bodied ways of life during the Paleocene, with crocodilians playing an important secondary role in some areas. The mammals were most successful, and they evolved explosively, their diversity rising from eight to 70 families.

The Radiation of Mammals: Molecular Studies

The fossil record suggests that there was "explosive radiation" among mammals in the early Cenozoic. This receives a ready explanation: the dinosaurs had been dominant in terrestrial ecosystems, world-wide, for over 100 million years. With their disappearance, new ecological roles suddenly became available for mammals (and birds), and dramatic adaptive radiation was a predictable response to that ecological opportunity. We see very few mammals in Cretaceous rocks, and they are all small.

But is that explosive radiation real? Perhaps the different groups of mammals had already diverged genetically, at small body size, long before the end of the Cretaceous, but were *ecologically released* after the K–T extinction. (Note that this question is exactly the same as the one we asked about the radiation of the metazoans in the late Precambrian [Chapter 4]).

How would we detect and describe a Cretaceous radiation of major mammal lineages? We could look more carefully at Cretaceous mammals, to try to find advanced characters among them. But the record is so poor that this approach has been very difficult. In any case, if the ancestors of, say, horses were mouse-sized, they would not look like, eat like, run like, or behave like horses, so they would also lack most of the skeletal characters that we use to recognize horses. Genetics is not ecology. An alternative approach is to look at molecular evidence.

Molecular geneticists have proposed that under certain circumstances, evolutionary changes in DNA and proteins can be selectively neutral, unaffected by natural selection. Such molecular changes should happen at a random rate that is fairly constant through time. In theory, such **molecular clocks** of evolutionary change, based on proteins, or specific genes, or DNA sequences from the nucleus or from mitochondria, may allow us to determine the times of divergence of *living* animals without ever having to look at or for their fossil ancestors.

The DNA clock suggests, for instance, that the common ancestor of apes and humans split from monkeys about 33 Ma. Gibbons split off about 22 Ma, followed by the orangutan at about 16 Ma. Finally the various living lineages of hominids diverged from one another between 10 Ma and 6 Ma. Protein clocks suggest more recent branching points.

It is increasingly clear that the clocks run at different speeds within different clades of animals. And in any case, neither protein changes nor DNA changes have happened with true clocklike regularity. Genes can be transferred from mitochondria to nucleus, for example, muddying the analysis. Arguments about clock rates often depend on complex arguments about the correct statistical methods to be used. It is clear that no simple analysis will turn out consistently reliable or even plausible dates for evolutionary divergences.

Molecular geneticists have discovered that some molecules and gene sequences are more "reliable" than others, and increasingly are driven to sophisticated analyses that will distinguish between "more reliable" and "less reliable molecules". If the assumptions are valid, the result will be more and more reliable reconstructions of evolutionary pathways.

While molecular geneticists have used the word "clock" so many times that it is built into their minds and their publications, one might whisper the thought that there may be no clock at all.

The plain biological fact is that it is dangerous to assume that amino acid or nucleotide substitutions, or mitochondrial genes, are selectively neutral. If evolution in these molecules responds to natural selection, then they will not show clock-like behavior. Change will be much more rapid at some times than others. One can average any kind of change over time, but the concept of a clock is plain wrong.

There is evidence of strong adaptive change in some proteins. **Lysozyme** is an enzyme that attacks bacterial cell walls, and it is used as a defensive chemical to resist bacterial invasion (for example, it is found in our tears and saliva). It plays a special role in digesting bacteria in the digestive tract of some mammals. The lysozymes in the stomachs of leaf-eating monkeys and cattle are almost identical, yet they are significantly different from the lysozymes of other mammals. Cattle and leaf-eating monkeys are very distantly related, but they both have a complex stomach with an anterior section where cellulose is broken down by fermenting bacteria, The bacteria are digested with the rest of the food in the normal, posterior stomach, with the aid of lysozyme. The lysozyme of leaf-eating monkeys and cattle is unusual because it can act at high acid levels and can resist attack by the digestive enzyme pepsin. Its amino acid structure has evolved by natural selection, not random change, independently and for good adaptive reasons in these two groups of mammals. One wonders how many other protein or DNA changes are adaptive.

Mitochondria play a vital role in physiology. Their genes mutate, of course, and mitochondrial diseases caused by such mutations can be catastrophic to the host animals that carry them. In this sense, mitochondrial genes are certainly subject to selection. But there is more. Mitochondria oxidize molecules, and dangerous oxygen radicals crop up in the reactions. Oxygen radicals have been implicated in ageing, and in many diseases. Just to take a simple example, animals belonging to a clade that evolved to larger size (so had an increased life span) would face a longer exposure to radicals leaking from mitochondria throughout their longer lives. One would predict strong selection as hosts and mitochondria both adjusted (by natural selection) to the new long life of the host.

In the same way, host animals and mitochondria would face important selection pressures if the host evolved a higher metabolic rate (involving increased oxygen flow through the mitochondria). Certainly birds and mammals, now endothermic, seem to have evolved different relationships with their respective mitochondria. Bird mitochondria leak fewer oxygen radicals into the tissues of their hosts, and birds live much longer than mammals of the same weight (rats versus parrots, for example). The hominoid primates (humans, chimps, gorillas) live much longer than other mammals of the same size, though the research has not yet been done on the relationship between us and our mitochondria.

 Obviously, molecules evolve, and the patterns of change will help greatly in understanding evolution. My point here is that molecular geneticists cling to the idea of a clock when they should not. Molecular change can be expressed on cladograms just like morphological change. **Cladograms** based on physical morphology do not use a time factor: they simply record similarities and differences. Time is introduced only into a **phylogram**, which expresses a different hypothesis (Chapter 3). Molecular differences should be integrated into paleontological observations without the concept of clock-like properties. And at that level, we will have access to a wealth of data that we can trust.

The major drawback of molecular paleontology is the fact that we can only collect evidence from living animals. If extinct clades played a major part in evolutionary pathways, molecular studies could not reveal it. Remember Figure 7.5. Paleontologists believe that they can reconstruct evolution with confidence because they have the fossil record to work with, including the ancestors and the extinct branches that cannot be studied by molecular geneticists. In the end, molecular evidence is calibrated from the fossil record, and molecular hypotheses are checked against the fossil record. (In fairness, I have to report that at least one genetically impressed anthropologist has wondered whether paleontologists should now be buried along with their fossils.)

It is important to remember that there is only one true evolutionary story, and we are trying to reconstruct it with all the tools available. Molecular and fossil evidence are both relevant; neither can give the complete story, and we must be careful to use each of them with full knowledge of the biases in each method.

So let us return to the radiation of the mammals. The Paleocene radiation of mammals that we see in the fossil record apparently occurred so fast that we cannot distinguish the very great number of branches that resulted in the major groups of mammals alive today. If those branches took place in the late Cretaceous, we can only hope to detect them by some fortunate fossil finds. Meanwhile, molecular geneticists are painting a vivid picture of deep, early, and unexpected branching events among Cretaceous placental and marsupial mammals. In this case, clock assumptions play a major role, and most paleontologists strongly reject the suggestion.

Figure 19.1 *Chriacus*, a Paleocene arctocyonid condylarth that looks very much like the living coati. (Restoration by E. Kasmer, under the supervision of K. D. Rose. Courtesy of Kenneth D. Rose of The Johns Hopkins University.)

The Radiation of Mammals: Molecular Results

Stripped of "clock" assumptions, many molecular results are reasonable (that is, they agree with fossil evidence!), which gives confidence in the methods. Thus, marsupials and monotremes always fall outside the groups that form the placental mammals. However, some results within placentals came as a surprise. Some of these are very exciting, because they give insights into the mammalian radiation that had not been discovered by standard morphological comparison .

For example, a large group of African mammals forms a clade separate from other placentals. The clade includes elephants, sea-cows, hyraxes, aardvarks, elephant shrews, and golden moles: a tremendous array of different body plans, sizes, and ecologies. It is unanimously supported by all available molecular evidence. It implies that Africa became isolated in Cretaceous times with a set of early placentals that evolved to fill all these ecological roles, separate from evolution on other continents. I shall return to this story in Chapter 20.

Some results are counter-intuitive. For example, whales form a clade with an early offshoot of the lineage that gave us artiodactyls—antelope, cattle, deer, pigs, and so on. Carnivores are linked with perissodactyls—horses, rhinos, tapirs, and so on. And all these four groups are apparently linked together as a clade. In contrast, zoologists and paleontologists have always linked artiodactyls and persissodactyls within a large group of herbivores, the ungulates. In addition, molecular evidence suggests that rodents may not be a true clade, but an artificial assemblage of little gnawing mammals that evolved much the same set of teeth, habits, and ecology. All these results need a lot of further analysis and debate.

Now let us return to the fossil record, which, remember, contains many extinct clades that played major roles in evolution and ecology.

The Paleocene

Paleocene mammals included recognizable ancestors of a great many living groups, including marsupials, shrews, rabbits, modern carnivores, elephants, primates, whales, and hedgehogs. South America was already isolated geographically, and ancestors of its peculiar fauna can be recognized there.

Among all this diversity, the **condylarths** were the dominant group of Paleocene mammals. They were generalized, rapidly evolving early ungulates. Most of them were herbivores of various sizes. But the arctocyonids had low, long skulls with canines and primitive molars, and were probably raccoonlike omnivores. *Chriacus* (Figure 19.1) had much the same size and body plan as the tree-climbing coati, but *Arctocyon* itself was the size of a bear and probably had much the same omnivorous ecology. Mesonychids were probably otter-like carnivores or scavengers, but some of them were good running predators on land. For those interested in the largest of anything, the mesonychid *Andrewsarchus* from the Eocene of Mongolia was the largest terrestrial carnivore/scavenger among mammals, with a skull nearly 1 meter (3 feet) long.

Paleocene mammals are generally primitive in their structure, but after a drastic turnover at the end of the epoch, many new groups appeared in the Eocene that survive to the present.

The Eocene

The turnover at the end of the Paleocene is partly related to climatic change and partly to free migration of mammals across the northern continents of Eurasia and North America. Roughly the same fossil faunas are found across the Northern Hemisphere in North America and Eurasia. South America, Africa + Arabia, India, and Australasia were island continents to the south of this great northern land area (Figure 19.2), and their faunal evolution is discussed separately in Chapter 20.

Many modern groups of mammals appeared very early in the Eocene, including rodents, advanced primates, and modern artiodactyls and perissodactyls. By the end of the Early Eocene, digging, running, climbing, leaping, and flying mammals were well established at all available body sizes. Above all, Eocene faunas record the evolution of many different groups of mammals into herbivores of all sizes. Many of these early herbivores were small or medium-sized, including the earliest known horse, *Hyracotherium*, but soon there were large-bodied herbivores that ranged up to 5 tons. In North America the large herbivores were **uintatheres**, followed by **titanotheres**; in South America they were **astrapotheres** (Figure 19.3); and in the Old World, especially in Africa, they were **arsinoitheres**.

Perissodactyls and artiodactyls appeared abruptly at the base of the Eocene in North America. Perissodactyls may have invaded, but it is likely that artiodactyls evolved in North America from an arctocyonid condylarth such as *Chriacus* (Figure 19.1). Small at first, both groups evolved long, slim, stiff legs and other adaptations for fast running.

Proboscideans (elephants and related groups) and sea cows, which belong to an African clade (p. 304), evolved along the African shores of the tropical Tethys Ocean that spread east-west between Africa and Eurasia. Many other herbivores evolved in isolation in South America.

Mammals did not evolve quickly into large carnivores. Some early carnivorous mammals, the mesonychids, arctocyonids, and creodonts,

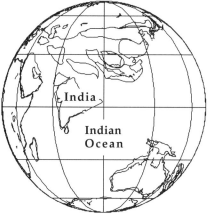

Figure 19.2 Eocene world geography at about 50 Ma, showing the Atlantic Ocean (above) and Indian Ocean (below). The northern land masses were close together, and land animals could walk freely from one to the other through the comparatively mild climates of the Eocene Arctic. India, Australasia, Africarabia, and South America could all be considered as island continents, because Australasia, South America, and Antarctica were linked only by a difficult passage through polar regions.

Figure 19.3   Many groups of mammalian vegetarians have evolved to about 5 tons body weight at different times and on different continents. This is *Astrapotherium* from the Early Cenozoic of South America. (After Riggs.)

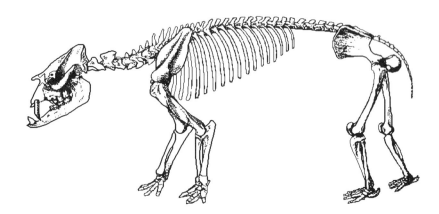

Figure 19.4   The giant carnivorous bird *Diatryma*. (Reconstruction by Bob Giuliani. © Dover Publications Inc., New York. Reproduced by permission.)

were probably the equivalents in size and ecology of hyenas, coyotes, and dogs. Larger mammals seems to have been omnivores or herbivores.

Along the tropical shores of Eurasia, the aquatic ancestors of whales were evolving into their oceanic life style. *Ambulocetus* was a swimming creature that can confidently be identified as a primitive cetacean, and probably had an ecology like a large sea lion. It swam in otter-like fashion, with a flexible spine, but it still had fairly effective limbs for moving about on shore, and had not yet evolved the tail flukes of later whales.

On land, the carnivorous mammals were outclassed in body size in the northern continents by large, flightless birds with massive heads and tearing and stabbing beaks, the diatrymas (Figure 19.4). Similar carnivorous birds called phorusrhacids (Figure 14.20) evolved independently in South America, and both groups were the dominant predators in their respective ecosystems for some time. At the same time, some crocodiles became important predators on land: for example, the pristichampsid crocodiles of Europe and North America evolved the high skulls, serrated teeth, and rounded tails of terrestrial carnivorous reptiles. One Eocene crocodile evolved hooves!

La Grande Coupure

Toward the end of the Eocene, many families on land and in the sea became extinct and were replaced by others. Naturally, the mammals that became extinct have come to be called archaic and the survivors modern, but this does not necessarily imply that there were major functional differences between them. The event has been called "La Grande Coupure" or "the great cut-off," and it has been well documented in Europe and Asia. Even so, the extinction was much less abrupt than the K–T event. Because it was gradual rather than catastrophic and was accompanied by changes in climate and ocean currents, agents here on Earth were probably responsible.

The Oligocene

As Antarctica became isolated and began to refrigerate, the Earth's climate began to cool on a global scale. It seems that the cooling took place in sharp steps, occasionally reversing for a while, so that there may have been a series of climatic events, each of which set up stresses on the ecosystems of the various continents. For example, a rapid cooling in southern climates in the mid-Oligocene seems to have had global effects, and there were some abrupt extinctions among North American mammals. Later events were even more severe, however.

The Later Cenozoic

In the Miocene the refrigeration of the Antarctic deepened, and its ice cap grew to huge size, affecting the climate of the world. Vegetation patterns changed, creating more open country, and a major innovation in plant evolution produced many species of grasses that colonized the open plains. The mammals in turn responded, and a grassland ecosystem evolved on many continents, continuing with changes to the present. The Savanna Story will receive separate treatment later in this chapter.

Climatic and geographical changes allowed exchanges of mammals between continents, often in pulses as opportunities occurred. A favorite example is *Hipparion*, a horse that migrated out of North America, where horses had originally evolved and spent most of their evolutionary history. It spread across the plains of Eurasia about 11 Ma, leaving its fossils as markers of a spectacular event in mammalian history.

By the end of the Miocene, the mammalian fauna of the world was essentially modern. Two further events demand special attention: the great series of ice ages that have affected the Earth over the last few million years (Chapter 24), and the rise to dominance of animals that greatly changed the faunas and floras of Earth—humans (Chapters 22 and 24).

Ecological Replacement: The Guild Concept

Although ancient mammal communities may have included some strange-looking animals, nevertheless certain ways of life are always present in a fully evolved tropical ecosystem. Plant life is abundant and varied, and provides food for browsers and grazers, usually medium to large in size. Small animals feed on high-calorie fruits, seeds, and nuts. Pollen and nectar feeding is more likely to support really tiny animals. Carnivores range from very small consumers of insects and other invertebrates to medium-sized predators on herbivorous mammals; scavengers can be any size up to medium. There may be a few rather more specialized creatures, such as anteaters, arboreal or flying fruit-eaters, or fishing mammals.

Easily categorized ways of life that have evolved again and again among different groups of organisms are called **guilds**, and their recognition helps to make sense of some of the complexity of mammalian evolution on several continents over more than sixty million years.

For example, the woodpecker guild includes many creatures that eat insects living under tree bark. Woodpeckers do this on most continents. They have specially adapted heads and beaks for drilling holes through bark, and very long tongues for probing after insects. But there are no woodpeckers on Madagascar, where the little lemur *Daubentonia*, the aye-aye, occupies the same guild. It has ever-growing incisor teeth, like a rodent, and instead of using beak and tongue like a woodpecker, it gnaws with its teeth and probes for insects with an extremely long finger. On New Guinea, where there are no primates and no woodpeckers, the marsupial *Dactylopsila* has evolved specialized teeth and a very long finger for the same reasons. Because these three species all belong to the same guild, understanding the adaptations of any one of them helps us to interpret other members. On the Galápagos Islands, the woodpecker finch *Camarhynchus* does not have a long beak but uses a tool, usually a cactus spine, to probe into crevices. In Australia, some cockatoos fill the woodpecker niche, but they rely on the brute strength of their beaks to rip off bark, and they have not evolved the sophisticated probing devices of the others. Another small mammal evolved woodpecker devices fifty million years ago. *Heterohyus*, from the Eocene of Germany, had powerful triangular incisor teeth, and the second and third fingers of each hand were very long, with sharp claws on the ends.

Some guilds are unexpected (to me). For example, there is a recognizable guild of small mammals that live among rocks. From marmots and rock hyraxes to chinchillas, pikas, and rock wallabies, small rock-dwelling mammals on several different continents look alike, behave alike, and even sound alike.

Of course, there is no guarantee that a guild will be occupied by only one major group. In the tropics today, small arboreal animals that feed at night are almost all mammals, but the daytime feeders are almost all birds. Most medium-sized predators and scavengers are mammals, but raptors are very effective at smaller body weights.

Cenozoic Mammals in Dinosaur Guilds

All Mesozoic mammals were small, and small body size implies that animals can play only a limited number of ecological roles, mainly insectivores and omnivores.

When dinosaurs disappeared at the end of the Cretaceous, some of the Early Paleocene mammals quickly evolved to take over many of their ecological roles, particularly as omnivores and vegetarians. Others continued to occupy the same small-bodied guilds that the Mesozoic mammals had occupied for a hundred million years. Even today, 90% of all mammal species weigh less than 5 kg (11 pounds).

Dinosaurs dominated many guilds in the Cretaceous, including that of large browsers. Most of them, such as the ceratopsians, hadrosaurs, and iguanodonts, weighed about 5 to 7 tons as adults. The K–T extinction wiped out all these creatures, and it was not until the late Paleocene that the guild was occupied again, by large mammals.

Although some birds are large herbivores (ostriches are omnivorous, but much of their food is browse), mammals are the dominant browsers and grazers today. Even at the very beginning of the Paleocene, the mammals were dominated not by insectivores but by the largely herbivorous early ungulates. Very late in the Cretaceous, some mammals evolved molars even more complex than tribosphenic molars. The new teeth permitted or even required complex jaw motions, but they allowed much more shearing and grinding than before. The capacity for grinding more and tougher food allowed mammals to turn to low-calorie vegetarian diets.

There seems to be something special about the 5- to 7-ton range for the largest land herbivores. This limit applied to all dinosaurs except for sauropods, and it has apparently applied to all mammals since, including living elephants and rhinos. Presumably there is some metabolic reason for this limit, associated with the fact that browse and forage is low in calories. The 5- to 7-ton size was approached by different mammalian groups in the different continents of the Paleocene and Eocene. The best record is in North America, where uintatheres and titanotheres followed the dinosaurs.

Uintatheres (Figure 19.5) were most successful in the northern continents, but they may have managed to cross from an original home in South America in the Paleocene. They had massive skeletons and gradually increased in size through the Paleocene and Eocene. They had large canine teeth modified into cutting sabers, but they were not carnivores. The large flattened molar teeth were used for grinding vegetation. (The sabers were probably for fighting between adults.) *Uintatherium* itself (Figure 19.5) was as large as a rhino.

These large herbivores were replaced in the large herbivore guild in North America, and later in Asia, by perissodactyls called **titanotheres** (or brontotheres). These were small in the Early Eocene, by the Middle Eocene they were large, and at the end of the Eocene they were very

Figure 19.5 *Uintatherium*, one member of a major group of large-bodied Paleocene and Eocene North American vegetarian mammals. (Reconstruction by Bob Giulani. © Dover Publications Inc., New York. Reproduced by permission.)

Figure 19.6 The Eocene titanothere *Brontops*. (Reconstruction by Bob Giulani. (© Dover Publications Inc., New York. Reproduced by permission.)

large indeed (Figures 19.6, 19.7). They evolved massive blunt horns as they became larger. The horns have been interpreted as ramming devices, but most of them have a shape and a position on the head that would have been much better designed for pushing and wrestling (Figure 19.7). Titanotheres became extinct at the end of the Eocene, and their guild was filled by the modern-looking rhinos in Eurasia and North America. Later, rhinos were joined in the guild by elephants, which had evolved in Africa but did not leave that continent until the Miocene.

Creodonts and Carnivores: Ecological Replacement or Displacement?

As we have seen, the larger-bodied carnivorous guild vacated by dinosaurs was occupied in the Paleocene by crocodiles and flightless birds. Many early mammals were insectivores and only became carnivores with the evolution of larger size. The success of the new early ungulates and their young meant prey for early carnivorous mammals, recognized in fossils by their biting and slashing teeth, powerful jaws, and strong, clawed feet. Most of these early hunters were creodonts. Creodonts included oxyaenids, which were probably ambush predators, and hyaenodonts, which were runners. Some mesonychids were also carnivores or scavengers. These mammals were the dominant small- and medium-sized carnivorous animals in Paleocene and Eocene ecosystems.

Some new small carnivorous mammals, the miacids, appeared in the Middle Paleocene but for 20 m.y. remained rather few in number and small in size. They lived alongside the larger carnivorous groups of creodonts and mesonychids. At the end of the Eocene all the larger predators except a few hyaenodonts became extinct, and miacids underwent an evolutionary radiation that produced all the modern types of carnivores (Figure 19.8): the mustelids (otters, weasels, and badgers), viverrids (mongooses and civets), canids (dogs), felids (cats), ursids (bears), and nimravids (sabertooths).

Figure 19.7 Titanotheres evolved to very large body size between the Early Eocene and the Early Oligocene. The upper diagram shows (A) an early small titanothere, *Eotitanops*, and (B) a gigantic late one, *Brontotherium*, in the act of displaying. The lower diagram shows *Brontotherium* walking naturally with the head held low. The huge double horns look to me like wrestling structures rather than ramming devices. (From Osborn.)

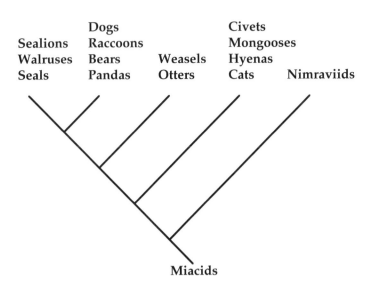

Figure 19.8 Phylogram of the Carnivora.

Did miacids suddenly evolve some character or characters that allowed them to outcompete the mesonychids and creodonts? Or did the disappearance of their competitors allow miacids to evolve into larger-bodied carnivores? In other words, was this change in the carnivore guild the result of replacement or competition? The extinction of creodonts has been attributed at various times to increased intelligence of modern carnivores over creodonts and/or to better running adaptations.

Leonard Radinsky attacked this question by studying the skull structure of living carnivores. The four largest carnivore groups—viverrids, canids (dogs), felids (cats), and mustelids (weasels)—have distinctly different skulls that reflect the different ways they bite and kill. Felids and mustelids have short, powerful jaws, and they kill small prey with a powerful bite at the back of the neck. Large cats kill larger prey with a powerful neck bite that strangles the prey. Canids, on the other hand, have long, slender jaws, and they kill small prey with a head shake. They can kill large prey only by tearing at them in packs. Radinsky hoped to be able to distinguish a new killing capability among miacids that would have given them a competitive edge over creodonts, but he could not see one. In short, there is no obvious key innovation behind the radiation of the modern Carnivora.

Without a key innovation, then, modern carnivores must have succeeded because their larger competitors disappeared for unknown reasons, along with other mammal groups, at the end of the Eocene at La Grande Coupure. This is an example of guild replacement, not of displacement. A final piece of evidence that supports this argument deals with the latest surviving creodonts, the hyaenodonts. The large running mesonychids disappeared at the end of the Eocene, and thereafter a surviving group of hyaenodonts became very large, replacing them. Modern carnivores did not take on the role of large predators until later, in the Miocene, with much larger cats and dogs.

The Savanna Story

Modern Savannas

Research by Samuel McNaughton and his colleagues on the savanna grazing ecosystem of East Africa revealed patterns that may also be true for other ecosystems in space and time.

Herbivores tend to graze off the tops of any plants they can reach, because the top of the plant contains the most tender, juicy parts, and is less well protected by any mineral or chemical compounds the plant produces. Grazing thus promotes the survival and evolution of plants that tend to grow sideways rather than upward. If grazing is continuous, such plants are selected because they lose less of their foliage. They are not shaded out by competitors that grow upward, because the grazing animals remove those competitors. Low plants tend to occupy a smaller area than high plants, so there is space for more plants in a grazed environment. This may often translate into more species as well as more individuals of one species.

McNaughton fenced off savanna areas to protect them from grazing. It turned out that grazed areas actually had much more available vegetation per cubic centimeter than fenced areas. Plants that are not grazed grow tall and airy, not low and bunched. This happens on lawns too, where mowing is artificial grazing. In areas that are grazed, therefore, food resources are densely packed. A grazing animal can get more food per bite than in ungrazed areas, and it feeds more efficiently.

For example, a cow needs a certain level of nutrition per mouthful in order to survive, considering the energy that is required to move, bite, chew, and digest that mouthful. If the cow lived on the Serengeti Plains of East Africa, it could not survive if it had to crop grassland that had grown more than about 40 cm (16 inches) high, but the same environment, already grazed down to 10 cm (4 inches) high, would provide a very rich food supply.

Grazed plants react in more sophisticated ways than by simply altering their growth habit. After some time, they coevolve with the grazers to produce different reproductive patterns and structures. For example, plants that can regrow from the base rather than the growing tip will be favored, as will plants that reproduce by runners. All this has important consequences. It implies that a grazing ecosystem is balanced evolutionarily so that the herbivores are controlling the type and density of their food resources, but at the same time the plant response forces certain behavioral patterns and perhaps social structures on the herbivores. The ecosystem will tilt out of balance unless the grazing pressure is maintained at a minimum level to keep the low-growing plants at an advantage over possible competitors. Grazing animals probably can't do this if they are solitary. Solitary grazers have two problems: they have to spend energy to defend a territory, and in open country they are liable to predation from running carnivores. Living in herds is an efficient solution to this problem, because it removes the need to spend energy on defense of a territory, it increases the chance of early warning of the approach of a predator, it allows group defense, and it provides a better guarantee of the heavy and continuous grazing that maintains a healthy ecosystem.

Furthermore, with a seasonal and local variation in food supply, it is easy to envisage the evolution of a set of grazing species, each specializing in a different part of the available food. In the Serengeti, for example, three different grazers eat grass and herbs. Zebra eat the upper parts of the blades of grass and the herbs, wildebeest follow up and eat the middle parts, and the Thomson's gazelle eats the lower portions. The teeth and digestive systems of each animal are specialized for its particular diet. Thus, a succession of animals grazes the plain at different times, each species modifying the plants in a way that permits its successor to graze more efficiently. A great diversity of grazers is encouraged: today there are ten separate tribes of bovid antelopes on the savannas of Africa.

Because these principles are so general, they have probably operated at least since grasslands spread widely in the Miocene. Furthermore, there were low-plant ecosystems even before the evolution of the first grasses at the end of the Oligocene. The first horses seem to have grazed in open country in the Paleocene, for example. If dinosaurs were warm-blooded, they probably faced similar problems related to feeding requirements per mouthful. Even if dinosaurs were cold-blooded, with lower metabolic requirements, they would still have faced similar problems.

Similar principles probably apply to browsers too. Obviously, the rules will be rather different, because the defense of many plants against browsing is to grow tall quickly. And finally, herbivores, whether they are grazers or browsers, are a food resource for predators and scavengers. The animals of the African savannas are in a delicate and interwoven ecological network.

McNaughton's work, which explicitly defined principles that many workers had guessed at previously, is likely to be a breakthrough not only in the understanding of modern savanna ecosystems but in the interpretation of past ones too.

Savannas in the Fossil Record

A major climatic change in the Miocene was apparently triggered by the refrigeration of the Antarctic and the growth of its huge ice cap. The cooler climate encouraged the spread of open woodland in subtropical latitudes, at the expense of thicker forests and woods. In California, for example, this occurred around 12 Ma, when the climate changed from wet summers to dry summers. There had been open woodland on Earth ever since the Permian, but the plants that grew in the open had been ferns and shrubs. The new feature of Miocene open country was the spread of grasses, with their high productivity.

Savanna ecosystems produce a great deal of edible vegetation, even though grasses have high fiber and low protein. Grasses are adapted to withstand severe grazing; they recover quickly after being cropped because they grow throughout the blade instead of mainly at its growing

Figure 19.9 The gigantic rhinoceros *Baluchitherium*. (Reconstruction by Bob Giuliani. © Dover Publications Inc., New York. Reproduced by permission.)

Figure 19.10 The giant Miocene camel *Aepycamelus*. (Reconstruction by Bob Giuliani. © Dover Publications Inc., New York. Reproduced by permission.)

tip. They have evolved tiny silica fragments, or **phytoliths**, that make them tough to chew and cause significant tooth wear in grazing animals.

The spread of grasses was perhaps encouraged at first by intense grazing pressure, but the whole savanna ecosystem quickly stabilized, no doubt through mechanisms like those suggested by McNaughton. There was a rapid and spectacular evolutionary response, especially from the mammals, which evolved many different grazing forms. This event and its continuation into plains ecosystems today is called The Savanna Story by David Webb, who has done the most to document it. The change in vegetation was worldwide, and although the North American evidence is most complete, similar trends can be traced on all the continents with subtropical land areas. On each continent, the savanna fauna evolved from animals that lived there before the major climatic change.

The animals that were particularly successful in savanna ecosystems were grazers or browsers on these open plains with only scattered woodland patches. Deer and antelope evolved to great diversity. Often their teeth evolved to become very long for their height, or **hypsodont**, with greatly increased enamel surfaces. Elephants, rodents, horses, camels, and rhinos, for example, all evolved jaws and teeth with adaptations for better grinding. Presumably, hypsodont teeth wore longer and permitted a grazer to chew tough fibers and resist the abrasion of phytoliths. Larger savanna animals also showed size and locomotory changes consistent with life in open country where there was nowhere to hide. They became taller and longer-legged, well adapted for running fast. Some of these Miocene plains animals were gigantic, and they included the largest land mammal that has ever lived, the Eurasian rhino *Baluchitherium* (Figure 19.9), as well as the tallest camel.

In the Miocene of North America, the grazers were at first native ruminants such as camelids and horses. Eurasian deer arrived and radiated during the Miocene. The Late Miocene savanna fauna of North America was very rich, peaking at about 50 genera of ungulates and large carnivores, dominated by hypsodont horses, camelids, and pronghorns.

Starting about 9 Ma, this North American savanna ecosystem suffered a series of shocks, including a great extinction that began about 6 Ma. New genera evolved and new immigrants arrived, but they did not come close to replacing the losses.

The extinction patterns are interesting. All nonruminant artiodactyls disappeared except for one peccary, *Platygonus*, which evolved shearing teeth and shifted toward a coarser, more fibrous diet. Grazing horses flourished, but browsing horses disappeared. Only hypsodont camelids survived, while short-toothed forms became extinct. The casualties included a giant camel *Aepycamelus*, 3.5 meters (12 feet) high (Figure 19.10), which was probably a giraffelike browser. Among the ruminants, the major survivors were the pronghorns, which are hypsodont.

The common ecological pattern is adaptation to coarse fodder and more open country, and it presumably reflects a change from savanna to steppe grassland. Perhaps rain shadow effects were produced by major uplift in western North America; but overall, global climates became colder in the Late Cenozoic.

Evolution by Improvement

The fossil record of mammals is so good that we can trace related groups of mammals through long time periods, and often across large areas and across geographic and climatic barriers. In many cases, we can see considerable evolutionary change in the groups, and because we understand the biology of living mammals rather well, we can interpret the changes confidently. Often the changes can be linked with specific biological functions and can be seen as allowing the animals to perform those functions in a more effective way. I shall deal with only one example: horses.

Horses

Horses are only one of a group of ungulates called **perissodactyls** (they usually have an odd number of toes on each foot). Tapirs and rhinos are also living perissodactyls (Figure 19.11), and titanotheres are extinct perissodactyls. While rhinos have had an interesting, eventful, and complex history, tapirs have evolved slowly enough to qualify as living fossils. This is especially interesting because horses have evolved so radically and so rapidly.

Perissodactyls appeared in numbers in the late Paleocene, replacing some of the earlier ungulates as small-bodied browsers. They show good adaptations for running, and their success can be explained in two ways: either the better running ability of the earliest perissodactyls made them more predator-proof than the earlier mammals they replaced, and/or better running ability was needed as Paleocene forest gave way to more open woodland in the early Eocene. The first perissodactyls had larger and more complex brains, too, and they began a radiation into different and often larger body types.

Hyracotherium, the first known horse and one of the first perissodactyls (Figure 19.12), was a browsing animal: the smallest species was about the size of a cat, but other species weighed up to about 35 kg (75 pounds). *Hyracotherium* first appeared in the Early Eocene of North America and Europe. Males were larger than females, with wider faces and longer canine teeth; presumably this was related to fighting among males. Large horses today fight with teeth and hooves, but at cat size or even at 35 kg, a kick does not have much impact, so *Hyracotherium* must have had a quite different fighting style. It was once thought that *Hyracotherium* lived in forests, but paleobotanical evidence shows that horses lived in open woodland right from the start.

The history of horses is perhaps the best known of any fossil vertebrate group. It was one of the earliest evolutionary sequences to be worked out well: paradoxically, the first good version was published in 1851 by Darwin's rival Sir Richard Owen, who named the earliest horse *Hyracotherium* and traced its lineage to the living *Equus.* Twenty years later, Darwin's friend Thomas Huxley was using an improved version of the horse story in his lectures to help promote Darwin's theory of evolu-

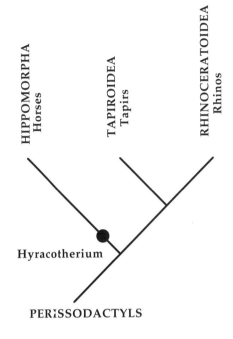

Figure 19.11 Phylogram of the major groups of living perissodactyls.

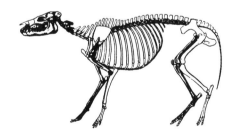

Figure 19.12 *Hyracotherium* was the earliest horse, but was also close to the ancestor of rhinos and tapirs (Figure 19.11). (After Cope.)

tion by natural selection. In 1876, however, Huxley visited the United States and immediately had to rewrite his lectures to incorporate the much richer North American fossil record of horses.

Hyracotherium had rather generalized teeth and probably a vegetarian diet to match. The teeth of *Mesohippus*, in the Early Oligocene, evolved to accentuate a combination of shearing and crushing, perhaps associated with succulent plants and seeds in its diet. Oligocene horses were rather bigger than their Eocene ancestors (perhaps 40–50 kg), and they had longer legs and feet. The number of toes dropped to three on each foot (*Hyracotherium* had four toes on its front feet), and the body frame of ribs and spine grew stronger with greater size.

With the cooling and climatic change in the Miocene, the decline in forested area, and the evolution of grasses as a major component in the larger open areas between woodland patches, there was more food for grazers as well as browsers. A number of Miocene horses survived as browsers, but others began to exploit the new food resource in earnest. Grazing exposes teeth to more abrasive plant material, and a grazer is more likely to get grit and sand in its teeth than a browser is; both conditions lead to greater tooth wear. The height of the molars of *Parahippus* was double that of earlier horses, the enamel was more complex, and the teeth contained more cement. Body size also doubled in *Parahippus*.

An explosive breakthrough of grazing horses came around 17–18 Ma, in the Middle Miocene, as many new species appeared. Species of *Parahippus* and *Merychippus* are good examples: they were the size of small ponies, with three toes on each foot (Figure 19.13, left). A whole set of evolutionary changes in *Merychippus* were adaptations for savanna grazing rather than savanna browsing. The teeth grew larger and longer, with deep roots and still more complex enamel, and they were supported by cement so they would last longer under the impact of a silica-rich diet. They were accommodated in a longer, larger jaw with a wide, flat muzzle, in a longer, deeper, heavier skull. Sophisticated analysis of jaw motions, revealed by wear patterns on its teeth, show that *Merychippus* chewed with a jaw action that differed distinctly from that of its predecessors, accentuating sideways slicing and shearing rather than shearing with downward crushing and compression.

Many of the new grazing horses were larger than their predecessors (100–150 kg), as low-quality forage encouraged larger gut capacity. The limbs became relatively longer and better designed for faster running, as safety came to depend more on speed than on stealth and camouflage. The gait became truly ungulate. Instead of running on a padded foot, as a tapir does and as earlier horses had done, *Merychippus* ran on a hoofed foot raised onto the toes, and the side toes were reduced in size and importance (Figure 19.13). Fast running requires mass to be reduced as much as possible at the ends of the limbs. The feet had powerful ligaments that could flex to reduce the stress of impact in running.

The new savanna habitats encouraged an increased diversity among horses. Between 18 Ma and 15 Ma there was a dramatic increase in the number of grazing horses in North America; in fact, species numbers increased until there were 18 grazing species altogether in North America.

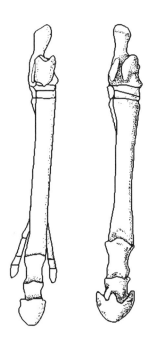

Figure 19.13 The hind feet of the Miocene horse *Merychippus* (left) and the modern horse *Equus* (right). Fast running requires that there should be as little mass as possible at the end of a limb, and *Merychippus* has evolved close to that ideal. *Equus* finally lost the side toes. (After Scott.)

Some of the new horses were large, others small, but all retained the improvements in skeletal and tooth characters that allowed them to graze in open grassland. A group of small, three-toed North American horses called **hipparions** evolved perhaps about 15 Ma in the Middle Miocene. *Cormohipparion* is the horse that found its way out of North America and invaded savannas throughout the Old World, leaving fossils so abundant that they form a valuable marker at about 11 Ma, the so-called **Hipparion Event** (Figure 19.14). Later hipparion species also found their way out of North America to create other, less dramatic markers in Old World fossil collections.

The Late Miocene North American horse *Pseudhipparion* holds the record for extreme tooth evolution. Although it was tiny, it had molar teeth that are proportionately higher than in any other horse known. Normally, teeth stop growing when their roots form. In living horses this happens when the molars are about half worn down. But in *Pseudhipparion* the tooth root did not form until very late in life, so the molars were ever-growing for most of its life. The same extended tooth development has evolved independently in rodents, rabbits, and pronghorns, and in some extinct grazing animals, for the same reason: to prolong the life of the molars and of the animal that uses them.

In North America, the adaptation of horses to open country was accelerated by the rise of the western mountains. The High Plains became steppe grassland rather than savanna. In the Pliocene, modern horses evolved from hipparions by reducing their toes to one (Figure 19.13, right), and *Equus* itself evolved perhaps 5 Ma. The **Equus Event** marks

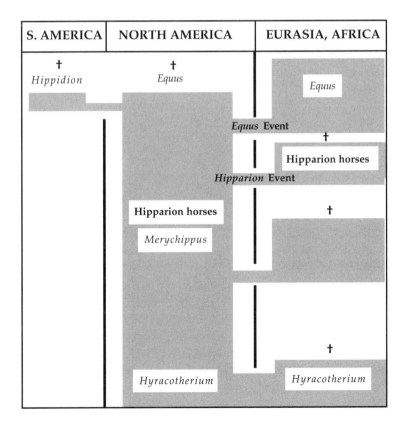

Figure 19.14 The biogeography of horses through time. Diagrams like this have been drawn since the 1880s. This version is based on the work of Bruce MacFadden (1992). Time is not to scale. The diagram shows how Huxley could tell a reasonable but incomplete story of horse evolution from evidence in Eurasia, and why he had to rewrite his lectures when he saw the North American evidence. At times, horses migrated out of North America across geographic barriers (solid vertical lines). Finally, New World horses became extinct, to be reintroduced from the Old World in 1492.

its invasion of the Old World (Figure 19.14). It reached India by 3 Ma and Europe by 2.6 Ma; zebras reached Africa at about the same time.

Horses today are reduced to only seven species worldwide, all of them native to the Old World. Przewalski's horse of Central Asia and the domesticated horse survive only under human protection. All so-called wild horses have escaped or have been released from domestication. One endangered species of wild ass survives in Africa and one in Asia, and three species of zebra remain in Africa. But this low diversity is a result of catastrophic extinctions in the last 2 m.y. At the end of the Pliocene, horses were abundant and diverse on all the continents except Australasia and Antarctica. Viewed at that time by a zoologist, their body plan and their adaptations would have been seen as remarkably successful. The most ironic twist to the horse story is, of course, that horses became extinct in the Americas in the Late Pleistocene (Figure 19.14 and Chapter 24) and were reintroduced only in 1492.

Review Questions

What led to the spread of grasslands over much of the lower and middle latitudes of Earth in Miocene times?

In any community of animals, there is always one that is the most powerful carnivore. Name one animal that was the top carnivore of its time. (You can't use *Homo sapiens*!)

Mammals and birds must have competed for ecological dominance after the end of the Cretaceous. Give an example of a bird that seems to have outcompeted mammals for ecological dominance.

Name one group of animals that took advantage of the great spread of Miocene grasslands, and name a continent that group lived in.

Further Reading

Easy Access Reading

Foote, M., et al. 1999. Evolutionary and preservational constraints on origins of biological groups: divergence times of eutherian mammals. *Science* 283: 1310–1314. Argument against deep Cretaceous roots for many mammal lineages.

Novacek, M. J. 1992. Mammalian phylogeny: shaking the tree. *Nature* 356: 121–125. Review discussing the difficulties of dealing with rapid radiations.

von Koenigswald, W., and H.-P. Schierning. 1987. The ecological niche of an extinct group of mammals, the early Tertiary apatemyids. *Nature* 326: 595–597.

Strauss, E. 1999. Can mitochondrial clocks keep time? *Science* 283: 1435–1438.

Thewissen, J. G. M., et al. 1994. Fossil evidence for the origin of aquatic locomotion in archaeocete whales. *Science* 263: 210–212, and comment, p. 180–181.

Wallace, D. C. 1999. Mitochondrial diseases in man and mouse. *Science* 283: 1482–1488.

Zimmer, C. 1997. The dolphin strategy. *Discover* 18 (3): 72–83.

Zimmer, C. 1998. The equation of a whale. *Discover* 19 (4): 78–84.

More Technical Reading

Benton, M. J. (ed.). 1988. *The Phylogeny and Classification of the Tetrapods. Volume 2: Mammals.* Oxford: Clarendon Press.

Bromham, L., et al. 1999. Growing up with dinosaurs: molecular data and the mammalian radiation. *Trends in Ecology and Evolution* 14: 113–118. Review of this controversy.

Cao, Y., et al. 1998. Conflict among individual mitochondrial proteins in resolving the phylogeny of eutherian orders. *Journal of Molecular Evolution* 47: 307–322.

Cooper, A., and R. Fortey. 1998. Evolutionary explosions and the phylogenetic fuse. *Trends in Ecology & Evolution* 13: 151–156. Section on Cretaceous mammals.

de Jong, W. W. 1998. Molecules remodel the mammalian tree. *Trends in Ecology & Evolution* 13: 270–275.

MacFadden, B. J. 1992. *Fossil Horses.* Cambridge: Cambridge University Press.

Rose, K. D. 1996. On the origin of the order Artiodactyla. *Proceedings of the National Academy of Sciences* 93: 1705–1709.

Savage, R. J. G., and M. R. Long. 1987. *Mammal Evolution: An Illustrated Guide.* London: Natural History Museum.

CHAPTER TWENTY

Geography and Evolution

Natural selection operates on individual organisms partly by their response to their environment. On a larger scale, the evolution of larger groups of organisms is strongly affected by major geographic effects. In this chapter I discuss some aspects of Cenozoic evolution that were affected by geography.

Australia

Australia is linked in people's minds with exotic creatures such as kangaroos and jillaroos, but they are only a part of the story of evolution on this isolated continent. Australian plants, insects, amphibians, reptiles, birds, and mammals are all unusual. Australia and New Zealand were part of Gondwanaland in Cretaceous times, joined to Antarctica in high latitudes. The climate was mild, however, and pterosaurs, dinosaurs, and marine reptiles have been found there. In early Cenozoic times the two land masses broke away from Antarctica and began to drift northward and diverge. In the process, both Australia and New Zealand became isolated geographically and ecologically from other land masses, and evolution among their faunas and floras led to interesting parallels with other continents.

Among amphibians, Australia has (or had) at least two species of frogs that brood young in their stomachs. Instead of the colubrid snakes and vipers that are abundant elsewhere, Australia has had a radiation of elapid snakes (cobras and their relatives) into 75 species, all of them virulently poisonous. The largest Australian predators are the salt-water

Figure 20.1 *Meiolania*, a very large extinct tortoise with horns and a clubbed tail, about 2 meters (6 feet) long, from Lord Howe Island off the east coast of Australia. (After Gaffney.)

crocodiles (the world's largest surviving reptiles), which lurk along northern rivers and shorelines, and large monitor lizards related to the Komodo dragon of Indonesia. Monitors are ambush predators, the largest living Australian monitor being 2 meters (over 6 feet) long. Smaller monitors dig for prey like the badgers of larger continents. In contrast, most Australian mammals are herbivores.

Extinct Australian reptiles included giant horned tortoises (Figure 20.1) that weighed up to 200 kg (450 pounds), a monitor 7 meters (23 feet) long that weighed perhaps a ton, and competed with large terrestrial crocodiles of about the same size and weight. The giant snake *Wonambi* was 6 meters (19 feet) long and must have weighed 100 kg (220 pounds). Extinct Australian birds included *Dromornis*, the heaviest bird that has ever evolved (Chapter 14).

Australia is the only continent with living monotremes. They have been in Australia since the Early Cretaceous (Chapter 15), and only one early Cenozoic monotreme tooth from southern Argentina shows that they once ranged more widely over Gondwanaland. The surviving monotremes are egg-laying mammals, including the duckbilled platypus and the echidna of Australia and New Guinea. Many aspects of monotreme biology are bizarre: for example, the platypus swims in muddy water with its eyes, ears, and nostrils tightly shut, searching for its crustacean prey with electrical sensors in its beak. Since monotremes have evolved to include the specialized platypus and ant-eating echidnas, it's likely that their fossil record will eventually show us many other surprises.

Marsupials now have been discovered in Eocene rocks in Antarctica and Australia, so it is likely they reached Australia from South America when the south polar region was much warmer than it is now, well before the refrigeration of Antarctica in Oligocene times. It was surprising to find in 1992 that a mysterious primitive *placental* mammal had also reached Australia by Eocene times. *Tingamarra* or its ancestors must

Figure 20.2 A gallery of Australian marsupials, all of which have placental ecological counterparts on other continents.

Kangaroos	Antelope
Wallabies	Rabbits
Wombats	Marmots
Phalangers	Squirrels
Koala	Sloths
"Mice"	Cats and weasels
"Moles"	Moles
Numbats	Anteaters
Tasmanian devil	Wolverine
†Diprotodonts	Rhinos, tapirs
†Marsupial "lion"	Large cats
†Tasmanian wolf	Dogs

also have walked across Antarctica to reach Australia and drop a right upper molar tooth into a billabong in southeast Queensland, but that's all the evidence we have right now. *Tingamarra*'s tooth shows us that marsupials did not evolve to dominate Australian mammalian faunas because they were somehow isolated from placental competition there.

By the Late Cenozoic, marsupials had evolved to fill most of the ecological roles in Australia that are performed by placental mammals on other continents (Figure 20.2). Wallabies and kangaroos are grazers comparable with antelope and deer, wombats are large burrowing "rodents" rather like marmots, and koalas are slow-moving browsers like sloths. The cuscus is like a lemur, and the numbat is a marsupial anteater. There are marsupial cats, marsupial moles, and marsupial mice, and at least six gliding marsupials can be compared with flying squirrels. The honey possum *Tarsipes* is the only nonflying mammal that lives entirely on nectar and pollen, which it gathers with a furry tongue. The small marsupial *Dactylopsila* of New Guinea has evolved specialized teeth and a very long finger to become a marsupial woodpecker (Chapter 19). The Tasmanian wolf and the Tasmanian devil are marsupial carnivores comparable in size and ecology to wolf and wolverine. They once ranged over the main continent of Australia. The Tasmanian devil is now confined to Tasmania, and the Tasmanian wolf is probably extinct.

The fossil record of extinct Australian marsupials is even more impressive. Entire families of marsupials are now extinct. Many were very large, including giant kangaroos and giant wombats that each weighed 200 kg or so (450 pounds). *Thylacoleo* was a Pleistocene carnivore whose name means the marsupial lion. It was indeed the size of a leopard, and had efficient stabbing and cutting teeth. It was better adapted for cutting off chunks of flesh than any living carnivore is (Figure 20.3). Diprotodonts were quadrupedal Pleistocene marsupials about the size of tapirs and rhinoceroses (Figure 20.4). They were the largest marsupials ever: the largest diprotodont was the size of a small elephant, almost 3 meters (10 feet) long and 2 meters (over 6 feet) high at the shoulder, weighing probably close to 2 tonnes. Discoveries of enormous numbers of Miocene bats and marsupials at Riversleigh, in Queensland will soon allow a better description of the radiation of these Australian mammals.

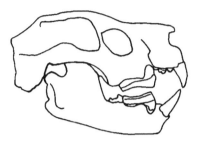

Figure 20.3 *Thylacoleo*, an extinct Australian marsupial carnivore the size of a leopard. (After Lydekker.)

Figure 20.4 A diprotodont, one of several large extinct quadrupedal marsupials in Australia. This one was the size of a tapir. (Reconstruction by Bob Giulani. © Dover Publications Inc., New York. Reproduced by permission.)

Figure 20.5 The biogeographic evolution of marsupials. An early (Cretaceous) radiation in Asia and North America was followed by a Late Cretaceous and Cenozoic radiation in South America, and Cenozoic dispersal through Antarctica to Australia, where a spectacular radiation occurred.

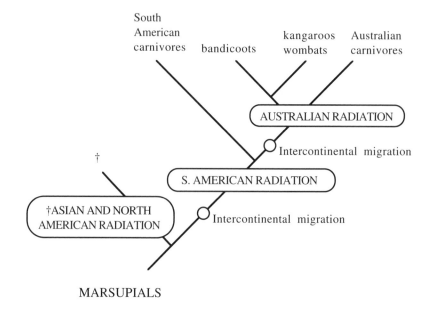

MARSUPIALS

People often talk of marsupials as primitive and inferior to placentals, and it's true that today they are outclassed in diversity and range by placentals. But marsupials do not always have inferior adaptations (Chapter 15). For example, a kangaroo is rather clumsy as it hops slowly around on the ground, using its tail as an extra limb in what is really a five-footed movement. It does use more energy than a placental at this speed. But at high speed a kangaroo is not only very fast (up to 60 kph, or 40 mph), but its incredibly long leaps are much more efficient than the full stride of a four-footed runner of the same weight.

Marsupials evolved in North America or Northern Asia (Figure 20.5). As we shall see later in this Chapter, there were many successful marsupials in South America throughout the Cenozoic. Marsupials most likely reached Australia from South America, presumably across Antarctica. This makes sense geographically, and in terms of relative similarity between the fossil faunas. Marsupials must have crossed Antarctica before it refrigerated, and the process must have excluded almost all placental mammals from Australia. The recent discovery of Eocene marsupials in Antarctica and Australia adds support to the idea. Placentals were rather limited in South American faunas at the time, which helps to explain why only one nonflying placental, *Tingamarra*, reached Australia.

Dromornithids (*mihirung* in aboriginal legend) are giant extinct Australian birds that must have evolved flightlessness and large body size (Figure 20.6). *Dromornis* was probably as large as *Aepyornis*, the elephant bird of Madagascar, and rivals it for the heaviest bird of all time. The living Australasian emu and cassowary are large ground-running ratites. Ratites are distributed on the southern continents that are remnants of Gondwanaland (Figure 14.24). We do not know yet whether the cassowary, emu, ostrich, and rhea are related in a clade, all descended from ancient ratites that lived in Gondwanaland, or whether they are descended from separate northern groups that each found their way to the southern continents (Chapter 14).

Figure 20.6 A dromornithid, or mihirung, a giant extinct bird from Australia, shown with a living emu for scale. These two birds may not be related: dromornithids may not be ratites.

The isolated position of Australia has meant that only very mobile birds and placental mammals (bats and humans) have reached it. Humans brought with them a host of other invaders, such as rats, cats, dogs, sheep, cattle, rabbits, cactus, fish, and cane frogs, with serious results for the Australian ecosystem. Captain Cook's first reaction to a kangaroo was to set his dog on it! More recently, other bizarre introductions have helped to restore a little of the damage—for example, the organism that causes the rabbit disease myxomatosis, and the dung beetles that keep Australian grasslands from being buried in cattle dung. The biogeographic story of Australia is still in an active phase.

New Zealand

New Zealand was part of Gondwanaland until the Cretaceous, and it had a normal fauna at that time. But it had no land mammals until humans arrived, and the rest of its prehistoric fauna suggests that migration into the region was difficult. The native fauna includes only four amphibian species, primitive frogs that hatch as miniature adults from the egg with no tadpole stage. There are only a few native reptile species: 11 geckos that all have live birth, 18 skinks, of which 17 have live birth, and the tuatara, which is an ancient and primitive reptile. New Zealand has no snakes and no normal lizards. The only native mammals are two species of bats.

The dominant prehistoric creatures of New Zealand were birds. The kiwis survive as nocturnal insectivores, but the major vegetarians were very large ratites, at least a dozen species of moas. The largest moa was 3.5 meters (11 feet) in height (Figure 14.24). Moas coevolved with New Zealand plants so that 10% of the native woody plants have a peculiar branching pattern called **divarication**—they branch at a high angle to form a densely growing plant with interlaced branches that are difficult to pull out or break. There are few leaves on the outside, and the largest, most succulent leaves are on the inside. But nine species of divaricating plants that grow more than 3 meters (10 feet) tall look more like normal trees once they reach that height, and other divaricating species grow more normally on small offshore islands. The only reasonable explanation of divarication is that it evolved as a defense against browsing moas, the largest of which was about 3 meters tall.

Other vegetarian guilds that were filled by small mammals on other land masses were partly occupied by moas and other birds and partly by huge flightless insects—enormous weevils and wetas (giant grasshoppers). It's not easy to identify the major prehistoric predators, but they were present. The largest surviving New Zealand birds (the kiwi, for example) are well camouflaged, although there is no obvious surviving predator on them. But extinct New Zealand raptors include a bird that was the largest goshawk that ever evolved (3 kg or 7 pounds in weight) and a huge extinct eagle that weighed about 13 kg (30 pounds).

In New Zealand, evolution's been slower

Vegetarian diversity's lower

Giant weevils eat seeds,

The wetas eat weeds,

But the grasses are cut by the moa.

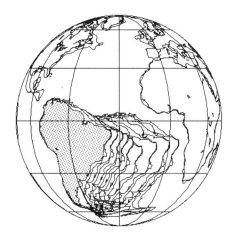

Figure 20.7 South America drifted west and then west-northwest during the Cenozoic, and for most of that time was an island continent accessible only to lucky or mobile immigrants.

South America

South America is in many ways more interesting than Australia for mammalian evolution because we know its history in more detail. South America split away from Africa in the Late Cretaceous (around 80 Ma) to become an island continent (Figure 20.7). In Cretaceous times the South American mammals and dinosaurs included unique forms belonging to primitive Jurassic groups that had become extinct everywhere else but continued to evolve in South America. Examples include the giant dinosaur *Megaraptor* and early mammals. Triconodonts, symmetrodonts, and a peculiar group of multituberculate mammals (Chapter 15) called gondwanatheres have all been collected from Cretaceous rocks in South America, yet therian mammals (marsupials and placentals) are not found.

Around the end of the Cretaceous, marsupials and placental herbivores arrived in South America, presumably from North America, and it is likely that South America provided a gateway to a route across Antarctica for at least the marsupials to reach Australia.

The climatic changes at the end of the Eocene seem to coincide with the arrival of a further few immigrants into South America: rodents and monkeys, tortoises, and colubrid snakes. The same climatic changes led to the spread of Oligocene grasslands over much of South America, and the early expansion of the South American placentals into a guild of open-country grazers.

Apart from these brief periods of immigration, Cenozoic evolution in South America took place in isolation for over 60 m.y. The strange South American mammals in particular are well known, and they divided up available ecological roles in the usual way. Charles Darwin noticed some peculiar fossil mammals in Argentina during his voyage on the *Beagle*, and later expeditions to Argentina have found hundreds of beautifully preserved Cenozoic fossils.

From Early Cenozoic times, the South American marsupials took on the roles of small insectivores (and still do). There is a living aquatic marsupial with webbed feet and a watertight pouch. *Argyrolagus* was a rabbit-sized marsupial that looked like a giant kangaroo rat. It hopped and had ever-growing molars for grazing coarse vegetation. The arrival of the placental rodents did not affect these small marsupials. One of the most successful marsupials in the world, even in the face of intense competition from placentals, is the small omnivorous opossum, *Didelphis*.

The placental grazers had evolved by the Miocene into a bewildering variety of forms ranging from rhino-sized to rabbit-sized. *Thoatherium* and *Diadiaphorus* (Figure 20.8) had an uncanny resemblance to horses, with long faces, horselike front teeth, grinding molars, straight backs, and slender legs ending in one or three toes. Some of their relatives looked like camels. Large vegetarians such as *Toxodon* had large grinding molars that grew through most of the life of the animal (Figure 20.9).

Armadillos, sloths, and anteaters are also characteristic South American mammals. Armadillos and their relatives evolved heavy body armor for protection and became highly successful opportunistic insectivores and scavengers. The Pleistocene armadillo *Glyptodon* was very large, probably a vegetarian, 1.5 meters (5 feet) long. It had a thick armored skullcap as well as body armor, and some glyptodont species had a spiked knob at the end of the tail (Figure 20.10). Glyptodonts were certainly too big to burrow like the smaller armadillos, and they had to be heavily armored and armed to survive out on the surface. Naturally, their skeleton was very strong to support all the weight of the armor.

Sloths now live in trees, eating leaves and moving with painful slowness. But remains of huge ground sloths have been found in South America, including one that must have been almost as large as an elephant. Anteaters evolved from the same group of ancestors but are now specialized to an amazing extent for eating termites, beginning by tearing apart their nests with tremendously powerful clawed forearms.

The most impressive South American creatures were the larger carnivores. None of them were placental mammals, and most were marsupials. This is not surprising, considering how savage the surviving little marsupial insectivores are, but it is unusual compared with other continents. Borhyaenids were basically like wolves, but were generally larger. *Proborhyaena* was as big as a bear and probably had a similar way of life. *Borhyaena* itself was a wolf-sized Miocene marsupial with canine teeth adapted for stabbing and molars that had evolved into meat-slicing teeth (Figure 20.11). It was a successful medium-sized carnivore, but it was the last of the large borhyaenid carnivores. They were replaced by invading placentals from the north and by giant predatory birds.

Thylacosmilids looked like large cats. *Thylacosmilus* was a marsupial sabertooth, but its savage stabbing canines were better designed than those of the placental sabertooth cats of North America. In *Thylacosmilus* the sabers were longer, slimmer, more securely anchored in huge, recessed tooth cavities extending far up the face; thus, they were better protected from damage than those of true cats (Figure 20.12). The sabers were ever-growing and self-sharpening, and they were backed by more powerful neck and head muscles. Presumably they were adapted to killing large (placental) ungulates by stabbing and slashing deep into the soft tissues of throat or belly. The cheek teeth were not as powerful as those of placental cats, however.

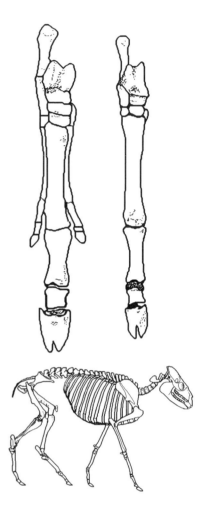

Figure 20.8 Above: The hind feet of two Cenozoic mammals from South America, *Diadiaphorus* (left), and *Thoatherium* (right), the thoat. Compare the hind feet of horses in Figure 19.13. These strikingly similar structures evolved in parallel in these two separate lineages of savanna running animals. Below: The skeleton of *Diadiaphorus*, showing that the resemblance to horses involved the entire body structure. (After Scott.)

Figure 20.9 *Toxodon*, a large vegetarian mammal from the Cenozoic of South America. (After Lydekker.)

Figure 20.10 A glyptodont, a giant, heavily armored extinct relative of living armadillos. This one was close to 3 meters (9 feet) long. (Reconstruction by Bob Giuliani. © Dover Publications Inc., New York. Reproduced by permission.)

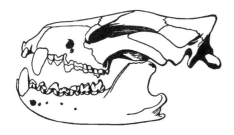

Figure 20.11 *Borhyaena*, a wolf-sized marsupial carnivore from the Cenozoic of South America. (After Sinclair.)

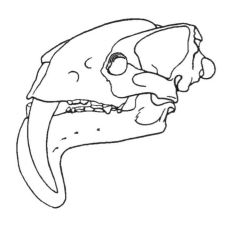

Figure 20.12 The skull, jaws, and teeth of the South American marsupial sabertooth *Thylacosmilus*. (From Riggs.)

These amazing marsupials had unusual competitors for mastery of the carnivorous guild, the phorusrhacids: flightless, ostrich-sized birds equipped with very powerful tearing beaks as well as foot talons (Figure 14.22). It seems that the phorusrhacids eventually gained the upper hand over the carnivorous marsupials.

South America had its own group of crocodiles, the sebecids. They apparently evolved in Gondwanaland in the Cretaceous, survived the K–T extinction, and radiated in the Early Cenozoic in South America to become powerful terrestrial predators. Unlike aquatic crocodiles, they had high, deep skulls and snouts. Other crocodilians in South America also evolved into unusual morphologies; for example, a duckbilled caiman is known from the Miocene of Colombia.

The South American ecosystem gained new immigrants in Oligocene times, around 25 Ma, with the arrival of rodents and primates, probably from Africa by way of islands in the widening Atlantic Ocean (Figure 20.7). Both groups radiated widely, the rodents into the various guinea pigs and their relatives, and the primates into the distinctive New World monkeys, evolving habits and characters in parallel with gibbons and Old World monkeys.

Other members of the Cenozoic South American fauna included more giant predators, the largest flying birds of all time, the **teratorns** (Chapter 14). The largest turtle of all time, *Stupendemys*, lived along the north coast.

This unique ecosystem suffered four tremendous shocks in ten million years and has almost completely disappeared. First, Antarctica froze up, with the result that the Humboldt Current, flowing most of the way up the west coast of South America, became much colder and stronger. Second, tectonic activity along the Pacific coast raised the Andes as a major mountain chain. Together, these two events drastically lowered rainfall over most of the continent, and much of the area turned from forest and well-watered plain to dry steppe. This led, in the later Miocene, to the extinction of many animals, including the terrestrial crocodiles and especially the large-bodied savanna herbivores.

Third, South America drifted northward towards Central and North America (Figure 20.7). By about 6 Ma, the gap was small enough to allow a few animals to cross it, more or less by accident. North American raccoons and some mice and rats crossed to the south, while two kinds of sloths crossed to the north. The effect of the competition was seen almost

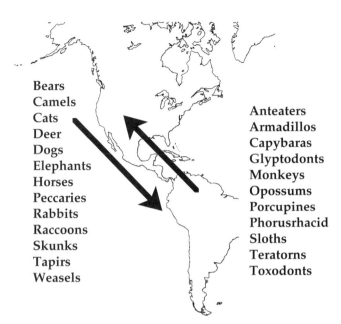

Figure 20.13 The Great American Interchange.

Bears
Camels
Cats
Deer
Dogs
Elephants
Horses
Peccaries
Rabbits
Raccoons
Skunks
Tapirs
Weasels

Anteaters
Armadillos
Capybaras
Glyptodonts
Monkeys
Opossums
Porcupines
Phorusrhacid
Sloths
Teratorns
Toxodonts

immediately. Many borhyaenids were replaced by raccoons, and the larg-est of them, the bearlike *Proborhyaena*, was replaced by a bear-sized rac-coon. Finally, at about 3 Ma, the last important sea barrier was bridged, and animals could walk from one continent to the other.

Ecological principles suggest what should happen when an exchange of animals takes place. A larger continent such as North America should contain a larger diversity of animals than its smaller counterpart, and the fossil record confirms that this was true just before the exchange. There-fore, if the same proportion of animals from each continent migrated to the other, one would expect more North American animals to go south than the reverse. If a continent can hold only so many families or genera of animals, then one would predict extinctions on each continent, but more in South America than in North America. The effect would be ac-centuated because North America was at least intermittently connected with Eurasia, and altogether this huge northern area of temperate open country held a great variety of savanna animals. In contrast, the area of savanna in South America was not as large as one would think, because the continent is widest in equatorial latitudes and narrows significantly to the north and south. South American savanna faunas might have been very vulnerable to invasion from the north.

The major exchange happened after 3 Ma. Camels, elephants, bears, deer, peccaries, horses, tapirs, skunks, rabbits, cats, dogs, kangaroo rats, and shrews entered South America. Monkeys, opossums, anteaters, sloths, armadillos, capybaras, toxodonts, porcupines, and glyptodonts mi-grated north, with the giant birds—a phorusrhacid and a few teratorns (Figure 20.13).

The South American immigrants to North America flourished there, and so did the successful North American immigrants that moved south. Overall, however, there was a net major extinction of South American groups (Figure 20.14). The large, native marsupial carnivores and most of the phorusrhacids seem to have been outcompeted by the cats and

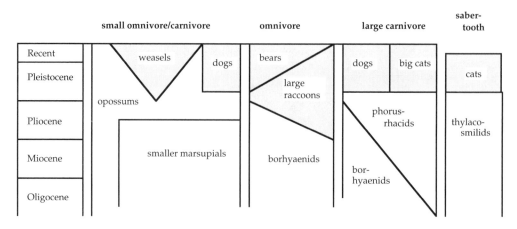

Figure 20.14 Balance sheet of the omnivore and carnivore guilds during the Great American Interchange. North American invaders are shown in stipple. Herbivore guilds show equally complete replacement. (Data from Larry G. Marshall.)

Figure 20.15 Africarabia drifted slowly northeast during the Cenozoic. Finally it collided with Western Asia in the Early Miocene along a line that is now the Zagros Mountains. The African continent then rotated slightly clockwise, splitting away from Arabia. At the end of the Miocene the northwest corner of Africa collided with Western Europe to close off the Mediterranean Sea as a vast lake that quickly dried up. Meanwhile the Red Sea opened up as Africa swung away from Arabia, and the great African Rift Valley was formed.

dogs from the North, and the remaining savanna browsers and grazers were almost all wiped out, perhaps outcompeted by the northern horses and camels, perhaps hunted out by the new predators. The sabertooth marsupial *Thylacosmilus* was replaced by a real sabertooth cat. Even earlier invaders suffered: the bear-sized raccoon was replaced by a true bear.

The geographical changes that had permitted the interchange also changed the climate of the Atlantic Ocean, and this in turn caused drastic changes in the land ecology of North and South America as the northern ice ages began in earnest around 2.5 Ma. Ice advances and retreats also caused great swings of vegetation patterns on time scales of 100,000 years, stressing the faunas in a way that had not happened before. Overall, the result was nearly as expected, and can be explained by current ecological theory, with the South American animals coming out much the losers. North American invaders survive in strength today in South America, including all the South American cats, the llamas, and dozens of rodents.

South American faunas suffered a fourth catastrophic extinction in the late Pleistocene. This time, similar extinctions took place in North America too, and we shall examine this in Chapter 24.

Africa

Africa (plus Arabia, so perhaps I should write Africarabia) was part of Gondwanaland until the Cretaceous, when it broke away from South America on the west and Antarctica and India on the east (Figure 20.15). From Late Cretaceous times onward, Africa and South America, and the animals and plants living there, had increasingly different histories.

Africa had dinosaurs much like those of the rest of the world during the Late Cretaceous, but there is no record of any African Cretaceous mammal. This may change soon, because molecular evidence suggests that there must have been spectacular evolution among strictly African mammals during the Cretaceous and Palaeocene. A clade of mammals, including seacows, elephants, hyraxes, aardvarks, tenrecs, and golden moles apparently evolved on the isolated continent of Africarabia, which had split from South America but was not close to Europe or Asia. The continent lay south of its present position, bounded on the north by the tropical Tethys Ocean .

Our first look at the fossil Cenozoic life of Africa comes from the Eocene rocks of Egypt, laid down on the northern edge of the continent. Shallow warm seas teemed with microorganisms whose shells formed the limestones from which the Pyramids and the Sphinx were carved.

Here we find early whales and sea cows, mammals that probably evolved adaptations for life in the sea in swamps and deltas along the shores of Tethys. **Moeritheres** were amphibious animals related to sea cows, and they are also related to elephants. *Moerithium* itself is an Eocene animal from Africa, and looked like a small, fat elephant with the ecology of a hippo. Other Eocene fossils from Egypt include some early primitive carnivores, the creodonts (also known from other continents).

By Oligocene times, Egypt was the site of lush deltas where luxuriant forest growth housed rodents, primates, and bats, all recent Eurasian immigrants. Piglike anthracotheres had crossed from Eurasia, but there were African groups too. There were eight different genera of Oligocene hyraxes, small- to medium-sized vegetarians that look much like rodents. *Arsinoitherium* was a large-bodied browser (Figure 20.16).

So Eocene and Oligocene African mammals are a mixture of native African groups that were evolving new characters and a few successful immigrants from Eurasia. But by Miocene times, Africarabia had drifted far enough north to bring it close to Eurasia, and finally the two continental edges collided (Figure 20.15). For much of the Late Cenozoic, a shallow seaway prevented free exchange of land animals, but there were two important times of uninterrupted migration, an early stage at about 18 Ma and another at about 15 Ma (Figure 20.17). The interchanges affected animal life throughout the Old World, almost on the same scale as the Great American Interchange. The seaway finally closed off about 12 Ma.

In the earlier event, 12 families of small mammals appeared in Africa in the Early Miocene, mostly insectivores and rodents from Eurasia. Ancestral ungulates (deer, cattle, antelope, and pigs) largely replaced the hyraxes at medium sizes, and rhinos and the first giraffes were large invaders. Cats arrived and began to replace the older creodonts. Going the other way, elephants walked out of Africa into Eurasia, in at least two

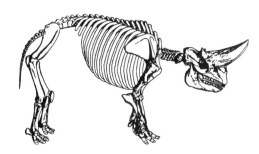

Figure 20.16 *Arsinoitherium*, a large browsing mammal weighing close to 5 tons, from the Oligocene of Africa. (Skeleton from Andrews. Reconstruction by Bob Giulani, (© Dover Publications Inc., New York. Reproduced by permission.)

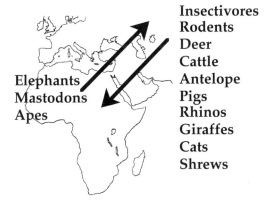

Insectivores
Rodents
Deer
Cattle
Antelope
Pigs
Rhinos
Giraffes
Cats
Shrews

Elephants
Mastodons
Apes

Figure 20.17 The Great Old World Interchange of animals between Africarabia and Eurasia in the Miocene.

major adaptive groups, mastodons and true elephants. Some large creodonts even reinvaded Eurasia from Africa.

In the second exchange around 15 Ma, a new set of African animals, including apes, quickly spread over the woodlands and forests of Eurasia. Hyenas and shrews migrated into Africa.

It is not clear whether the continental collision itself altered world climate, or whether climate was affected more by major events in the Southern Hemisphere. Whatever the cause or causes, the Miocene change from forest to savanna was partly responsible for the success of the large number of grazing animals listed above.

By the end of the Miocene, more immigrants had appeared in Africa: small animals, including many bats, and the three-toed horse *Hipparion*. Meanwhile, hippos evolved in Africa, and the antelope and cattle that had arrived earlier evolved into something close to the incredible diversity we see today in the last few game reserves.

Africa and Eurasia have been connected by land since the Miocene, but this does not automatically imply free exchange of animals. For example, the Mediterranean Sea was closed off at its western end at the end of the Miocene, around 6 Ma, and it dried into a huge salty desert like a giant version of Death Valley. Only a few animals could have crossed this barrier. Later, the development of desert conditions in the Sahara formed another fearsome barrier to animal migration for most of the past few million years. Today, North African animals are more like those of Eurasia than those of sub-Saharan Africa.

The Early Pleistocene saw a large extinction in Africa, with one-third of the mammals becoming extinct. But they were replaced by newly evolving species, so total diversity remained high. Africa apparently did not feel the effects of the ice ages too drastically. In contrast, when European animals were pushed southward by the advancing cold and ice, they could not cross the Mediterranean and the Sahara, and many became extinct. Human hunting activities have affected Africa less than other continents, perhaps because humans evolved gradually there and the animals had time to adjust to them. On other continents human impact was much more sudden and severe. Animals formerly widespread over the world are now confined to Africa or nearly so (rhinos, lions, cheetahs, hyenas, and wild horselike species—zebras). Protected there by the geographical, climatic, and historical events of the Late Cenozoic, many creatures survived relatively successfully in Africa until this century.

Islands and Biogeography

Strict geographic barriers prevent land plants and animals living on islands from moving easily to other land areas, and potential invaders also must cross barriers. This means that island faunas and floras tend to evolve in greater isolation than those with wider and more variable habitats. Of course, this is true at any scale, whether we look at small islands or continent-sized ones. Islands past and present can teach us a great deal about evolution. It is no accident that Darwin was particularly enlightened by his visit to islands like the the Galápagos, and Wallace by his years in Indonesia.

We have seen some of the vagaries of continental faunas over a time scale of tens of millions of years, but it is worth looking at cases where smaller-scale events on smaller islands over smaller lengths of time show the rapidity and power of natural selection in isolated populations.

The Raptors of Gargano

In 1969, three Dutch geologists were exploring the Mesozoic limestones of the Gargano Peninsula in southern Italy (Figure 20.18). Sometime in the Early Cenozoic, this block of land was raised above sea level and caves and fissures formed in the limestone. In Early Miocene times the Gargano area was cut off from the mainland by a rise in sea level to form an island in the Mediterranean Sea of the time. Land animals living there were isolated on the island as the sea rose. Over only a few million years, animals occasionally fell into fissures in the limestone, where they were covered by thin layers of soil and preserved as fossils. Today, the limestones are quarried for marble, and the bones can be found in the pockets of ancient soils exposed in the quarry walls.

No large animals were isolated on Gargano as it was cut off. The only large reptiles were swimmers (turtles and crocodiles) and the only mammalian carnivore was also a swimmer, a large otter with rather blunt teeth that probably ate shellfish most of the time and would not have hunted on land.

Because there were no land carnivores, small mammals evolved quickly into spectacular forms. Small rabbit-like pikas were abundant, and gigantic dormice evolved on the island. Giant hamsters were eventually outcompeted by true rats and mice. Some of the Gargano mice grew to giant size, with skulls 10 cm (4 inches) long, and many evolved fast-growing teeth as complex as those of beavers. They probably chewed very tough material. *Hoplitomeryx* is a small deer which evolved horns instead of antlers (Figure 20.19).

If there were no cats, dogs, or other terrestrial carnivores, how were the rodents kept under control? By disease and starvation? And why did *Hoplitomeryx* evolve spectacular horns, if there were no carnivores to fight off? The horns were too lethal to have been used for fighting between individuals of the species.

The answer to these questions seems to have been raptors—birds of prey. A giant buzzard, *Garganoaetus*, was as large or larger than a golden eagle. Presumably it hunted by day. It would have been perfectly capable of taking a small or young *Hoplitomeryx*, and the horns may have evolved to protect the back of the neck against raptors. Normally, small deer hide in vegetation, but Gargano was a bare, limestone island, with no cover by day. At night, the owls took over: the largest barn owl of all time evolved on Gargano.

Figure 20.18 The geographic setting of the Gargano Peninsula, in Italy, which formed an island in Miocene times.

Figure 20.19 A deer with horns instead of antlers, *Hoplitomeryx*, evolved in geographic isolation on the Gargano Peninsula in Miocene times. (After Leinders.)

Giant Pleistocene Birds on Cuba

Cuba had a strange set of animals isolated on it during the ice ages. Pleistocene mammals have been found in enormous numbers in limestone cave deposits, and we have a reasonable idea of the unusual ecology the island must have had. In particular, there were enormous numbers of ground sloths and rodents, and insectivores were very common. Tens of thousands of mouse jaws have been found in one cave, and another site yielded over two hundred ground sloths. In addition, large numbers of fossil vampire bats imply that there were large numbers of warm-blooded animals for them to prey on. Similar but less spectacular fossils have also been found on Puerto Rico and Hispaniola.

There are practically no carnivorous mammals in these deposits, and as at Gargano, we are forced to wonder what kept the animal populations in check. The answer here too seems to be raptors. In the caves with the animal bones there are also great numbers of the bones of small birds. This suggests that the cave deposits are mainly the accumulations of owl pellets and bat colonies. But the size of the bones indicates that the owls were producing pellets much larger than normal owls do.

In 1954 a gigantic fossil owl was discovered, large enough to have preyed upon juvenile ground sloths. Later that year, a fossil eagle bigger than any living species was found. A fossil vulture as large as a condor, and a fossil barn owl as large as the species at Gargano, fill out a picture of a set of predators quite alien to our experience today.

The gigantic owl *Ornimegalonyx* must have stood a meter high. It may not have been a powerful flyer, because its breastbone looks weak relative to the rest of the skeleton. But with its tremendous beak and claws, it could have preyed successfully on rodents and young sloths. By day the giant eagle would have performed the same function—it is larger than the monkey-eating eagle of the tropical forest today. Presumably the giant vulture fed from the carcasses of giant ground sloths, and the other large owls added to the flying nocturnal predators.

The whole ecosystem became extinct towards the end of the Pleistocene on Cuba and on all the other Caribbean islands. We don't know enough of the geological history of Cuba to suggest that human intervention caused these extinctions.

Other Biogeographic Islands

Islands in the biogeographic sense do not have to be small pieces of land separated by water. Mountain-dwelling species can be isolated by stretches of plains country around them, so that in some parts of Africa each mountain system has its own spider fauna. Lake faunas are separated by watersheds. Woodland faunas can be separated by open stretches of prairie. Even in the ocean, shallow-water animals find deep ocean basins just as much of a barrier to them as land masses are. Marine organisms can often be thought of as living on water islands surrounded by seas of land.

For example, in the Late Oligocene, around 30 Ma, walruses and sea lions evolved around the edges of the North Pacific, while seals evolved around the edges of the North Atlantic. The two groups did not intermingle until much later in the Cenozoic, presumably because they were separated by the barrier of the north-south land masses of the Americas.

From Gondwanaland to Antarctica

Gondwanaland began to split apart significantly in the Cretaceous, and the continental blocks of Africarabia, India, and South America began to move northward, leaving Antarctica near the pole. Warm currents still flowed far south in the Early Cenozoic, and Antarctica had a temperate climate until the end of the Eocene, perhaps 37 Ma. Plant vegetation, spores and pollen of ferns and deciduous hardwoods, molluscs, and small mammals are part of the fossil record of moist temperate conditions. Isotope data from the southern oceans indicate a high-latitude temperature of perhaps 12°C (54°F), the same as those of the ocean floors at the time. Equatorial surface temperatures were moderate, about 18°C (64°F), perhaps because of efficient heat exchange with the rest of the oceans.

Around the end of the Eocene, Antarctica began to cool significantly, and its climate became arctic. Some ice formed on East Antarctica, and sea ice formed around it (glacial rocks were dropped into nearshore Oligocene sediments from floating ice). As pure ice crystals freeze out of normal sea water, they leave behind very cold, salty, dense seawater. If a great quantity of cold, salty water is formed, it does not lose its identity by diffusion into neighboring water masses. It stays as a dense mass and sinks to the seafloor, where it is far removed from any heat source. Huge quantities of water are formed in this way around Antarctica today. As **Antarctic Bottom Water** or **ABW** (Figure 20.20), it creeps along the ocean floor wherever gravity takes it, displacing any warmer (lighter) water and keeping the seafloor temperature close to zero. In modern oceans, ABW flows as far as the North Atlantic and North Pacific ocean floors, and it is stopped in the Indian Ocean only by the land masses of Asia that block off the northern end. Even so, the oceans are now very cold for most of their depth, and these deep, cold waters form a major part of the ocean environment, the **cryosphere**.

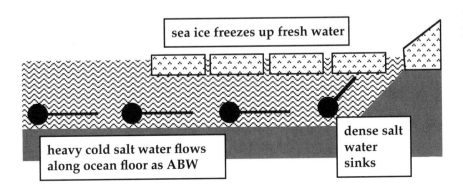

Figure 20.20 The formation of Antarctic Bottom Water is driven by the low air temperatures that freeze the surface waters into sea ice, leaving behind a very cold, dense brine that sinks to the ocean floor.

sea ice freezes up fresh water

heavy cold salt water flows along ocean floor as ABW

dense salt water sinks

ABW spread out northward for the first time in the Early Oligocene, bringing with it the origin of the cryosphere. A major environmental crisis overtook seafloor organisms. High-latitude ocean temperatures and all ocean-floor temperatures dropped dramatically down to about 4°C (39°F). Ocean current patterns changed drastically with the introduction of deep polar water flow, and they affected global climate, biology, and ecology in the Terminal Eocene Event.

For much of the Oligocene, Antarctica continued to have a polar climate with small glaciers, but it did not build up a huge ice cap. By the middle Oligocene, about 32 Ma, the split between Australia and Antarctica was wide enough to allow deep-water currents to pass between them. This change in ocean currents set off an important cooling event that affected continents as far away as North America, where several groups of mammals, including the brontotheres, became extinct quite abruptly. Finally, at the end of the Oligocene, South America had moved far enough northward to allow deep-water flow through the Drake Passage, completing a deep-water pathway all around Antarctica in the Shrieking Sixties, the latitude belt between 60° and 70° S (Figure 20.21). The development of this ocean current (today's Antarctic Circumpolar Current) meant that Antarctica became almost completely isolated from warming winds and currents, and a large ice sheet slowly built up on East Antarctica in the Middle Miocene, perhaps 15 to 13 Ma .

This was the first significant buildup of ice on land since the Late Paleozoic glaciation of Gondwanaland, and enough water froze up in the Antarctic ice sheet to drop world sea level by about 60 meters (200 feet). The freezing of Antarctica was a major climatic event, and it must have helped to cool the rest of the world even if it did not cause the entire temperature drop that occurred at the time. The Middle Miocene ice buildup had dramatic effects on land organisms, certainly in South America and perhaps on other continents too. The collision of Eurasia with Africa, which slightly preceded the ice buildup, added to the ecological impact by changing oceanic circulation in the tropical belt.

Finally, in events we will look at in more detail in Chapter 23, the world descended into an ice age around 2.4 to 2.5 Ma.

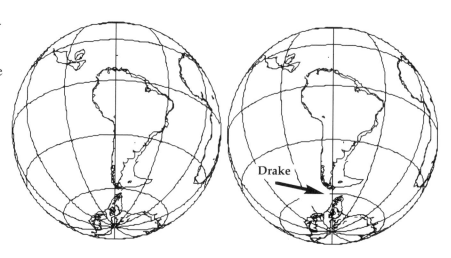

Figure 20.21 South America drifted slowly away from Antarctica and opened the deep-water Drake Passage between them. These two maps show the geography of the area at 40 Ma (left) and 20 Ma (right). The change is rather subtle, but its effects were dramatic. The opening of the Drake Passage permanently changed the climate of the Antarctic region and of the entire globe.

I have stressed events in and around Antarctica as if that continent were the sole determinant of Earth's climate. This is too simple. Solar heating is concentrated in the tropics, which are the hot ends of the circulation currents in the oceans and atmosphere. The poles are the cold ends of the currents. The flow of heat from hot to cold drives weather and climate, so events either in the tropics or at either pole could have important climatic effects. But at present it is the Antarctic that communicates freely by oceanic currents with the world's oceans. Antarctic Bottom Water covers the floors of almost all the oceans, and Antarctica holds over 90% of the world's ice on and around its surface, in an ice sheet that averages over 2 km thick and reaches twice that thickness in some places. It is the persistently cold pole, and it is a major control on overall global climate. The North Pole, on the other hand, is much more unstable. In Earth's present geography, conditions in the Northern Hemisphere can and do fluctuate dramatically and drive shorter-period changes in climate such as the advances and retreats of the Pleistocene ice sheets.

Review Questions

Give an example of parallel evolution in which different groups of animals on different continents, or at different times, evolved to have much the same way of life, i.e. occupied the same guild.

List three examples of creatures on the island continent of Australia that evolved unusual ways of life compared with creatures on other continents.

List three examples of creatures on the Cenozoic island continent of South America that evolved unusual ways of life compared with creatures on other continents.

What happened when South America drifted north and became joined to North America by way of the Panama or Central American isthmus?

Dinosaurs are pretty much the same from every continent. But if you go to the zoo, you find that big land mammals today are practically all different from one continent to the next, and are even very different from one part of a continent to another. Why the difference?

Describe very briefly the fauna of South America about 5 million years ago, when it was an isolated continent.

What is so important about the discovery of a fossil marsupial in Antarctica?

Further Reading

Easy Access Reading

Diamond, J. M. 1990. Biological effects of ghosts. *Nature* 345: 769–770. The prehistoric ecology of New Zealand.

Flannery, T. F. 1995. *The Future Eaters: An Ecological History of the Australasian Lands and People.* New York: George Braziller. Part I is the story of Australasian life before the arrival of humans.

Godthelp, H., et al. 1992. Earliest known Australian Tertiary mammal fauna. *Nature* 356: 514–516.

Marshall, L. G. 1988. Land mammals and the Great American Interchange. *American Scientist* 76: 380–388.

Marshall, L. G. 1994. The terror birds of South America. *Scientific American* 270 (2): 90–95.

Pascual, R., et al. 1992. First discovery of monotremes in South America. *Nature* 356: 704–706.

Springer, M. S., et al. 1997. Endemic African mammals shake the phylogenetic tree. *Nature* 368: 61–64.

More Technical Reading

Archer, M., et al. 1989. Fossil mammals of Riversleigh, Northwestern Queensland: preliminary overview of biostratigraphy, correlation and environmental change. *Australian Zoologist* 25: 29–65.

Flynn, J. J., and A. R. Wyss. 1998. Recent advances in South American mammalian paleontology. *Trends in Ecology & Evolution* 13: 449–454.

Rasmussen, D. T., and E. L. Simons. 1988. New Oligocene hyracoids from Egypt. *Journal of Vertebrate Paleontology* 8: 67–83. Hyraxes formed an Oligocene mini-radiation.

Rich, P. V., and G. F. van Tets (eds.). 1985. *Kadimakara*. Victoria, Australia: Pioneer Design Studio. The extinct animals of Australia.

Richardson, K. C., et al. 1986. Adaptations to a diet of nectar and pollen in the marsupial *Tarsipes rostratus* (Marsupialia: Tarsipedidae). *Journal of Zoology, London A* 208: 285–297.

Savage, R. J. G., and M. R. Long. 1987. *Mammal Evolution: An Illustrated Guide*. London: British Museum (Natural History).

Simpson, G. G. 1980. *Splendid Isolation*. New Haven: Yale University Press. South American mammals.

Stanhope, M. J., et al. 1998. Molecular evidence for multiple origins of Insectivora and for a new order of endemic African insectivore animals. *Proceedings of the National Academy of Sciences* 95: 9967–9972.

Stanhope, M. J., et al. 1998. Highly congruent molecular support for a diverse superordinal clade of endemic African mammals. *Molecular Phylogenetics and Evolution* 9: 501–508. A classic case of "shingling" (publishing the same discovery twice): essentially the same science as reported in their other paper.

Stehli, F. G., and S. D. Webb (eds.). 1985. *The Great American Biotic Interchange*. New York: Plenum.

Webb, S. D. 1991. Ecogeography and the Great American Interchange. *Paleobiology* 17: 266–280.

There is no convenient summary for Gargano. Specialist papers on the fauna have appeared (in English) in the Dutch journal *Scripta Geologica*. Two papers in Volume 77 (1986) contain references that lead to all the others.

South America had quite a bunch

Of sabertooths looking for lunch.

Their marsupial grab

Packed a great deal of stab,

But there wasn't much munch in their crunch!

From an aphorism by Michael Archer

Primates

We are particularly interested in our own ancestry. After all, the recent evolution of primates has produced humans, the most widespread, powerful, and potentially destructive biological agent on Earth.

Most living primates are small, tropical, tree-dwelling animals that eat high-calorie food, mainly insects. This is particularly true of groups that retain the most primitive primate characters. Taken at face value, this suggests that primate ancestors searched for insects, fruit, seeds, or nectar on small branches, high in trees, and in smaller bushes. Evolutionary evidence supports this scenario, because primates are most closely related to three other mammal groups that also live in trees—tree shrews, dermopterans or "flying lemurs," and fruit bats. This group of small arboreal mammals had invaded forest habitats by the end of the Cretaceous, and several lines must have survived the K-T extinction.

Primates have large eyes, turned forward to give excellent stereoscopic vision. The combination of large eyes and stereoscopic vision may have evolved in primates—as in cats, owls, and fruit bats—to help search for food at night by sight rather than smell, because it allows the animal to judge the distance of a food item without moving its head. Stereoscopic vision promotes agility and coordination, especially when an animal has hands and feet adapted for grasping and fine manipulation, with pads and nails rather than paws and claws. Grasping feet and hands allow primates to forage along narrow branches, and live prey or other food can be reached or seized by a hand or hands rather than by a lunge with the whole body and head. Compare the coiled strike of snakes and the tongue strike of chameleons (Figure 21.1), frogs, and toads, which all do the same thing in different ways.

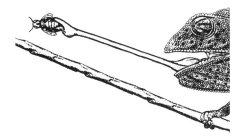

Figure 21.1 Catching small, agile prey without moving the entire body.

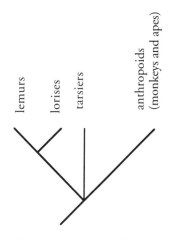

Figure 21.2 Cladogram of living primates. The branches between the groups are very deep in time.

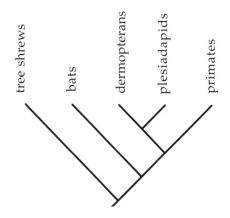

Figure 21.3 Cladogram of some tree-dwelling mammal groups: primates and their nearest relatives. This cladogram, one of several possibilities, shows primates and dermopterans as closest allies.

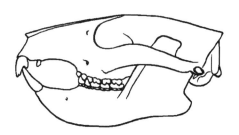

Figure 21.4 A typical plesiadapid, one of a radiation of early primatelike animals that included animals with rodentlike ecology. (After Gingerich and Krause.)

All primate fetuses show rapid growth of the brain relative to the body, so they are born with relatively larger brains than other mammals. Gestation time is long for body size, and primates have small litters of young that develop slowly and live a long time. Primates evolved high learning capacity, complex social interactions, and unusual curiosity. The evolution of curiosity is useful in searching for food, and high learning capacity, memory, and intelligence help individuals to make correct responses in a complex, ever-changing environment.

Living primates are often divided into two groups: small-brained, small-bodied animals called **prosimians**, and the relatively large-brained **anthropoids** (monkeys and apes). However, prosimians contain two clades, each with a long evolutionary history: the **tarsiers** of Southeast Asia on the one hand, and the **lorises** of Africa and the **lemurs** of Madagascar on the other. The evolutionary relationship among these primate groups is unclear (Figure 21.2).

Earliest Primates: Are They Dermopterans?

A clade of small tree-dwelling mammals had radiated by the end of the Late Cretaceous. The surviving descendants of that clade are bats, primates, tree shrews, and the only surviving species of dermopteran, the colugo or "flying lemur" of the Indonesian rain forest (Figure 21.3). Because they were such close relatives, it might be difficult to distinguish the earliest small primates from the earliest members of the other groups. All these animals were small and probably ate nectar, gum, pollen, seeds, insects, and fruit in the canopy forest.

One of the major groups of small forest mammals in the Paleocene of North America and Europe was the **plesiadapids**. We have only a few plesiadapid bones other than skulls and teeth, but as far as we can tell, they were more closely related to dermopterans than primates (Figure 21.3). Some smaller plesiadapids look as if they had evolved the complex gliding of their living relative, the colugo. Larger plesiadapids were rather heavy in build, probably like marmots or large squirrels, small-brained and rather small-eyed, with teeth that were adapted for cropping vegetation (Figure 21.4).

The larger plesiadapids became extinct at the close of the Paleocene. Because they looked and probably lived like large rodents, they must have competed to some extent with Paleocene multituberculates (Chapter 15). The virtual extinction of larger plesiadapids and multituberculates in North America coincides exactly with the arrival from Asia of rodents and true primates. This looks like a case of extinction by competition.

The Prosimians

Living lemurs are confined to Madagascar, and must have reached that island from Africa (Figure 19.2). They flourished in that island refuge.

Molecular evidence suggests that lemurs reached Madagascar in Eocene times. No anthropoid primates reached Madagascar until humans did so about 2000 years ago. The actual fossil record of lemurs goes back only as far as the Miocene, but that is enough to document a startling radiation into at least 45 species of lemurs on the island, adapted to a great variety of life styles.

Living lemurs (Figure 21.5) are specialists at vertical clinging and leaping, in which the front limbs are used for manipulating, grasping, and swinging, while the hind limbs are powerful for pushing off. Most lemurs are medium-sized (weighing a few pounds), and are omnivorous, eating fruits and leaves. A few lemurs are small: the mouse lemur weighs only 50 grams or so (about 2 ounces). The largest lemur, *Archaeoindris*, reached around 200 kg, the size of a gorilla, and became extinct only recently; as an adult it must have been a ground dweller. The recently discovered extinct lemur *Palaeopropithecus* was adapted for moving slowly in the forest canopy in the same way as the South American sloth, while *Megaladapis* was probably rather like the Australian koala in its ecology.

Lorises are small, slow-moving, nocturnal hunters of insects. They live in the African forest and are most closely related to lemurs.

Tarsiers, in comparison, are small, agile primates, adapted to eating small animals and insects. They live only in Southeast Asia today, but were much more widespread in the past. Essentially, tarsiers are living fossils, with ancestors in the early Cenozoic that seem to have had much the same anatomy and way of life.

Figure 21.5 The black lemur, a living prosimian primate from Madagascar. (From Meyers.)

Omomyids and Adapids

Because living prosimians are geographically and ecologically restricted, we might expect to gain more insight into early primate evolution from their extinct relatives, the **omomyids** and **adapids**.

As we have seen, plesiadapids dominated North American Paleocene forest habitats. Rodents and true primates arrived in North America only at the end of the Paleocene, probably from Asia, and their arrival seems to have resulted in the disappearance of multituberculates (Chapter 15) as well as the plesiadapids. These Eocene immigrants included two distinctly different groups of true primates. The **omomyids** were small, alert, active nocturnal insect eaters in the forest (Figure 21.6). More than 20 genera of omomyids are known, and they obviously played a significant role in Eocene forest communities. The **adapids** include *Diacronus*, from Paleocene rocks in South China, a plausible ancestor for the Early Eocene *Cantius* from western North America. Adapids look like small lemurs in limb structure, and are related to them. They presumably moved in the same way (Figure 21.7).

Eocene primates apparently lived in the forests all across the Northern Hemisphere, in Eurasia and North America. North American rocks have been more intensively studied: more than 30 species of primates are known from early Eocene rocks of North America alone, though almost all of them are known from skulls, jaws, and teeth, and only three have

Figure 21.6 Life restoration of the small Eocene omomyid *Tetonius*, from North America, as an alert tarsierlike animal. By L. Kibiuk, under the supervision of K. D. Rose. (Courtesy of Kenneth D. Rose of The Johns Hopkins University.)

Figure 21.7 *Notharctus*, an adapid from the Eocene of North America. It probably had an ecology like that of its living relatives, the lemurs (Figure 21.8). Like many vegetarian primates, *Notharctus* was rather small-brained (for a primate). (Negative 319565. Courtesy of the Department of Library Services, American Museum of Natural History.)

been preserved with a reasonable proportion of the body skeleton. All omomyids were small, but many adapids evolved toward larger body size and turned from branch-stalking for insects to a swinging and leaping mode of movement, and to a diet that included much more plant material as well as animal prey, with some eating fruit and others leaves. We do not know whether the adapids or the omomyids (or some other prosimian group still to be found) were closest to the basal primate.

Eocene primates were most likely adept at four-footed climbing and leaping from branch to branch in three dimensions, using the full grasp of hands and feet for catching and holding small branches (Figures 21.6, 21.7). All the different ways in which lemurs, monkeys, gibbons, great apes, and humans move could have evolved from this generalized style shared by early primates. The arm-swinging or brachiating of gibbons could have arisen by emphasizing the arms in movement. The careful, multilimbed climbing of orangutans in trees, the agility of monkeys, the four-footed scrambling and shambling of heavy apes, and the trotting of baboons on the ground could each have evolved by using all the limbs equally. The bipedal walking and running of australopithecines and humans could have been achieved by accentuating the role of the hind limbs in powerful pushing and of the forelimbs in grasping and handling. The primate phylogeny of Figure 21.8 can be used to map this locomotory radiation.

The northern continents drifted slowly northward during the Eocene, and their climates slowly cooled. Finally, at the end of the Eocene, the primates disappeared from northern latitudes. Refugee adapids reached Southeast Asia in the Late Eocene, and as the ancestors of lemurs and lorises, they must have crossed Africa to reach Madagascar. By Oligocene times there were practically no primates left in northern continents.

Primates did not evolve in North America—they invaded from Asia. And later primate evolution toward humans seems to have occurred in Africa. So recent research has begun to focus on China and Africa in the search for the early primate radiation, and the ancestry of anthropoids.

There were already primate-like creatures in Africa in Paleocene times. *Antiatlasius* is known only from a few isolated teeth. It was probably tarsier-sized, but we know practically nothing else about it. Although this evidence is fragmentary, anthropoids may have evolved very early (though still prosimian in their status and biology), and in Africa.

Eosimias ("the dawn monkey") from the Middle Eocene of China, is tiny, about the same size as the smallest living anthropoid, the pygmy marmoset of South America. From incomplete specimens, *Eosimias* has been claimed as the earliest, most primitive anthropoid yet discovered;

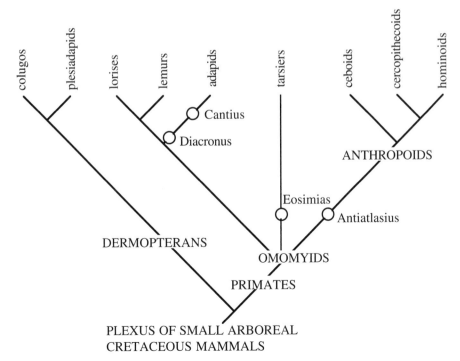

Figure 21.8 One of several possible phylograms of primate evolution. Radiations of omomyids and adapids in northern continents have provided much of the evidence for early primate evolution, and we may expect revisions as more evidence turns up in Africa and Asia.

yet strong counterarguments suggest that it is a tarsier or close tarsier relative, and therefore not relevant to anthropoid ancestry.

The Origin of Anthropoids

The living higher primates, or **anthropoids** (monkeys and apes), have evolved into a variety of life styles and habitats that extends from the huge herbivorous gorillas to the tiny gum-chewing marmosets of South America. Living anthropoids are divided into three evolutionary groups: **cercopithecoids** (Old World monkeys), **ceboids** (New World monkeys), and **hominoids**, which include gibbons, apes, and humans. Apart from *Eosimias* (see above), all available evidence points to the Eocene of Africa as the time and place for the origin and initial radiation of anthropoids.

The Late Eocene Primates of Egypt

In Late Eocene and Oligocene times, the Fayum district of Egypt, not far from Cairo, lay on the northern shore of Africarabia as it drifted slowly northeastward (Chapter 20). Thousands of fossilized tree trunks, some of them more than 30 meters (100 feet) long, show that tropical forests of mangroves, palms, and lianas grew along the levees of a lush, swampy delta. Water birds such as storks, cormorants, ospreys, and herons were abundant, as they are today around the big lakes of central Africa. Fishes, turtles, sea snakes, and crocodiles lived in or around the water, and early relatives of elephants and hyraxes foraged among the rich vegetation. The primates presumably ate fruit in the trees. The same fauna has been discovered as far south as Angola, so the Fayum animals were widespread around the coasts of Africa in Eocene and Oligocene times.

More than 2000 specimens of 19 species of fossil primates have now been collected from the Fayum deposits, most of them on expeditions led by Elwyn Simons. The primates include tarsiers, a loris, and an extinct line of adapids. But the others are anthropoids, all of which look like tree-climbing fruit and insect eaters. The most abundant and best-known early Fayum anthropoid is *Catopithecus* (Figure 21.9).

All the Fayum anthropoids that are well enough known to compare individual sizes are sexually dimorphic. Males are larger and had much larger canine teeth than females, implying that males displayed or fought for rank, and that the animals had a complex social life that included groups of females dominated by a single male. Among living primates, it is generally the larger-bodied species that have these characters, especially in the Old World. The Fayum anthropoids show that size is not important in evolving these sex-linked characters, and they also suggest that these traits may well be basic to anthropoids. Simons and his colleagues suggest that they arose when anthropoids became active in daylight: group defense may be linked with the social structure.

Other Fayum primates include **parapithecids**, a family whose name indicates that although they have some advanced characters, they do not

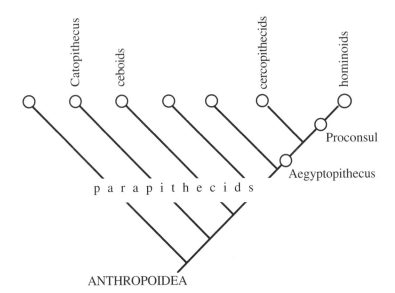

Figure 21.9 Phylogram of the higher primates with evidence included from the fossils of the Fayum. Compare with Figure 21.8. A miscellaneous group of anthropoids from the Fayum, the parapithecids, include the ancestor of the ceboids (New World monkeys), and of *Aegyptopithecus*, which has characters that allow it to have been the ancestor of both Old World monkeys (cercopithecids) and the hominoids. *Proconsul* is a possible ancestor of all hominoids. This scheme is consistent with biogeographic evidence that suggests the ceboids separated from the Old World primates around the end of the Eocene.

look like the direct ancestors of monkeys or apes. They are small, weighing only up to 3 kg (7 pounds). Parapithecid skulls are rather like those of Old World monkeys, but the rest of the skeleton looks primitive, more like that of South American monkeys. The parapithecids are a miscellaneous group that includes the basal genera for all later anthropoids (Figure 21.9). Parapithecids probably had a basic style of primate ecology, eating fruit in the trees. *Apidium*, for example, seems to have been adapted for leaping and grasping in trees.

Later Fayum anthropoids were larger, monkey-sized primates with an adult weight of 3 to 6 kg (7 to 14 pounds). They had shorter snouts than the earlier Fayum primates but were not as flat-faced as later hominoids. *Aegyptopithecus* (Figure 21.10) is the best known of these later Fayum fossils. Its heavy limb bones suggest that it was a powerfully muscled, slow-moving tree climber, ecologically like the living howler

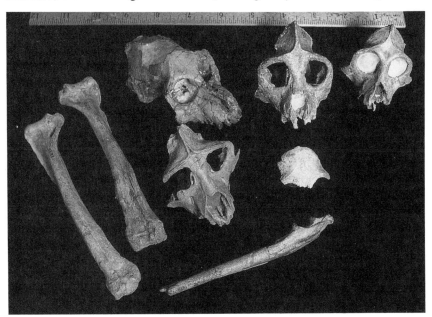

Figure 21.10 *Aegyptopithecus*, a little anthropoid from the Late Eocene of Egypt. It is the closest species we have yet found to the common ancestor of Old World primates (Figure 21.9). (Photograph courtesy of Elwyn L. Simons of Duke University.)

monkey of South America. It had many primitive characters, but its advanced features were more like those of apes than monkeys. Its brain was large for its body size, for example, and its foot bones were like those of Miocene apes. It had powerful jaws for its size, too. It may well be the common ancestor of all higher primates in the Old World: the cercopithecoids, or Old World monkeys, and *Proconsul* and the line leading to hominids (Figure 21.9).

The New World Monkeys

Primates reached South America by Oligocene times, and evolved there in isolation, never again influenced by exchange and contact with other primate groups. No prosimian or apelike primate has ever been found in South America. Instead, the New World primates evolved to fill the ecological niches that monkeys and gibbons occupy in Old World forests.

New World primates, the **ceboids**, have some unique characters: prehensile tails that can be used as a fifth limb, and four more teeth than Old World primates, cercopithecoids. Ceboids are related to cercopithecoids only in that they are both anthropoids (Figure 21.10); many of their monkey-like characters evolved independently. Ceboid color vision uses different nerve pathways than that of cercopithecoids and apes, for example, and it evolved once only, very early. It's tempting to speculate that it evolved as the immigrant (nocturnal) ancestors of ceboids began to adapt to living actively by day.

The earliest ceboids, *Dolichocebus* from Patagonia and *Branisella* from Bolivia, are both Late Oligocene in age, perhaps 26 or 27 Ma. *Dolichocebus* is very much like the living squirrel monkey, and represents an ideal ancestor for it. In the same way, some Miocene South American monkeys are much like living spider monkeys and howler monkeys. A genuinely modern-looking owl monkey is known from Miocene rocks of Colombia at 12 to 15 Ma. Many of today's South American monkeys therefore qualify as living fossils. Either they evolved early and rapidly, or they have a longer fossil record still to be discovered. All of them are tree dwellers. South American primates did not evolve into terrestrial ways of life as Old World primates did, even though there have been extensive savannas in South America since the Miocene.

Ceboid ancestors probably evolved from African immigrants that crossed the widening Atlantic in early Oligocene times. For want of better information I have shown them as diverging from early Fayum anthropoids like *Aegyptopithecus* (Figure 21.9). More fossils must be discovered before we can say much about this divergence.

We know more about the emergence of Old World monkeys. The small fossil *Victoriapithecus* from the Miocene of East Africa is an ideal ancestor for this group. Interestingly, it is small even for a monkey at 3–5 kg (7 to 11 pounds), and seems to have had a semi-terrestrial ecology rather than the tree-dwelling habit that one might have expected.

Emergence of the Hominoids

Living hominoids include **hylobatids** (gibbons), **pongids** (the remaining species of Asian ape, the orangutan), **panids** (African apes, the chimps and gorillas) and **hominids** (only humans survive in this family). The physical, molecular, and genetic structure of living hominoids has been studied closely. Humans, gorillas, and chimps are very similar in genetic makeup and in protein chemistry, much closer than they are in body structure, but the orangutan differs significantly, and gibbons even more.

Hominoids almost certainly evolved from some African genus like *Aegyptopithecus*. The DNA clock suggests that the common ancestor of all hominoids split from monkeys about 33 Ma. The gibbons split off about 22 Ma, followed by the orangutan lineage at about 16 Ma. Finally the various living lineages of panids and hominids diverged from one another between 10 Ma and 6 Ma (Figure 21.11). Protein clocks suggest more recent branching points.

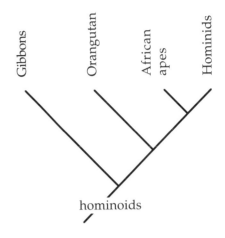

Figure 21.11 The cladogram of living hominoids that is suggested by molecular and fossil evidence. The timing of the branching points is uncertain, and is not likely to be settled soon. Extinct groups are not shown on this cladogram.

Miocene Hominoids

About 20 Ma, Africarabia formed a single land mass that lay south of Eurasia and was separated from it by the last remnant of the Tethys Ocean. African animals were evolving largely in isolation from the rest of the world, and some groups, including the hominoids, were confined to Africarabia at this time, though they were widespread across it.

Early Miocene faunas of Africa were dominated by elephants and rhinos at large body size, primitive deer and hyraxes at medium size, and insectivores common at small sizes. The environment was forest, broken by open grassland and woodland. Primates of all kinds flourished, although it is difficult to describe their ecology and habits because body skeletons are not as well known as skulls. But prosimians and monkeys were rare, while hominoids were diverse and abundant. We have over 1000 hominoid fossils from the Early Miocene of Africa, most dating from 19 to 17 Ma and most from East Africa.

The dominant hominoids were the apelike **dryomorphs**. Like living African apes, they had relatively small cheek teeth with thin enamel, implying a soft diet of fruits and leaves, and a way of life foraging and browsing in trees like most living monkeys. (True monkeys were scarce at this time, remember.) Dryomorphs varied in weight from a large species of *Proconsul* in which males weighed about 37 kg (80 pounds) down to *Micropithecus* at about 4 kg (9 pounds), and their locomotion varied accordingly. *Micropithecus* and *Dendropithecus* were not as well adapted for arm-swinging as living gibbons, but they were lightly built and relied more on brachiating than did other Miocene primates.

The best-known and most important form is *Proconsul* itself (Figure 21.13). There were several species of this animal by 18 Ma. The most complete specimens are from a small species that weighed only about 9 kg (20 pounds) but had a baboon-sized brain; that is, its brain was larger

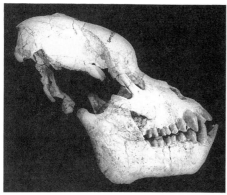

Figure 21.12   Top: Reconstruction of *Proconsul*, the Miocene dryomorph from East Africa that could be the common ancestor of later hominids. Parts in black are reconstructed; other bones are known. Bottom: The best-known skull of *Proconsul*. The jaw does not project so much if the skull is tilted to the position it had in a quadrupedal pose in life (compare top picture). (Courtesy of Alan Walker of Pennsylvania State University.)

relative to body size than that of living monkeys. Its skeleton was a mixture of primitive characters that are also found in monkeys, gibbons, and chimps; altogether, these characters indicate a rather basic quadrupedal, tree-climbing, fruit-eating primate that could be the ancestor of all later hominoids. *Proconsul* had advanced hominoid characters of the head and jaws, though most of its body skeleton remained unspecialized. A large *Proconsul* may have spent a lot of time in deliberate climbing or on the ground, like a living chimp. Like them, it was probably versatile in its movements, and capable of occasional upright behavior.

Morotopithecus is a large ape from Uganda, probably as old as 20 Ma. Although we do not have a skeleton as complete as *Proconsul*, *Morotopithecus* is clearly large (40 to 50 kg, or 100 pounds), and the pieces we have are more advanced than the same pieces of *Proconsul*. In other words, *Morotopithecus* is probably close to the direct ancestry of all later hominoids. It was probably a rather heavy slow climber, hanging in trees and eating fruit.

Africarabia drifted northward during the Miocene (Chapter 20) and finally collided with Eurasia to form an irregular mountain belt from Iran to Turkey. The collision interrupted tropical oceanic circulation and set off climatic changes. Temperatures cooled in East Africa, and almost all the northern continents experienced dramatic changes in faunas and floras. Forests became much more open, and grasses evolved to form wide expanses of savanna.

An exchange of animals between Africarabia and Eurasia added to the ecological turmoil of the times (Chapter 20). In that process, African hominoids successfully invaded Eurasian plains and woodlands.

In Africa the dryomorphs remained in the forests, which were thinned or diminished by cooling temperatures. They came under increasing pressure from the evolving monkeys. At 15 Ma, *Victoriapithecus* is clearly a monkey rather than an ape. It still has features of a common ancestor with apes, but its anatomy could also make it the common ancestor of all living Old World monkeys. Monkeys have increased in abundance and diversity so that today they rather than apes dominate the remaining forests of Africa and Asia.

Some late dryomorphs reached Eurasia, but they were apparently numerous only in Europe. *Dryopithecus* is a European fossil, known from Spain to Hungary. Perhaps because it lived in a cool region, *Dryopithecus* was bigger and stronger than most dryomorphs. It shows adaptations for branch-swinging with the trunk more or less vertical, which gave it some of the characters of the living orang. Its skull is not like orangs, however, and the position of *Dryopithecus* is arguable, depending whether one thinks skull characters or trunk characters are more important.

Sivapithecids

At about 14 Ma, new hominoids appeared alongside the dryomorphs: the **sivapithecids** were the dominant group in East Africa. The earliest sivapithecid is *Kenyapithecus*, dated about 14 Ma. It is generalized enough to

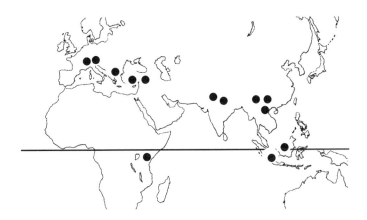

Figure 21.13 The geographical distribution of sivapithecids. Miocene specimens from Kenya, Hungary, Greece, Turkey, India, and China have all been given different names. The giant Pleistocene form *Gigantopithecus* and the living orangutan, *Pongo*, are later sivapithecid descendants.

be a descendant of *Proconsul* and/or *Afropithecus*, and to be the ancestor of all later large apes: sivapithecids + pongids on one hand, and panids + hominids on the other.

Sivapithecids are apes, known from a wide area that stretches from East Africa to Central Europe, and eastward as far as China (Figure 21.13). Sivapithecids probably evolved from an Early Miocene ancestor that lived in Africa, possibly *Afropithecus*, which lived in both African and Arabian parts of Africarabia. *Sivapithecus* itself was an Asian ape.

We have a great number of sivapithecid fossils, but they are mostly jaws, skulls, and isolated teeth; few body or limb bones are well known. Thus we can reconstruct sivapithecid heads rather well, but we know little about their body anatomy, posture, or locomotion.

Many different names have been applied to sivapithecids in their various countries of discovery. Hungarian, Turkish, Kenyan, Indian, Chinese, and Greek specimens were all given different names, for example. Part of the problem of naming sivapithecids is that there is a good deal of variation between individuals. As in orangutans, the skulls of males are much larger and broader than those of females.

All sivapithecids had thick tooth enamel and powerful jaws, suggesting that their diet required prolonged chewing and great compressive forces on the teeth. In living primates with thick enamel, such as orangutans or mangabeys, teeth and jaws like these are correlated with a diet of nuts, or fruits with hard rinds. One can hear an orangutan cracking nuts a hundred meters away! Perhaps sivapithecids diverged from the dryomorph diet of soft leaves and fruits to exploit a food source that had so far been available only to pigs, rodents, and bears.

Nut eating can be an activity of tree or ground dwellers, or creatures making the evolutionary and ecological transition from woodland to open ground. We cannot yet tell whether sivapithecids were foraging for fruit and nuts in the trees (with an arboreal life like that of orangutans) or under the trees (with adaptations for ground living).

One late sivapithecid was adapted to live entirely on the ground. The huge ape *Gigantopithecus* lived in southern and eastern Asia from about 7 Ma well into the Pleistocene. It had huge grinding teeth and weighed several hundred pounds (Figure 21.14). It probably lived on very coarse vegetation, as an ecological equivalent of the giant ground sloth of the American Pleistocene, or the Asian giant panda, or the African mountain

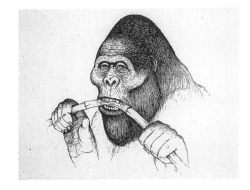

Figure 21.14 Reconstruction of the giant Asian sivapithecid ape *Gigantopithecus*. (© Stephen Nash and Russell Ciochon. Courtesy of Russell Ciochon, University of Iowa.)

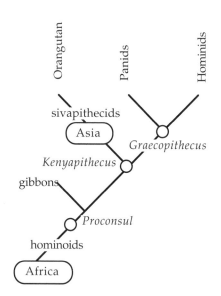

Figure 21.15 A phylogram that shows the sivapithecids fitting into hominoid evolution as ancestors of the orangutan.

gorilla. *Gigantopithecus* survived in Asia as recently as 300,000 or 250,000 years ago. It was certainly contemporaneous with *Homo* in Eastern Asia, and its bones, teeth, and jaws may be responsible for Himalayan folklore about the abominable snowman, or yeti.

It is clear now that sivapithecids have nothing to do with human ancestry but are instead ancestors of the living Asian ape, the orangutan. In Africa, dryomorphs evolved toward hominids (Figure 21.15). The molecular clock suggests 17 Ma for the divergence, and the new Kenyan fossils seem to agree with that estimate.

After about 11 Ma, migration between Africa and Eurasia was essentially cut off. Thereafter hominoids evolved independently in Eurasia and Africa, eventually leaving the sivapithecids in Asia and the hominids in Africa. The African fossil record of hominids is horribly incomplete during the critical time after 11 Ma when they radiated to become separate lineages: we simply haven't found these fossils yet.

Between 8 and 5 Ma, the climate of Eurasia slowly changed to encourage even more open grasslands instead of woodland and forest. Then the history of Eurasian apes became one of struggling survival rather than innovation and evolution. European dryomorphs disappeared around 8 Ma, and the only remaining sivapithecids were the East Asian animals that led to *Gigantopithecus* and the orangutan. This means that Eurasia is not the continent in which to search for direct human ancestry (Figure 21.15). It is the African story that we must now follow.

By 7 to 5 Ma, the African forest had become dominated by monkeys, who displaced the dryomorphs ecologically and presumably restricted all the surviving forest apes. This is the time interval in which we can look for dramatic finds in the near future.

Review Questions

It is generally agreed that we retain characters that were evolved by our primate ancestors when they lived in trees. List and explain those characters.

On which continent did most primate evolution take place? Discuss the evidence carefully in answering this question.

Draw a phylogram that illustrates the geography of primate evolution.

Further Reading

Easy Access Reading

Badgley, C., et al. 1988. Paleoecology of a Miocene, tropical, upland fauna: Lufeng, China. *National Geographic Research* 4: 178–195.

Beard, K. C., et al. 1996. Earliest complete dentition of an anthropoid primate from the late Middle Eocene of Shanxi Province, China. *Science* 272: 82–85. The latest paper on *Eosimias* from China.

Collura, R. V. and C.-B. Stewart. 1995. Insertions and duplications of mtDNA in the nuclear genomes of Old World monkeys and hominoids. *Nature* 378: 485–489. Some problems in molecular clocks.

Dean, D., and E. Delson. 1992. Second gorilla or third chimp? *Nature* 359: 676–677.

Gebo, D. L., et al. 1997. A hominoid genus from the Early Miocene of Uganda. *Science* 276: 401–404. *Morotopithecus*.

Godinot, M., and M. Mahboubi. 1992. Earliest known simian primate found in Algeria. *Nature* 357: 324–326.

Kay, R. F., et al. 1997. Anthropoid origins. *Science* 275: 797–804, and discussion, v. 278, pp. 2134–2136. Strange methods were used to get from the data to the cladogram.

Kelley, J., and Q. Xu. 1991. Extreme sexual dimorphism in a Miocene hominoid. *Nature* 351: 151–153, and comment, pp. 111-112. *Lufengpithecus.*

Moyá-Solá, S., and M. Köhler. 1996. A *Dryopithecus* skeleton and the origins of great-ape locomotion. *Nature* 379: 156–159, and comment, pp. 123–124.

Olson, S. L., and D. T. Rasmussen. 1986. Paleoenvironment of the earliest hominoids: new evidence from the Oligocene avifauna of Egypt. *Science* 233: 1202–1204.

Setoguchi, T., and A. L. Rosenberger. 1987. A fossil owl monkey from La Venta, Colombia. *Nature* 326: 692–694.

Simons, E. L. 1995. Skulls and anterior teeth of *Catopithecus* (Primates: Anthropoidea) from the Eocene and anthropoid origins. *Science* 268: 1885–1888, and comment, p. 1851.

Stewart, C.-B., et al. 1987. Adaptive evolution in the stomach lysozymes of foregut fermenters. *Nature* 330: 401–404, and comment, p. 315.

Tattersall, I. 1993. Madagascar's lemurs. *Scientific American* 268(1): 110–117.

Walker, A., and M. Teaford. 1989. The hunt for *Proconsul*. *Scientific American* 260(1): 76–82.

More Technical Reading

Benefit, B. R. 1999. *Victoriapithecus*: the key to Old World monkey and catarrhine origins. *Evolutionary Anthropology* 7: 155–174.

Boissinot, S., et al. 1998. Origins and antiquity of X-linked tri-allelic color vision systems in New World monkeys. *Proceedings of the National Academy of Sciences* 95: 13749–13754. The unusual color vision of New World monkeys evolved once (presumably shortly after their arrival).

Ciochon, R., et al. 1990. *Other Origins: The Search for the Giant Ape in Human Prehistory.* New York: Bantam Books.

Ciochon, R., et al. 1996. Dated co-occurrence of *Homo erectus* and *Gigantopithecus* from Tham Khuyen cave, Vietnam. *Proceedings of the National Academy of Sciences* 93: 3016–3020.

Fleagle, J. G., and R. F. Kay. (eds.). 1994. *Anthropoid Origins.* New York: Plenum.

Krause, D. W. 1986. Competitive exclusion and taxonomic displacement in the fossil record: the case of rodents and multituberculates in North America. *Contributions in Geology of the University of Wyoming, Special Paper* 3: 95–117.

Leakey, R. E. F., et al. 1988. Morphology of *Afropithecus turkanensis* from Kenya. *American Journal of Physical Anthropology* 76: 289–307.

Martin, R. D. 1990. *Primate Origins and Evolution.* London: Chapman and Hall.

Rasmussen, D. T., et al. 1998. Tarsier-like locomotor specializations in the Oligocene primate *Afrotarsier*. *Proceedings of the National Academy of Sciences* 95: 14848–14850.

Rose, K. D. 1990. Postcranial skeletal remains and adaptations in early Eocene mammals from the Willwood Formation, Bighorn Basin, Wyoming. *Geological Society of America Special Paper* 243: 107–133.

Schwartz, J. H. 1990. *Lufengpithecus* and its potential relationship to an orang-utan clade. *Journal of Human Evolution* 19: 591–605.

Simons, E. L. 1987. New faces of *Aegyptopithecus* from the Oligocene of Egypt. *Journal of Human Evolution* 16: 273–289.

Simons, E. L., et al. 1999. Canine sexual dimorphism in Egyptian Eocene anthropoid primates: *Catopithecus* and *Proteopithecus*. *Proceedings of the National Academy of Sciences* 96: 2559–2562.

Van Couvering, J. A., and J. A. Harris. 1991. Late Eocene age of Fayum mammal faunas. *Journal of Human Evolution* 21: 241–260.

Yoder, A. D., et al. 1996. Ancient single origin for Malagasy primates. *Proceedings of the National Academy of Sciences* 93: 5122–5126.

Evolving Toward Humans

We know practically nothing of the evolution of the hominoid lineages that led to gorillas and chimps. Molecular and genetic evidence suggests that our closest living relatives are chimps, with gorillas a little further away. Our own lineage, the **hominids**, probably separated from that of chimps around 5 or 6 Ma. Even so, our DNA is at least 98.5% identical to that of chimps. Obviously the 1.5% that is different has led to great changes in our bodies, brains, and behavior.

Over time, there have been perhaps a dozen species of hominids, but we, as *Homo sapiens*, are the only surviving one. As many as six earlier species of *Homo* have become extinct, and another six hominid species have usually been placed in the genus *Australopithecus*. A newly discovered species seems to require a new generic name (*Ardipithecus*), even though it is likely to be ancestral to one or perhaps all of the species of *Australopithecus*. (Be warned that almost everything I write in this chapter is being argued over by paleoanthropologists. I have tried, as usual, to select what I think are the most likely hypotheses.)

I shall refer to all the hominid species that are not *Homo* as **australopithecines**. These hominids lived in Africa, south of the Sahara Desert, from perhaps 4.3–4.3 Ma to about 1.4 Ma or a little later. We have enough evidence to reconstruct a vivid picture of australopithecine life.

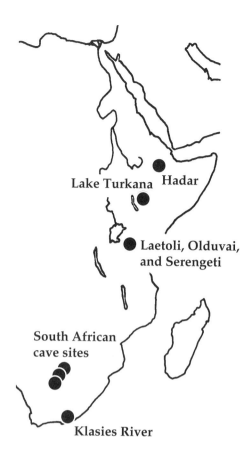

Figure 22.1 Some important hominid-bearing regions and localities in Africa.

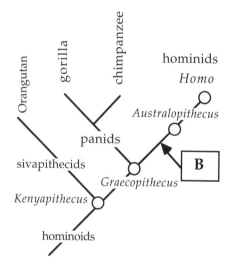

Figure 22.2 A hypothesis (phylogram) for the separation of hominids from Asian and African apes. B marks the evolution of bipedalism, which could also be said to define the origin of the family of hominids, and its subgroup the australopithecines.

Australopithecines

The New Australopithecines

The earliest two species of australopithecines are *Ardipithecus ramidus*, described in 1994 from rocks in Ethiopia dated at 4.3–4.4 Ma, and *Australopithecis anamensis*, described in 1995 from rocks in Kenya dated at 4.1–4.2 Ma. *Ardipithecus ramidus* is the most primitive (i.e., ape-like) australopithecine yet found, and was probably a forest dweller. The slightly younger *Australopithecus anamensis* has a jaw that is even more ape-like, but its arm and leg bones suggest upright (bipedal) posture and locomotion. It would make a good ancestor for later *Australopithecus*: in fact some of the specimens currently identified as *A. afarensis* may actually belong to *A. anamensis*. *A. anamensis* is rather large, perhaps 50 kg, 110 pounds. New specimens still being cleaned, and new information will pour out in the next few years. However, it is fair to say that there are no major evolutionary surprises (yet) in these new fossils.

Footprints at Laetoli

The East African Rift splits Africa from Ethiopia to Zambia and Malawi. Among its unusual geological features are volcanoes that sometimes erupt carbonatite ash, which is composed largely of a bizarre mixture of calcium carbonate and sodium carbonate. One of these volcanoes, Sadiman, stood near the Serengeti Plain, in northern Tanzania (Figure 22.1), in Late Pliocene times. After carbonatite ash eruptions, the sodium carbonate in the ash dissolves in the next rain, and as it dries out the ash sets as a natural cement. Any animals moving over the damp surface in the critical few hours while it is drying will leave footprints that can be preserved very well. As long as the footprints are covered up quickly (for example, by another ash fall), rainwater percolating through the ash will react with the carbonate to make a permanent record.

Sadiman erupted one day about 3.6 Ma, towards the end of the dry season. Ash fell on the plains near Laetoli, 35 km (20 miles) away, and hominids walked across it, leaving their footprints along with those of other creatures. The vital point about the tracks is that the hominids were walking fully erect, long before hominid jaws, teeth, skull, and brain reached human proportions, shape, or function.

Why would a hominid become bipedal? Most suggestions are related to carrying things with the hands and arms (infants, weapons, tools, food), to food gathering (seeing longer distances, foraging over greater ranges, climbing vertically, reaching high without climbing at all), to defense (seeing longer distances, throwing stones, carrying weapons), to better resistance to heat stress (less sweat loss and better cooling), or to staying within reach of rich food resources by migrating with the great plains animals (carrying helpless young over long distances). These are all reasonable suggestions, but all are difficult to test.

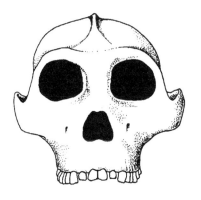
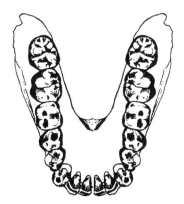

Figure 22.3 The skull, jaw, and teeth of australopithecines. Left: The skull of *Australopithecus robustus*. (After Howell.) Right: The teeth and jaw of *A. afarensis*, idealized after the specimen AL–400. (Adapted from Johanson and Maitland.)

Probably the trend toward the use of the forelimbs for gathering food and the hind limbs for locomotion began among tree-dwelling primates long before *Australopithecus*. But this thought is based mainly on my own experience in picking and eating fruit, and the realization that forelimbs are more effective for that job than teeth and jaws alone. The final achievement of erect bipedality on the ground was probably an extension of previous locomotion and behavior, rather than something completely different. A phylogram for the evolution of bipedalism and the evolution of the hominids is shown in Figure 22.2.

It's almost certain that the footprints at Laetoli were made by a species of *Australopithecus*. Australopithecines can be summarized by the slogan "Brains of apes and bodies of men," but, of course, much more can be said about them. The footprints from Laetoli show that they walked upright. All australopithecines are similar below the neck, apart from size differences, so they all probably moved in much the same way. Their movements were probably not exactly like ours, but their leg and hip bones indicate that they walked and ran well. At the same time, the limb joints and toes suggest that australopithecines spent a lot of time climbing in trees as well as walking upright on the ground.

Australopithecines were smaller than most modern people. They were all around 30–50 kg (65–110 pounds) as adults (males heavier than females) but their bones were strongly built for their size. The skull was even stronger, and very different from ours (Figure 22.3, left). The relative brain size was about half of ours, even allowing for the smaller body size of *Australopithecus*, but the jaw was heavy and the teeth, especially the cheek teeth, were enormous for the body size. The canine teeth were large and projecting, especially in males. The whole structure of the jaws and teeth suggests strength (Figure 22.3, right).

The small size of the brain and the thickness of the skull may be linked with another feature that separates us from *Australopithecus*. The birth canal in the pelvis of australopithecines is wide from side to side, but narrow from front to back, so that there may have been a special mode of delivery for even the small-brained babies that australopithecines had (Figure 22.4). In *Homo* the birth canal is rounder (Figure 22.4), presumably to accommodate the passage of a baby with a very large head (and a very large brain). If so, a larger brain was important enough that this visible difference in skeletal anatomy was evolved in *Homo*.

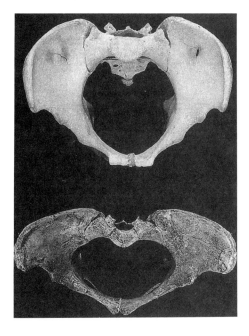

Figure 22.4 The difference in structure between the pelvis of a modern woman (*Homo sapiens*, upper photograph), and the reconstructed pelvis of the female australopithecine Lucy (*Australopithecus afarensis*, lower photograph). The modern pelvic canal is much rounder, presumably to accommodate the birth of a larger-brained baby. (© 1988 C. Owen Lovejoy. Courtesy of Owen Lovejoy, Kent State University.)

Like ourselves, australopithecines were built for trotting endurance rather than blinding speed, an adaptation that would be better suited to foraging widely in open woodland than to skulking in forests, provided that *Australopithecus* was not easily picked off by large sprinting carnivores. (Reasons may have been related to group defense, which allows baboon troops to roam freely on the ground, or to early possession and use of defensive tools.) *Australopithecus* had long arms and fingers that were capable of sensitive motor control. In a tall biped walking upright on the ground, the arms would have been free for carrying, throwing, and manipulating.

Australopithecus afarensis

The best-known collections of early australopithecines are from Laetoli and from Hadar in Ethiopia (Figure 22.1). Each district has produced spectacular finds. At Laetoli there are the footprints, plus remains of at least 22 individuals; at Hadar, bone fragments from at least 35 individuals are preserved rather better. All the specimens belong to one species, *Australopithecus afarensis*, which was probably descended from *A. anamensis*, and was ancestral to all the later species of *Australopithecus* and to *Homo* as well (Figure 22.5).

Hadar lies in the Afar depression, a vast arid wilderness in northeast Ethiopia (Figure 22.1). At 3 to 4 Ma it was the site of a lake fed by rivers tumbling out of winter snowfields on the plateau of Ethiopia. The australopithecines lived and are fossilized along the lake edges. Delicately preserved fossils such as crab claws and turtle and crocodile eggs suggest that the australopithecines had rich protein foods available to them, and skeletons of hippos and elephants suggest that there was rich vegetation in and around the lake edges. All the Hadar specimens are dated at about 3.2 Ma, so they are considerably later than the Laetoli australopithecines.

The best-preserved Hadar skeleton is the famous Lucy. Lucy was small by our standards, a little over 1 meter (42 inches) in height. She was full-grown, old enough to have had arthritis. Her brain was small at

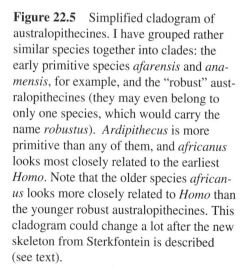

Figure 22.5 Simplified cladogram of australopithecines. I have grouped rather similar species together into clades: the early primitive species *afarensis* and *anamensis*, for example, and the "robust" australopithecines (they may even belong to only one species, which would carry the name *robustus*). *Ardipithecus* is more primitive than any of them, and *africanus* looks most closely related to the earliest *Homo*. Note that the older species *africanus* looks more closely related to *Homo* than the younger robust australopithecines. This cladogram could change a lot after the new skeleton from Sterkfontein is described (see text).

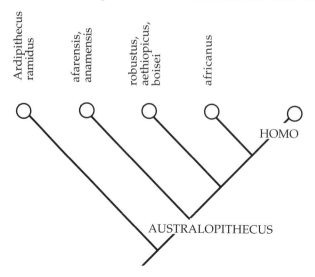

about 385 cc, compared with 1300 cc for an average human. Her large molar teeth suggest that *A. afarensis* was a forager and collector eating tough fibrous material.

Baboons sleep in high places—trees or high rocks—and are great opportunists in taking whatever food is available. They live and forage in troops and have a cohesive social structure that gives them effective protection from predators even though they are fairly small as individuals. But *Australopithecus* walked upright, whereas baboons trot on four limbs. Ecologically, *Australopithecus* may have been a super-baboon. Walking upright, with its arms free for carrying, it may have been a more effective forager than a baboon, which can carry only what it can put into its mouth and stomach. Perhaps the requirements and advantages of efficient troop foraging and defense encouraged tight social cohesion among australopithecines, long before tools permitted technological advances.

Australopithecus in South Africa

Isolated caves scattered over the high plains of South Africa are mined for limestone, and hominid fossils have been found in the limestone (Figure 22.6). But cave deposits are difficult to interpret and date accurately. Roof falls and mineralization by percolating water have disturbed the original sediments, and few of the radioactive minerals in cave deposits allow absolute dating. Thus there have been problems in relating South African hominid fossils to their well-dated East African counterparts.

New research will soon change that. The cave deposits are now being dated with a new technique that works on changes in the Earth's magnetic field. And a dramatic new find at the Sterkfontein cave late in 1998 revealed a largely complete skeleton of an australopithecine that dates from well before 3 Ma.

To date, the best-known early australopithecine from South Africa is *Australopithecus africanus* (Figure 22.6). Although it was about the same body weight as *A. afarensis*, *A. africanus* was taller but more lightly built and had a larger brain, perhaps 450 cc. The teeth and jaws continued to be large and strong, with molars twice as large as chimpanzee molars, suggesting that the diet remained mainly vegetarian. However, new evidence from isotopes in the teeth suggest that *A. africanus* ate vegetarian animals as well, possibly catching small animals, or scavenging meat from carcasses. Tooth wear suggests an average life span for *A. africanus* of perhaps 20 years, maybe with a maximum of 40 years, about the same as a gorilla or chimpanzee. The arms were relatively long compared with *A. afarensis*, suggesting that *A. africanus*, though perfectly erect and able to walk and run on the ground, spent a good deal of time in trees.

If the new South African fossil is *A. africanus*, we will have even more information about it; if it is a new species, it could alter our ideas about australopithecine biology and evolution in ways that I cannot predict as I write this section. It will probably take two years to extract, study, describe, and interpret the new find: the delay is aggravating, but that is the nature of paleontology.

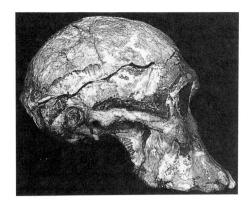

Figure 22.6 A skull of *Australopithecus africanus* from the cave deposits at Sterkfontein in South Africa. This specimen was discovered by deliberately dynamiting the cave limestone, which is very solid. When the smoke cleared after one blast, the skull was found blown almost in two, showing the brain cavity still empty and lined with lime crystals. The skull has been rejoined along the line of breakage. (Negative 2A 333. Courtesy of the Department of Library Services, American Museum of Natural History.)

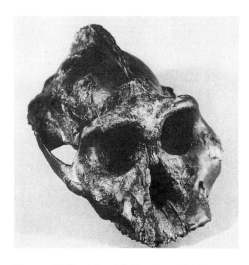

Figure 22.7   Alan Walker found this robust australopithecine skull in northern Kenya. First called the "Black Skull", it is now usually called *Australopithecus aethiopicus*. (Courtesy of Alan Walker of Pennsylvania State University.)

Robust Australopithecines

Australopithecines with heavily built skulls are called **robust** to distinguish them from those with lightly built skulls (such as *A. africanus*) which are called **gracile**. The best example of a robust skull is the oldest one, the so-called *Black Skull* (Figure 22.7) from the Turkana Basin of northern Kenya (Figure 22.1) dating from about 2.5 Ma. The Black Skull is usually called *Australopithecus aethiopicus*. It has a skull much heavier and stronger than *A. afarensis*, although the brain was no larger and the body was not very different. The jaw extended further forward, the face was broad and dish-shaped, and there was a large crest on the top of the skull for attaching very strong jaw muscles (Figure 22.7). The molar teeth of the Black Skull are as large as any hominid teeth known, about four or five times the size of ours. Yet the front teeth of robust australopithecines are small.

Later robust forms have been found all over East and South Africa between 2.5 Ma and 1.4 Ma. In East Africa they are usually called *Australopithecus boisei* (the famous Zinj of Louis Leakey), and in South Africa they are called *A. robustus* (Figure 22.3, left). There are enough fossils to suggest that robust australopithecines changed over the million years of their history, evolving a larger brain (perhaps 500 cc rather than 400 cc) and a flatter face.

The robust australopithecines are certainly an ecological group. The large jaw weight and the huge molars, with their very thick tooth enamel, were ecological adaptations that indicate great chewing power and a diet of coarse fibre. However, almost all the characters that are used to define robust australopithecines are connected with the huge teeth, and the modifications of the jaws and the face during growth that are required to accommodate the teeth. So any australopithecine population that evolved huge teeth would have come to look "robust." Therefore the robust australopithecines may not be an evolutionary group. They may be three separately evolved species; they may be three related species; or they may be variants of the *same* species (which would have to be called *robustus*). Some specialists prefer to give robust australopithecines their own generic name, *Paranthropus*, but I have not used this name (Figure 22.5). The robust australopithecines could easily have evolved from *A. afarensis*.

Australopithecus garhi, and Butchering Tools

An astonishing find was reported in early 1999. Rocks 2.5 million years old in Ethiopia yielded enough pieces of two or three skeletons to allow the description of a new species of *Australopithecus*, *A. garhi*. Furthermore, the same bed yielded evidence of the use of stone tools for butchering meat and smashing bones.

Australopithecus garhi is a normal gracile australpithecine, except that it has very large teeth for the size of its jaw and skull. The skull is far too primitive for it to belong to *Homo*, and its brain size is only about 450 cc. However, given its age, location, and the features of its skeleton,

A. garhi would be a good ancestor for *Homo*.

There are many animal bones in the same rock bed, and some of them have been sliced and hammered in ways that betray intelligent butchering. Most likely, the butchers used their tools carefully, because there were no suitable rocks nearby, and all tools had to be carried in (and carried out for further use).

Before 1999, it had generally been thought that the defining characters of *Homo* versus *Australopithecus* included a larger brain and the use of tools. Unless there are fossils of *Homo* still to be found in this locality, we now must accept that *A. garhi* was making, carrying, and using tools effectively. Perhaps the great ecological advantages gained by the invention of butchering tools encouraged exactly the changes in the *Australopithecus garhi* lineage that led quickly to increased brain size, reduced tooth size, and the status of first *Homo*.

Once again, apparently major transitions disappear as we collect more fossils: we have seen this for the transition between birds and dinosaurs, between cynodonts and mammals, and now between australopithecines and *Homo*.

The Appearance of *Homo*

Homo appeared in the fossil record with the cooling and drying of tropical climates that accompanied the beginnings of the great glaciations. Perhaps as early as 2.4 Ma, hominids with greatly increased brain size and correspondingly reduced teeth and jaws appeared in Africa. They are sufficiently like ourselves in jaws, teeth, skull, and brain size to be classed as *Homo*. The earliest definite specimen of *Homo*, with an age of at least 2.33 Ma, consists only of an upper jaw, so has not been given a species name. The most familiar species of early *Homo* is *Homo habilis* (Figure 22.8). *Homo rudolfensis* is known from East Africa around 2 Ma, largely from skulls, and is often given a separate name because people think that the differences between it and the lighter-skulled *H. habilis* are too great for them to have belonged to one species (others think they are sexual dimorphs, with *rudolfensis* as the male). This story is bound to change as we find more fossils. Meanwhile I will call them "early *Homo*" or *Homo habilis*.

Early *Homo* was small by modern standards, perhaps just over a meter (4 feet) tall, but was at least as heavy as contemporary robust australopithecines at about 30–50 kg (65–110 pounds). The difference in brain size is striking, however. The brain size was about 650 cc, considerably larger than the brain of an australopithecine. It may have reflected a new level of brain organization in early *Homo*.

We have only a few sets of bones of *H. habilis*, but there is enough evidence from hands, legs, and feet to suggest that *H. habilis* was adapted for spending a lot of its time climbing in trees. New skull characters and brain size are major reasons for assigning *habilis* to *Homo* rather than *Australopithecus*, though even this is debated.

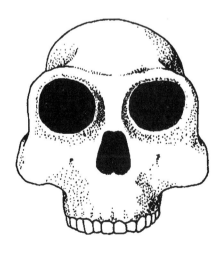

Figure 22.8   Homo habilis: a reconstruction of specimen KNM-ER1470. (Redrawn after Howell.)

Figure 22.9 An Oldowan artifact shown about half-size. Perhaps it was some sort of tool in itself, but maybe it was used as a core for chipping off more useful smaller tools. (After Gowlett.)

We have a good record of the tools that were used by *Homo habilis* (and probably *Australopithecus garhi* before it). They are called **Oldowan** tools because they were first identified by the Leakeys in Olduvai Gorge (Figure 22.1). They are often large and clumsy-looking objects with simple shapes, and not all of them were useful tools in themselves. Instead, many objects may be the discarded centers (cores) of larger stones from which useful scraping and cutting flakes had been removed by hammering with other stones (Figure 22.9).

Oldowan tools demonstrate the use of stone in a deliberate, intelligent way, and the flakes were probably made and used for cutting up food items. For example, an excavation in the Turkana Basin turned up the skeleton of a hippopotamus lying near an ancient river bed. Cobbles naturally occurring close by on a gravel bank in the river had been broken to produce simple tools. Marks on the hippo bones showed that they had been scraped, and that the tendons and ligaments had been cut, to allow meat to be taken from the carcass. There was no indication that the hippo had been killed by the tools.

Nicholas Toth has reproduced and used Oldowan-style artifacts from East African rock types. He showed that the toolmaker was sophisticated in selecting appropriate rocks and making the most of them. Toth's experiments on fresh carcasses of East African animals show that Oldowan axes, flakes, and cores are excellent tools for slitting hides, butchering carcasses, and breaking bones for marrow. Toth was also able to determine that *Homo habilis* was right-handed!

Some Oldowan sites were visited many times. They contain accumulations of bones, stones, and tools, brought to the site over periods of years. Flakes were made on site from stones that had been carried there. This may not indicate a systematic return to a homesite, but it does indicate an intelligent return to sites that perhaps were particularly suitable for food processing and tool making.

From Super-Baboon to Super-Jackal

Was early *Homo* a hunter or a scavenger? This may be a non-question, because all hunters will eat a fresh carcass, and all scavengers will cheerfully kill a helpless prey if they can. Evidence from Turkana and Olduvai suggests that early *Homo* was a scavenger on large carcasses but hunted small- and medium-sized prey. Thus early *Homo* may have had the ecology of a super-jackal, foraging in groups over long distances in search of large, fresh carcasses killed by other predators. Rhinos, hippos, and elephants have thick and leathery hides, difficult for vultures, jackals, and hyenas to pierce, but stone tools allowed *Homo* to make short work of dismembering a large carcass. Between carcass finds of large animals, early *Homo* may have foraged for leopard kills of medium-sized animals, left hanging in trees. *Homo* may also have been an opportunistic hunter of small- to medium-sized prey that was brought to central sites for butchering, and also a forager searching for fruits, berries, grains, roots, grubs, locusts, and lizards. *Australopithecus africanus* may have eaten small animals as well: early *Homo* (and possibly *A. garhi*) used tools to make that opportunistic way of life more efficient.

The concept is exciting. A new ecological niche opened up, or became much more profitable, with the invention of tools and the ability to use them intelligently. Visiting American anthropologists with no previous experience in the African bush were able to learn quickly how to find large carcasses of animals killed in woodlands and smaller carcasses cached in trees by leopards; it is perfectly reasonable to expect that early *Homo* could have done so too. Simultaneous or consequent changes in diet, brain size, and possibly even social structure are consistent with the apparently rapid advances in skull characters, but not body anatomy, in early *Homo*, and its replacement of gracile australopithecines. Perhaps *Homo* did not compete ecologically with the surviving australopithecines (the robust forms). Certainly robust australopithecines and *Homo* co-existed for over a million years in the same environments.

One can imagine how early *Homo* could have improved its competitive ability by exploring and exploiting the possibilities of tool use. Weaponry would naturally follow from tool use during scavenging. Food and infants could be transported safely from place to place with carrying devices. Increasing behavioral complexity would probably act to increase the value of brain growth and learning ability, and perhaps we may speculate (but not too wildly) about the increasing value of, or need for, sophisticated communication within and among social groups of humans.

Hominids and Cats in South Africa

Most hominid fossils found in South Africa have come from caves (Figure 22.1), many of which had steep or vertical entrances. It is unlikely that the hominids lived in the caves. Instead, the piles of bones there probably fell into the caves from above. In addition to hominid skeletons (mostly australopithecines), the fossils include bones of rodents, hyraxes, antelopes, baboons, two species of hyenas, leopards, and three extinct species of stalking sabertooth cats, one as big as a lion and two the size of a leopard.

C. K. Brain realized that the hyrax skulls in the cave deposits are all damaged in a peculiar way. Leopards always eat hyraxes completely, except for the fur, the gut, and the skull and jaws, and as they get at the brain and tongue they leave characteristic tooth marks on the skull, just like those on the fossil hyraxes. Cheetahs today can eat the backbones of baboons but not those of antelopes. Fossils from Swartkrans cave include many antelope vertebrae but none from baboons. Only baboon skulls are found. Furthermore, it looks as if the cave fossils were selected by size. There are very few juvenile baboon skeletons at Swartkrans, but many juvenile australopithecines. Some of the primate skulls show teeth marks that look exactly like those made today by leopard canines. Some of the fossil antelopes are bigger than those killed today by leopards, and sabertooth cats may have been responsible for them.

Leopards today like to carry their prey up trees. On the bare plains of South Africa, a cave entrance is one of the few places that seedlings can find safe rooting away from browsers, fire, and winter frost. Brain suggests that Pliocene and Pleistocene sabertooth cats killed prey and carried them to safe places to eat, undisturbed by jackals and hyenas, in trees growing at the entrances of caves such as Swartkrans and Sterkfontein. Uneaten parts of the carcasses fell into the caves, away from the hyenas, and were buried and preserved as soil, debris, and limestone deposits filled the cave.

Hominoids may have been a preferred meal for sabertooths for a long time. Numerous and well-preserved fossils of sivapithecids and gibbons in South China, at about 6 Ma, consist almost entirely of skulls and skull fragments, with few other bones of the skeleton, and there are large and impressive sabertooth canines in the same beds.

The large cats were the dominant carnivores in South Africa when the cave deposits were formed, and we can imagine them stalking and killing fairly large prey animals, including *Australopithecus robustus*. But there are relatively few fossils of early *Homo* at Swartkrans, Sterkfontein, or among other early cave deposits, suggesting that *Homo* was either rare or comparatively safe from big cats by virtue of habits, intelligence, or defensive methods and weapons. *Homo* did not have to be immune from big cat predation, just well defended enough that the big cats hunted other prey most of the time. *Homo,* of course, eventually replaced *Australopithecus* in South Africa.

There are many bones of *Australopithecus africanus* in the rock bed Member 4 at Sterkfontein, but no tools. Member 5, which overlies it, contains many tools, including choppers and diggers, animal bones with cut marks on them, and a few fossils of *Homo habilis*. The contrast between these two levels is striking in every aspect of their fossil record. As Brain sees it, the replacement of big cats by *Homo* as the dominant predators in South Africa was a major step toward human control over nature, and the beginning of our rise to dominance over the planet.

Homo erectus

Some extraordinary changes took place in the African plains ecosystem, beginning about 1.5 Ma. It is tempting to associate them with the appearance of new species of human, *Homo erectus*. An excellent specimen of *H. erectus* was discovered in 1984 west of Lake Turkana in Kenya, in sediment dated about 1.5 Ma. Although the body had been trampled by animals, so that the bones were broken and spread over 6 or 7 meters, careful collecting recovered an almost complete skeleton. The skeleton came from a boy 11 or 12 years old who stood 1.6 meters (65 inches) high (Figure 22.10). It is therefore possible that adult males stood close to 1.8 meters (6 feet) tall. The nose was enlarged and projected, as in modern humans but unlike australopithecines or *H. habilis*. This character suggests that *H. erectus* was adapting to greater exposure to dry air, for longer times and during greater activity.

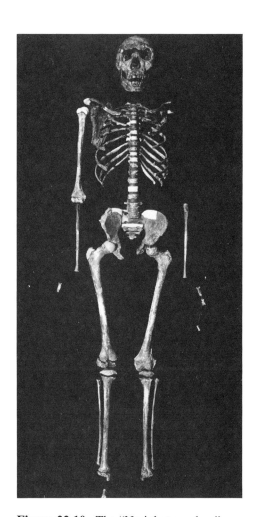

Figure 22.10 The "Nariokotome boy", KNM-ER-15000, the best specimen of early *H. erectus,* found in 1984 near Lake Turkana in Kenya. (Courtesy of Alan Walker of Pennsylvania State University.)

Homo erectus was strongly built, and was a specialized walker and runner with large hip and back joints capable of taking the stresses of a full running stride. There is less evidence of tree-climbing ability than there is in early *Homo*, though *H. erectus* would have been no worse at it than we are. *H. erectus* is also advanced in skull characters. The skull is thick and heavy by our standards, but brain size had increased to around 900 cc.

Quite suddenly, at about 1.4–1.5 Ma, all over East Africa, *Homo erectus* is found associated with a completely new set of stone tools. The **Acheulian** tool kit is much more effective than the older Oldowan, but experiments by Nicholas Toth show that Acheulian tools required much greater strength and precision to make and use than Oldowan tools. Acheulian craftsmen shaped their stone cores into heavy axes and cleavers at the same time as they flaked off smaller cutting and scraping tools. Most Acheulian tools are well explained as heavy-duty butchering tools (Figure 22.11). And around this time, robust australopithecines, *A. africanus*, and sabertooth cats all became extinct in South Africa. All other species of early *Homo* were already gone from Africa, and by 1 Ma, the last robust australopithecines and the last two species of sabertooth cats were gone too.

It is tempting to correlate all these events with the achievement of some dramatically new level of intellectual, physical, and technical ability in *Homo erectus*. *H. erectus* was much bigger than any preceding human. Most paleontologists believe that the evidence from anatomy, from tools, and from animal remains found with *H. erectus* suggests that this was the first effective human hunter of large animals. Alan Walker has suggested that the entire ecosystem of the African savanna was reorganized as *Homo erectus* came to be a dominant predator instead of a forager, scavenger, and small-scale hunter. African kill sites with butchered animals suggest a sophisticated level of achievement. Some experts even suggest that we should re-define the origin of *Homo* to the appearance of *Homo erectus*.

Homo erectus was the earliest species of *Homo* to migrate outside Africa. The earliest specimens of *H. erectus* from Java may be as old as 1.8 Ma, though this date is contested. If the date is correct, it implies that *H. erectus* left Africa almost as soon as it evolved there, before its African branch invented the Acheulian tool kit.

Other specimens of *Homo erectus* from the Middle East and from China have an age around 1.0 Ma. *H. erectus* may have reached as far east as the island of Flores, in Indonesia, before 750,000 BP (years before present), a feat which involved two sea crossings, of 15 and 12 miles. There are no fossils, only a few tools, but the story fits with the fact that three major animals became extinct quite suddenly on Flores around 900,000 BP: a pygmy elephant, a giant tortoise, and a giant lizard related to the Komodo dragon.

The Asian specimens of *H. erectus* had their own versions of stone tool-making styles. Specimens from Java had a brain size just under 1000 cc, but brain size had reached 1100 cc by the time of "Peking Man," who occupied caves outside Beijing between 500,000 and 300,000 BP. The

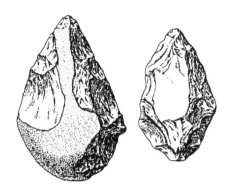

Figure 22.11 An Acheulian tool kit, well designed for heavy-duty butchering. Shown about half size. (Redrawn after Mary Leakey: I have "freshened up the edges.")

Caves are wonderful places for lairs

For sabertooth tigers and bears

But "Try and eject us!"

Said Homo erectus,

"We need this place for our heirs!"

successful long-term occupation of North China by these people indicates that they had solved the problems of surviving a challenging seasonal climate.

Fossils of *H. erectus* occur over a time range of a million years and a
geographic range of thousands of miles. Some paleontologists insist that
the variation among them is no more than it is among *Homo sapiens* today, and therefore that all the specimens should be placed into one species, *Homo erectus*. Here is the problem. If hominids were spread widely
but thinly over the Old World, the widely separated populations would
tend to diverge in some characters, both anatomical and cultural. Compare the diversity of modern humans, who are much more densely and
continuously distributed. Geneticists argue that on theoretical grounds it
would be unlikely for a hominid species to extend from South Africa to
Southeast Asia, so some paleontologists call the African *erectus* specimens *Homo ergaster*. (One could look at a modern analog and see that
leopards extend over such a range today.)

It's difficult to separate reality from theory, and it's impossible to
use a naming scheme that will make everyone happy. Even readers of
National Geographic have to deal with messy alternative schemes, each
of which has enthusiastic and often intolerant proponents. I have done
the best I can in choosing a scheme to describe here.

It looks as if *Homo erectus* was the first species to control fire. There
is good evidence for camp fires in a South African cave at Swartkrans,
dating from at least 1 Ma, and in various caves from France to Beijing, in
cool northern latitudes.

We know from the shape of the pelvis that *H. erectus* babies were
born as helpless as modern human babies are, and it is clear that the brain
grew a lot after birth, as our brains do. This implies an long period of
care for a baby that probably could not walk for several months. That is
an enormous price to pay for a larger brain, and would only have been
evolutionarily worthwhile (selected for) if there was a large pay-off for
learning and intelligence.

All these lines of evidence imply a complex and stable social structure for *Homo erectus*, though details are certainly not available. I will
make one comment of my own. The cooperation required to build, start,
control, maintain, and transport a fire is very high. It is difficult (for me)
to imagine a campfire without conversation. But as soon as any hominid
evolved language, that would have begun the novel process of transferring abstract information and knowledge directly and immediately from
individual to individual, replacing indirect methods such as taking and
showing, demonstrating and copying, or sharing the same real experience. The ability to short-circuit the processes of teaching and learning
must have allowed much more knowledge to be transmitted, absorbed,
and retained in a society, with obvious advantages to all its members.
Later modifications in educational techniques have served mainly to accelerate the transfer of information and to make it possible at a distance,
by the invention of writing and reading, numbers and alphabets, schools,
printing, telephones, broadcasting, and so on. Many experts believe that
language is a very recent invention, essentially by *Homo sapiens*. That

may be true for the complex activity that all modern humans are so good at. But language must have evolved, like every other characteristic of modern humans. I suspect that the information explosion set off by modern electronics is only the latest in a series that started around a campfire a million years ago. This book and the Macintosh I wrote it on are direct continuations of that tradition.

After *Homo erectus*

After *Homo erectus*, the story of human evolution becomes very messy. As fossil and molecular evidence has accumulated, anthropologists have erected finer and finer subdivisions for species. In early 1999 we have a majority opinion that postulates several species of *Homo* have evolved during the last million years, all but one of which have become extinct. I believe that I can resolve some of the issues to make a better story, as I write this final version in June 1999. Remember that I am a mere paleontologist, not an expert in paleanthropology. As always, you are free to construct your own interpretation of the story, and as always, new data will cause us all to re-think and re-interpret in the future.

According to the current majority story, *Homo erectus* did not evolve into *Homo sapiens*. *H. erectus* may have survived in Java until only 30,000 BP, according to some new (but controversial) dates. If so, *erectus* must have been contemporary with modern humans, who had already walked across Java on their way to Australia by around 50,000 BP. The current story is that around 500,000 to 300,000 BP, while *H. erectus* continued to thrive in Asia, a more advanced species arose in Africa perhaps a million years ago, and spread northward into Europe. This species, described from Spain, has been called *Homo antecessor*. In Europe, *H. antecessor* evolved by about 500,000 BP into *Homo heidelbergensis*, described originally from Germany. Around 400,000 BP, *H. heidelbergensis* was making beautifully crafted hunting spears in Germany. They are <u>throwing</u> spears, up to 3.2 m long (10 feet), carved to angle through the air like modern javelins, and the spears are associated with butchered horses and other bones from elephant, rhino, deer and bear.

These European populations evolved into Neanderthals, *Homo neanderthalensis*, a group of humans who were strongly adapted for cold-climate living along the fringes of the Ice-Age tundra from Spain to Central Asia.

Meanwhile, about 300,000 BP an African population of *antecessor* developed a new type of stone tool technology known as Middle Stone Age/Middle Paleolithic technology. One of these African populations evolved into fully modern *Homo sapiens*, perhaps around 200,000 BP to 100,000 BP, during the worst glacial period, when human populations may have been stressed and fragmented.

During the last interglacial (around 120,000 BP) modern humans are found in South Africa, and outside Africa for the first time (in Israel). Modern humans may have traversed around the shores of the Indian

Ocean to Southeast Asia, but they were apparently not yet able to sweep aside other species. Neanderthals were firmly entrenched in Europe, and *Homo erectus* in Asia. As the interglacial ended and the climate turned cool again, Neanderthals re-occupied the Middle East around 70,000 BP, taking the place of modern humans.

All this changed around 45,000 years ago. Modern humans now swept all competing species from the Old World: the Neanderthals in Europe, and *Homo erectus* from Southeast Asia. This was ethnic cleansing: the fossil record suggests there were no survivors this time.

Details are argued: for example, some people do not accept *H. antecessor* as a species (it was described only in 1997), and would simply include it in *heidelbergensis*.

But the fundamental scenario is fiercely debated, and it is important. At least some anthropologists argue that a widespread *Homo erectus* evolved into a widespread *Homo sapiens* across the entire Old World, and the "species" that have been named are subspecies or variants that do not imply inability to interbreed. There are claims of transitional skeletons between *erectus* and *sapiens* in at least four different regions: Indonesia, China, Africa, and the Middle East.

If that scenario is true, there would have to be gene flow between the continents, to keep the scattered populations within the same interbreeding gene pool. Yet the DNA structure of living humans shows a distinct division into African and non-African types. Furthermore, the variation in modern human DNA is very restricted. A single breeding group of chimpanzees in the Taï forest in Africa has more variability than the entire human race today. This suggests very strongly to geneticists that all modern humans are descended from an ancestral population that was not only small—say 10,000 or so—but was small for a long time.

If this estimate is correct, there is no way that 10,000 people could have interbred and inhabited more than a relatively small area, even if they were wandering hunter-gatherers. The only possible conclusion (if the assumptions and calculations are correct) is that all living humans are exclusively descended from ancestral *Homo sapiens* who evolved (from *Homo heidelbergensis*) in a suitably small population and region in Africa, and spread from there throughout the Old World.

The difference between "African" and "non-African" DNA is explained if a small founder population left Africa, carrying with them only a small sample of the genetic variation found in Africa as a whole. These founder populations then expanded as they occupied Eurasia, growing into a large population with a distinctly non-African DNA structure. This explanation was first favored in the late 1980s, when it was called the **Mitochondrial Eve** or **Out of Africa** hypothesis. The original presentation of the hypothesis was flawed (it was based on poor use of the computer program that processed the data), but in fact it is likely to be true.

The more pressing questions involve the timing, and what happened to the human populations of the rest of the Old World.

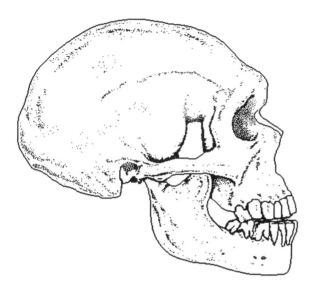
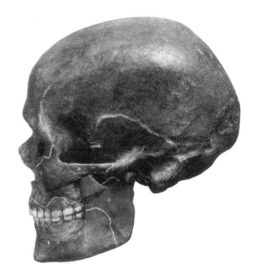

Neanderthals—A Live Issue?

The people we call Neanderthals lived between 90,000 and 30,000 BP in Europe and along the mountain slopes on the northern edges of the Middle East, as far east as Iraq. They were named after a site in the Neander Valley in Germany. Neanderthals had a way of life that was distinctly sophisticated in living sites, tools, and behavior.

Neanderthals differ from living humans in having big faces with large noses, large front teeth, and little or no chin (Figure 22.12, left). These characters are connected: Neanderthal front teeth show heavy wear, as if they used their incisors for something that demanded constant powerful pressure (softening hides by chewing, as Eskimos used to do?). Human faces are plastic, especially in early growth, and either by use or by genetic fixation, the frontal chewing of Neanderthals seems to have encouraged growth of the facial bones to support the front teeth against the skull. The nasal area was essentially swung outward from the face, so that the nose was even bigger and more projecting than it is in most hominids. Neanderthal brain size, at 1450 to 1500 cc, was at least equal to that of living humans and sometimes greater. Another special Neanderthal character was a very strong, stocky body with very robust bones, which may have helped conserve body heat in a cold climate and/or may reflect a lifestyle that required great physical strength. Most Neanderthal fossils are found in deposits laid down in the harsh climates of the next-to-last ice age.

Most Neanderthal tools are made in a style called **Mousterian**. They include scrapers, spear points, and cutting and boring tools (Figure 22.13) made from flakes carefully chipped off a stone core. Marks on Neanderthal teeth suggest that they stripped animal sinews to make useful fibers by passing them through clenched teeth, just as Australian aboriginals do. But perhaps the most enlightening Neanderthal finds are their ceremonial burials. Bodies were carefully buried, with grave offerings of tools and food. Enormous quantities of pollen were found with the body

Figure 22.12 Skulls of a Neanderthal (left) and a Cro-Magnon (right). Their brain sizes were about equal, but Cro-Magnons left by far the more sophisticated tools, art, and artifacts. (Neanderthal redrawn, idealized, and simplified from Trinkaus; Cro-Magnon skull from negative 310705, courtesy of the Department of Library Services, American Museum of Natural History.)

Figure 22.13 A Neanderthal tool kit, shown about half size. (After Bordes.)

of Shanidar IV, a Neanderthal man buried in Iraq. The pollen came from seven plant species in particular. All seven have brightly colored flowers, all seven bloom together in the area in late April, and all have powerful medicinal properties. It is difficult to avoid the conclusion that Shanidar IV was carefully buried with garlands of healing herbs chosen from early summer flowers, suggesting an intense concern for the abstract world.

Who were the Neanderthals? Neanderthal DNA (recovered from the original fossil from the Neander Valley) is distinctly different from that of any living human population. Neanderthals became adapted to life in the cold climate along the edges of the ice sheets from Western Europe to central Asia, by evolving characters of their own. The more geographically isolated they were, the more extreme their Neanderthal characters became, until they became visibly different from both *heidelbergensis* and from the *sapiens* populations that were evolving in Africa. The current interpretation of these facts is that Neanderthals were not *Homo sapiens*. They are not related to us, but are a separate and extinct species, *Homo neanderthalensis*.

One has to be careful, however. A human population was isolated on the island of Tasmania for 10,000 years after sea level rose as the great ice sheets melted. I do not know of a DNA study of this population (there are no full-blooded Tasmanians left). Perhaps their DNA also would have been "different" from that of all other humans at some level, yet they were still able to interbreed with invading Europeans even as they were hunted down with extreme prejudice.

In the Middle East, Neanderthals seem to have alternated with *Homo sapiens*, with a fluctuating border between them. It looks as though both peoples in the region were making the same Mousterian tools, which have been identified as far south as the Sudan, but none of the <u>fossils</u> are intermediate, suggesting that the two did not interbreed. Neanderthals lived in the Middle East in cooler, wetter times, while *Homo sapiens* lived there in hotter, drier times. Each was fitted to a particular climatic zone in which the other could not compete; neither was "superior" during those tens of thousands of years.

Neanderthals disappeared from the Middle East about 45,000 BP, then from Eastern and Central Europe, and finally from northwest Europe (France and northern Spain). The last dated surviving Neanderthals held out in upland France until about 38,000 BP and in southern Spain and Portugal until about 30,000 BP.

European Neanderthal sites typically contain less standardized tools, made only from local stone and flint, but the very last Neanderthals in Western Europe showed a distinctly more advanced culture. Their tools are called Châtelperronian. These tools show clear evidence of the old Mousterian style, but they also have some similarities to the Aurignacian tools that the newly arrived, fully modern humans (the Cro-Magnon people, Figure 22.12) were using at about the same time in Western Europe. The last Neanderthal sites in France also contain simple ornaments, and it is tempting to suggest that Neanderthals may have copied some of the CroMagnon technology and art. This suggestion is controversial, because Neanderthal artifacts are not found at Cro-Magnon sites, or the reverse.

Some anthropologists think that the late Neanderthals were evolving their own advanced culture just before their demise (perhaps in the same way that Aztecs and Incas were evolving new political systems just as the Spanish arrived). Neanderthal and Cro-Magnon peoples overlapped in Europe for 10,000 years or so, a very long time in historical terms.

The skeletons at Châtelperronian sites are Neanderthal. A beautifully preserved skull of a late Neanderthal baby from France shows a structure of the inner ear that is quite different from that in modern humans. Yet as I write, there are newspaper reports of the discovery of a 4-year old child in Portugal who seems to combine the teeth and jaws of a modern human with the strong bones of a Neanderthal: and the site is dated several thousand years after Neanderthals supposedly became extinct. This skeleton needs close examination. If it is an intermediate, then we can throw away the concept that *Homo neanderthalensis* (and its predecessors?) are separate from *Homo sapiens*. They would represent only isolated populations, comparable with the way that Eskimo or Lapps or Ainu were isolated from other humans long enough to evolve distinctive characters.

There is evidence that Cro-Magnons and Neanderthals were quite different socially. There is not only the evidence of art and symbolism; Cro-Magnons may well have been better big-game hunters, and some people even argue that Cro-Magnons had true language, while Neanderthals did not. But we are quite different socially from Papua New Guineans: that means nothing genetically. We need a better explanation. What superiority did *Homo sapiens* have? Was it weaponry? or social cohesion? It was not language as such: Neanderthals could certainly speak.

As Cro-Magnon people took over the cold and forbidding European peninsula, they were only a local population on the northwest fringe of the human species, but they are important because they have yielded us the best-studied set of fossils, tools, and works of art from the depths of the last Ice Age (Figure 22.14).

Cro-Magnon sites yield richer and more sophisticated art and sets of tools, and more complex structures and burials than Neanderthals. In particular, Cro-Magnons made stone tips for projectiles. Neanderthals may have had arrows and spears, but if they did, they were not stone tipped. Mousterian tools are more frequently wood-working tools than those of Cro-Magnons, who worked more with bone, antler, and stone.

Cro-Magnons and their contemporaries in Eastern Europe, the Gravettians, left evidence of a capacity for habitat destruction that is typically modern in style. From Russia to France, sites contain the remains of thousands of horses and hundreds of woolly mammoths. Cro-Magnons were also responsible for the magnificent cave paintings of extinct Ice Age animals, drawn by people who saw them alive (Figure 22.14), and they made (and presumably played) bone flutes. Cro-Magnons were painting on cave walls 30,000 years ago, and Gravettians were firing small terracotta figurines in kilns more than 25,000 years ago. Cro-Magnons and their contemporaries had a tremendous ecological impact on the world (Chapter 24).

Figure 22.14 Portraits of a woolly mammoth and a woolly rhinoceros, by unknown CroMagnon artists. (After drawings made by de Breuil from cave paintings in France.)

Assessing the New Evidence

Here is a way to combine the evidence we now have. We living humans are so genetically uniform compared with chimpanzees that there must have been some special event or events sometime in our history. Humans would potentially be capable of interbreeding with a much more genetically diverse set of partners than we currently have available to us. In the recent analysis of over 1000 humans and chimpanzees, the Neanderthal genes cluster close to those of modern humans, though outside them. In other words, we should not be surprised by the recent discovery of a human who shows every sign of being having Neanderthal characters. Instead, we should be surprised that there are not more of them.

The molecular evidence, then, suggests that ancestral humans were capable of interbreeding with other humans that are currently classified as separate species. If that list included *Homo neanderthalensis*, and we now have evidence of that, it must also have included *Homo heidelbergensis* and *Homo antecessor*, who were not as derived (specialized) as Neanderthals.

This has two important implications. First, it gives a picture of a broadly distributed species over the past million years plus (we have to call it *Homo sapiens*) in Africa, Europe, and Western Asia. The species was genetically varied (like chimpanzees) and included geographic subgroups (let us call them subspecies), to which we can give names (antecessor, etc.), and for which we can envisage different cultures. There is considerable evolution within this widespread group, and a lot of it was regional. But they never lost the ability to interbreed, so that around 30,000 BP, arriving modern humans were still able to produce offspring with Neanderthals. We still do not know whether *Homo sapiens* were able to breed with the *Homo erectus* they encountered in Eastern Asia, though that would not be a surprise either.

This scenario underlines the origin of living humans in Africa, and makes it reasonable to envisage one or more sets of African groups emerging from Africa with some new technology or behavior that allowed them to invade and take over the rest of the world.

It also raises the specter of ethnic cleansing. The other species of *Homo* disappeared, and at some point, even *Homo sapiens* lost a great deal of the genetic variability that one would normally expect to see in a hominid species. I think it is unlikely that some natural catastrophe affected *Homo sapiens* but not the other surviving apes. It is much more likely that the branches of human descent were pruned by other humans.

Migration-with-ethnic-cleansing may not be one's favorite image of the human race, but we have to deal with evidence. Certainly genocide occurred in the past (read the Bible or the newspaper). Events in Bosnia and Kosovo come to mind as I write, but Rwanda, Kurdistan, Uganda, and Nazi Germany should not be forgotten either, and the history of any continent has horrific examples. Within living memory, cannibal villagers in Papua New Guinea would try to kill off all members of a target community, because survivors would be likely to retaliate.

Evolution among Humans Today

Given the vastly different biological and ecological environments of the species of *Homo* since 2.4 Ma, it's likely that the selective pressures on soft-part anatomy and behavior have been as intense as those on skeletal features. There is clear evidence among living humans of regional evolution to suit the particular environment; for example, some of the characters of soft anatomy and body proportions are strongly linked to climate in many human groups. Nose shape is strongly correlated with humidity. Eskimos are endomorphic to resist body cooling. Peoples living at high altitude in Asia and South America have adapted physiologically to low oxygen levels. There are extremely ectomorphic, dark-skinned tropical people, pink-skinned people in northern Europe, and so on. Testis size varies markedly among human groups, as does the frequency of twinning. Such features must have evolved under intense regional selection, combined with the slow spread and mixing of genes at a time when human groups could not travel great distances. Such differences are visibly diminishing in certain modern populations (Hawaii, Brazil, and California come to mind).

The evolution of behavior cannot be assessed very well from the fossil record, but the variation in social structure within hominoid species is very great and suggests radical behavioral evolution, at least over the past 10 m.y. New research on mating patterns among animals suggests that much of human sexual anatomy and sexual behavior may be linked with the evolutionary breakthrough that began with early *Homo* and the use of tools to achieve ecological dominance.

It is sometimes claimed that natural selection no longer acts on modern humans because our surroundings are so artificial. Most people are now more insulated against diseases, environmental fluctuations, and accidents than humans were only a few centuries ago. Yet selection still operates strongly even in the most "advanced" societies. The genes for sickle-cell anemia are now generally harmful, instead of being favored in malarial regions. The genes that predispose non-Europeans to diabetes and gall-bladder cancer are more easily triggered into action on a westernized ("coca-colonized") diet, whereas they had little or no effect in their original environment. Among the Micronesian population of Nauru, nearly two-thirds of adults have diet-induced diabetes by the age of 60, and the Pima and Maricopa Indians of Arizona have even higher total rates of diabetes.

The white Afrikaners of South Africa have some of the highest concentrations of genetic defects in the world, because so many of them are descended from only a few families of Dutch immigrants. Only 20 families passed on their names to nearly a million Afrikaners. At least 30,000 Afrikaners (all descended from one immigrant Dutch orphan, Araiaantje Adriaansse, sent out in 1688 to marry an early settler, Gerrit Janz van Deventer), carry genes for *porphyria variegata*. This is a hemoglobin abnormality that used to be only moderately debilitating, but is dangerous today because medicines such as barbiturates induce a severe and often lethal reaction in carriers of the genes.

Finally, it appears that although the maximum human lifespan has not changed very much, average life expectancy has. At least some human societies now depend strongly on physical, social, and intellectual nurturing of children long past physical maturity. Characteristics that are normally juvenile attributes in primates, such as imagination, curiosity, play, and learning, are now encouraged in early adult years. The trend is presently social. There is no evidence yet of any evolutionary feedback creating delayed physical maturity or increased mental capacity. It is not yet clear whether this will occur and alter human biology as well as human culture. Potentially, an increased learning period could have enormous consequences for us and for every living thing on Earth.

Review Questions

Which hominid was the first to walk upright?

Discuss where to draw the line that separates *Australopithecus* from *Homo*. Name the characters that you would choose to distinguish the two genera.

Which pair of these are the closest relatives?

ourselves (humans)

African apes (chimps and gorillas)

Asian apes (orangutans)

As far as we know, which hominid was the first to make stone tools? (One point for the genus name, one for the species.)

What character makes hominids—*Australopithecus* and *Homo* —different from their sister group the apes?

We know that the first hominid walked upright. But how do we know?

It is a debatable point whether *Homo erectus* was a good hunter or not. I certainly believed so when I wrote the chapter. Give one argument that says *erectus* was a good hunter.

Draw maps and diagrams to illustrate the complexity of human evolution after *Homo erectus*.

Identify the hominid associated with this limerick:

> He butchered and hammered the dead,
> We assume he was very well fed,
> More astonishing still,
> He learned this new skill
> With a very small brain in his head.

Further Reading

More Accessible Reading

Abbate, E., et al. 1998. A one-million-year-old *Homo* cranium from the Danakil (Afar) depression of Eritrea. *Nature* 393: 458–460.

Agnew, N., and M. Demas. 1998. Preserving the Laetoli footprints. *Scientific American* 279 (3): 44–55.

Appenzeller, T. 1998. Art: evolution or revolution? *Science* 282: 1451–1454.

Asfaw, B., et al. 1992. The earliest Acheulean from Konso-Gardula. *Nature* 360: 732–734.

Asfaw, B., et al. 1999. *Australopithecus garhi*: a new species of early hominid from Ethiopia *Science* 284: 629–635, and comment, pp. 572–573. See also de Heinzelin, 1999.

Ayala, F. J. 1995. The myth of Eve: molecular biology and human origins. *Science* 270: 1930–1936.

Bahn, P. G., and J. Vertut. 1998. *Journey Through the Ice Age*. Berkeley, California: University of California Press. The art work of the Cro-Magnons.

Bahn, P. G. 1998. Neanderthals emancipated. *Nature* 394: 719–720. Neanderthals at Arcy-sur-Cure evolved their culture independently.

Balter, M. 1996. Cave structure boosts Neandertal image. *Science* 271: 449.

Bermudez de Castro, J. M., et al. 1997. A hominid from the Lower Pleistocene of Atapuerca, Spain: possible ancestor to Neandertals and modern humans. *Science* 276: 1392–1395, and comment, pp. 1331–1333. *Homo antecessor*.

Blumenschine, R. J., and J. A. Cavallo. 1992. Scavenging and human evolution. *Scientific American* 267 (4): 90–96.

Brain, C. K. 1981. *The Hunters or the Hunted? An Introduction to African Cave Taphonomy*. Chicago: University of Chicago Press.

Cartmill, M. 1998. The gift of gab. *Discover* 19 (11): 56–64. Origin of language.

de Heinzelin, J., et al. 1999. Environment and behavior of 2.5-million-year-old Bouri hominids. *Science* 284: 625–629.

Diamond, J. M. 1992. Diabetes running wild. *Nature* 357: 362–363. Nauru.

Diamond, J. M. 1992. *The Third Chimpanzee*. New York: HarperCollins.

Diamond, J. M., and J. I. Rotter. 1987. Observing the founder effect in human evolution. *Nature* 329: 105–106. South African Afrikaner genetics.

Gibbons, A. 1993. Pleistocene population explosions. *Science* 262: 27–28.

Gibbons, A. 1998. Ancient island tools suggest *Homo erectus* was a seafarer. *Science* 279: 1635–1637. In Indonesia.

Gibbons, A. 1998. Which of our genes makes us human? *Science* 281: 1432–1434.

Gore, R. 1997. The dawn of humans. *National Geographic* 191 (2): 72–99 [February]; 191 (5): 84–109 [May]; 192 (1): 96–113 [July]; and 192 (3): 92–99 [September].

Holden, C. 1998. No last word on language origins. *Science* 282: 1455–1458.

Johanson, D., and B. Edgar. 1996. *From Lucy to Language*. New York: Simon & Schuster.

Klein, R. G. 1989. *The Human Career: Human Biological and Cultural Origins*. Chicago: University of Chicago Press. Comprehensive text with 65 pages of references. New edition in preparation, 1999.

Kunzig, R. 1997. The face of an ancestral child. *Discover* 18 (12): 88–101. *Homo antecessor* from Spain.

Larick, R. and R. L. Ciochon. 1996. The African emergence and early Asian dispersals of the genus *Homo*. *American Scientist* 84: 538–551.

Leakey, M., and A. Walker. 1997. Early hominid fossils from Africa. *Scientific American* 276 (6): 74–79. *A. anamensis*.

Lewin, R. A. 1997. *Bones of Contention: Controversies in the Search for Human Origins*. 2d ed. Chicago: University of Chicago Press.

Lewin, R. A. 1998. *The Origin of Modern Humans*. 2d ed. New York: Scientific American Library.

Li, T., and D. A. Etler. 1992. New Middle Pleistocene hominid crania from Yunxian in China. *Nature* 357: 404–407. Apparent transition between *erectus* and *sapiens* in China.

Lieberman, D. E. 1998. Sphenoid shortening and the evolution of modern human cranial shape. *Nature* 393: 158–162.

Lovejoy, C. O. 1988. Evolution of human walking. *Scientific American* 259 (5): 118–125.

McCollum, M. A. 1999. The robust australopithecine face: a morphogenetic perspective. *Science* 284: 301–305, and comment, pp. 230–231.

Mellars, P. 1998. The fate of the Neanderthals. *Nature* 395: 539–540. News from a conference in August 1998.

Pinker, S. 1994. *The Language Instinct*. New York: William Morrow and Company. Chapter 11 deals with the origin of language among humans.

Schick, K. D., and N. Toth. 1993. *Making Silent Stones Speak*. New York: Simon and Schuster.

Semaw, S., et al. 1997. 2.5-million-year-old stone tools from Gona, Ethiopia. *Nature* 385, 333–336, and comment, pp. 292–293. Oldowan tools at 2.5–2.6 Ma.

Shreeve, J. 1995. *The Neanderthal Enigma: Solving the Mystery of Modern Human Origins*. New York: William Morrow and Company. A masterpiece of scientific writing. It discusses human origins, not just Neanderthals. Paperback edition, Avon Books, 1996.

Shreeve, J. 1995. Sexing fossils: a boy named Lucy? *Science* 270: 1297–1298.

Shreeve, J. 1995. The Neanderthal peace. *Discover* 16 (9): 71–81.

Sillen, A., and C. K. Brain. 1990. Old flame. *Natural History* 99 (4): 6–10. Fire at Swartkrans.

Sponheimer, M., and J. A. Lee-Thorp. 1999. Isotopic evidence for the diet of an early hominid, *Australopithecus africanus*. *Science* 283: 368–370, and comment, p. 303. Meat?

Stringer, C., and C. Gamble. 1993. *In Search of the Neanderthals*. London: Thames and Hudson. Very readable; stresses Stringer's own views, of course.

Stringer, C., and R. McKie. 1997. *African Exodus: the Origins of Modern Humanity*. New York: Henry Holt.

Suwa, G., et al. 1997. The first skull of *Australopithecus boisei*. *Nature* 389: 489–492, and comment, pp. 445–446. *A. boisei* and *A. robustus* may be the same.

Swisher, C. C., et al. 1994. Age of the earliest known hominids in Java, Indonesia. *Science* 263: 1118–1121, and comment, pp. 1087–1088.

Swisher, C. C., et al. 1996. Latest *Homo erectus* of Java: potential contemporaneity with *Homo sapiens* in Southeast Asia. *Science* 274: 1870–1874.

Tattersall, I. 1995. *The Last Neanderthal*. New York: Macmillan.

Tattersall, I. 1997. Out of Africa again... and again? *Scientific American* 276 (4): 60–67.

Tattersall, I. 1998. *Becoming Human: Evolution and Human Uniqueness*. New York: Harcourt Brace.

Thieme, H. 1997. Lower Palaeolithic hunting spears from Germany. *Nature* 385: 807–810, and comment, pp. 767–768. Throwing spears from 400,000 BP.

Toth, N. 1987. The first technology. *Scientific American* 256 (4): 112–121.

Toth, N., et al. 1992. The last stone ax makers. *Scientific American* 267 (1): 88–93.

Trinkaus, E., and P. Shipman. 1993. *The Neandertals: Changing the Image of Mankind*. New York: Alfred A. Knopf.

Walker, A., and P. Shipman. 1996. *The Wisdom of the Bones: in Search of Human Origins*. New York: Alfred A. Knopf. Superb account of the way an outstanding scientist works and thinks, with a vivid reconstruction of *H. erectus*.

Ward, R., and C. Stringer. 1997. A molecular handle on the Neanderthals. *Nature* 388: 225–226. Brief review of the work on Neanderthal DNA.

White, T. D., et al. 1993. New discoveries of *Australopithecus* at Maka in Ethiopia. *Nature* 366: 261–265, and comment, p. 207.

White, T. D., et al. 1994. *Australopithecus ramidus*, a new species of early hominid from Aramis, Ethiopia. *Nature* 371: 306-312 [description], and v. 375, p. 88 [new name, *Ardipithecus*].

Wolpoff, M., and R. Caspari. 1997. *Race and Human Evolution: A Fatal Attraction*. New York: Simon and Schuster. Argues for a multi-regional evolution of *Homo sapiens*.

Wood, B., and M. Collard. 1999. The human genus. *Science* 284: 65–71. Astonishingly illogical, and trivial. The main point is that early species of *Homo* are rather like the australopithecines (that they evolved from)!

More Technical Reading

Aiello, L., and P. Wheeler. 1995. Brains and guts in human and primate evolution: the expensive organ hypothesis. *Current Anthropology* 36: 199–221.

Bermúdez de Castro, J. M., et al. 1999. A modern human pattern of dental development in Lower Pleistocene hominids from Atapuerca-TD6 (Spain). *Proceedings of the National Academy of Sciences* 96: 4210–4213. In *Homo antecessor*.

Brantingham, P. J. 1998. Hominid-carnivore coevolution and invasion of a predatory guild. *Journal of Anthropological Archaeology* 17: 327–353.

Churchill, S. E. 1998. Cold adaptation, heterochrony, and Neandertals. *Evolutionary Anthropology* 7: 46–61.

Clarke, R. J. 1998. First ever discovery of a well-preserved skull and associated skeleton of an *Australopithecus*. *South African Journal of Science* 94: 460–463.

Eyre-Walker, A., et al. 1999. How clonal are human mitochondria? *Proceedings of the Royal Society of London B* 266: 477–483.

Fifer, F. C. 1987. The adoption of bipedalism by the hominids: a new hypothesis. *Human Evolution* 2: 135–147. Stone throwing.

Foley, R. 1998. The context of human genetic evolution. *Genome Research* 8:339–347.

Hagelberg, E., et al. 1999. Evidence for mitochondrial DNA recombination in a human population of island Melanesia. *Proceedings of the Royal Society of London B* 266: 485–492.

Harpending, H. C., et al. 1998. Genetic traces of ancient demography. *Proceedings of the National Academy of Sciences* 95: 1961–1967.

Hey, J. 1997. Mitochondrial and nuclear genes present conflicting portraits of human origins. *Molecular Biology and Evolution* 14: 166–172.

Hoffecker, J. F. 1998. Neanderthals and modern humans in Eastern Europe. *Evolutionary Anthropology* 7: 129–141. Neanderthals abandoned the area in the face of the last Ice Age, and it was empty until occupied by modern people.

James, S. R. 1989. Hominid use of fire in the Lower and Middle Pleistocene. *Current Anthropology* 30: 1–26.

Kramer, A. 1991. Modern human origins in Australasia: replacement or evolution? *American Journal of Physical Anthropology* 86: 455–473. Evolution, says Kramer.

Mellars, P. 1996. *The Neanderthal Legacy*. Princeton, N. J.: Princeton University Press. Neanderthal culture.

Mountain, J. L. 1998. Molecular evolution and modern human origins. *Evolutionary Anthropology* 7: 21–37. Discusses the methods and assumptions. See Eyre-Walker et al., 1999 and Hagelberg et al., 1999: some of the assumptions may not hold.

Stanley, S. M. 1992. An ecological theory for the origin of *Homo*. *Paleobiology* 18: 237–257.

Stoneking, M., et al. 1997. Alu insertion polymorphisms and human evolution: evidence for a larger population size in Africa. *Genome Research* 7: 1061–1071.

Treves, A., and L. Naughton-Treves. 1999. Risk and opportunity for humans coexisting with large carnivores. *Journal of Human Evolution* 36: 275–282.

Wallace, D. C., et al. 1985. Dramatic founder effects in Amerindian mitochondrial DNAs. *American Journal of Physical Anthropology* 68: 149–155.

Ward, C., et al. 1999. The new hominid species *Australopithecus anamensis*. *Evolutionary Anthropology* 7: 197–205.

Weiss, K. M., et al. 1984. A New World Syndrome of metabolic diseases with a genetic and evolutionary basis. *Yearbook of Physical Anthropology* 27: 153–178. Diabetes etc.

Wheeler, P. E. 1991. The influence of bipedalism on the energy and water budget of early hominids. *Journal of Human Evolution* 21: 117–137.

Zietkiewicz, E., et al. 1998. Genetic structure of the ancestral population of modern humans. *Journal of Molecular Evolution* 47: 146–155.

CHAPTER TWENTY-THREE

The Ice Age

Earth's climate has gone through dramatic changes, from very warm to icy, and we would like to know why those changes happen. The question is important for three reasons.

First and most vital, climate is one of the most important physical factors of the environment for all organisms, and climatic changes have almost certainly been major factors affecting the evolution of life. We saw in Chapter 5 that plate tectonic movements can change oceanic and continental geography, and then how geographic changes can modify seasonal climatic patterns and affect the ecology and evolution of organisms in major global events. There were major effects on life as world geography changed with the breakup of Pangea in the Late Mesozoic and Early Cenozoic. Some effects resulted directly from geographic isolation, but others were mediated through the indirect effects of geography on climate. Many puzzles of later Mesozoic evolution may be resolved when we can reconstruct paleoclimates more accurately. Certainly this has been the result of concentrated research programs on Cenozoic paleoclimates.

Second, we are living through an ice age now, and have been for the past 2.5 million years or so. We happen to live during a particularly warm stage in it, but there is no sign that it is over. Great ice sheets expanded and covered much of the northern continents then retreated again. They have done so at least 17 times in the past two million years. Yet ice ages have been rare during Earth's history. So we face two problems in explaining ice ages. How do ice ages begin in the first place? And why should ice sheets advance and retreat so dramatically during an ice age?

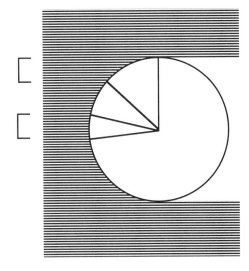

Figure 23.1 Above: A light shining on a sphere never illuminates its surface evenly, because the light rays hit the surface at different angles. The Sun's rays are always at a low angle in polar regions, even in midsummer. Below: Two identical packets of solar rays shine on the Earth. One shines on the tropics, where the rays are vertical and concentrated on a small area. The other shines on high latitudes, where the rays are at a low angle and are spread over a greater area. In this diagram the northern packet is spread over about five times as much area as the tropical packet, so high-latitude solar heating per unit time is at best only 20% of that in the tropics.

Third, by using fossil fuels we are pouring vast amounts of carbon dioxide and methane into the atmosphere, yet we have no real idea what this might do to Earth's climate. 1998 was the hottest year since we began keeping accurate records, so we might guess that we are on the wrong track!

The controls on climate are really quite simple, though the ways they operate in detail are too complex even for modern computers. We will begin with the simplest factors and try to explain how climate works, how we can reconstruct ancient climates, how Earth's climate changes, and how climatic changes affect life.

The Earth is basically a sphere in space receiving radiation from a distant but powerful source, the Sun (Figure 23.1), and radiating some of that energy back into space. If solar radiation were all absorbed and evenly distributed, it would maintain the whole Earth at about 7°C (45°F). But some radiation is reflected back into space, and the rest is not evenly distributed. The sun is always at a high angle in the tropics and a low angle at the poles, even in the middle of summer (Figure 23.1). Thus the tropics are always warmer than the poles, usually about 30°C compared with 0°C.

Earth is tilted on its axis of rotation, however, so the northern and southern hemispheres alternately receive greater and lesser radiation as Earth orbits the sun (Figure 23.2), giving us **seasons**, which are more extreme away from the tropics.

The equator and the poles receive about the same time per year in sunlight, even though the amount of radiation actually received is very different at equator and poles (Figure 23.1). Day and night are about equal length all year on the equator. But at the poles, there is a long period of winter darkness and a long period of midnight sun in the summer.

The Earth's heat does not stay where it is received. Earth has two enormous fluid masses on its surface, the ocean and the atmosphere, and the temperature differences generated by uneven solar radiation set up convection currents that act to redistribute heat.

Heat flows poleward from the tropics in warm winds in the atmosphere and warm currents in the oceans. But atmospheric and ocean circulation are both affected by the distribution of land and sea, so the system is complex. Nevertheless, there are some basic principles involved. In the atmosphere, surface winds are generated in predictable ways by pressure differences that result from solar radiation. Those surface winds are deflected by Earth's rotation into characteristic patterns, and in turn the winds blow over the ocean surface and help to generate ocean current systems.

Because Earth's climate operates on known principles, we can in theory calculate the climate for any particular place on Earth. There are so many variables, however, that this is not yet possible. What scientists do instead is to formulate models of climate on a simplified Earth, beginning with a simple sphere. As the model is made more complex, it should be able to predict with greater and greater accuracy what the real Earth is like. In the process, climatologists should be able to improve the equations that lie behind the calculations.

Modern computer models of climate are called general climate models (GCMs). They require a lot of supercomputer time and are expensive to run. GCMs can now simulate present-day climate reasonably well, if one feeds in the patterns of solar radiation and its reflection back into space, the distribution of land, sea, and sea ice, and the position, area, and height of major mountain ranges and ice caps. But GCMs are not yet very good at modelling ancient climates: facts from the rock record are more reliable than a computer simulation of that same paleoclimate. Recently 15 different GCMs made startlingly different simulations of the "greenhouse" effect of increasing the amount of carbon dioxide in the atmosphere. This means that we don't know which GCM to trust as we try to simulate ancient climate. We need to be able to check computer models of past climate against facts of the rock record.

The ice caps of Antarctica and Greenland are made from snow that has fallen on the polar regions every year for millenia. Each year's snow makes an individual layer, allowing researchers to drill down into the ice cap and study past climate over the specific, countable years recorded by the ice. Ancient snowfalls trapped air and dust that remain in the ice as gas bubbles and solid particles, and these too can be studied.

In the 1980s and early 1990s, new drilling techniques produced the highest-quality cores yet to be drilled in the Antarctic and Greenland ice caps, sampling ice up to 200,000 years old. Although the ice cores sample only the most recent part of geological time, they have revolutionized our thinking about ice ages in particular and climatic change in general.

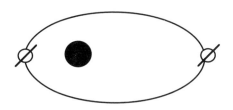

Figure 23.2 Earth's axis is tilted. As it orbits around the sun, solar radiation is concentrated on one hemisphere and then the other, giving Earth seasons.

Ice Ages and Climatic Change

Ice ages are not common events in Earth's history. There was a widespread ice age toward the end of the Precambrian, perhaps even a "snowball Earth," at about 600 Ma (Chapter 4). In Late Ordovician times, when northern Gondwanaland was over the South Pole, a great ice sheet spread over most of North Africa and probably further, triggering enough changes in marine life to mark the end of the Ordovician Period and the beginning of the Silurian. Gondwanaland drifted across the South Pole during the rest of the Paleozoic Era, with a particularly important glacial period in South America at the end of the Devonian. A small ice sheet lay over South Africa in the Early Carboniferous, but large-scale glaciation once again spread over most of Gondwanaland in the Late Carboniferous and Early Permian. Traces of this event, in the form of scratched rock surfaces and piles of glacial rock debris, are widespread in South Africa, South America, Australia, India, and Antarctica. A Northern Hemisphere glaciation occurred in Siberia at the end of the Permian Period. But afterward there was no major ice age for well over 200 m.y., until the present one began. Paleoclimatic evidence suggests that the Earth's surface cooled over the past 60 m.y., until finally the planet dropped into the present ice age.

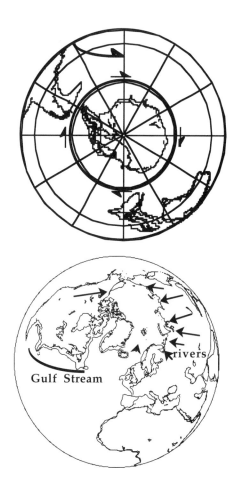

Figure 23.3 Top: The South Pole is refrigerated because it is isolated from the warmth of the tropics by the "Shrieking Sixties," a belt of latitudes in which west winds circulate endlessly. The bold circle is shown at about 58° S. Warm currents, such as the one that sweeps southward in the South Atlantic, cannot reach Antarctica. Bottom: The North Pole is refrigerated because it is isolated from the warmth of the tropics by land masses with only narrow or shallow seaways between them. Only the narrow passage between Iceland and Britain allows warm water into the Arctic Ocean. Note also that the Arctic Ocean receives a great deal of freshwater runoff from the great rivers of Canada and Siberia.

The only external factors that could generate major climate change are astronomical processes—**changes in Earth's orbit** or **changes in solar radiation**. Such changes occur, but they are probably too small to generate major climate change by themselves. They cause fluctuations in climate, however. An **asteroid impact** could conceivably trigger a climate change, but only for a short time and only if conditions were already just right to start *and maintain* a change over considerable time (Chapter 18).

It seems that we must look for mechanisms here on Earth for major climatic changes. Two processes can affect the amount of solar radiation that Earth retains. Some solar radiation is reflected back into space (the **albedo effect**), and a change in the amount of heat reflected would cool or warm the Earth. Gases in the atmosphere, especially carbon dioxide and methane, are very effective in absorbing solar radiation (**the greenhouse effect**), and changes in the amounts of such gases could strengthen or weaken the amount of solar radiation.

In discussing the major climate changes we call ice ages, we will look specifically for ways to set up a cold period, or ice age; by reversing the argument, we could look for ways in which Earth's climate could become globally very warm. The basic preconditions for climate change on Earth are simple. For an ice age, there must be a lot of snowfall in areas where it will build up rather than remelt. Such a situation can occur if Earth's global geography is arranged in the right way. An ice age, or any other climate change, can be encouraged or discouraged by the geographic changes that result from plate tectonic movements.

Geography: Isolating the Poles

The poles do not receive much heat by the time currents have traveled over thousands of kilometers, and they receive even less heat if anything blocks or obstructs the free circulation of winds and currents.

Earth's present poles are isolated from warming currents in two different ways. South of South America, Africa, and Australia there is a belt of latitudes (the Shrieking Sixties) with no land masses at all. By chance, this latitude corresponds with a westerly wind belt that funnels waves, storms, and ocean currents endlessly round the Southern Ocean. No heat-bearing winds or currents ever reach south of the Sixties to bring warmth to Antarctica, which is therefore permanently refrigerated (Figure 23.3). The huge Antarctic ice sheet has built up almost from sea level, so the South Pole stands on 3 km (close to 2 miles) of ice, and Antarctica holds 90% of the world's ice even though its snowfall is very low. Thus the South Pole is isolated by water barriers and climatic barriers that result from the geography of the Southern Hemisphere.

The North Pole, on the other hand, is almost completely surrounded by land. The Bering Strait, between Alaska and Siberia, is narrow and very shallow. The Davis Strait, between Canada and Greenland, is largely blocked by the islands of the Canadian Arctic, and only a narrow seaway between Iceland and Britain allows water and Russian submarines in or out of the Arctic Ocean. The Arctic Ocean is therefore largely insulated from warm ocean currents (Figure 23.3).

In addition, the Arctic Ocean has a net surplus of water. The Canadian Shield drains mainly to the north, via the Mackenzie River and a large number of smaller rivers flowing to Hudson Bay. Several enormous rivers drain much of European Russia and almost all of Siberia to the Arctic Ocean (Figure 23.3). The polar ocean receives little solar energy, so there is little evaporation. With a large supply of fresh water and little water loss, the Arctic Ocean provides a significant net outward flow of surface water to the main world ocean. Its surface waters are less salty than the average ocean water, so the surface water freezes easily into pack ice, and the central Arctic Ocean is covered by permanent floating ice that reflects radiation back into space. In winter there is an even larger seasonal growth of sea ice. The Arctic, then, is abnormally cold today for reasons that also result from the quirks of continental movement that molded the regional geography.

The differences between Earth's north and south polar regions are instructive in another way. Antarctica has had an ice sheet for perhaps 15 m.y., but Earth has not been in an ice age for all of that time. The reason is that Antarctica is an island continent surrounded by ocean, and its ice sheets and glaciers cannot spread northward over large areas of lowland and shallow sea. Instead, Antarctic ice reaches the ocean in giant, flowing sheets that break off into icebergs as they reach deep water. Giant ice shelves lie in protected shallow gulfs like the Ross Sea and the Weddell Sea, but they are small compared with the area of the Antarctic continent.

Thus there is a natural limit to the extent of Antarctic glaciation, set by the same continuous Southern Ocean that insulates the continent from warm-water currents. Although the initial growth of the Antarctic glaciers played an important part in cooling the Earth over the past 40 m.y. (by reflecting heat back to space, and beginning the flow of Antarctic Bottom Water on the ocean floors), it did not meet all the conditions for an ice age because the ice was confined geographically.

The Arctic Ocean has some permanent pack ice, and Greenland has a thick ice sheet that is geographically restricted, just like the Antarctic ice sheet, because it sits on an island land mass. Greenland is the only large land mass north of 70° N latitude. But just to the south lie the northern coasts of Eurasia and North America, which provide possible bases for large ice sheets if conditions for ice accumulation should change slightly (Figure 23.3). Thus the geographical arrangement of land masses in the northern continents set the stage for the present ice age and for the ice advances and retreats that have punctuated it.

Reflecting Solar Radiation: The Albedo Effect

The **albedo** of a surface specifies how much energy is reflected from it. A perfect reflector has an albedo of 1, and a perfect absorber has an albedo of 0. A surface reflects more—it has a higher albedo—if it is relatively smooth and if radiation strikes it obliquely. For example, the albedo of water with waves on its surface is only about 0.05; therefore, only about 5% of solar radiation is reflected back from the ocean if the sun is right overhead (in the tropics, for example). That is why Earth's oceans look dark on satellite photographs. In higher latitudes, where the sun is lower in the sky, more solar radiation is reflected back from a water surface, up to 12% or so in polar seas. On land, the situation is complicated by vegetation cover. About 15% is reflected off vegetation at any latitude. But deserts reflect a much higher percentage, and snow and ice reflect up to 80%, so deserts and ice sheets look light on satellite images.

The best way to cool the Earth by albedo change would be to have land at high latitudes where snow and ice can build up into very large ice sheets, reflecting back any sunshine into space. In reverse, a change in climate that melted ice off high-latitude land masses would effectively break up an ice age. Furthermore, if an ice age is to *begin*, this must be a *new* situation.

In summary, ice ages are encouraged or discouraged by the geography of land and water masses on Earth. But geography changes through the action of plate tectonics. Changes in geography also act to vary the albedo of the Earth, the scale and activity of ocean currents, and the distribution of heat to different regions, all of which affect climate. In general, ice ages require large areas of land in high latitudes. The poles should be isolated from warm water. Finally, to lock the Earth into a long glacial period, there must be room for large continental ice sheets to spread out and provide high reflectivity to large regions. Geography thus controls whether or not Earth's heat is well distributed, and whether or not polar ice sheets can form. Plate tectonics controls continental distributions, and the necessary conditions for generating an ice age may arise from time to time just by the motions of the plates.

Climatic Fluctuation: Glacial Advances and Retreats

Earth's major climate changes contain distinct fluctuations. For example, dramatic advances and retreats of ice can occur even while Earth is locked into an ice age. Huge areas of the northern continents are covered by debris dropped by ice sheets during 40 or so glacial advances and retreats during the past two million years. Large temperature fluctuations are recorded by microfossils in seafloor sediments. World sea level has fluctuated by more than 70 m (220 feet) as 5% of Earth's water has alternately been frozen into ice sheets and melted away. Such changes in sea level are recorded worldwide in sedimentary deposits far from the ice sheets. For example, islands and atolls in the Atlantic and Pacific Oceans have been repeatedly exposed and reflooded.

Ancient rocks also show evidence of frequent and important change in sea-level. For example, regular cycles of limestone, sandstone, and coal formation in the Carboniferous rocks of North America and Europe resulted from the cyclic rise and fall of sea level, and those rocks were deposited in tropical latitudes far from the glaciations of Gondwanaland that indirectly generated them. Many other cases of climatic cycling have been identified, even at times when Earth had no ice sheets, so we should look for a general cause for them, unconnected with ice sheets as such.

The Astronomical Theory

The **astronomical theory of ice ages** was suggested more than a century ago, was worked out by hand by Milutin Milankovitch in the 1920s, and was refined by computer calculations in the 1970s. It has been confirmed by evidence from microfossils that record temperature fluctuations in the oceans. The Milankovitch theory suggests that slight variations in Earth's orbit around the sun and in the tilt of the Earth's axis make significant differences to climate (see Box 23.1).

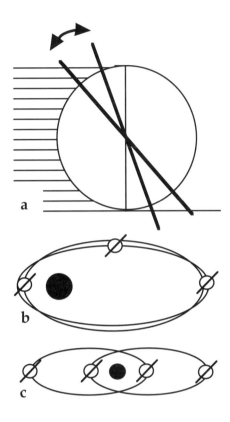

Figure 23.4 Some parameters of Earth's orbit can change over time, and as they do so they affect Earth's climate. (a), Greater or lesser tilting of Earth's axis causes stronger or weaker seasons. (b), The eccentricity of the Earth's elliptical orbit means that one pole almost always has greater seasonal effects than the other; changes in eccentricity weaken or strengthen that effect. (c), The precession of the elliptical orbit alternates the eccentricity effect between the poles.

Box 23.1 THE COMPONENTS OF THE MILANKOVITCH THEORY

Tilt. Increased or decreased tilting of the axis, which varies between 22° and 24.5°, increases or decreases the effect of the seasons in a cycle of about 41,000 years (Figure 23.4a).

Precession. Earth's orbit is not a circle but an ellipse, with the sun at one focus (Figure 22.4b). One pole is closer to the sun in its winter, while the other is closer in its summer. Thus, at any time, one pole has warm winters and cool summers, while the other pole has warm summers and cool winters. However, the slow rotation or precession of the Earth's orbit around the sun alternates the effect between the two poles in cycles of 19,000 or 23,000 years (Figure 23.4c).

Eccentricity. Earth's orbit varies so that it is more elliptical at some times than at others, strengthening or weakening the precession effect. This variation in orbital eccentricity affects climate in cycles of about 100,000 years and about 400,000 years. Of course, when eccentricity is low (when the orbit is closer to being circular), the precession effect is much lessened.

Each of the long-term changes in Earth's orbit is regular, but they interact in complex ways. They sometimes strengthen and sometimes cancel one another, altogether producing cycles that recur at intervals of tens or hundreds of thousands of years. They have astrophysical causes and affect orbital characters all the time, whether Earth is in an ice age or not.

Most important, Milankovitch cycles can trigger the advance and retreat of ice sheets, if conditions for an ice age are already present. Computer models of ice advances and retreats agree well with data from the geological record. For most of the present ice age, glacial advances and retreats happened at 41,000 year intervals, and since about 800,000 years ago, the cycles have occurred every 100,000 years. Milankovitch models also predict that ice sheets should melt much more rapidly than they build up, and the geological record confirms this. The models suggest that the present mild climate on Earth is very unusual for our geography. Interglacial periods with reduced northern ice sheets are very short in comparison with glacial periods with large ice sheets.

The theory Milankovich built

Has been proved right up to the hilt

Its three major components,

Say its proponents,

Are precession, eccentricity, tilt.

Figure 23.5 Top: Optimal conditions for melting ice sheets in the Northern Hemisphere. The Earth's orbit takes it farther from the sun in the nothern winter (right), so that winters are cold and dry and there is not much snow. In the following summer the Earth is closer to the Sun, the summer is hot, and a lot of melting occurs. Bottom: Optimal conditions for ice sheets to grow in the Northern Hemisphere. The winters are warm (and wet, so that there is a lot of snow). The following summers are cool, so little melting occurs. The eccentricity of the orbit is much exaggerated in this diagram to make the point, but the principle applies.

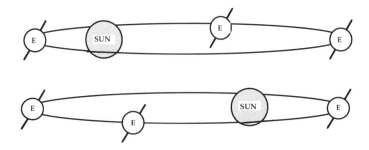

When northern summers are hot, it is very difficult for ice sheets to form or to be maintained (Figure 23.5, top). The winters are cold, and there is not much snowfall during cold dry weather patterns, yet there is a lot of melting during the hot summer that follows. Conversely, the best conditions for generating northern ice sheets occur when the Northern Hemisphere has cool summers and warm winters (Figure 23.5, bottom). Large amounts of snow build up from warm wet storms in the winter, but are not melted the following cool summer. The largest snowfalls occur on continental edges where storms strike inland after passing over warm ocean water, so medium-sized continents close to warm oceans (such as Europe, Greenland, and North America) accumulate more snow and ice than large continents such as Asia.

Probably, then, the stage is set for major climatic change by continental movements (plate tectonics, an internal Earth mechanism). Some small event such as a major eruption or an asteroid impact triggers the change. Then Earth's geography locks the planet into the new climate, until plate tectonic movements change the geography sufficiently to favor a major change once again. While all this goes on, the Milankovitch cycles themselves continue to modulate the climate.

Smaller Cycles and Very Rapid Climatic Fluctuation

One of the great surprises of recent paleoclimatic work has been the discovery that there are shorter climatic fluctuations within Milankovitch cycles. Some of these may be harmonics, as in music, but there is no reason that climatic changes should all be *cyclic*. (Climatologists tend to be physicists, who like to do Fourier analysis, so are predisposed to looking for cycles.) Climatic change could be fractal, or even stochastic! Another, greater surprise is that glacial climates can change with dramatic speed, while interglacial climates apparently do not. In glacial times, especially as glacial times are ending, temperatures can apparently increase or decrease on the scale of several degrees in ten years, and longer-term fluctuations can raise or lower temperatures several degrees for a thousand years at a time. Several such events are recorded in the Greenland and Antarctic ice caps, and in smaller mountain ice sheets and glaciated areas from New Zealand and Chile to North America and Europe. They are global, and they are intense, yet they are too short-lived to be caused by Milankovitch effects. The events may well be related to major crises that could affect great ice sheets—but how? The best suggestion is that these rapid events have their origins in the ice sheets round the North Atlantic, with the effects spread globally as ocean circulation is disturbed. But this line of research is very new and we have a lot to learn about it.

Could rapid changes in climate occur in warm times too—now, for example? At the very least, we should be very concerned that the Earth's climatic system can be more unstable than we thought. Such changes have not occurred in recorded human history, but the Earth is clearly capable of generating them. What if the Antarctic ice cap began a crisis next year? The effects on human society would be dramatic. A great deal of effort is going into this line of research.

Now it is time to apply all this theory to Earth's latest ice age, the Pleistocene Ice Age in which we are all living.

The Pleistocene Ice Age

Earth has been locked into an ice age since about 2.7 Ma, but its effects have been most marked in the Northern Hemisphere. As far as we can tell, a tropical plate tectonic movement set it up (Figure 23.6). At the end of the Miocene, about 4.5 Ma, Central America approached its present position as a bridge between North and South America, blocking a major equatorial current that flowed from the Atlantic to the Pacific. Some of that water was deflected northward into the Atlantic in a much strengthened Gulf Stream current. By the Late Pliocene, perhaps 3 Ma, animals were walking freely between North and South America, and all the Atlantic equatorial water was deflected north into the Gulf Stream.

The closing of Panama primed the Northern Hemisphere for major growth of ice sheets by bringing warmer water, more evaporation, and more precipitation to the North Atlantic. This also increased rainfall and snowfall into Eurasia as far as Siberia, where the great Russian rivers flow into the Arctic Ocean. Data from ocean sediment and from Lake Baikal in Siberia show that this was happening as early as 4.5 or 4.6 Ma.

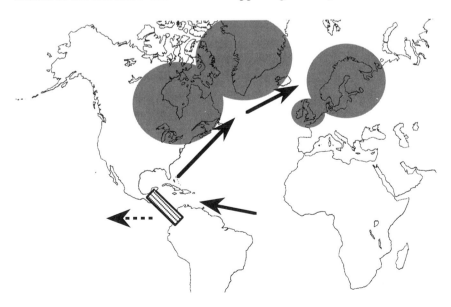

Figure 23.6 The possible plate tectonic trigger for the northern ice ages. At about 3 Ma, the normal westward flow of warm Atlantic tropical water was blocked by the new land bridge of Central America, and the Great American Interchange took place (Chapter 19). Warm, equatorial Atlantic water, which formerly flowed into the Pacific, now heated up even more as it circulated around the Caribbean, and then flowed out northeast past Florida to strengthen the Gulf Stream. Warm currents flowed into the North Atlantic and caused greater snowfall around its northern edges. Huge ice caps quickly built up in Labrador, Greenland, and Scandinavia.

The result was that the Arctic Ocean received more river run-off, and became fresher and easier to freeze. Arctic sea ice began to build up, reflecting heat back into space by the albedo effect.

The final trigger was apparently pulled as the Earth's obliquity increased. From 4.5 Ma to about 3 Ma, the tilt of the axis was not extreme. After about 2.7 Ma, it varied more, and the tilt increased slightly, making the seasons more extreme and kicking the Earth into an ice age as soon as the Milankovitch cycle reached a phase in which the Northern Hemisphere had warm, wet winters and cool summers (Figure 23.5). Canada, Greenland, and northwest Europe built up thicker and larger snowfields. The new snowfields reflected more sunlight back into space, and large ice sheets built up and expanded still further, until the Northern Hemisphere was gripped in a glacial climate that we now call the "Ice Age."

Thus the northern glaciations that began at the end of the Pliocene were centered on huge new ice sheets in exactly the North Atlantic regions supplied with moisture by the Gulf Stream (Figure 23.6). It may have taken only 10,000 years to build northern ice sheets thousands of feet thick. At the same time there was severe cooling in the Southern Hemisphere.

Climate and Geography During the Ice Ages

Once ice sheets built up, they altered climatic patterns in the North Pacific and North Atlantic. Sea surfaces in the North Atlantic froze as far south as New York and Spain (Figure 23.7). Warm Gulf Stream waters were diverted eastward toward North Africa, instead of bringing warm, moist climate to western Europe as far north as Scandinavia. When this diversion occurred, as it did in every glacial period, North Atlantic ice buildup was slowed, and the ice sheets of the Northern Hemisphere became ice deserts. The jet stream was split and diverted far south of the North American ice sheet, and around its polar edge. With only limited

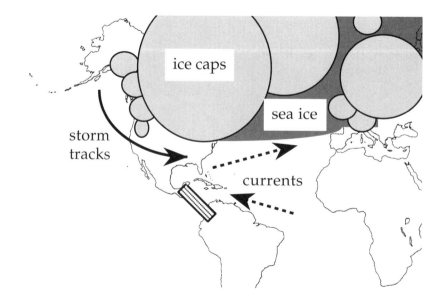

Figure 23.7 The winter climate of the North Atlantic 18,000 years ago. This sketch is based on reconstructions by the CLIMAP project. The fully developed ice sheets cooled the North Atlantic so that sea ice extended as far south as New York and Spain. Winter storm tracks were diverted far south in North America, forming large lakes in what is now the western and southwestern desert country. Warm winds and currents in the Atlantic were also far south of their present latitude. They brought moisture to what is now the Sahara Desert, leaving much of Europe as ice sheet and tundra.

snowfall, the ice sheet became vulnerable to rapid melting the next time the astronomical cycle reversed seasonal conditions.

As ice sheets build up, water is frozen out of the oceans, and sea level drops. At first, this aids ice sheet growth, for several reasons. The ice sheets are higher above sea level and are therefore colder, encouraging more ice to form. As the sea retreats from the shallow continental shelves, the newly exposed land provides more area for ice sheets to spread, and they add to cooling by their increased albedo. The increased area of each land mass experiences greater climatic extremes.

Later in ice buildup, however, the glaciated regions sink under the weight of the ice sheets, flooding their edges and converting parts of the ice sheets into floating ice shelves that are more unstable than land-based ice sheets. This makes the ice sheets vulnerable to rapid melting and breakup when the astronomical cycle reverses.

The last great glacial advance was at its maximum somewhere around 20,000 ± 2000 BP (depending on uncertainties in radiocarbon age dates). In North America, ice advanced as far as New York, St. Louis, and Oregon, then shrank back to Greenland and the Canadian Arctic islands. Ice scour removed great blocks of rock and transported them for hundreds of miles.

The North American ice sheets diverted the jet stream southward, taking the main storm track with it (Figure 23.7). The western United States became much wetter than it is today, so that great freshwater lakes formed from increased rainfall and from meltwater along the front of the ice sheet. River channels were blocked by ice to the north, and as the southern edges of the ice sheets flowed far enough southward to melt, almost all of the melt water drained south to the Gulf of Mexico down a giant Mississippi River, instead of eastward to the Atlantic.

Ice sheets have edges where they melt. Gigantic lakes can form around ice sheets, some of them held back by ice dams. Anywhere huge lakes lie behind ice dams, enormous floods can be released by ice melting, overflow, or earthquake. In North America, gigantic floods poured over Washington and Oregon from glacial Lakes Bonneville and Missoula. To the east, floods from glacial Lake Agassiz on the Canadian prairies drained catastrophically into the Great Lakes or into the Arctic by way of Hudson Bay.

As the North American ice sheets began to melt and retreat, water flow down the Mississippi to the Gulf of Mexico must have increased enormously. The water draining from the melting North American ice sheet changed the seawater composition of the Gulf of Mexico as it poured southward down the Mississippi in enormous quantities beginning about 14,000 BP, perhaps at ten times its current flow. Finally, the edge of the ice sheet retreated northward until the Great Lakes began to drain to the Atlantic instead, first down the Hudson Valley and then the St. Lawrence.

More subtle effects occurred in warmer latitudes. Increased rainfall in the Sahara during ice retreat formed great rivers flowing to the Nile from the central Sahara; they were inhabited by crocodiles and turtles, and rich savanna faunas lived along their banks.

Review Questions

A geographic change led to the beginning of the Ice Age that has recently affected the northern continents. What was that geographic change? how long ago did it happen? and how did it set off the Ice Ages?

Why is Antarctica a frozen continent?

Describe or draw the astronomical situation that could set up the melting of northern ice sheets.

Describe or draw the astronomical situation or the climatic conditions that could lead to an advance of the northern ice sheets over much of the northern continents.

Further Reading

Easy Access Reading

Broecker, W. S., and G. H. Denton. 1989. What drives glacial cycles? *Scientific American* 262 (1): 49–56.

Broecker, W. S., et al. 1989. Routing of meltwater from the Laurentide Ice Sheet during the Younger Dryas cold episode. *Nature* 341: 318–321.

Burton, K. W., et al. 1997. Closure of the Central American Isthmus and its effect on deep-water formation in the North Atlantic. *Nature* 386: 382–385.

Cess, R. D., et al. 1993. Uncertainties in carbon dioxide radiative forcing in atmospheric general circulation models. *Science* 262: 1252–1255.

Chorlton, G., et al. 1983. *Planet Earth: Ice Ages.* Alexandria, Virginia: Time-Life.

COHMAP members. 1988. Climatic changes of the last 18,000 years: observations and model simulations. *Science* 241: 1043–1052.

Covey, C. 1984. The Earth's orbit and the Ice Ages. *Scientific American* 250(2): 58–66.

Driscoll, N. W., and G. H. Haug. 1998. A short circuit in thermohaline circulation: a cause for Northern Hemisphere glaciation? *Science* 282: 436–438.

Haug, G. H., and R. Tiedemann. 1998. Effect of the formation of the Isthmus of Panama on Atlantic Ocean thermohaline circulation. *Nature* 393: 673–676.

Imbrie, J., and K. P. Imbrie. 1979. *Ice Ages: Solving the Mystery.* Short Hills, New Jersey: Enslow Publishers.

Jouzel, J., et al. 1993. Extending the Vostok ice-core record of palaeoclimate to the penultimate glacial period. *Nature* 364: 407–412.

Kerr, R. A. 1996. Ice rhythms: core reveals a plethora of climate cycles. *Science* 274: 499–500. News report.

Olsen, P. E. 1986. A 40-million-year lake record of early Mesozoic orbital climatic forcing. *Science* 234: 842–848.

Radok, U. 1985. The Antarctic ice. *Scientific American* 253 (2): 98–105.

Raymo, M. E. 1998. Glacial puzzles. *Science* 281: 1467–1468. If you feel you understand Milankovitch cycles, this short review leads you into some subtleties in current research.

Stommel, H., and E. Stommel. 1979. The year without a summer. *Scientific American* 240(6): 176–186.

Humans and the Ice Age

Amazingly, the severe fluctuations of climate do not appear to have affected ice-age plants or animals very much. Glacial advances and retreats, though they were rapid on a geological time scale, were slow enough to allow communities to migrate north and south with the ice sheets and the climatic zones and weather patterns affected by them. Communities close to mountain glaciers were able to adjust to advances and retreats by simply moving up and down in altitude. Tropical rain forests were very much reduced, but the habitat did not disappear, and their fauna and flora survived well. Tropical savannas were more extensive during the drier times that accompanied glaciations.

Life and Climate in the Ice Ages

The most interesting effects were controlled by changes of sea level that occurred with every glacial advance and retreat. Each major glaciation dropped world sea level by 120 meters or so (about 400 feet), exposing much more land area and joining land masses together. Each new melting episode reflooded lowlands to recreate islands.

Most continents carry examples of creatures stranded by flooding and the warming that occurred during and after the last ice retreat. In the Sahara Desert, for example, there are cypress trees perhaps 2000 years old. They set seed that never germinates because the climate is now much drier than it used to be. Ancient rock paintings of giraffes and antelope confirm the evidence of the cypress trees. Giraffes migrated south to the savannas; cypresses are confined to the north around the Mediterranean; and the Sahara Desert is a fearsome barrier to biological exchange.

A few creatures were trapped in geographical cul-de-sacs and wiped out. Advancing ice sheets, not St. Patrick, wiped out snakes from Ireland,

and snakes have not yet been able to cross the Irish Sea to recolonize the island. The Loch Ness Monster is impossible because Loch Ness was frozen under a mile of ice at 18,000 BP. The forests of western Europe were trapped between ice sheets from Scandinavia and wiped out. After the ice sheets melted, western Europe was recolonized by deciduous hardwoods; elsewhere, in North America, Scandinavia, and Siberia, the great boreal forests are dominated by conifers.

In the seas, many species of warm-water molluscs that retreated southward with the first great cooling episode were trapped in marginal seas around the North Atlantic: in the Caribbean, the Mediterranean, and the North Sea. They could not find a clear way of retreat southward, and they suffered great extinctions. But most of these events occurred in early ice advances, and the creatures that survived the first few glacial episodes were well adapted to Pleistocene conditions and suffered little extinction during subsequent glaciations.

We have good evidence of the plant and animal life of the Pleistocene. Enormous bone deposits in Alaska and Siberia and fossils found in caves, sinkholes, and tar seeps have provided excellent evidence of rich and well-adapted ecosystems on all continents.

There have been much greater changes in terrestrial animals and plants during and after the last glaciation than in any previous one, and the effects have often varied with the size of the land area. The land areas sometimes showed dramatic changes. For example, Alaska and Siberia were joined across what is now the Bering Strait, and Greenland was joined to North America, to form one giant northern continent. Australia was joined to New Guinea, and Indonesian seas were drained to form a great peninsula jutting from Asia (Figure 24.1). Box 24.1 compares the

Figure 24.1 World geography at 18,000 BP. The largest ice sheets are shown in heavy stipple. Areas of land and sea, covered with tundra and sea ice respectively, are shown in lighter stipple. I have indicated some important geographic changes to coastlines. For example, much of the seafloor off Southeast Asia, Australia, and the West Indies was dry land at the time. (Simplified from maps produced by the CLIMAP project, and by Denton and Hughes.)

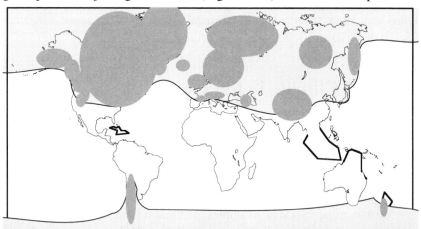

BOX 24.1 WORLD GEOGRAPHY, NOW AND AT 18,000 BP — THE MAJOR LAND MASSES

THE MAJOR LAND MASSES NOW		MAJOR ICE AGE LAND MASSES
1. Eurasia plus Africa	6. New Guinea	1. One major world continent
2. The Americas	7. Borneo	2. Antarctica
3. Antarctica	8. Madagascar	3. Meganesia: Australia + New Guinea
4. Australia	9. Baffin Island	4. Madagascar
5. Greenland	10. Sumatra	5. New Zealand

major land masses now and at 18,000 BP at the height of the last glaciation; it shows how drastically seawater barriers were removed to join land masses together.

Continental Changes

On major continents, the larger birds and mammals of the Pleistocene were most unlike their modern counterparts. Just before the last ice advance, North America had mastodons, mammoths, giant bison, ground sloths, sabertooth cats, horses, camels, and dozens of other large mammals. Eurasia had most of these, plus giant deer and woolly rhinos. The giant ape *Gigantopithecus* roamed the Himalayan slopes (Chapter 21). The moas of New Zealand and the elephant birds of Madagascar are well known (Chapter 20), but Australia had giant ground birds as large as these and a dozen giant marsupials. All these creatures are now extinct.

The catastrophic extinctions occurred at different times on different continents. In each case, the mammals and birds were part of flourishing ecosystems. For example, North America has a very good fossil record of large Pleistocene mammals. Twenty genera became extinct in the 2 m.y. before the last ice sheet melted, then 35 genera were lost in less than 3000 years! Radiocarbon dates for the extinction cluster around 11,000 BP; where the record is good, the extinctions look sudden. North American extinctions are listed (in part) in Box 24.2.

Some ice-age animals, such as the woolly mammoth and the woolly rhinoceros, were specifically adapted to life in cold climates. They were much hairier than their living relatives, and they have been found in areas of tundra that were very close to the ice sheets at the time. Woolly mammoths were sometimes killed by falling into ice crevasses. Their bodies have been discovered still frozen in permafrost in Siberia, preserved well enough to tell us quite a lot about their way of life (Figure 24.2). Gallons of frozen stomach contents show that woolly mammoths ate sedges and grasses and browsed tundra trees such as alder, birch, and willow. The tusks of adults were well shaped for clearing snow from forage in winter. Woolly mammoths were adapted specifically to tundra, and we have evidence of their reaction to ice advances and retreats: the only change was in size. Siberian woolly mammoths were about 20%

**BOX 24.2. HERE IN NORTH AMERICA LIE THE MORTAL REMAINS
OF DOZENS OF SPECIES OF EXTINCT CREATURES,
ALL OF WHICH DIED OUT ABOUT 11,000 BP.**
Requiescant in pace   †   RIP

† Six species of large ground sloth
† Three species of tapirs
† Two species of sabertooth cats
† North American cheetah
† North American lion
† Giant turtle

† Giant bison
† Mastodon
† Two species of mammoths
† Four species of camels
† Giant anteater
† Giant condor

† Seven species of deer, moose, and antelope
† Ten species of North American horses
† Eight species of cattle and goats
† Two species of large bears
† Dire wolf

Figure 24.2 The discovery of the famous frozen Beresovka mammoth in Siberia in 1900. (Taken from an old magic-lantern slide.)

larger during the warm interglacials than they were at the coldest times.

Large Pleistocene mammals were well able to withstand climatic change as well as climatic severity. Their large body sizes gave them low metabolic rates, so they could live on rather poor-quality food. As adults, they were largely free from the danger of predation by carnivores. Yet it was the large mammals and birds that became extinct, while smaller species did not suffer as much. The plants the large mammals ate are still living, and so are the small birds, mammals, and insects that lived with them. In the oceans, nothing happened to large marine mammals.

In North and South America, the extinctions took place in a short time toward the end of the ice age, very close to 11,000 BP. This was an unusually cold, dry time, so it has been easy for North American geologists to argue that the extinction was related to climate change.

But that explanation, even if true, covers only some of the American extinctions and does not apply at all to the rest of the world. For example, the giant ground sloths of Arizona were browsers and ate semidesert scrub that was available in the area before, during, and after they died out. Other things being equal, we should prefer another hypothesis if it explains more data more simply.

There is no question that climatic change around 11,000 BP was rapid. Yet the very same species of animals had already survived a dozen or more similar events. There is nothing climatically unique about the last ice retreat. The previous ice retreat, about 125,000 BP, was just as sudden but caused no extinctions. So if climatic change did not result in the extinctions, what did? The problem of Pleistocene extinctions has been debated ever since ice-age animals were discovered, and an 800-page book devoted to the problem in 1984 showed that there was continuing major disagreement. The strongest evidence supports an idea put forward in its current form mainly by Paul Martin. Martin's idea is called the **overkill hypothesis** because he stresses one human behavior in particular—hunting. In every case, invading humans were skilled hunters, encountering animals that had never seen humans before. Martin listed seven major lines of evidence (Box 24.3). All the pieces of evidence,

---

BOX 24.3 PAUL MARTIN'S EVIDENCE IN FAVOR OF THE OVERKILL HYPOTHESIS

1. Large mammals and ground-living birds were affected most. North America lost 35 genera, and South America lost even more.

2. Extinctions occurred in different areas at very different times.

3. Extinct animals were not replaced.

4. Extinctions were closely linked in time and space with human arrival.

5. Large mammals survived best in Africa and Asia. Extinctions were much more severe in the New World (Australasia and the Americas).

6. Where extinctions are well dated, they were sudden: North America and New Zealand are the best examples.

7. There are very few places where mammal remains occur with human remains or human artifacts. This implies that co-existence was brief.

---

rgues Martin, are consistent with the idea that the sudden arrival of human invaders in an ecosystem was responsible for the extinctions. Other corollaries of human arrival may play an important part, so Martin's idea should not be judged entirely on the hunting overkill that he stresses most. To test his idea, we can draw on data from the only three major continents that were colonized suddenly by humans: North and South America, and Australia.

The Americas

Human Arrival

Humans crossed into North America from Siberia at a time when the Bering Strait region was a dry land area, **Beringia**. In the depths of the Ice Ages, Beringia was a frigid, barren waste swept by violent winds blowing dust and sand from the edge of the ice sheet. As the ice began to melt and retreat, however, a varied Arctic vegetation became established in Beringia and began to attract and support a fauna of large ice-age mammals, including woolly mammoths, horses, camels, sheep, deer, musk-oxen, and ground sloths. Even then, Beringia was separated from the rest of North America by the ice sheets of the Canadian Shield and the Rocky Mountains, which flowed together in what is now Alberta (Figure 24.1). An ice-free corridor to the south opened up into the rest of the Americas only as the main Canadian ice sheet retreated eastward. The important event in human migration is not when people reached Beringia, but when they broke past the ice barriers to the temperate and tropical Americas to the south.

Did humans reach the Americas only as the last glacial period ended, or had they done so during another period of opportunity long before? There is now compelling evidence that humans were living in Monte Verde, in southern Chile, at 12,500 BP or before. Most likely, these people had arrived by boat along the western American coast: there is scattered evidence of very early American fisherfolk at sites in British Columbia, southern California, and Peru. The coastal people seem to have eaten shellfish, seabirds, and small fish. As far as we can tell, they had very little effect on the American continental ecosystems, and they may not have left many descendants. The American continental ecosystems did not receive full human impact until around 11,000 BP, with the arrival of big-game hunters.

Genetic evidence suggests that surviving native Americans are distinct from Eurasians in DNA structure. This is probably the result of a strong genetic founder effect, in which a small group of people managed to enter the Americas through some kind of bottleneck in a single invasion of invaders whose descendants dominated the eventual population of the New World. Native Americans eventually paid a terrible price for this: the founder effect meant that they were genetically homogeneous, and when Europeans arrived with new diseases in 1492, any susceptibility to a particular disease was practically universal among the population. Imported diseases decimated native Americans, where a more varied gene pool would have suffered less catastrophic results.

But when did the great wave of continental colonists arrive (after the coastal fisherfolk)? A distinctive and short-lived tool and weapon culture, the Clovis culture, spread rapidly across North America from Washington to Mexico. All the radiocarbon dates for Clovis sites in the western United States cluster around 11,000 BP. Tools found in Beringia shortly after 12,000 BP look like precursors to the tools carried by Clovis people across the North American plains. The invaders were already skillful hunters of large mammals across the northern tundra.

Soon after 11,000 BP, there is evidence of human occupation throughout the Americas. By then, the Clovis artifacts had been replaced by more advanced and more regionally varied sets of tools, and people had adopted varied life styles appropriate to the regions.

Large Animals

Along with the arrival of Clovis culture in the Americas, we see evidence of large-animal hunting. The **PaleoIndians** hunted mammoths and mastodons. There are cut marks on mastodon bones found close to the edge of the ice sheet near the Great Lakes, and it seems that humans butchered the carcasses into large chunks and cached them for the winter in the frigid waters under shallow, ice-covered lakes, just as Inuit do today in similar environments. We can tell that the PaleoIndians' favorite hunting season for mastodons was late summer and fall, whereas natural deaths occurred mainly at the end of winter when the animals were in poor condition. A mammoth skeleton from Naco had eight Clovis points in it. Two juvenile mammoths and seven adults were killed with Clovis tools near Colby, Wyoming, and the way the bones are piled suggests meat-caching there too. The last mammoths in the Great Lakes area, the last native North American horses, and ground sloths and mountain goats in the Southwest all died out around 11,200 BP. In west Texas, mammoths, horses, and bison are common before 11,000 BP, but after that we find only bison bones. The skeleton of a giant extinct turtle found in Little Salt Spring in southwest Florida had a sharpened wooden stake jammed

Life and Climate During the Ice Ages

between the shell and the breastbone. The turtle had apparently been killed and cooked on the spot in its shell. There are ground sloths, bison, and a young elephant at about the same level, but afterward it seems that the PaleoIndians ate white-tailed deer.

Megaherbivores and Medium-Sized Animals

There was extinction among medium-sized mammals, although one would expect some of them (camels, horses, and deer, for example) to have been resistant to extinction because of their speed, agility, and rapid reproduction, even in the face of expert hunters. The answer to this puzzle may be found in ecosystems that include **megaherbivores** (very large herbivorous mammals more than 2000 kg [2 tons] in weight). On the plains of Africa, for example, the largest animals, elephant and rhino, can have drastic effects on vegetation. Elephants destroy trees and turn dense forest into open woodland by opening up clearings in which smaller browsing animals multiply. Eventually the elephants turn any local habitat into grassland. They then migrate to another woodland habitat, leaving the trees to recover in a long-term ecological cycle that can take decades to complete. White rhinos graze high grass so effectively that they open up large areas of short grassland for smaller grazing animals.

Thus, in the long run, megaherbivores keep open habitats in which smaller plains animals can maintain large populations. Where elephants have been extinct for decades, the growth of dense forest is closing off browsing and grazing areas, and smaller animals are also becoming locally extinct. Many of the problems in African national parks today occur because they are not big enough to allow these cycles of destruction and migration to take place naturally.

But what would happen if megaherbivores were completely removed from an ecosystem—by hunting, for example? Megaherbivores breed slowly and cannot hide. They would be particularly vulnerable to skillful hunters. Norman Owen-Smith proposed that the disappearance of Pleistocene megaherbivores (Figure 24.3) soon led to the overgrowth of many habitats, reducing their populations of smaller animals too. Thus, even if early hunters hunted or drove out only a few species of megaherbivores, they could have forced ecosystems so far out of balance that extinctions would then have occurred among medium-sized herbivores too, especially if hunters were forced to turn to the latter as prey when the megaherbivores had gone.

There may be more subtle effects of removing large herbivores. Plants sometimes coevolve with herbivores that disperse their seeds. Very large herbivores are likely to encourage the evolution of large, thick-skinned fruits, and a sudden extinction would leave the fruits without dispersers. Even today, guanacaste trees of Central America produce huge crops of large fruits, most of which lie and rot. Daniel Janzen suggested that these fruits coevolved with large elephants (gomphotheres), which became extinct with the other large American mammals.

Figure 24.3 Pleistocene megaherbivores. We don't have to rely on the inferences of paleontologists to tell us that there were megaherbivores in Pleistocene ecosystems. They were observed and illustrated by competent ecologists of the time. (Same as Figure 22.15.)

> They fall from the branches to wait,
>
> But they're 12,000 summers too late.
>
> You can smell them for miles,
>
> They're rotting in piles,
>
> The fruits that the gomphotheres ate.

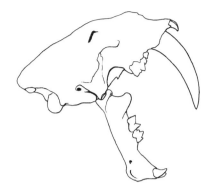

Figure 24.4 The large sabertooth cat *Smilodon* occurs in large numbers among the Pleistocene fossils of the La Brea tar pits in Los Angeles. It is the State Fossil of California. It was a major predator in the Pleistocene ecosystems of North America.

Figure 24.5 The huge long-horned bison of the Pleistocene North American plains, *Bison latifrons.* (Reconstruction by Bob Giuliani. © Dover Publications Inc. Reproduced by permission.)

Predators and Scavengers

Predator species such as the sabertooth cats (Figure 24.4) and the North American lion could have been reduced to dangerously low levels by the removal of their prey by overkill; there is no need to think in terms of the direct, systematic overkill of predator species that modern humans often carry out. In turn, scavengers may also depend on populations of large mammals to provide the carcasses they feed on. Thus, the giant teratorn known from the La Brea tar pits (Figure 14.25) is extinct, and the so-called "California" condor once nested from the Pacific coast to Florida. Pleistocene caves high on vertical cliffs in the Grand Canyon of Arizona contain bones, feathers, and eggshells of this condor, along with the bones of horses, camels, mammoths, and an extinct mountain goat. The condor vanished from this area at the same time as the large mammals did, presumably because its food supply largely disappeared.

Survivors

What about the surviving large mammals in North America? It turns out that many of them were originally Eurasian and crossed into the Americas late in the Pleistocene. Thus, bear, moose, musk-oxen, and caribou had been exposed to human hunting in Eurasia before 11,000 BP. There were no North American extinctions after 8000 BP at the latest, presumably when a new stable balance had evolved. There were separate regional cultures in the Americas by this time, but there were no new significant extinctions even though tools and weapons had improved.

Bison were a special case. They were American natives, and although the immense long-horned bison (Figure 24.5) became extinct, the smaller bison survived in great numbers. Perhaps the removal of larger competitors encouraged this success. Moreover, bison survived in the face of intense and wasteful hunting by PaleoIndians, whose methods were by no means as ecologically sound as their descendants sometimes claim. A well-studied site in Canada reveals that PaleoIndians, hunting on foot with stone weapons, would stampede whole herds of bison along preplanned routes that led to a cliff edge or **buffalo-jump**. The bison would then be finished off and butchered at the cliff base. The method naturally resulted in the deaths of many young animals; only about a quarter of the animals killed were full-grown. The site, appropriately named Head-Smashed-In, had been used for more than 5000 years when it became obsolete as firearms reached the tribes in the nineteenth century. Given that there are dozens more buffalo-jump sites stretching across the Great Plains from Alberta to Texas, most of them still to be investigated, it seems less surprising that humans would be capable of the overkill that the fossil record suggests, and more surprising that the bison adjusted so successfully for so long to human predation pressure.

Australia

Australia suffered more severe extinctions than any other continental sized land mass. It lost every terrestrial vertebrate larger than a human. It lost a giant horned turtle as big as a car, and its giant birds, the dromornithids (Chapter 20). It lost its top predators, including *Megalania*, the largest terrestrial lizard that ever evolved, 7 meters (24 feet) long. *Megalania* was related to the living Komodo dragon but weighed more than eight times as much. Other predators were a huge terrestrial crocodile, a carnivorous kangaroo, and a 5-meter (16-foot) python. Australia lost about 20 large marsupials, including all the diprotodonts, huge four-footed vegetarians the size of tapirs; a wombat the size of a cow; and the largest kangaroo of all time, *Procoptodon*, a browser 3 meters high that was the ecological equivalent of a tapir or ground sloth. Only a few large animals survived in Australia, but small animals were less affected.

The extinctions are dated to around 50,000 BP, a time that coincides, as far as we can tell, with human arrival in Australia. The slow-running giant dromornithid bird *Genyornis* disappeared from habitats where fast-running emus survived, and there seems to be a memory of the dromornithids in aboriginal legend as the mihirung.

The extinctions coincide roughly with a change in vegetation associated with increased burning. This was not a time of climatic change, so the increased burning may have been generated by the early Australians. They were migrating into a dry-country ecosystem that was unfamiliar to them because they came from the moister tropical ecosystems of New Guinea. They had to learn slowly how to adapt to drier Australian conditions, just as Europeans had to do tens of thousands of years later. One of the easiest ways of clearing Australian vegetation is to burn it: burning makes game easier to see and hunt, and Australian aborigines today have complex timetables for extensive seasonal brush burning that has dramatic effects on regional ecology. Most likely, then, the Australian extinctions were the direct result of human invasion, through the introduction of large-scale burning as well as hunting.

On the basis of this evidence from three continents, it looks as if Martin's hypothesis of human impact is stronger than any other. Now we will look at smaller land masses that were subjected to much the same human impact and see how they fared rather differently.

Island Extinctions

Island animals can often evolve into unique sets of creatures, and geographical changes that connect previously isolated areas can have dramatic and damaging efects on species and communities (Chapter 20). Human arrival has often had a catastrophic effect on island faunas. The world may have 25% fewer species than it did a few thousand years ago, and most of those extinctions took place on islands. For example, native Tasmanians killed off a unique penguin sometime after the 13th century, 600, years before they in turn were wiped out by European settlers.

Figure 24.6 One of the species of moa in New Zealand. (After Frohawk.)

BOX 24.4 Case Study: New Zealand

A thousand years ago, New Zealand was an isolated set of islands without land mammals (except for two species of bats). Birds were the dominant vertebrates, and the largest were the moas, giant flightless browsing birds the size of ostriches (Chapter 20) (Figure 24.6). The moas and other native creatures survived as glacial periods came and went, yet they became extinct within a few hundred years of the arrival of the Polynesian Maori people after 1000 AD.

There seem to have been two main reasons for the extinctions, and all of them are connected with human arrival. First, evidence of hunting is clear and appalling. Midden sites that extend for acres are piled with moa bones, with abundant evidence of wasteful butchering. The bones are so concentrated in some places that they were later mined to be ground up for fertilizer. The middens contain bones of eleven of the twelve extinct species of moas, and they also contain bones of tuataras, very primitive reptiles. Second, the Maori brought rats with them, which ate insects directly, killed off reptiles by eating their young, and exterminated birds by robbing their nests. The tuataras (Chapter 10), the giant flightless wetas (insects that had been the small-bodied vegetarians of New Zealand (Chapter 20), and many flightless birds including the only flightless parrot, the kakapo, were practically wiped out by rats. There were many other more subtle ecological changes. A giant eagle that probably preyed on moas died out with them, for example. And when the moa were gone, the Maori took up serious cannibalism, because humans were the largest remaining protein packets.

Half of the original number of bird species in New Zealand were extinct before Europeans arrived, and the new settlers only acted to increase the changes in New Zealand's landscape and biology. Forests were cleared even faster, and new mammals were introduced. European rats were the worst offenders against the native birds, but cats, dogs, and pigs were also destructive, rabbits destroyed much of their habitat, and deer competed with browsing birds. The tuatara now lives only on a few small, rat-free islands, and the kakapo survives precariously in remote areas where it is threatened by wild cats. Bird populations are still dropping in spite of efforts to save them.

As a microcosm of the problem, consider the Stephen Island wren, the only flightless songbird that has ever evolved. This species had already been exterminated from New Zealand by the Polynesian rat before European arrival. The entire remaining population of this species, which was by then confined to one island, was eaten by Tibbles, a cat brought to the island in 1894 by the keeper of a new lighthouse.

A convict colony established by the British wiped out an endemic seabird on Norfolk Island, between Australia and New Zealand. Several small, unique native birds fared better for a while on Lord Howe Island, further north: they lived alongside the early settlers until a shipwreck allowed rats to reach the island in 1918. Within a few years five species had completely disappeared.

On Madagascar, large lemurs, giant land tortoises, and the huge flightless elephant birds (Chapter 14) disappeared after the arrival of humans somewhere between 0 and 500 AD. Here too, large forest areas were cut back and burned off to become grassland or eroded, barren wasteland. No native terrestrial vertebrate heavier than 12 kg (25 pounds) survived after 1000 BP. Dates are not well known, and the extinction may have taken 1000 years instead of being sudden. It's clear that human arrival was part of a "recipe for disaster" (as David Burney has called it). The desperate erosion and poverty of much of the countryside of Madagascar today underlines the fact that humans are still involved in self-destructive deforestation, in spite of the evidence all around them of its horrific after-effects.

A panda-sized marsupial lived in New Guinea in the Late Pleistocene, and although it is now extinct, the plants it ate are still flourishing.

New discoveries of the bones of extinct flightless birds in Hawaii suggest that devastating extinctions followed the arrival of Polynesians. The Hawaiian Islands are famous for honeycreepers, which evolved there into many species like Darwin's finches on the Galápagos Islands. But there were 15 more species of honeycreepers before humans arrived. Two-thirds of the land birds on Maui were wiped out by the Polynesian arrival, probably by a combination of hunting, burning, and the arrival of rats. As in New Zealand, further extinctions that followed the European arrival were severe but not as drastic, probably because the bird fauna had already been so depleted.

The same process is recorded on almost all the Pacific islands in Melanesia, Polynesia, and Micronesia. Almost all of them, apparently, had species of flightless birds that were killed off by the arriving humans and/or their accompanying rats, dogs, pigs, and fires. As many as 2000 bird species may have been killed off as human migration spread across the ocean before the arrival of Europeans. It may not be an accident that Darwin was inspired by the diversity of the Galápagos Islands: these were never occupied before European discovery in 1535, and human impact was relatively slight until whalers arrived in strength around 1800.

Several islands in the Mediterranean Sea—Cyprus and Crete are good examples—held fascinating evolutionary experiments after the ice ages. There were pigmy elephants and pigmy hippos, giant rodents, and dwarf deer. These mammals had evolved on these isolated islands in much the same way as did the fauna of Gargano during Miocene times (Chapter 20). Many of the island animals disappeared as Neolithic peoples discovered how to cross wide stretches of sea and colonized the islands several thousand years ago.

Europeans killed off the dodo on Mauritius (Figure 24.7) and several species of giant tortoises there and in the Galápagos; deliberate burning, and the goats, pigs, and rats they brought, completed a great deal of destruction of native plants, birds, and animals. Europeans supervised the destruction of Steller's sea cow in the North Pacific. They killed off the great auk of the North Atlantic, the huia and other small birds of New Zealand (Box 24.4), and an unknown number of species of birds of paradise in New Guinea, all to satisfy the greed of egg and feather collectors.

Figure 24.7 The dodo of Mauritius. This giant flightless pigeon was hunted out by meat-hungry European sailors. We now have only some dried-out museum scraps and a few drawings. (From Lydekker.)

Figure 24.8 *Sylviornis*, the extinct Du of New Caledonia. The bird stood about a meter (3 feet) high. (Redrawn from a reconstruction by F. Poplin and C. Mourer-Chauviré; the head ornament is based on oral legend of the Melanesians of New Caledonia.)

> The Du is a bird that's now long lost
>
> It laid eggs in a pile of warm compost,
>
> But the chicks that did hatch
>
> Were too easy to catch
>
> And became Melanesian pot-roast.

They drove fur-bearing mammals close to extinction worldwide.

Irrespective of race, color, and creed, it seems, human arrival in the midst of a fauna and flora unused to hunting pressure, to extensive burning, or to rats, cats, and pigs, has spelled disaster. One common factor is naiveté. Charles Darwin described the complete lack of fear of humans of the Galápagos animals and birds, an almost universal feature of creatures never exposed to human hunting; and modern observers like Tim Flannery have recorded the same behavior. New Zealand's inhabitants had never seen a land mammal before the Polynesians arrived.

The Du

Very large bird bones were discovered a few years ago on the Isle of Pines, off New Caledonia. The fossil was named *Sylviornis*, and although it was large, it was not a ratite, but a very large flightless megapod. Megapods are a family of birds that includes the mallee fowl of Australia and ranges through eastern Indonesia and Australasia. No bird anywhere near that size now lives on New Caledonia. *Sylviornis* is therefore extinct, and its remains are dated at about 3500 BP, when humans had already reached its island.

The Melanesians of the Isle of Pines tell stories of a giant red bird, the Du (Figure 24.8), which did not sit on its single egg to hatch it. Although the Melanesians did not know it at the time, this behavior is unique to megapod birds, which lay their eggs and cover them with rotting vegetation to keep them at an even warm temperature. The male keeps close control of the egg temperature by adjusting the compost heap on an almost hourly basis for weeks at a time. The Du, then, was *Sylviornis*, and the legend shows that it was known to early man.

The Isle of Pines has large areas covered by large and mysterious mounds, never associated with original human artifacts. The mounds are the right size to be Du hatching mounds, still preserved in enormous numbers. They give some idea of the numbers of the Du, and they illustrate the massive disaster that overtook the bird at a time when there was no significant climatic change in its environment.

Experienced Faunas

We have already seen that many survivors in North America had been used to hunting pressure in Eurasia. Humans developed their hunting skills in the Old World, and although there were extinctions of large mammals there, they were spread out over longer times than the New World extinctions were. For example, in Africa *Homo heidelbergensis* butchered giant baboons, hippos, and the extinct elephant *Deinotherium*. The remains of 80 giant baboons have been found at one site dated at about 400,000 BP. At Torralba, in Spain, the remains of 30 elephants, 25 horses, 10 wild oxen, 6 rhinos, and 25 deer were found on one site.

It therefore fits the overkill or human-impact hypothesis that the most important local extinctions in the Old World took place in habitats that modern humans were invading in strength for the first time. These invasions took place along the edges of the ice sheets, and even then humans are implicated in the disappearance of only a few large mammals of the northern Eurasian tundra, especially the woolly mammoths and the woolly rhinoceros.

It looks as if mammoths became extinct as advanced hunting techniques allowed humans to range closer to the ice sheets. For example, the advance of ice sheets toward the peak of the last glaciation seems to have driven the Gravettian people (Chapter 22) out of the northern Carpathian Mountains of Central Europe toward the south and the east, where they discovered and invaded the mammoth steppe of Ukraine for the first time. The Gravettians were already using mammoth bones as resources. At Predmost in the Czech Republic, a site dating from just before the coldest period of the last glaciation (28,000–22,000 BP) contains the bones of at least 1000 woolly mammoths. These people routinely buried their dead with mammoth shoulder bones for tombstones.

The pattern in these extinctions was always the same. The large mammals were hunted out of the optimum part of their range, and then the last survivors hung on in the inhospitable (usually northern) parts of their range until newly invading humans or climatic fluctuations killed them off. For example, woolly mammoths, woolly rhinoceroses, and giant deer, along with horses, elk, and reindeer, reinvaded Britain from Europe after the ice sheets began to retreat and birch woodland and parkland spread northward. Mammoths flourished in Britain until 12,800 BP at least, but then human artifacts appeared at 12,000 BP, and the largest animals of the tundra fauna quickly disappeared.

The giant deer is sometimes called the Irish elk, partly because it is best known from Ireland. It was not an elk but a deer the size of a moose, with the largest antlers ever evolved, more than 3 meters (10 feet) in span (Figure 24.9). It was adapted for long-range migration and open-country running, and its diet was the high-protein willow vegetation on the edges of the northern tundra. It once ranged from Japan to France, but it did not reach North America, where the moose is an approximate ecological counterpart. The giant deer disappeared from Eurasia in a sequence that started in Eastern Siberia and proceeded westward. It survived longest in Western Europe and finally, after the ice sheets melted and sea level rose, it was confined to the island of Ireland. The giant deer flourished there in a warm period until about 11,000 BP, but it then died out in a cold period, possibly because it was unable to retreat southward to a better climate. Humans did not reach Ireland from Europe and Britain until the climate warmed again, around 8000 BP.

Mammoths, which had lived much farther south, were confined to the tundra north of the Black Sea by 20,000 BP. We have an interesting record of life around 15,000 BP on the plains of Russia and Ukraine. Several dozen living sites were built on low river terraces by people from the Gravettian culture. Each major site contains the remains of several large buildings whose foundations and lower walls were made entirely

Figure 24.9 The giant Pleistocene deer of Eurasia, *Megoceros*, usually called the Irish elk. Males carried the largest antlers ever evolved. (Reconstruction by Bob Giuliani. © Dover Publications Inc. Reproduced by permission.)

The Ukrainian tundra was bleak

Till they found a solution unique,

But lack of supplies

Soon caused the demise

Of their mammoth construction technique.

from mammoth bones. The buildings were large, 4–7 meters (13–23 feet) across and up to 24 sq m (240 sq ft) in area. The foundation was made of the heaviest bones, carefully aligned. Skulls at the base were followed by jaws and then long bones, with the resulting pattern providing an aesthetic geometric arrangement as well as sound architecture. The roofs were probably lighter structures made from branches, hides, or sod. (Only a thousand years ago, the Inuit of Greenland were using whale skulls and whale ribs in the same way, roofing the dwellings with sod.)

One group of four buildings was built using bones from at least 149 mammoths. It's not clear whether the mammoths had been killed, or whether bones from old skeletons had been collected. Zoia Abramova has remarked that it would have been easier to obtain bones from living mammoths than to dig them out of permafrost. Others disagree, arguing that dwelling sites may even have been chosen because they were close to massive mammoth bone accumulations. Since no one involved in the arguments has completed even one of these tasks, we are not likely to get any agreement soon. Certainly, the sites in this unique area give some idea either of the numbers of mammoths that once roamed the plains, or of the hunting efficiency of these stone-age peoples, or both.

Inside, the Gravettians laid down clear river sand as a floor, built hearths (fuelled by mammoth bone), and remained secure and warm during the winter. They made finishing touches to stone tools (leaving behind their antler hammers and chipped flakes), skinned animals (leaving behind the bones), ground ochre for dye, and sewed with ivory needles.

The mammoth-bone dwellings are surrounded by pits dug into the permafrost that were probably used to store meat long-term. Lewis Binford has made a close and vivid comparison between the whale- and caribou-hunting Inuit of Arctic North America today and the tundra dwellers of the mammoth steppe. Many features of their buildings and the food storage pits are closely similar, implying that the ancient Gravettians were effective hunters even if they also used and re-used old bones.

Older generations of Inuit did not hesitate to attack a 25-ton bowhead whale from flimsy boats, though their modern descendants prefer motor boats and assault rifles. Ice-age hunters may or may not have attacked an 8-ton mammoth directly. They may have used sophisticated, low-risk techniques such as stampeding or trapping. At any rate, it doesn't take a great deal of imagination to see why man and mammoth could not have coexisted for long in this mammoth steppe country that had been the main range of the species. (The same Gravettian people had built mammoth-bone dwellings in Central Europe several thousand years before, when mammoths still lived there.)

Some woolly mammoths survived in the permafrost areas of northern Siberia until perhaps 10,000 BP, in an area where humans arrived late. Even then, there was still one mammoth refuge left, in the Wrangel Islands, a small group of low-lying islands off the north coast of Siberia. Forage was poor, and the last mammoths were small, perhaps 2 tons instead of the 6 tons of their ancestors. The last woolly mammoths died out in the Wrangel Islands only 3000-4000 years ago, at a time when there were large cities in the ancient civilizations of Eurasia and the Egyptian pyramids were already old. We are not sure what killed off these last piti- ful survivors of the great mammoths, but humans reached the Wrangel Islands about that time.

There is a myth that primitive peoples live in ecological harmony with the plants and animals around them, and that it has been only with the arrival of modern civilization that major ecological imbalances have arisen. We have seen several examples that explode this myth, and there are many more.

Ancient peoples have destroyed their own civilizations on islands with delicate ecosystems. The Easter Islanders who built their famous enormous stone statues on the island also deforested their fertile, produc- tive land until it became a barren waste and they became a wretched band of refugees surviving by shoreline scavenging and primitive fish- ing. But sophisticated peoples on continents have harmed themselves too. The Anasazi Indians, who built a complex civilization on the Colora- do Plateau, stripped their environment of trees until the erosion and silta- tion that followed ruined their irrigation projects and they disappeared as a significant people. (Tim Flannery calls these sorts of self-destructive societies "The Future Eaters.") But are we doing any better?

The World Today

The Spanish introduced cattle to Argentina in 1556; by around 1700 there were about 48 million head of wild cattle on the plains. By 1750 they had been all but exterminated by a comparatively sparse human population with primitive firearms. This is even more incredible than the North American slaughter of about 60 million bison a century later with much more effective rifles, and it is more evidence in support of the plau- sibility of the overkill hypothesis.

Stripping tropical forest from hillsides not only removes the plants and animals that are best adapted to life there, but it results in erosion that removes the few nutrients left in the soil, destroying any agricultural value the land may have. It also results in much greater run-off and downstream flooding, which destroys or silts up rivers, irrigation chan- nels, and fields downstream, harming ecosystems and productivity there too. This scenario has been played out now in Ethiopia, Madagascar, and Haiti in horrific proportions; it is happening throughout the rain forests of Brazil and Indonesia, it is destroying the reservoirs that provide water for the Panama Canal, and yet we do not seem to have learned the lesson.

One can argue that humans at 11,000 BP, perhaps even at 500 BP, did not know enough ecology, did not have enough recorded history, did not know enough archeology or paleontology, and did not have enough of a global perspective to realize the consequences of their impact on an ecosystem. But that is not true today. We have the theory and the data to know exactly what we are doing. We transport species to new continents and islands without proper ecological analysis of their possible impact. We approach our environment sometimes with stupidity, sometimes with greed, but usually with both. We know very well that the tropical regions of the world are a treasure house of species, many of them valuable to us and many of them undescribed. Yet we deliberately introduce alien carnivorous fish into tropical lakes, ruining fisheries that have been stable for centuries. We know that clearing tropical forests will quickly destroy the low level of nutrients in the soil and will render those areas useless for plant growth. Yet we go ahead anyway, sometimes for a quick profit on irreplaceable timber, sometimes to achieve a few years' agricultural cropping before the land is exhausted. We poach gorillas and shoot animals for trophies. Indonesians and Malaysians clear tropical forests to supply the wood for the eleven billion disposable chopsticks used each year in Japan. Africans destroy rhinos to supply Asians with useless medicines and Yemenis with dagger handles; Africans poach elephants for ivory that ends up as ornaments and trivia in Japan, Europe, and North America. Fishing grounds have been plundered world-wide. Tanzanian and Filipino fishermen use dynamite sticks, killing off the reefs their fish depend on and any hope they have of catching anything next year. Filipinos capture tropical fish for American aquariums and Chinese restaurants and "medicine" shops by dosing their reefs with cyanide and catching the few survivors. The Japanese and Norwegians catch whales under the guise of "research"—after all, they've already paid for the ships and want to get their money back. Giant clams are poached from marine preserves all over the South Pacific to feed the greed of Chinese "gourmets". North Americans complain about the destruction of tropical rain forests for export to Japan, while their own lumber companies are felling the last of the old Douglas fir and redwood forests of the northwest for export to Japan. Italians and French shoot millions of little songbirds each year for "sport." The French test nuclear weapons on South Pacific atolls, and British and French nuclear waste has made seaweed radioactive all the way to the Arctic Ocean. All industrial nations continue to destroy forests and lakes with acid rain, though we know how to prevent the pollution that causes it. Ignorance is not the problem in any of these cases: poverty, greed, and arrogance are to blame.

We could live perfectly well—in fact, we could have a vastly increased quality of life—without disturbing the equilibrium of marmosets, gorillas, orangutans, chimps, whales, and all the other endangered species, if we took a grip on our own biology. What we need is a sense of collective responsibility and enlightened self-interest. It's a difficult message to get across because evolution and society both tend to reward the short-term goals of individuals rather than the long-term welfare of communities or societies.

It is in the interest of everyone, for themselves and for their children, to make our future secure not just for survival but for quality of life. If we don't solve our problems by our own voluntary actions, natural selection will do it for us. If we can learn anything from the fossil record, it is that extinction is the fate of almost every species that has ever lived on this planet. There is no automatic guarantee of success. Every individual in every generation is tested against the environment. We have the power and the knowledge to control our environment on a scale that no other species has ever done. So far, we have used those abilities to remove thousands of other species from the planet. If we destroy our environment to the point where the human species fails the test, becoming either extinct or less than human, it will serve us right. But the greater tragedy would be our legacy, because we'll destroy much of the world's life along with ourselves. I believe that any rational God would have intervened long ago to prevent the wholesale destruction of so many of His creatures. We have only ourselves and one another to blame and to rely on.

The anthropologist David Pilbeam has written that we have only just begun to tap the potential of the human brain. He had better be right.

Review Questions (Chapter 24)

If you were to visit North America a million years ago, during the Ice Ages, what would be your overall impression of the land mammals that you saw?

How did climate changes during the Ice Age affect life?

What's the evidence that humans were at least partly responsible for the extinction of large Ice Age animals?

Describe the science behind this limerick:

> The Ukrainian tundra was bleak,
> Till they found a solution unique.
> Then lack of supplies
> Soon caused the demise
> Of their mammoth construction technique.

Identify the organism featured in this limerick:

> The morning was hardly propitious
> When sailors discovered Mauritius.
> They captured the lot,
> Stewed then up in a pot,
> And pronounced them extinct, but delicious.

Review Questions (For the book)

THOUGHT QUESTIONS

What's the evidence for evolution? How do we know that the fossils we see in the rocks evolved one from another, and were not the result of separate creations?

Caves are very good places for preserving bones. But even though one can sometimes make very good collections of fossils from caves, the bones may not give a reliable sample of the ecology of the time. Why not?

Fact 1. A large herbivore needs gut bacteria to digest plant cellulose. Fact 2. Gut bacteria work best in a narrow range of temperature. Use these two facts to comment on the history of large vegetarian land vertebrates.

There are no obvious mechanisms for temperature control in early amphibians, or in later ones for that matter. Why, then, did their descendants the reptiles, the mammals, probably the dinosaurs and certainly the birds, bother to evolve thermoregulation?

Large vegetarian animals on Earth seem to be around 5 tons. First, list four examples of 5-ton vegetarians, extinct or living. Then please suggest reasons why there might be an optimum size for large vegetarians.

We know that continental movements affect life. Select one example where continental movements affected the history of life, and describe it.

What are turbinates, and why are they important in assessing the history of land vertebrates?

Now think turbinates again. How could sauropods grow to such enormous sizes, making them by far the largest land vertebrates of all time?

Further Reading

Ice Age Life and the Great Extinction

Bahn, P. G. 1993. 50,000-year-old Americans of Pedra Furada. *Nature* 362: 114–115. News report on this controversial site.

Barnosky, A. D. 1986. "Big game" extinction caused by late Pleistocene climatic change: Irish Elk (*Megaloceros giganteus*) in Ireland. *Quaternary Research* 25: 128–135.

Clausen, C. J., et al. 1979. Little Salt Spring, Florida: a unique underwater site. *Science* 203: 609–614.

Diamond, J. M. 1989. Quaternary megafaunal extinctions: variations on a theme by Paganini. *Journal of Archaeological Science* 16: 167–175.

Dillehay, T. D. 1999. The Late Pleistocene cultures of South America. *Evolutionary Anthropology* 7: 206–216.

Fisher, D. C. 1984. Mastodon butchery by North American PaleoIndians. *Nature* 308: 271–272.

Frison, G. C. 1989. Experimental use of Clovis weaponry and tools on African elephants. *American Antiquity* 54: 766–783.

Frison, G. C. 1998. Paleoindian large mammal hunters on the plains of North America. *Proceedings of the National Academy of Sciences* 95: 14576–14583.

Geist, V. 1986. The paradox of the giant Irish stags. *Natural History* 95 (3): 54–65.

Gladkih, M. I., et al. 1984. Mammoth-bone dwellings on the Russian Plain. *Scientific American* 251 (5): 164–175.

Guthrie, R. D. 1990. *Frozen Fauna of the Mammoth Steppe.* Chicago: University of Chicago Press. Paperback, excellent reading and excellent science.

Harris, J. M., and G. T. Jefferson (eds.). 1985. *Rancho La Brea.* Los Angeles: Los Angeles County Natural History Museum.

Kershaw, A. P. 1986. Climatic change and Aboriginal burning in north-east Australia during the last two glacial/interglacial cycles. *Nature* 322: 47–49.

Kurtén, B. 1976. *The Cave Bear Story.* New York: Columbia University Press.

Kurtén, B., and E. Anderson. 1980. *Pleistocene Mammals of North America.* New York: Columbia University Press.

Lewin, R. 1998. Young Americans. *New Scientist*, 17 October 1998.

Lister, A., and P. G. Bahn. 1994. *Mammoths.* New York: Macmillan.

Martin, P. S. 1990. 40,000 years of extinction on the "planet of doom." *Palaeogeography, Palaeoclimatology, Palaeoecology (Global and Planetary Change)* 82: 187–201. Includes summary of Pacific island extinctions.

Martin, P. S., and R. G. Klein (eds.). 1984. *Pleistocene Extinctions.* Tucson: University of Arizona Press. Includes Martin's overkill hypothesis.

Meltzer, D. J. 1993. Coming to America. *Discover* 14 (10): 90–97. Monte Verde.

Miller, G. H., et al. 1999. Pleistocene extinction of *Genyornis newtoni*: human impact on Australian megafauna. *Science* 283: 205–208, and comment, pp. 182–183.

Owen-Smith, N. 1987. Pleistocene extinctions: the pivotal role of megaherbivores. *Paleobiology* 13: 351–362.

Pielou, E. C. 1991. *After the Ice Age: The Return of Life to Glaciated North America.* Chicago: University of Chicago Press. Paperback.

Piperno, D. R., et al. 1990. Paleoenvironments and human occupation in late-glacial Panama. *Quaternary Research* 33: 108–116. Fire-clearing of forests probably began here about 11,000 BP.

Poplin, F., and C. Mourer-Chauviré. 1985. *Sylviornis neocaledoniae* (Aves, Galliformes, Megapodiidae), oiseau géant éteint de l'Ile des Pins (Nouvelle-Calédonie). *Geobios* 18: 73–107. The Du.

Sandweiss, D. H., et al. 1998. Quebrada Jaguay: early South American maritime adaptations. *Science* 281: 1830–1832, and comment, pp. 1775-1777.

Soffer, O., et al. 1997. Cultural stratigraphy at Mezhirich, an Upper Palaeolithic site in Ukraine with multiple occupations. *Antiquity* 71: 48–62.

Soffer, O., and N. D. Praslov (eds.). 1993. *From Kostenki to Clovis: Upper Paleolithic—PaleoIndian Adaptations.* New York: Plenum Press. Chapters by Olga Soffer and Lewis Binford.

Sutcliffe, A. J. 1985. *On the Track of Ice Age Mammals.* London: British Museum (Natural History).

Vartanyan, S. L., et al. 1993. Holocene dwarf mammoths from Wrangel Island in the Siberian Arctic. *Nature* 362: 337–340, and comment, pp. 283–284.

The Modern World (the last 5000 years or so)

Anderson, A. J. 1991. *Prodigious Birds: Moas and Moa Hunting in Prehistoric New Zealand.* Cambridge: Cambridge University Press.

Bahn, P. G. and J. Flenley. 1992. *Easter Island, Earth Island.* London: Thames and Hudson. See also summary of this work by Diamond 1995.

Black, F. L. 1992. Why did they die? *Science* 258: 1739–1740. Diseases that decimated Native Americans.

Burney, D. A. 1993. Recent animal extinctions: recipes for disaster. *American Scientist* 81: 530–541.

Cheke, A. S. 1987. An ecological history of the Mascarene Islands. Pages 5–89 in A. W. Diamond (ed.). *Studies of Mascarene Island Birds*. Cambridge: Cambridge University Press. The gruesome story of the end of the dodo of Mauritius, the solitaire of Rodriguez, and so on.

Culotta, E. 1995. Many suspects to blame in Madagascar extinctions. *Science* 268: 1568–1569. News update.

Diamond, J. M. 1986. The environmentalist myth. *Nature* 324: 19–20.

Diamond, J. M. 1989. The present, past and future of human-caused extinctions. *Philosophical Transactions of the Royal Society of London B* 325: 469–477.

Diamond, J. M. 1991. Twilight of Hawaiian birds. *Nature* 353: 505–506.

Diamond, J. M. 1992. Twilight of the pygmy hippos. *Nature* 359: 15.

Diamond, J. M. 1994. The last people alive. *Nature* 370: 331–332. Societies that self-destructed.

Diamond, J. M. 1995. Easter's end. *Discover* 16 (8): 62–69. The destruction of Easter Island by its own inhabitants. Required reading!

Flannery, T. F. 1995. *The Future Eaters*. New York: George Braziller. The history of Meganesia (Australasia and associated islands). Has much deeper significance than simply a regional history.

Flannery, T. F., 1996. *Throwim Way Leg: Adventures in the Jungles of New Guinea*. London: Weidenfeld and Nicolson. Enthralling account of an Australian zoologist in New Guinea discovering how easy it is for people to watch species go extinct without either noticing or caring very much.

Goldschmidt, T. 1996. *Darwin's Dreampond: Drama in Lake Victoria*. Cambridge, Mass.: MIT Press. The devastation of the astounding biology of Lake Victoria.

Gould, S. J. 1996. The dodo in the caucus race. *Discover* 17 (11): 22–33.

King, C. M. 1984. *Immigrant Killers*. Oxford: Oxford University Press. Mostly about New Zealand but with summaries of other island extinctions.

Montevecchi, B. 1994. The great auk cemetery. *Natural History* 103 (8): 6–9.

Quammen, D. 1996. *The Song of the Dodo*. New York: Scribner.

Reeves, B. O. K. 1983. Six milleniums of buffalo kills. *Scientific American* 249 (4): 120–135.

Steadman, D. W. 1995. Prehistoric extinctions of Pacific island birds: biodiversity meets zooarchaeology. *Science* 267: 1123–1131.

Steadman, D. W., et al. 1999. Prehistoric birds from New Ireland, Papua New Guinea: extinctions on a large Melanesian island. *Proceedings of the National Academy of Sciences* 96: 2563–2568. Another sobering assessment of the scale of prehistoric extinction on islands.

Stearns, B.P., and S. C. Stearns. 1999. *Watching, From the Edge of Extinction*. New Haven: Yale University Press. Cases of imminent extinction.

Invertebrate Paleobiology

This Appendix is not meant to be a substitute for a full account of invertebrate paleontology. Typically, invertebrate zoology and invertebrate paleontology are taught as complete courses in colleges and universities, for the very good reason that it takes that kind of depth of treatment before there is any real understanding of the topic. So texts on these subjects are several hundred pages long (three examples sitting on my desk are 382, 488, and 713 pages long).

The average person and the average student are not familiar with invertebrates, especially the marine animals that make up the bulk of the fossil record: they are the creatures that are squashed under our shoes as we scramble about the seashore. Even so, these groups were part of the great metazoan radiation, and they show the results of evolutionary processes just as much as the more familiar vertebrates do. Some groups have much more complete fossil records than vertebrates provide us, but they are difficult to read without some knowledge of the anatomy, physiology, and ecology of the group.

I have chosen to illustrate some of the fascinating aspects of invertebrate paleobiology by brief introductions to a chosen few fossil groups: five phyla, and eight subgroups of those phyla. I chose them because one can read a great deal of biology from their fossil remains. One can reconstruct the anatomy and ecology fairly well, and sometimes the behavior and physiology as well. It is not a coincidence that I have published research on five of the eight groups: I was attracted to them because of the prospect of delving deeply into the biology of their fossil members.

I did not include sponges, corals, or gastropods in my chosen groups. Sponges generally fall to pieces after death, and good fossils are very rare. Corals preserve well, but a fossil coral is only the carbonate cup that the living animal sat on, and it is difficult to read much of the biology of the animal from that. (My friend and former student John Pandolfi would probably disagree with that gloomy assessment!)

Gastropods are fascinating and varied in today's seas, but again, much of their biology is in the soft parts that are not preserved or reflected in the shell. I admire people who do their best to reconstruct the paleobiology of gastropods! So, on to the mini-essays...

UNDERSTANDING INVERTEBRATE PALEOBIOLOGY

To an anatomist or systematist, it is vitally important to know what animals are. To an ecologist and, I suggest, a paleobiologist, it is important to know what animals do (or did). So my discussion of fossil invertebrates is made from a functional approach. This approach, often called functional morphology, is out of fashion at the moment. Stephen J. Gould trashed functional morphology in an influential paper about 20 years ago, on the grounds that it was mere "adaptive story-telling." While zoologists and vertebrate paleontologists went on their merry way, analyzing the ways animals worked, invertebrate paleontologists largely stopped being paleobiologists, and worked on systematics, diversity counts, and descriptive anatomy rather than functional morphology. So read on: you may enjoy or detest this approach.

The most important fossil invertebrates in terms of abundance lived in the sea and had hard parts: this bias of the fossil record (Chapter 2) gives us a very long and very full history of the creatures that lived, interacted, and evolved in the shallow seas of our planet.

As each major group of invertebrates evolved hard parts (Chapter 4), it was enabled to perform a certain way of life better than before. The variety of invertebrate body plans suggests the variety of those ways of life, but we can only infer what those ways of life were by careful testing of hypotheses. The easiest method is to compare a fossil invertebrate with a living relative, noting similarities and differences. If no living relative is close enough, then there may be a living analog: a creature that looks as if it does the same sort of thing with a different set of hard parts. And if all else fails, one must analyse the fossil on its own terms, applying "reverse engineering" to try to reconstruct how it was formed as it grew, and what it did as it grew.

As we saw in Chapter 4, sponges and cnidarians are rather simple metazoans, but most invertebrate phyla are derived "worms", three-dimensional animals that are more differentiated. They have complex body parts, and those parts are integrated in intricate ways to form the functioning animal. Members of any given phylum can and do evolve into a variety of ways of life. Part of the challenge of invertebrate paleobiology is to recognize, describe and classify the range of anatomy within a given group. But to put it into a paleoecological context, it is important to understand what the animal did in its ecosystem. This is where functional analysis plays its role.

Some of the anatomy and function of an animal is related strictly to its soft parts: digestion and reproduction are simple examples. Snails (gastropods) have an enormous range of soft-part morphology contained within a relatively restricted set of shell shapes. On the other hand, some groups of animals have a rather uniform set of soft parts inside a great variety of shell forms: brachiopods, cephalopods, and echinoids are examples. For paleobiologists, the challenge lies in inferring the soft-part anatomy of a fossil from the hard parts that remain, and then putting together a reconstruction of the ecology of the creature. Historically, some reconstructions have been better than others, and it was some of the poor-quality efforts that led Gould to deride the whole enterprise.

If hard parts are internal, they serve primarily as support structures and/or bases for muscle attachment. Our own skeletons are good examples. In sponges, tiny pieces of mineral or organic solids help to prevent a large sponge from collapsing under its own mass, and from waving around like a lump of Jello in water currents. Essentially, hard parts allow sponges to grow to large size. In cnidarians, the hard parts of, say, corals lie under the animal tissue, forming a spiny or cup-like substrate into which the coral contracts to avoid being eaten, and a solid base that grows to project the little animal upward into food-rich water currents. Neither sponges nor cnidarians have to have hard parts, but those species that have evolved them find them useful in the way they exploit their environments.

If hard parts are external, they may provide protection, support, and many other advantages, but they also enclose the soft parts of the animal, making it more difficult for the animal to gather food and oxygen. Much of the morphology of shelled invertebrates involves solutions to the problem of interacting with the environment as needed, while retaining the advantages of living inside a hard skeleton.

In the following sections, I feature selected groups of invertebrates, describing their functional anatomy, and outlining their evolutionary history. Beware! The standard reference book, the *Treatise on Invertebrate Paleontology* (*Treatise* from now on) consists of literally dozens of volumes. I am skimming very lightly indeed here, just as I am in the rest of this book.

PHYLUM ARTHROPODA: TRILOBITES

Arthropods are covered by an external skeleton, which consists of an organic solid called chitin, often reinforced by calcite (calcium carbonate). Living arthropods include insects, mites, and spiders, and are ubiquitous on land; living marine arthropods include crustaceans and are ubiquitous in the ocean. I shall feature only trilobites, which were among the first metazoans to evolve hard parts. Looking back, trilobites seem simple arthropods, but that is only a comparative judgment in relation to the complexity of all the living groups.

Trilobites were covered with a protective exoskeleton, made up of pieces jointed together. The "head-shield" or **cephalon** was a massive structure that gave a solid cover over the first several segments. Even the eyes were solid parts of the cephalon, made of translucent material to allow light to shine through to the light receptors underneath (of course that is true of insect eyes too).

Behind the cephalon lay the **thorax**. It was much more flexible, because each piece was a separate sheet, covering only one segment of the thorax. Behind the thorax was a tail-shield, or **pygidium**, again one solid piece covering several segments.

Typically, a trilobite fossil consists only of this exoskeleton, but fossils like those of the Burgess Shale (Chapter 4) show us the soft parts that lay underneath. The exoskeleton provided a firm base for attaching walking legs driven by powerful muscles: each segment of the trilobite bore one pair of legs. Because the trilobite had to respire under its exoskeleton, each pair of walking legs also carried a pair of small gills. Trilobites had no claws or specialized appendages like those of crustaceans for dealing with food objects, so they must have eaten mud, or at best, soft prey like worms. The mouth lay under the middle of the cephalon, and a large stomach largely filled the center of the cephalon, with an intestine running the length of the trilobite's central axis to its pygidium. There are trace fossils that were made by trilobites walking and digging shallow pits in the soft sea-floor.

Trilobites were extinct by the end of the Permian, and were rather rare after the extinction at the end of the Devonian (Chapter 6). They are abundant and visible fossils in Cambrian rocks, dominating most museum collections from that period. They became more diverse into the Ordovician, as did most marine groups, but declined in relative diversity and abundance after that.

They continued to evolve, however, and some of the most remarkable trilobites are Devonian.

Cambrian trilobites were no doubt able to use the flexibility of the thorax to roll up in defense, but the very small pygidium probably did not fit very well against the underside of the cephalon when they did this. The exception was a group of tiny trilobites, the agnostids, which folded up like a snap purse, and had cephala and pygidia that are very difficult to tell apart in many specimens.

Ordovician and later trilobites had much larger pygidia, and many of them folded up with a very tight match between pygidium and cephalon, making an almost perfect seal. Only tiny spaces for respiration were left open, and these were often protected by spines.

A trilobite cannot grow unless it sheds its old solid exoskeleton and grows a new, larger one. Crustaceans do this, of course, though insects do not. For trilobites, this meant special adaptations for molting. The cephalon was the most difficult piece to shed, because its edges curled around to form a protective frame on the underside of the head. Most trilobites evolved a line of weakness across the cephalon so that it would break easily at molting time, and that line of weakness typically curved around the eye, which had to molt off the transparent outer lens cover without damaging the permanent photoreceptors in the process.

An individual trilobite that survived to large size had left many (sometimes dozens) of cast-off molted exoskeletons as potential fossils. Baby trilobites had no thorax, and they added a new thoracic segment every time they molted, so we have been able to collect and understand many of the juvenile stages, not confusing them with fossils of adult but naturally small trilobites.

Some trilobites had wonderful vision. The Devonian group of phacopid trilobites had many lenses in each eye, but each lens transmitted a perfect picture, rather than the one-spot-of-light that each insect lens transmits. The result is that while a bee or a beetle sees one picture per eye, a phacopid trilobite saw dozens of pictures per eye. While one could argue about the reason for this adaptation (in "adaptive story-telling"), its existence is real and could not have been recognized unless paleontologists had looked at the eye and asked why the structure was so complex. Research put together by several paleontologists suggests that each eye could see in stereo!

PHYLUM BRACHIOPODA: BRACHIOPODS

Brachiopods have a shell consisting of two **valves** that enclose the animal completely. That means that brachiopods can be very well protected, but cannot feed without opening the shell. The feeding organ, the **lophophore**, consists of a cartilage-like base that supports an array of flexible **filaments** (the best analogy is a comb). Each filament carries **cilia** that beat in such a way as to bring water currents through the space between the filaments. As they pass through the gap, any food particles (microorganisms) are caught and passed down the filament to its base and then to the mouth, while unwanted particles (say grains of silt) are allowed to pass. The brachiopod gathers oxygen from the water current, too.

The lophophore lies completely inside the shell. The brachiopod opens its shell a little to feed, almost always along a hinge between the valves. The lophophore is arranged in a complex shape so that the overall pumping action of the cilia draws new water in through the sides of the shell opening, through the filaments, while filtered water is pushed out of the shell along the midline. This system probably dictates the fact that brachiopods are symmetrical around this midline, in both shell shape and internal anatomy.

Brachiopods take in very high-quality food, which they can digest almost completely. So most brachiopods have a blind-ending gut, occasionally regurgitating small packets of waste material. The shell interior is almost completely occupied by the lophophore, with only a tiny body at the back of the shell near the hinge, and a very thin layer of tissue, the **mantle**, which lines the inside of the shell and secretes the crystals that form it.

Most of the body mass consists of the muscles that open and close the shell, and many fossil brachiopod shells carry the scars of those muscles, so that we can reconstruct the internal anatomy accurately. The muscles are attached either to the internal surface of the valves, or raised on solid calcite platforms that give them better leverage as they open and close the shell.

Many brachiopods also built a fragile calcite framework to help hold the lophophore in place, and we can sometimes find that structure in fossils and reconstruct the water pathway through the filter of the lophophore.

Once the larval brachiopod settles on the sea floor after a brief time in the plankton, it does not move any more. Most living brachiopods are fixed to the sea-floor or to other shells by a **pedicle**, a leathery holdfast that emerges through an opening at the back of the shell. These brachiopods can alter the attitude of the shell by contracting or relaxing the muscles attached to the base of the pedicle. Other brachiopods secrete a cement that attaches one of the valves permanently in one place in one position.

Brachiopod anatomy has not changed much since the Paleozoic, so we have a reasonable understanding of brachiopod ecology through time. Brachiopods occur in early Cambrian rocks, but they diversified dramatically in the Ordovician and were extraordinarily abundant on Paleozoic sea-floors. They suffered severely in the Permian extinction, and although the survivors had a subdued recovery, brachiopods are minor components in only a few marine settings today and are absent from most. However, they have been very well studied because we have 540 million years of fossil record for the group. Their history is hard to document in a few sentences (it took three volumes of the *Treatise* even back in 1965).

As a gross generalization, Paleozoic brachiopods included several groups with flat shapes with straight hinges. Many of them must have had a way of life lying freely on the sea-floor like little saucers, or cemented to rocks and reefs rather like some oysters do today. Other Paleozoic brachiopods were rounded, with pedicles attaching them to hard objects, often other brachiopods!

Shallow-water brachiopods, especially reef-dwellers, were hard hit in the extinction at the end of the Devonian (Chapter 6), but recovered to be spectacular members of Permian faunas. However, the Permian extinction wiped out most brachiopods, especially (again) those that lived in Permian tropical environments such as reefs. The survivors of the Permian extinction tended to be rounded brachiopods with pedicles: in other words, hard-substrate specialists. In addition, most living brachiopods live in cool waters (higher latitudes and deep waters), and only survive in the tropics in sheltered reef caves.

Today, brachiopods are specialists in nutrient-poor environments. They have low metabolic rates, very small bodies for the size of their lophophores, and may be able to survive in low-nutrient environments where more energetic creatures would starve. That may not have been true of the many Paleozoic groups that are now extinct, but their anatomy suggests that it was.

PHYLUM BRYOZOA: BRYOZOANS

Bryozoa (bryozoans) are lophophorates: that is, they feed and respire by using a row of filaments arranged around the mouth. This feature means that they are related to brachiopods, but the two phyla are otherwise very different. Essentially, however, bryozoans can be understood by regarding them as miniaturized brachiopods.

All bryozoans are tiny, and they always form **colonies**, made up of a set of individuals (called **zooids**). The colony is founded by a larva settling on the sea-floor, and all other individuals in the colony are produced by successive budding of new zooids. So all zooids in a colony are clones.

Although some zooids in some bryozoans are specialized as reproductive individuals, most zooids have a lophophore for feeding. The filaments, arranged in a ring, generate a current that pulls water downward into the lophophore, and particles in that current are swept toward the mouth. The particles are accepted or rejected there.

Bryozoan zooids usually secrete calcite tubes or boxes as protective skeletal covers. Much of the morphology of a zooid deals with the way that the lophophore is extended into the water for feeding, and retracted for safety. Different bryozoan groups extrude and retract the lophophore in different ways: some use a muscle system, others use sophisticated hydraulic systems, using internal coelomic fluid or external seawater as a working fluid. Some bryozoans have hinged lids for their zooids.

A zooid is so small that it does not need any special respiratory or circulatory system. At the same time, its small size makes it very vulnerable to predation, and the tiny lophophore cannot generate a powerful feeding current. Bryozoans therefore provide a good example of the value of modular construction: a colony is much more than a set of individual zooids.

Because the zooids are clones, the bryozoan colony can be built to promote the welfare of the whole colony, rather than that of the individual zooids. For example, many bryozoans build a calcite skeleton for the colony that may seal off (and kill) some early-formed individuals, because the colony as a whole benefits.

A colony of zooids, each secreting a tiny calcite **zoecium**, builds a structure that can be large and strong, providing more protection from predators, stability against wave action, and resistance to sediment abrasion than could have been built by a single zooid. In particular, there is a hydrodynamic advantage for a colony. Many zooids, each one pulling water into its lophophore, can generate a feeding current over the colony that is much more powerful than simple addition would suggest: thus a bryozoan colony benefits from size (and age).

There is a cost, however: the powerful colony-wide feeding current is only effective if all the zooids pump at the same time, and only if the zooids are arranged so that their pumping does not cancel out. In other words, there must be a *colony-wide* pumping system, reflected in the way that the individual zooid are physically arranged.

All these factors result in morphologies that are adaptive not only for in individual zooids, and also on a colony-wide basis. For example, tissue connects one zooid with another (they did bud from one another in the first place). Such extra-zooidal tissue can be adapted to secrete calcite outside the regular system of zooids, for example, to form additional pads of calcite to make the colony stronger; to form root-like structures that can anchor the colony on a soft substrate, or cement it to a hard one; or to form cross-struts between the branches of a colony.

The shape and structure of the colony may be adaptive to the environment: thus a colony may be a low encrusting form in rough water, or a tall, fragile, branching structure in quiet water.

Bryozoans are known from Ordovician times to the present. **Stenolaemate** bryozoans are dominant from the Ordovician to the Cretaceous. They have rather simple tube-shaped zooids, and the lophophore is extruded by a simple ring of muscles. Though they also have a long history, **gymnolaemate** bryozoans radiated in the Cretaceous as part of the Mesozoic Marine Revolution (Chapter 5). They have box-like zooids with hinged lids and a hydraulic apparatus for operating the lophophore, and in general show greater specialization among the zooids of a colony.

Bryozoans were a major component of many shallow marine faunas from Ordovician to Recent. But they are difficult to identify for the average paleontologist, and tend to occur as large piles of shell debris, broken off the original colonies, rolled around and worn down to irregular fragments. Only the comparatively rare cases of excellent preservation have allowed the detailed work on functional analysis mentioned here.

PHYLUM MOLLUSCA: BIVALVES

Molluscs are shelled creatures that have diverged so much from one another that we cannot draw, even in theory, a satisfactory common ancestor. The first molluscs appear in the early Cambrian, but several distinct groups had already evolved by that time, and the lineages that gave rise to them are not preserved.

Bivalves, as the name suggests, are molluscs that live inside a shell made of two valves secreted (rather as in brachiopods) by a thin internal sheet of tissue, the **mantle**. Like brachiopods, bivalves must open and close the shell to feed, respire, reproduce, and so on. Unlike brachiopods, the bivalve shell is not opened and closed by opposing muscle systems. Instead, a bivalve closes its shell with powerful **adductor** muscles that tie the valves together. As the valves close, they compress an elastic **ligament** that lies close to a hinge line at the back of the shell. When the adductor relaxes, the ligament springs the valves open. (That is why dead bivalves have the valves gaping open.)

The way the hinge functions is probably quite complex, but difficult to analyze in living bivalves. The hinge structures must guide the valves from the back so their edges meet precisely at the front. Various styles of hinge **teeth** have evolved in different bivalve groups to do this, and some of them apparently define clades very well; other styles of hinge have evolved more than once.

Bivalves respire with a gill that looks superficially like a brachiopod lophophore. The **gill** is arranged so that beating cilia generate a water current that is drawn into the shell, processed, and then pumped out in another direction, to avoid recycling the water. However, bivalve gills are more complex in structure than lophophores, and their filaments are often interconnected to make the pump stronger. The result is that bivalves can generate a much more powerful water current than any brachiopod. There are different types of gill among living bivalves, and although they are useful in classification, they never fossilize.

Most living bivalves filter the water current for food, thus living on plankton or tiny suspended debris particles. Particles are sorted partly at the gills, and partly by fleshy lobes (**labial palps**) near the mouth. Some bivalves use large labial palps to feed directly, extending the palps outside the shell to gather mud and food from the sediment.

Most living bivalves have a **foot**, a fleshy organ which can extend outside the shell into the sediment. By using the foot, bivalves can move slowly, digging into the sediment for a hold, and pulling themselves along. Usually, the foot is used for burrowing into the sediment until the shell is largely or completely under the surface (presumably for protection). The procedure is complex, because each cycle of digging, holding, and pulling requires the opening and partial closing of the hinge, coordinated between the adductor muscles and the muscles that operate the foot.

Burrowing into the sediment cuts the bivalve off from direct access to sea water (and oxygen and food), so the edges of the mantle are often extended into tubes (**siphons**) that project upward to the surface and guide water currents down into the shell cavity and back out again. Sometimes a bivalve may live in a deep burrow, feeding and respiring through siphons that are much longer than its shell. They can be retracted into the bivalve's burrow but not into the shell. Bivalves that do not burrow, but attach themselves to hard substrates, may still have siphons to guide the water currents in and out.

The powerful gills and the siphon system have allowed bivalves to explore shell shapes that are not possible for brachiopods. The relative amount of burrowing activity can be read in the sophistication of the hinge and the size of the adductor muscle and ligament. The relative development of the siphons can be read in the presence or absence, and size, of muscle attachment lines on the shell. The shell shape reflects life style (burrowing, digging, attached, and so on). So for bivalves, as for brachiopods, we can put together a reasonable ecological history of the group.

We do not understand why bivalves are not as abundant and diverse in Paleozoic seas as the brachiopods. It appears that they did not explore the burrowing way of life extensively in the Paleozoic, though many of them were shallow burrowers. The Paleozoic extinctions did not affect them greatly, however, and after the Permian extinctions, the bivalves radiated in Mesozoic seas while the brachiopods did not recover well.

The greatest radiation of bivalves, in Cretaceous times, is linked with adaptations that allowed them to burrow deeper and faster, especially the evolution of a powerful foot and long and complex siphons. Today, bivalves are abundant in all sea-floor environments, from shoreline to deep sea, from Equator to Pole, and even in hot springs along the volcanic mid-ocean rifts.

PHYLUM MOLLUSCA: CEPHALOPODS

Living cephalopod molluscs include soft-bodied forms like octopus and squid, whose fossil record is very poor. Fossil cephalopods include many extinct shelled forms, which have only a handful of survivors that include the pearly nautilus.

Cephalopods are the most sophisticated of molluscs in terms of intelligence, mobility, and complex behavior, though these are attributes that are practically impossible to detect or test in the fossil record.

The most obvious soft-body character of cephalopods is the **head**, with large eyes and a mouth that contains a powerful beak. Cephalopods are predatory carnivores for the most part, and they use sophisticated swimming to search for their prey. The prey is caught and handled with the **arms**, typically 8 or 10, that surround the mouth.

The body contains the gut and reproductive organs, and large **gills** to provide the oxygen for locomotion. The gills lie in a large **mantle cavity**, and they are force-fed water in a pumping action that involves muscles, not simply cilia. Cephalopods have made use of that respiratory pumping for locomotion, to "jet" water out of the mantle cavity as needed, giving a powerful acceleration to the body. The water is ejected through a moveable **funnel**, so that the animal can accelerate in a wide range of directions. This jetting is sometimes used in emergency, or in pursuit, but it requires a lot of energy, so many cephalopods swim slowly, using fins or arms instead.

Nautilus and almost all fossil cephalopods are preserved because they secrete a chambered shell that sits behind the body. The animal lives in the last-formed chamber, and the others are progressively abandoned and largely sealed to form the beautiful coiled shells of nautiloids and ammonites (often prominently displayed in museums and rock shops).

The chambers are not entirely sealed, however. They continue to play a vital role in the biology of the animal. A tube called the **siphuncle** connects all chambers with the back of the body. As they are formed during growth, the chambers are originally filled with body fluid, but typically that is pumped out through the siphuncle under the control of the animal, to maintain the overall density of shell + body close to that of seawater. This gives the animal neutral buoyancy, and by pumping fluid in or out of chosen chambers, it can balance its body in the right attitude in the water. All this is as vital to a *Nautilus* as it

is to a submarine or a SCUBA diver. The anatomical and biochemical control system is complex and sophisticated, and that must have been as true for Ordovician nautiloids as it is for living ones.

The earliest cephalopods (from the late Cambrian) were small nautiloids, and their shells show that they were still in the process of perfecting the buoyancy and balance system (it naturally becomes more complex as the animal evolves to large size).

By the Ordovician, however, nautiloids had evolved to very large size. Many were long and straight, and they secreted shell within the chambers to form permanent balance weights at the back of the shell that could not be removed but only increased. We classify the Paleozoic straight nautiloids on characters associated with their different solutions to the buoyancy and balance problem.

Many Middle Paleozoic nautiloids had evolved the trick of coiling the shell into a spiral. This probably made balance easier, and attitude more stable in the water, but the system needed more sensitive adjustment that could not be solved by simply dumping shell into the chambers. We classify these nautiloids on structures associated with the siphuncle.

Ammonoids evolved from nautiloids in the Middle Paleozoic. Ammonoids had sophisticated chamber walls that met the outer shell along complex, convoluted lines that make for easy categorization of families and genera. Although they were never abundant in the Paleozoic, ammonoids recovered quickly from the P–Tr extinction whereas nautiloids recovered slowly and at low diversity. Successive groups of ammonoids culminated in a flourishing radiation of **ammonites** in Jurassic and Cretaceous times. Ammonites are so abundant, and evolved so quickly, that they are the most important large invertebrates used to subdivide Mesozoic time in marine sediments (into ammonite zones that can be recognized over vast regions).

The **belemnites** were squid-like cephalopods, also abundant in Mesozoic seas, and well preserved because they secreted hard posterior calcite weights for balance.

Ammonites and belemnites became extinct at the K–T boundary, a difficult event to understand given their dominance in Cretaceous rocks. A few surviving belemnoids (relatives of belemnites) radiated into living squid; the long-surviving nautiloids are represented by *Nautilus*; and no-one can trace the ancestry of octopus.

PHYLUM ECHINODERMATA

Echinoderms live inside calcite boxes (**tests**) made up of many little plates that enclose the body almost completely. There is no hinge (like brachiopods or bivalves), no huge opening that contains a head and arms (like cephalopods), and no protected underside (like that of trilobites). Echinoderms live inside their boxes, and communicate with the outside world through little holes. This means that they have a good fossil record, because the calcite boxes preserve well, and the holes preserved in them can tell us a lot about the biology of extinct echinoderms.

The mouth is small, and while it may be underneath or on top of the animal, it can take in only small items of food. These vary from plankton and particles filtered from the water, to shreds of algae and diatom scum from rocks, to organic mud from the sea floor. This means in turn that fecal matter is not great, even for sediment feeders, so the anal opening is also small.

Like many marine invertebrates, echinoderms reproduce by spawning quantities of eggs and sperm into the water, so there is usually an identifiable pore or set of pores used for this purpose.

Echinoderms communicate with the outside world by using little hydraulic pipes called **tube-feet**. They are offshoots of a **water-vascular system** of larger internal pipes containing body fluid under pressure. Each tube-foot extends to the exterior through holes or slits in the test. It is actuated by hydraulic pressure and muscles set into its walls, and often carries a tool at the end: a sucker, a pincer, or a set of cilia.

Tube-feet have evolved to carry out dozens of necessary tasks in different echinoderms. They may have pincers to hold small objects or to nip at predators. They may carry body fluid in and out of the shell to be re-oxygenated from sea water. They may carry cilia that generate water currents for respiration or feeding. They may move sediment so that the echinoderm can burrow.

The most familiar example of an echinoderm is a starfish. The box that is the body is extended into "arms" that carry great numbers of tube-feet, especially on the underside. Little suckers on the ends of the tube-feet allow the starfish to walk along the seafloor and to grasp prey. The tube-feet are arranged in groups, but each tube-foot is individually controlled, and can be extended or retracted by increasing or decreasing the hydraulic pressure of the body fluid inside it.

Many starfish catch and open prey using tube-feet. Other echinoderms use tube-feet to rummage in the sea-floor sediment, pick up food items, and push them into the mouth. Other echinoderms filter food from the water by extending arms that carry tube-feet specialized for generating water currents, filtering them, and carrying the food back along special grooves to the mouth. With all this flexibility in potential range of design, echinoderms have evolved into many different body patterns in sea-floor ecosystems. Usually, but not always, the echinoderm body plan is pentameral, so the calcite box has pentameral symmetry.

Echinoderms did have thin tissues extending outside their boxes. The calcite of the box is porous so that tissue runs through it, to add more calcite to the test, to grow the edges of the plates as the box increases in size, and to repair any damage. Often there are spines or arms on the outside, and these too are sheathed in the very thin tissue that forms them. Muscles may move the spines. Always, however, the tissue exposed to the outside is minimized, to make it uneconomical for potential predators to feed on.

Because of the inherent limitations of living inside a box, echinoderms are particularly good at living on small particles: eating sea-floor sediment, and filtering food from the water. The major Paleozoic groups seem to have been largely filter-feeders, and most of them lived attached to the seafloor, often with a body raised on a stalk to take advantage of water currents. The Blastozoa and Crinozoa explored variations on this way of life.

The Echinozoa, especially the Echinoidea or sea-urchins, radiated greatly during the Mesozoic to become mobile feeders on and in the sea-floor. Some graze on algae, macerating them into a paste that is easy to ingest. Others select food particles from seafloor sediment.

The Asterozoa include starfish and brittle-stars, and most of them are bottom-feeders. Many starfish have extended the echinoderm repertoire of ecology by becoming carnivores, seizing and pulling open bivalves with their tube-feet. Because of the limitations of their feeding structures, these starfish have to extrude the stomach to digest the prey.

Several other Paleozoic echinoderm groups were fascinating but never numerous or diverse. Some remain to be properly interpreted in functional terms.

PHYLUM ECHINODERMATA: BLASTOZOANS

Echinoderms occurred in early Cambrian rocks as the enigmatic helicoplacoids that are not well understood (Chapter 4). However, by late Cambrian and early Ordovician times, an array of echinoderms had already evolved into rather specialized and ecologically diverse groups.

The blastozoans and crinozoans evolved two different ways of filter-feeding from plankton. Both groups had generally globular bodies, and most of them attached themselves to the sea floor, yet raised themselves off it by forming some sort of stalk. The simplest stalk consists of a series of cylindrical calcite disks formed one after the other on the underside of the body, held together by skin, ligament, and/or muscle. (Muscle is required if the animal is to have any control over the attitude of the stalk, but that might not be necessary—almost all plants work well without significant control.) The more sophisticated crinozoans clearly had better control over their stalks than blastozoans, because the disks are shaped, with facets that allowed movement, and there are scars where ligaments and/or muscles ran up and down the stalk.

The stalk raised the body up off the sea-floor to give access to food in the plankton. Cystozoans and blastozoans both projected structures into the water, and collected food with them. However, the feeding and respiratory systems were different in the two groups, and the characters associated with them allow us to differentiate between and within them.

Blastozoans had **brachioles** rather than arms. That means that the brachioles did not have tubes from the water-vascular system running along them, and that means that brachioles did not bear tube-feet. Instead, they probably pumped water and trapped food items by cilia and mucus alone, and probably relied on water currents around them for major water flow. The brachioles were arranged in sets to give effective water filtering, and to allow efficient transport from their bases into the body.

Because there was no water-vascular system in the brachioles, blastozoans had to have a respiratory system that was separate from the feeding system. Holes in the test allowed tube-feet to project, but for efficient oxygen exchange, random arrangements of such holes are not sufficient. Various special systems characterize the various blastozoans.

The Diploporita (Ordovician to Devonian) had pairs of pores running through the plates of the test. Presumably a tube-foot or similar structure passed body fluid up one pore and down the other; or passed seawater down one and up the other. At first the pore-pairs were distributed over the test, but later they were arranged in close groups, organized so that water currents flowing over the test made sense (no reuse of deoxygenated water).

The Rhombifera (Ordovician to Devonian) had the pairs of pores lying close together and parallel, so that a lot of oxygen exchange could be accomplished in a localized but very efficient part of the test. The groups are set in a pattern that forms a diamond shape, giving the name to the group. Once we understand the rhombs, it is clear that the respiratory currents were efficiently directed all around the test (no reused water).

The Blastoidea or blastoids (Silurian to Permian) are much more sophisticated. First, the brachioles are more tightly organized, and arranged so that one can reconstruct water currents (for feeding) directed downward on to particular zones on the top of the test. Right by the bases of the brachioles, and in the direct path of this water, are entrances to an internal set of chambers with convoluted folds. These chambers are internalized respiratory structures (**hydrospires**) where water was used for respiration, then funneled out through rather large holes that flush it away from the test and away from the incoming food- and oxygen-laden water. This elegant arrangement makes for elegant little fossils that are very abundant in many Late Paleozoic rocks. The blastoids survived to the end of the Permian while other blastozoans became extinct by the end of the Devonian. It is tempting to speculate that the blastoids survived longer than the others because of their elegant solution to the problems of feeding and respiration.

PHYLUM ECHINODERMATA: CRINOZOANS

Crinozoa (**crinoids**) have a body plan and ecology basically like blastozoans: that is, they filter-feed from sea water, raising the test on a stalk and using arms to gather food. However, there are significant differences.

Most important, the arms of a crinoids are real **arms**, like those of starfish: they contain the water-vascular system. They carry tube-feet, so have the potential to be better at generating water currents and gathering food items. And they can serve a dual purpose, because they can also serve as respiratory structures.

There is a further benefit to the crinoid because of this dual function: no other part of the test needs to have significant openings in it. As we saw in blastozoans, the Diploporita and Rhombifera had many holes for their pore-pairs, even if they were concentrated into rhombs in the Rhombifera. The blastoids organized their respiratory structures close to the brachioles, but they still had openings in the test for respiratory currents. In crinoids, those currents are structly external, and respiration takes place on the arms. The crinoid test is strong and thick, often rigid, with well-defined patterns of large plates that persist throughout the history of major clades. All the major remaining perforations, the mouth, the anus, and the pore for the gonads, are in the roof of the test, protected by the arms when they are folded down.

The arms of crinoids can and do branch, while the brachioles of blastozoans apparently could not. So crinoids may grow dozens of arms that form a wide, dense filter across water currents. Some crinoids had patterns of arm branching that gave them mathematically optimal gathering structures.

Crinoids have a geological record that extends from Ordovician to today (with a possible Cambrian form too). They occur in dense clusters on the deep ocean floor, in tropical reefs, and in many fossil localities. The disks from their stems formed great banks of debris on the sea-floor in Carboniferous times. Paleozoic crinoids seem to have adapted quite tightly to filtering at different levels in the water, so that one can infer a layered or "tiered" structure to crinoid communities, with long-stalked crinoids filtering high, and short-stalked crinoids feeding lower in the water column.

The Paleozoic crinoids were filter-feeders attached by stalks to the sea floor. They were devastated in the P–Tr extinction, with only a handful of species surviving to become the ancestors of later forms. The post-Permian crinoids are "articulate": they have very flexible arms that are often used for feeble swimming, or for crawling along the sea floor to find an optimum site for filter-feeding. (Of course, mobile crinoids have no stalks.) Modern stalked crinoids are often deep-water forms, while shallow-water crinoids move around. Some Cretaceous rock beds have hundreds of stalkless crinoids on them, suggesting that these crinoids floated as plankton in the open sea.

PHYLUM ECHINODERMATA: ECHINOIDS

Formally speaking, the Echinozoa are a clade of echinoderms, most of which live freely, not attached to the seafloor. By far the most important of these groups is the Echinoidea, or sea urchins.

"Regular" echinoids have a test that is generally rounded, close to globular in some cases. The mouth is centered in an opening on the underside (the **peristome**, at the south pole of the creature), and columns of plates radiate from the edges of the peristome in five-fold symmetry, running all the way past the midline to the top (the north pole), where the anus is situated. These plates are in alternating bands called **ambulacra** and **interambulacra**. The ambulacral plates are pierced by holes that allow tube-feet to emerge to the outside of the test for varied functions associated, as usual, with locomotion, feeding, respiration, and so on. The interambulacral plates act as strong spacers between the ambulacra. All the plates may carry spines, but the spines are often larger on the interambulacral plates, which themselves tend to be larger and stronger than ambulacral plates, and can stand more leverage on them.

Regular echinoids rely mainly on spines for locomotion by levering on the substrate. That style of movement works best on hard substrates, and in turn requires long tube feet to hold the echinoid down while the spines push. There are all kinds of variations in the density, type, and arrangement of spines to suit the specific habitat of the echinoid, and in turn, the spine morphology is reflected in the type, size, and arrangement of the ambulacral and interambulacral plates that the spines are set on. The echinoid test looks deceptively simple, but a lot of interesting research projects remain to be done!

Regular echinoids have calcite plates inside the peristome that are connected by muscles and tendons, allowing them to act together as jaws that can scrape, grind, and tear at pieces of food. Regular echinoids are bottom feeders, scraping or tearing algae, and eating small pieces of food that are picked up and passed to the mouth by feeding tube-feet. Food is passed along the gut, and feces are passed out through the anus at the top of the test.

The earliest echinoid is Ordovician, but echinoids are neither numerous nor varied in Paleozoic rocks. All Paleozoic echinoids are regular, and came very close to extinction at the end of the Permian. Only a few species survived.

During the Triassic and afterward, regular echinoids radiated. One group, the cidaroids, remained very much like their Paleozoic ancestors, but others evolved to multiply their ambulacral plates, thus increasing the density of the tube feet in the test surface.

The most exciting of all echinoderm evolutionary radiations occurred in the Jurassic and Cretaceous within the "irregular echinoids." These echinoids broke away from the radial, globular body form of regular echinoids, and in doing so were able to explore radically new ways of life.

Irregular echinoids are bilaterally symmetrical. One ambulacrum is always set facing "forward," and the anus migrates in the other direction so that it lies in the posterior half of the test, often right at the posterior end. On the underside, the mouth is set toward the front of the test, and irregular echinoids tend to move "forward" in and on the sea-floor.

Most irregular echinoids are burrowers, using spines and tube-feet to move sediment aside in a complex and sophisticated pattern of movement. The ambulacra naturally evolve to place long, large respiratory tube-feet in the upper (northern) half of their path from pole to pole, and this concentration is often shown very clearly by tight petal-shaped patterns of pores near the top of the test. At the same time, tube-feet near the mouth become specialized for picking up food particles from the sediment, and tube-feet near the anus may build a special "sanitary burrow" that extends backward from the test. A diet of selected soft organic fragments may allow the jaw to be greatly reduced or even lost. The whole test may become shaped to lay out the respiratory and feeding structures more effectively, and to allow deeper and deeper burrowing. The more complex the burrowing adaptations are, the more visible they are in the test shape, and in the patterns of pores and spine bases on the plates. Several classic examples of increasing adaptation to burrowing have been documented in echinoid clades.

Some irregular echinoids have adapted to shallow burrowing in sand. "Sand dollars" evolved in the Cenozoic, and have very flattened tests with numerous short spines. They often show distinct grooves on the underside, which are used to transport food along the under surface of the test to the mouth.

Glossary

ABW. A mass of water produced along the edge of Antarctica as ocean water is severely cooled. ABW then drops to the sea floor and spreads over much of the ocean floor.

Acanthodians. Small fishes that probably represent the most primitive of the jawed fishes or gnathostomes.

Acoels. Group of flatworms, possibly the closest living group to the ancestor of three-dimensional metazoans.

Acritarchs. Fossil protists that appear in the fossil record around 1600 Ma. They are the most abundant fossils of Proterozoic times. They are probably the resting stages or cysts of photosynthetic protists.

Actinistia, see Coelacanths

Actinopterygians. Ray-fin fishes. A clade of jawed bony fishes, with fins stiffened with many tiny radiating bones.

Adapids. Major extinct clade of early prosimian primates.

Adenine. One of the bases of nucleic acid.

Adenosine triphosphate, see ATP

Aetosaurs. Group of thecodonts, typically 4-legged predators.

African apes, see panids

Agnatha. Jawless fishes.

Aistopods. A small group of early aquatic amphibians that lost their limbs early in amphibian evolution.

Alanine. A base of nucleic acid.

Albedo. In climatology, the reflectance of the Earth's surface.

Algae. In informal use, "seaweeds". Photosynthesizing protists in aquatic habitats.

Alligators, see crocodilians

Alvarezsaurids. Small Cretaceous clade of bird-like dinosaurs (or dinosaur-like birds), very close to the transition between these groups.

Amber. Tree resin that is hardened to the point that it can be preserved in rocks through geologic time. Often contains beautiful fossils.

Amino acids. Important class of organic molecules, often formed abiotically. Proteins are polymerized amino acids.

Ammonia. Compound of nitrogen and hydrogen, thought to be important in the origin of organic compounds and of life.

Amniotic egg. Vertebrate egg that contains an embryo and an internal support system for it; often surrounded by a shell and always by a membrane.

Amoeba. A genus of heterotrophic protist that never forms a skeleton. Often featured in introductory biology classes.

Amphiaspids. Clade of early jawless fishes.

Amphibians. Major clade of vertebrates. In cladistic terms, tetrapods; in informal use, tetrapods that are not amniotes.

Amphioxus. Informal name for *Branchiostoma*, a living chordate that looks similar to the ancestors of vertebrates.

Anapsid. Term for the skull of some reptiles. Anapsid skulls have no hole behind the eye socket.

Angara(land). Name for an ancient Paleozoic continent that consisted primarily of Northern Siberia.

Angiosperms. Flowering plants. Defined by their evolution of highly protected seeds, rather than their evolution of flowers.

Ankylosaurs. Small clade of medium-sized, heavily armored ornithischian dinosaurs.

Anomalocarids. Very large Cambrian predators from the Burgess Fauna, possibly the sister group of arthropods.

Antarctic Bottom Water, see ABW.

Anteaters. Mammals specialized to tear open ant hills and eat ants. They are not a clade: this habit evolved at least twice.

Anteosaurs. Small clade of carnivorous therapsids.

Anthracosaurs. Group of early amphibians, probably close to the ancestry of reptiles.

Anthropoids. The higher primates: monkeys and apes.

Antiarchs. Clade of placoderms: small, armored, slow.

Archaea, see Archaebacteria

Archaean Era. Period of Earth history from 3800 Ma to 2500 Ma, a definition that takes geologists from the earliest known rocks to the beginning of the Proterozoic Era.

Archaebacteria. Clade of prokaryotes that are now separated from other bacteria. They may be the most primitive living cells, and include methanogens.

Archaeocyathans. Group of Cambrian sponge-like fossils. They are most likely genuine sponges that became extinct.

Archosauromorphs. Major clade of diapsid reptiles that may include turtles as well as rhynchosaurs and archosaurs.

Archosaurs (Archosauria). Major clade of diapsid reptiles that includes dinosaurs, pterosaurs, crocodilians, and birds.

Arctocyonids. Early mammals that probably had a raccoon-like ecology. Possibly the ancestral group for ungulates.

Armadillos. Primitive living mammals whose ancestry lies in South America with the large armored glyptodonts.

Arsinoitheres. Large-bodied extinct herbivorous mammals that evolved in Africa in early Cenozoic times.

Arthrodires. Clade of large, powerfully-swimming placoderm fishes.

Arthropods, Arthropoda. The jointed animals, a major phylum of metazoans that includes insects, crustaceans, spiders, trilobites, and so on. Arguably has always been the metazoan phylum with the greatest number of species, since its appearance in the Cambrian.

Artiodactyls. A large group of mammals. Most are herbivores and have an even number of toes: they include cattle, sheep, deer, antelope, etc.

Asteroids. Large bodies of rock orbiting the Sun in interplanetary space, most of them probably formed during the very early history of the Solar System.

ATP. Adenosine triphosphate, an organic molecule that stores a great deal of energy in its structure. Biologically, that energy can be released in a reaction that underlies the energy cycle of living things.

Australopithecines. The primate group that lies at the base of the clade of hominids: they include human ancestors.

Autotrophy. The ability to feed oneself, typically but not universally by photosynthesis.

Bacteria. Informally, those organisms that have no cell nucleus (prokaryotes). Increasingly, however, the term is restricted to the group of prokaryotes that are not Archaea.

Banded iron formations, see BIF

Basalts, flood, see Flood basalts

Beringia. Name for the land surface that connected Alaska and Siberia at times when sea level was low.

BIF. Banded iron formations. Major rock type formed dominantly between 2500 Ma and 1800 Ma, probably by the oxidation of iron in sea water.

Biogeography. Study of the geographic distribution of organisms.

Bivalves. Major clade of molluscs. Includes familiar groups such as oysters, clams, mussels, and so on.

Borhyaenids. Medium- to large-bodied carnivorous marsupials that flourished in the Cenozoic in South America.

Brachiopods. Metazoan phylum of filter-feeding shelled invertebrates. Abundant in the Paleozoic Era, rare today.

Brontotheres, see Titanotheres

Bryozoans. Phylum of tiny filter-feeding metazoans, typically colonial. They are well preserved in the fossil record because they calcify the little tubes and boxes they live in.

Bulldozer hypothesis. The concept that filter-feeders of the Paleozoic Fauna did not succeed well after the Permian extinction because of the disturbance of soft seafloors by strongly burrowing organisms.

Cambrian Fauna. Those marine metazoan groups that flourished particularly in the Cambrian period.

Cambrian radiation. Term to summarize the observation that many diverse groups of invertebrates appear very suddenly in the fossil record early in the Cambrian.

Carbon-14. An unstable isotope of carbon, used for absolute age dating in archeological contexts. Useless for dating in time periods before 50,000 years ago.

Carnivora. Formal name for a clade of advanced mammals that are typically but not universally carnivorous. Includes cats, dogs, weasels, otters, etc.

Carnosaurs. Informal name for a group of very large carnivorous dinosaurs. Not a clade.

Carrier's Constraint. Term to summarize the idea that vertebrates with sprawling limbs cannot breathe while they run.

Caseids. Group of herbivorous pelycosaurs.

Cassowary. Large ground-running bird of Australasia, related to emus and other ratites.

Cats, sabertooth. Large cats that have extremely long stabbing canine teeth. The sabertooth condition has evolved several times independently among large cats.

Catalyst. A chemical that helps a reaction to go faster, though it is not consumed in that reaction.

Caterpillars. Informal term for the larvae of insects such as moths and butterflies. They are major herbivores in some ecosystems.

Ceboids. New World monkeys.

Cell membrane. The specialized coating that surrounds a cell.

Cellulose. A polymer of sugar molecules, used in many plant cells to construct a cell wall.

Cenozoic Era. Period of Earth history from 65 Ma to the present; follows the Mesozoic Era.

Centipedes. Clade of carnivorous arthropods.

Cephalaspids. Clade of early jawless fishes.

Cephalochordates. Clade of soft-bodied chordates that includes the amphioxus, and is probably most closely related to vertebrates.

Cephalopods. Clade of molluscs, characterized by the evolution of a well-developed head; typically swimming carnivores.

Ceratopsians. Clade of ornithischian dinosaurs: the horned dinosaurs.

Ceratosaurs. An early clade of theropod dinosaurs.

Cercopithecoids. The Old World monkeys.

Chameleons. Group of insect-eating lizards, mainly tropical.

Champsosaurids. Small clade of aquatic Mesozoic and early Cenozoic reptiles, superficially crocodile-like but not related to them.

Chert. Mineral formed from impure silica, typically in sediments.

Chlorophyll. Molecule that traps light, and is used in photosynthesis to help build energy-packed molecules.

Chloroplasts, see Plastids

Choana. Technical term for nostril in early vertebrates.

Choanoflagellates. Protists that form small colonies of cells, each with a flagellum. These colonies behave like simple miniature sponges, and may resemble sponge ancestors.

Chondrichthyes. The cartilaginous fishes: sharks and rays.

Chordates. Major phylum of metazoans, includes vertebrates.

Circulatory system. Any anatomical system that acts to transport fluids around an organism. In vertebrates, refers particularly to the blood system.

Clade. A set of organisms grouped together because they (and only they) are thought to have evolved from a common ancestor.

Cladistic classification. A biological classification that tries to reflect the evolutionary steps that occurred during the history of a selected group of organisms.

Cladogram. A diagram that portrays the relative differences and similarities within a selected group of organisms.

Clam. Informal name for some bivalves, typically those that burrow into the seafloor.

Classification. In biology and paleontology, the science of naming and organizing creatures into natural groups.

Clays. Minerals produced by the chemical breakdown of other minerals. They are typically soft, absorbent, chemically active substances, so have been suggested as important agents in the origin of life.

Clone. A cell or organism produced by asexual production, so that it retains the precise genetic identity of its parent.

Clovis. Name applied to the people and culture of big-game hunters once thought to be the first human occupants of the Americas.

Cnidarians. Major metazoan phylum that contains jellyfish, corals, sea pens, and so on. Has very simple sheet-like structure of tissue; also characterized by nematocysts or stinging cells.

Coal. Rock formed almost entirely of carbon, dominantly composed of the remains of ancient plants and plant debris.

Coelacanths. Clade of sarcopterygian (lobe-fin) fishes that evolved in Devonian times and still has two living species.